WILEY SERIES IN COMPUTATIONAL STATISTICS

Consulting Editors:

Paolo Giudici
University of Pavia, Italy

Geof H. Givens
Colorado State University, USA

Bani K. Mallick
Texas A&M University, USA

Wiley Series in Computational Statistics is comprised of practical guides and cutting edge research books on new developments in computational statistics. It features quality authors with a strong applications focus. The texts in the series provide detailed coverage of statistical concepts, methods and case studies in areas at the interface of statistics, computing, and numerics.

With sound motivation and a wealth of practical examples, the books show in concrete terms how to select and to use appropriate ranges of statistical computing techniques in particular fields of study. Readers are assumed to have a basic understanding of introductory terminology.

The series concentrates on applications of computational methods in statistics to fields of bioinformatics, genomics, epidemiology, business, engineering, finance and applied statistics.

A complete list of titles in this series appears at the end of the volume.

Bayesian Modeling
Using WinBUGS

Bayesian Modeling Using WinBUGS

Ioannis Ntzoufras

Department of Statistics
Athens University of Economics and Business
Athens, Greece

WILEY

A JOHN WILEY & SONS, INC., PUBLICATION

Library of Congress Cataloging-in-Publication Data is available.

Ntzoufras, Ioannis, 1973-
  Bayesian modeling using WinBUGS / Ioannis Ntzoufras.
    p. cm.
  Includes bibliographical references and index.
  ISBN 978-0-470-14114-4 (pbk.)
  1. Bayesian statistical decision theory. 2. WinBUGS. I. Title.
  QA279.5.N89 2009
  519.5'42—dc22   2008033316

Printed in the United States of America.

10 9 8 7 6 5 4 3

To Ioanna and our baby daughter

CONTENTS

PREFACE

Since the mid-1980s, the development of widely accessible powerful computers and the implementation of Markov chain Monte Carlo (MCMC) methods have led to an explosion of interest in Bayesian statistics and modeling. This was followed by an extensive research for new Bayesian methodologies generating the practical application of complicated models used over a wide range of sciences. During the late 1990s, BUGS emerged in the foreground. BUGS was a free software that could fit complicated models in a relatively easy manner, using standard MCMC methods. Since 1998 or so, WinBUGS , the Windows version of BUGS, has earned great popularity among researchers of diverse scientific fields. Therefore, an increased need for an introductory book related to Bayesian models and their implementation via WinBUGS has been realized.

The objective of the present book is to offer an introduction to the principles of Bayesian modeling, with emphasis on model building and model implementation using WinBUGS . Detailed examples are provided, ranging from very simple to more advanced and realistic ones. Generalized linear models (GLMs), which are familiar to most students and researchers, are discussed. Details concerning model building, prior specification, writing the WinBUGS code and the analysis and interpretation of the WinBUGS output are also provided. Because of the introductory character of the book, I focused on elementary models, starting from the normal regression models and moving to generalized linear models. Even more advanced readers, familiar with such models, may benefit from the Bayesian implementation using WinBUGS .

Basic knowledge of probability theory and statistics is assumed. Computations that could not be performed in WinBUGS are illustrated using R. Therefore, a minimum knowledge of R is also required.

This manuscript can be used as the main textbook in a second-level course of Bayesian statistics focusing on modeling and/or computation. Alternatively, it can serve as a companion (to a main textbook) in an introductory course of a Bayesian statistics. Finally, because of its structure, postgraduate students and other researchers can complete a self-taught tutorial course on Bayesian modeling by following the material of this book.

All datasets and code used in the book are available in the book's Webpage: `www.stat-athens.aueb.gr/~jbn/winbugs_book`.

IOANNIS NTZOUFRAS

Athens, Greece
June 29, 2008

ACKNOWLEDGMENTS

I am indebted to the people at Wiley publications for their understanding and assistance during the preparation of the manuscript. Acknowledgments are due to the anonymous referees. Their suggestions and comments led to a substantial improvement of the present book. I would particularly like to thank Dimitris Fouskakis, colleague and good friend, for his valuable comments on an early version of chapters 1–6 and 10–11. I am also grateful to Professor Brani Vidakovic for proposing and motivating this book. Last but not least, I wish to thank my wife Ioanna for her love, support, and patience during the writing of this book as well as for her suggestions on the manuscript.

I. N.

ACRONYMS

ACF	Autocorrelation
AIC	Akaike information criterion
ANOVA	Analysis of variance
ANCOVA	Analysis of covariance
AR	Attributable risk
BF	Bayes factor
BIC	Bayes information criterion
BOA	Bayesian output analysis (R package)
BP	Bivariate Poisson
BOD	Biological oxygen demand (data variable in example 6.3)
BUGS	Bayesian inference using Gibbs (software)
CDF	Cumulative distribution function
COD	Chemical oxygen demand (data variable in example 6.3)
CODA	Convergence diagnostics and output analysis software for Gibbs sampling analysis (R package)
CPO	Conditional Predictive Ordinate
CR	corner (constraint)
CV	Cross-validation

CV-1	Leave-one-out cross-validation
DAG	Directed acyclic graph
DI	Dispersion index
DIBP	Diagonal inflated bivariate Poisson distribution
DIC	Deviance information criterion
GLM	Generalized linear model
GP	Generalized Poisson
GVS	Gibbs variable selection
ICPO	Inverse conditional predictive ordinate
i.i.d.	Independent identically distributed
LS	Logarithmic score
MAP	Maximum a posteriori
MP model	Median probability
MCMC	Markov chain Monte Carlo
MCE	Monte Carlo error
ML	Maximum likelihood
MLE	Maximum-likelihood estimate/estimator
NB	Negative binomial
OR	Odds ratio
PBF	Posterior Bayes factor
PD	Poisson difference
p.d.f.	Probability density function
PO	Posterior model odds
PPO	Posterior predictive ordinate
RJMCMC	Reversible jump Markov chain Monte Carlo
RR	Relative risk
SD	Standard deviation
SE	Standard error
SSVS	Stochastic search variable selection
STZ	sum-to-zero (constraint)
TS	Total solids(data variable in example 6.3)
TVS	Total volatile solids (data variable in example 6.3)
WinBUGS	Windows version of BUGS (software)
ZI	Zero inflated
ZID	Zero inflated distribution
ZIP	Zero inflated Poisson distribution
ZINB	Zero inflated negative binomial distribution

| ZIGP | Zero inflated generalized Poisson distribution |
| ZIBP | Zero inflated bivariate Poisson distribution |

CHAPTER 1

INTRODUCTION TO BAYESIAN INFERENCE

1.1 INTRODUCTION: BAYESIAN MODELING IN THE 21ST CENTURY

The beginning of the 21st century found Bayesian statistics to be fashionable in science. But until the late 1980s, Bayesian statistics were considered only as an interesting alternative to the "classical" theory. The main difference between the classical statistical theory and the Bayesian approach is that the latter considers parameters as random variables that are characterized by a prior distribution. This prior distribution is combined with the traditional likelihood to obtain the posterior distribution of the parameter of interest on which the statistical inference is based. Although the main tool of Bayesian theory is probability theory, for many years Bayesians were considered as a heretic minority for several reasons. The main objection of "classical" statisticians was the subjective view point of the Bayesian approach introduced in the analysis via the prior distribution. However, as history had proved, the main reason why Bayesian theory was unable to establish a foothold as a well accepted quantitative approach for data analysis was the intractabilities involved in the calculation of the posterior distribution. Asymptotic methods had provided solutions to specific problems, but no generalization was possible. Until the early 1990s two groups of statisticians had (re)discovered Markov chain Monte Carlo (MCMC) methods (Gelfand and Smith, 1990; Gelfand et al., 1990). Physicists were familiar with MCMC methodology from the 1950s. Nick Metropolis and his associates had developed one of the first electronic supercomputers (for those days) and had been testing their theories in physics using Monte Carlo techniques. Implementation of the MCMC methods in combination with the rapid evolution of personal computers made the new computational tool popular within a few years. Bayesian statistics suddenly became fashionable, opening new highways for statistical research. Using MCMC, we can now set up and estimate complicated models that describe and solve problems that could not be solved with traditional methods.

Since 1990, when MCMC first appeared in statistical science, many important related papers have appeared in the literature. During 1990–1995, MCMC-related research focused on the implementation of new methods in various popular models [see, e.g., Gelman and Rubin (1992), Gelfand, Smith and Lee (1992), Gilks and Wild (1992), Dellaportas and Smith (1993)]. The development of MCMC methodology had also promoted the implementation of random effects and hierarchical models.

Green's (1995) publication on reversible jump Markov chain Monte Carlo (RJMCMC) algorithm boosted research on model averaging, selection and model exploration algorithms [see, e.g., Dellaportas and Forster (1999), Dellaportas et al. (2002), Sisson (2005), Hans et al. (2007)]. During the same period, the early versions of BUGS software appeared. BUGS was computing-language-oriented software in which the user only needed to specify the structure of the model. Then, BUGS was using MCMC methods to generate samples from the posterior distribution of the specified model. The most popular version of BUGS (v.05) was available via the Internet in 1996 [manual date August 14, 1996; see, Spiegelhalter et al. (1996*a*)]. Currently WinBUGS version 1.4.3 [1] is available via the WinBUGS project Web-page (Spiegelhalter et al., 2003*d*). Many add-ons, utilities, and variations of the package are also available. The development of WinBUGS had proved valuable for the implementation of Bayesian models in a wide variety of scientific disciplines. In parallel, many workshops and courses have been organized on Bayesian inference, data analysis, and modeling using WinBUGS software. WinBUGS is a key factor in the growing popularity of Bayesian methods in science.

Development, extensions, and improvement of MCMC methods have also been considered in statistical research since the mid-1990s. Automatic samplers, which will be directly applicable in any set of data, are within this frame of research and have led to the slice sampler (Higdon, 1998; Damien et al., 1999). Various samplers designed for model and variable evaluation have been also produced; for a comprehensive review, see Sisson (2005). Perfect sampling (Propp and Wilson, 1996; Møller, 1999) and population-based MCMC methods (Laskey and Myers, 2003; Jasra et al., 2007) can also be considered as interesting examples of the more recent development of MCMC algorithms.

Finally, more recent advancements in genetics have given new impetus to Bayesian theory. The generally large amount of data (in terms of both sample size and variable size) have rendered the more traditional methods inapplicable. Hence Bayesian methods, with the help MCMC methodology, are appropriate for exploration of large model and parameter spaces and tracing the most important associations; see, e.g., Yi (2004).

This book focuses on building statistical models using WinBUGS. It aims to assist students and practitioners in using WinBUGS for fitting models starting from the simpler generalized linear-type models and progressing to more realistic ones by incorporating more complicated structures in the model.

The present chapter provides a comprehensive, short introduction to Bayesian theory. Only the most essential elements of the specific topic are emphasized. Nevertheless, since this is intended as a comprehensive overview for students and practitioners, detailed illustration using examples is also provided with emphasis on data analysis.

---

[1] Version 1.4.1 appeared on September 22, 2004; version 1.4.2, on March 13, 2007; version 1.4.3, on August 6, 2007

1.2   DEFINITION OF STATISTICAL MODELS

One of the most important issues in statistical science is the construction of probabilistic models that represent, or sufficiently approximate, the true generating mechanism of a phenomenon under study. The construction of such models is usually based on probabilistic and logical arguments concerning the nature and function of a given phenomenon.

Assume a random variable Y, called *response*, which follows a probabilistic rule with density or probability function $f(y|\boldsymbol{\theta})$, where $\boldsymbol{\theta}$ is the parameter vector. Consider an independent, identically distributed (i.i.d.) sample $\boldsymbol{y} = [y_1, \dots, y_n]^T$ of size n of this variable; where the $\boldsymbol{A}^T$ denotes the transpose of a vector or matrix $\boldsymbol{A}$. The joint distribution

$$f(\boldsymbol{y}|\boldsymbol{\theta}) = \prod_{i=1}^{n} f(y_i|\boldsymbol{\theta})$$

is called the *likelihood* of the model and contains the available information provided by the observed sample.

Usually models are constructed in order to assess or interpret causal relationships between the response variable Y and various characteristics expressed as variables $X_j, j \in \mathcal{V}$, called *covariates* or *explanatory variables*; j indicates a covariate or model term and $\mathcal{V}$ is the set of all terms under consideration. In such cases, the explanatory variables are linked with the response variables via a deterministic function and part of the original parameter vector is substituted by an alternative set of parameters (denoted by $\boldsymbol{\beta}$) that usually encapsulate the effect of each covariate on the response variable. For example in a normal regression model with $\boldsymbol{y} \sim N(\boldsymbol{X\beta}, \sigma^2 \boldsymbol{I})$ the parameter vector is given by $\boldsymbol{\theta}^T = [\boldsymbol{\beta}^T, \sigma^2]$.

1.3   BAYES THEOREM

Let us consider two possible outcomes A and B. Moreover, assume that $A = A_1 \cup \cdots \cup A_n$ for which $A_i \cap A_j = \emptyset$ for every $i \neq j$. Then, *Bayes' theorem* provides an expression for the conditional probability of A_i given B, which is equal to

$$P(A_i|B) = \frac{P(B|A_i)P(A_i)}{P(B)} = \frac{P(B|A_i)P(A_i)}{\sum_{i=1}^{n} P(B|A_i)P(A_i)}.$$

In a simpler and more general form, for any outcome A and B, we can write

$$P(A|B) = \frac{P(B|A)P(A)}{P(B)} \propto P(B|A)P(A)$$

This equation is also called *Bayes' rule*, although it was originally found by Piere-Simon de Laplace (Hoffmann-Jørgensen, 1994, p. 102).

The above rule can be used for inverse inference. Assume that B is the finally observed outcome and that by A_i we denote possible causes that provoke B. Then $P(B|A_i)$ can be interpreted as the probability that B will appear when A_i cause is present while $P(A_i|B)$ is the probability that A_i is responsible for the occurrence of B that we have already observed.

Bayesian inference is based on this rationale. The preceding equation, which at a first glance is simple, offers a probabilistic mechanism of learning from data (Bernardo and Smith, 1994, p. 2). Hence, after observing data $(y_1, y_2, \dots, y_n)$ we calculate the posterior distribution $f(\boldsymbol{\theta}|y_1, \dots, y_n)$, which combines prior and data information. This posterior distribution is the key element in Bayesian inference.

Example 1.1. Suppose that in a case–control study, we trace 51 smokers in a group of 83 cases of lung cancer and 23 smokers in the control group of 70 disease-free subjects. The prevalence rate (estimate of the proportion of the disease at the population) of lung cancer is equal to 1%. The aim is to calculate the probability that a smoker will develop lung cancer.

From this example we can estimate the following probabilities:

$$P(\text{smoker}|\text{case}) = \frac{51}{83} = 0.615, \quad P(\text{smoker}|\text{control}) = \frac{23}{70} = 0.329$$

and $P(\text{case}) = 0.01$. From the Bayes theorem we calculate

$$
\begin{aligned}
P(\text{case}|\text{smoker}) \quad &= \quad \frac{P(\text{smoker}|\text{case})P(\text{case})}{P(\text{smoker}|\text{case})P(\text{case}) + P(\text{smoker}|\text{control})P(\text{control})} \\
&= \quad \frac{0.615 \times 0.01}{0.615 \times 0.01 + 0.329 \times 0.99} = 0.0185
\end{aligned}
$$

Hence the probability of a smoker to develop lung cancer is equal to 1.85% (approximately 2 people over 100).

Using similar arguments we can calculate the probability of a nonsmoker to develop the disease, which is equal to 0.0099 and the relative risk (RR) is equal to

$$RR = \frac{P(\text{case}|\text{smoker})}{P(\text{case}|\text{nonsmoker})} = \frac{0.0185}{0.0099} = 1.87 \ .$$

Therefore, the probability for a smoker to develop lung cancer is 87% higher than the corresponding probability for nonsmokers.

1.4 MODEL-BASED BAYESIAN INFERENCE

Bayesian statistics differ from the classical statistical theory since all unknown parameters are considered as random variables. For this reason, *prior distribution* must be defined initially. This prior distribution expresses the information available to the researcher before any "data" are involved in the statistical analysis. Interest lies in calculation of the *posterior distribution* $f(\boldsymbol{\theta}|\boldsymbol{y})$ of the parameters $\boldsymbol{\theta}$ given the observed data $\boldsymbol{y}$. According to the Bayes theorem, the posterior distribution can be written as

$$f(\boldsymbol{\theta}|\boldsymbol{y}) = \frac{f(\boldsymbol{y}|\boldsymbol{\theta})f(\boldsymbol{\theta})}{f(\boldsymbol{y})} \propto f(\boldsymbol{y}|\boldsymbol{\theta})f(\boldsymbol{\theta}).$$

The posterior distribution embodies both prior and observed data information, which is expressed by the prior distribution $f(\boldsymbol{\theta})$ and the likelihood

$$f(\boldsymbol{y}|\boldsymbol{\theta}) = \prod_{i=1}^{n} f(y_i|\boldsymbol{\theta}),$$

respectively. Throughout this book, the marginal probability or density function of random variable X evaluated at x is denoted by $f(x)$, while the corresponding conditional probability or density function of random variable X evaluated at x given that $Y = y$ is denoted by $f(x|y)$. With this notation, no distribution parameters are represented. If the random variable X follows a specific distribution D with parameters $\boldsymbol{\theta}$, the notation $f_D(x\,;\boldsymbol{\theta})$ is used to denote the corresponding probability or density function evaluated at $X = x$. For example, $f_N(x\,;\mu_0,\sigma_0^2)$ denotes the density function of a normal distribution with mean equal to μ_0 and variance equal to σ_0^2 evaluated at $X = x$.

Specification of the prior distribution is important in Bayesian inference since it influences the posterior inference. Usually, specification of the prior mean and variance is emphasized. The prior mean provides a prior point estimate for the parameter of interest, while the variance expresses our uncertainty concerning this estimate. When we a priori strongly believe that this estimate is accurate, then the variance must be set low, while ignorance or great uncertainty concerning the prior mean can be expressed by large variance. If prior information is available, it should be appropriately summarized by the prior distribution. This procedure is called *elicitation* of prior knowledge. Usually, no prior information is available. In this case we need to specify a prior that will not influence the posterior distribution and "let the data speak for themselves". Such distributions are frequently called *noninformative* or *vague prior distributions*. A usual vague improper prior distribution is $f(\boldsymbol{\theta}) \propto 1$, which is the uniform prior over the parameter space. The term *improper* here refers to distributions that do not integrate to one. Such prior distributions can be used without any problem provided that the resulting posterior will be proper. A wide range of "noninformative" vague priors may be used; for details, see Kass and Wasserman (1995) and Yang and Berger (1996). In this book, we use the term *low information prior* for proper prior distributions with large variance. Such priors contribute negligible information to the posterior distribution.

Summary measures such as the moments of the posterior distribution can be used for inference concerning the uncertainty of the parameter vector $\boldsymbol{\theta}$. To be more specific, measures of central location such as the posterior mean, median, or mode can be used as point estimates, while the $q/2$ and $1-q/2$ posterior quantiles can be used as $(1-q)100\%$ posterior credible intervals.

We can use the Bayes rule to infer for any parameter of interest $\boldsymbol{\theta}$ even when the observed data are collected sequentially at different timepoints (e.g., in prospective studies). Before any data are available, we use only the prior distribution $f(\boldsymbol{\theta})$ for inference. When a set of data $\boldsymbol{y}^{(1)}$ is observed, we can use the posterior distribution $f\left(\boldsymbol{\theta}|\boldsymbol{y}^{(1)}\right) \propto f\left(\boldsymbol{y}^{(1)}|\boldsymbol{\theta}\right)f(\boldsymbol{\theta})$. When a second set of data is available, we can use the posterior from the first instance as a prior and incorporate the new data in a new updated posterior distribution. Hence the updated posterior distribution will be given by

$$\begin{aligned} f\left(\boldsymbol{\theta}|\boldsymbol{y}^{(1)},\boldsymbol{y}^{(2)}\right) &\propto f\left(\boldsymbol{y}^{(2)}|\boldsymbol{\theta}\right)f\left(\boldsymbol{\theta}|\boldsymbol{y}^{(1)}\right) \\ &\propto f\left(\boldsymbol{y}^{(2)}|\boldsymbol{\theta}\right)f\left(\boldsymbol{y}^{(1)}|\boldsymbol{\theta}\right)f(\boldsymbol{\theta}). \end{aligned}$$

This equation can be generalized for data collected in t different time instances using the equation

$$\begin{aligned} f\left(\boldsymbol{\theta}|\boldsymbol{y}^{(1)},\ldots,\boldsymbol{y}^{(t)}\right) &\propto f\left(\boldsymbol{y}^{(t)}|\boldsymbol{\theta}\right)f\left(\boldsymbol{\theta}|\boldsymbol{y}^{(1)},\ldots,\boldsymbol{y}^{(t-1)}\right) \\ &\propto \prod_{k=1}^{t} f\left(\boldsymbol{y}^{(k)}|\boldsymbol{\theta}\right)f(\boldsymbol{\theta}). \end{aligned}$$

Where it is obvious that Bayesian theory provides an easy-to-use mechanism to update our knowledge concerning the parameter of interest $\boldsymbol{\theta}$.

In order to complete the definition of a Bayesian model, both the prior distribution and the likelihood must be fully specified. Having specified these two components, we then focus on describing the posterior distribution using density plots and descriptive measures.

We may divide the whole procedure into four stages: model building, calculation of the posterior distribution, analysis of the posterior distribution, and inference — final conclusions concerning the problem under consideration. In the first stage we need to consider a model (likelihood/parameters/prior) with reasonable assumptions. In the second stage, we calculate the posterior distribution of interest with the appropriate method of computation. Then we focus on the posterior analysis using descriptive measures, figures, and credible intervals. Finally, we draw conclusions concerning the problem which we are dealing with.

More specifically, in the first stage (model building) we may follow the procedure described below:

1. Identify the main variable of the problem (called response Y) and the corresponding data y.

2. Find a distribution that adequately describes Y.

3. Identify other variables that may influence the response variable Y (called *covariates* or *explanatory variables*).

4. Build a structure for the parameters of the distribution (using deterministic functions).

5. Specify the prior distribution (select the distributional family and specify the prior parameters; select between using a noninformative prior or incorporating preceding known information and/or experts' opinion in our prior distribution).

6. Write down the likelihood of the model.

In the second stage we identify first the method of calculation of the posterior distribution (analytically, asymptotically, or using simulation techniques) and then implement the selected method to estimate the posterior distribution.

Concerning the analysis of the posterior distribution, we may proceed with (all or some of the) the measures proposed below:

1. Visually inspect the marginal posterior distributions of interest. Possible plots that can be obtained are as follows.

    a. Marginal posterior density or probability plots if analytical or asymptotic methods are used

    b. Marginal posterior histograms (or density estimates) for continuous variables and bar charts for discrete or categorical variables

    c. Boxplots of the marginal posterior distributions

    d. Bivariate posterior plots (e.g. contour plots) to identify and study correlations

2. Calculate posterior summaries (means, medians, standard deviations, correlations, quantiles) and 95% or 99% posterior credible intervals

3. Calculate the posterior mode and the area of highest posterior density (where possible)

This description is only indicative and can be enriched with more details and further analysis. One further important issue is the implementation of diagnostic tests or checks concerning the appropriateness of the adopted model. Various techniques may be used to check whether the assumptions of the model are valid and whether the fit of the model is adequate, to test specific hypotheses leading to different conclusions, and to compare different models that may represent totally different scientific scenarios. All these procedures may lead to a new revised model and hence represent an important part of model building. We refer to this important aspect in Chapters 10 and 11.

Another important issue is the robustness of the posterior distribution. We can assess how robust the posterior distribution is to the selection of the prior distribution via *sensitivity analysis*, in which we assess changes in the posterior distribution over different prior distributions. When prior information is available, sensitivity analysis focuses on the structure of the prior distribution; when noninformative priors are used, it focuses on how different choices of prior parameters may influence the posterior inference.

Another important aspect is prediction. Bayesian theory provides a realistic and straightforward theoretical frame for the prediction of future observations through the *predictive distribution*. The predictive distribution is equivalent to the fitted (or expected or predicted) values in classical theory with the difference that now we directly deal with a "distribution". This distribution is also used for checking the assumptions as well as the fit of the model. We will refer at this important issue in Chapter 10.

1.5 INFERENCE USING CONJUGATE PRIOR DISTRIBUTIONS

Usually the target posterior distribution is not analytically tractable. In the past, intractability was avoided via the use of *conjugate prior distributions*. These prior distributions have the nice property of resulting to posteriors of the same distributional family. Extensive illustration of conjugate priors is provided by Bernardo and Smith (1994).

A prior distribution that is a member of the distributional family D with parameters $\boldsymbol{\alpha}$ is conjugate to the distribution $f(\boldsymbol{y}|\boldsymbol{\theta})$ if the resulting posterior distribution $f(\boldsymbol{\theta}|\boldsymbol{y})$ is also a member of the same distributional family. Therefore

$$\text{if } \boldsymbol{\theta} \sim \mathcal{D}(\boldsymbol{\alpha}) \text{ then } \boldsymbol{\theta}|\boldsymbol{y} \sim \mathcal{D}(\widetilde{\boldsymbol{\alpha}}),$$

where $\boldsymbol{\alpha}$ and $\widetilde{\boldsymbol{\alpha}}$ are the prior and posterior parameters of D. In many simple cases, the posterior parameters are expressed as weighted means of the prior parameters and maximum-likelihood estimators. In this section we focus on simple models with one or two parameters. Special attention is given to distributions that belong to the exponential family.

1.5.1 Inference for the Poisson rate of count data

Let us assume a set of discrete count data $\boldsymbol{y}$ in which we wish to estimate their mean λ. Assuming a Poisson distribution with mean λ for the data, we write

$$y_i \sim \text{Poisson}(\lambda) \text{ for } i = 1, \dots, n.$$

In this simple example, the parameter of interest is the Poisson rate λ, therefore $\boldsymbol{\theta} = \lambda$, while the likelihood is given by

$$f(\boldsymbol{y}|\lambda) = \prod_{i=1}^{n} \frac{e^{-\lambda}\lambda^{y_i}}{y_i!} = \frac{e^{-n\lambda}\lambda^{\sum_{i=1}^{n} y_i}}{\prod_{i=1}^{n} y_i!}.$$

Let us now consider a gamma prior distribution for λ with parameters a and b and density function

$$f(\lambda) = \frac{b^a}{\Gamma(a)} x^{a-1} e^{-b\lambda}.$$

In this setup the prior parameter vector is given by $\boldsymbol{a} = (a, b)$.

The resulting posterior distribution is equal to

$$f(\lambda|\boldsymbol{y}) \quad \propto \quad f(\boldsymbol{y}|\lambda)f(\lambda) \propto \frac{e^{-n\lambda}\lambda^{\sum_{i=1}^{n} y_i}}{\prod_{i=1}^{n} y_i!} \times \frac{b^a}{\Gamma(a)}\lambda^{a-1}e^{-b\lambda}$$

$$\propto \quad e^{-(n+b)\lambda}\lambda^{n\bar{y}+a-1},$$

where $\bar{y}$ is the sample mean. Since the density of any gamma distribution with parameters a and b, denoted by gamma(a, b), is proportional to $x^{a-1}e^{-bx}$, we reach the conclusion that

$$\lambda|\boldsymbol{y} \sim \text{gamma}(n\bar{y} + a, n + b), \tag{1.1}$$

that is, the posterior distribution is a gamma distribution with parameters $\tilde{\boldsymbol{a}} = (n\bar{y} + a, n + b)^T$. Therefore the gamma distribution is conjugate to the Poisson distribution.

The posterior mean of λ is given by

$$E(\lambda|\boldsymbol{y}) = \tilde{\mu}_\lambda = \frac{n\bar{y} + a}{n + b},$$

while the posterior variance is given by

$$V(\lambda|\boldsymbol{y}) = \tilde{\sigma}_\lambda^2 = \frac{n\bar{y} + a}{(n + b)^2}.$$

These quantities can be rewritten in the following form

$$E(\lambda|\boldsymbol{y}) = \left(\frac{n}{n+b}\right)\bar{y} + \left(\frac{b}{n+b}\right)\left(\frac{a}{b}\right) = w\bar{y} + (1-w)\left(\frac{a}{b}\right) = w\bar{y} + (1-w)E(\lambda)$$

$$V(\lambda|\boldsymbol{y}) = \frac{n^2}{(n+b)^2}\left(\frac{\bar{y}}{n}\right) + \frac{b^2}{(n+b)^2}\left(\frac{a}{b^2}\right) = w^2\left(\frac{\bar{y}}{n}\right) + (1-w)^2 V(\lambda),$$

where $w = n/(n+b)$. The posterior mean is expressed as a weighted average of the prior of the sample mean (which is the maximum-likelihood estimator).

The usually selected "low information" prior is a gamma distribution with low and equal prior parameters such as $a = b = 10^{-3}$. This prior is convenient since its mean is equal to one while the variance is given by $1/a$, which becomes large (expressing prior "ignorance") for low values of a. Moreover, when $a = b \to 0$, then $w \to 1$, and the posterior mean of λ coincides with the sample mean $\bar{y}$.

1.5.2 Inference for the success probability of binomial data

Let us consider a set of binomial data y_i that express the number of successes over N_i attempts (for $i = 1, 2, \ldots, n$). Hence $y_i \sim \text{binomial}(\pi, N_i)$, resulting to a likelihood given by

$$f(\boldsymbol{y}|\pi) = \prod_{i=1}^{n}\left\{\binom{N_i}{y_i}\pi^{y_i}(1-\pi)^{N_i-y_i}\right\}$$

$$= \prod_{i=1}^{n}\binom{N_i}{y_i}\pi^{\sum_{i=1}^{n} y_i}(1-\pi)^{\sum_{i=1}^{n} N_i - \sum_{i=1}^{n} y_i}$$

$$= \prod_{i=1}^{n}\binom{N_i}{y_i}\pi^{n\bar{y}}(1-\pi)^{N-n\bar{y}},$$

where $N = \sum_{i=1}^{n} N_i$ is the total number of the Bernoulli experiments in the sample.

If we consider a beta prior distribution with parameters $\boldsymbol{a} = (a, b)^T$, denoted by beta$(a, b)$, and density function

$$f(\pi) = \frac{\Gamma(a)\Gamma(b)}{\Gamma(a+b)} \pi^{a-1}(1-\pi)^{b-1},$$

then the resulting posterior is also a beta distribution since

$$
\begin{aligned}
f(\pi|\boldsymbol{y}) \quad &\propto \quad f(\boldsymbol{y}|\pi)f(\pi) \\
&\propto \quad \prod_{i=1}^{n} \binom{N_i}{y_i} \pi^{n\overline{y}}(1-\pi)^{N-n\overline{y}} \times \frac{\Gamma(a)\Gamma(b)}{\Gamma(a+b)} \pi^{a-1}(1-\pi)^{b-1} \\
&\propto \quad \pi^{n\overline{y}+a-1}(1-\pi)^{N-n\overline{y}+b-1} \Leftrightarrow \\
\pi|\boldsymbol{y} \quad &\sim \quad \text{beta}(n\overline{y} + a, N - n\overline{y} + b) \quad\quad\quad (1.2)
\end{aligned}
$$

with posterior parameters $\widetilde{\boldsymbol{\alpha}} = (n\overline{y} + a, N - n\overline{y} + b)^T$ and posterior mean and variance

$$
\begin{aligned}
E(\pi|\boldsymbol{y}) \quad &= \quad \widetilde{\mu}_\pi = \frac{n\overline{y} + a}{N + a + b}, \\
V(\pi|\boldsymbol{y}) \quad &= \quad \widetilde{\sigma}_\pi^2 = \frac{(n\overline{y} + a)(N - n\overline{y} + b)}{(N + a + b)^2(N + a + b + 1)}.
\end{aligned}
$$

Similarly to the Poisson case, the posterior mean can be also expressed as a weighted average of the prior and the sample proportion since

$$
\begin{aligned}
E(\pi|\boldsymbol{y}) \quad &= \quad \left(\frac{N}{N+a+b}\right)\left(\frac{n\overline{y}}{N}\right) + \left(\frac{a+b}{N+a+b}\right)\left(\frac{a}{a+b}\right) \\
&= \quad w\left(\frac{n\overline{y}}{N}\right) + (1-w)\left(\frac{a}{a+b}\right)
\end{aligned}
$$

where $w = N/(N + a + b)$, $n\overline{y}/N = \sum_{i=1}^{n} y_i/N$ is the sample proportion and $a/(a + b)$ is the mean a beta prior distribution with parameters a and b.

A beta distribution with equal and low parameter values can be considered as a low-information prior (e.g., $a = b = 10^{-3}$). Other choices that are usually adopted are the beta$(\frac{1}{2}, \frac{1}{2})$ or the uniform distribution denoted by $U(0, 1)$ [i.e., a beta$(1, 1)$ distribution]. The latter can be consider as a low-information prior distribution since it a priori gives the same probability to any interval of the same range. Nevertheless, this prior will be influential when the sample size is low. This might not necessarily be a disadvantage since, for small sample size, the posterior will also reflect the low available information concerning the parameter of interest π.

1.5.3   Inference for the mean of normal data with known variance

When we are interested in the calculation of the posterior distribution of the mean of a response variable which takes all values in the set of real numbers, then a frequently used (and in several cases reasonable) assumption is that y_i follows a normal distribution. In this case the likelihood is given by

$$f(\boldsymbol{y}|\mu, \sigma^2) \quad = \quad \prod_{i=1}^{n} \left\{ \frac{1}{\sqrt{2\Pi\sigma^2}} \exp\left(-\frac{(y_i - \mu)^2}{2\sigma^2}\right) \right\}$$

$$= (2\pi\sigma^2)^{-n/2} \exp\left(-\frac{1}{2\sigma^2} \sum_{i=1}^{n}(y_i - \mu)^2\right)$$

$$= (2\pi\sigma^2)^{-n/2} \exp\left(-\frac{1}{2\sigma^2} \left(\sum_{i=1}^{n} y_i^2 + n\mu^2 - 2n\mu\bar{y}\right)\right).$$

Using a normal $N(\mu_0, \sigma_0^2)$ prior distribution with $f(\mu|\sigma^2) = f_N(\mu; \mu_0, \sigma_0^2)$, the posterior distribution is given by

$$\begin{aligned}
f(\mu|\sigma^2, \boldsymbol{y}) \;\; &\propto \;\; f(\boldsymbol{y}|\mu, \sigma^2) f(\mu|\sigma^2) \\
&\propto \;\; f(\boldsymbol{y}|\mu, \sigma^2) f_N(\mu; \mu_0, \sigma_0^2) \\
&\propto \;\; (2\pi\sigma^2)^{-n/2} \exp\left(-\frac{1}{2\sigma^2}\left(\sum_{i=1}^{n} y_i^2 + n\mu^2 - 2n\mu\bar{y}\right)\right) \\
&\quad \times (2\pi\sigma_0^2)^{-1/2} \exp\left(-\frac{1}{2\sigma_0^2}(\mu - \mu_0)^2\right) \\
&\propto \;\; \exp\left(-\frac{1}{2\sigma^2}(n\mu^2 - 2n\mu\bar{y}) - \frac{1}{2\sigma_0^2}(\mu^2 - 2\mu\mu_0)\right) \\
&\propto \;\; \exp\left(-\frac{1}{2}\left\{\left[\frac{n}{\sigma^2} + \frac{1}{\sigma_0^2}\right]\mu^2 - 2\mu\left[\frac{n\bar{y}}{\sigma^2} + \frac{\mu}{\sigma_0^2}\right]\right\}\right) \\
&\propto \;\; \exp\left(-\frac{1}{2}\left[\frac{n\sigma_0^2 + \sigma^2}{\sigma^2\,\sigma_0^2}\right]\left\{\mu^2 - 2\mu\left[\frac{n\bar{y}\sigma_0^2 + \mu_0\sigma^2}{n\sigma_0^2 + \sigma^2}\right]\right\}\right).
\end{aligned}$$

Note that σ^2 and σ_0^2 are different known quantities that correspond to the variances of the random variable Y and the prior distribution of μ, respectively [i.e. $V(Y_i) = \sigma^2$ and $V(\mu) = \sigma_0^2$].

The posterior described above is a normal distribution with mean and variance given by

$$E(\mu|\boldsymbol{y}) = \tilde{\mu} = \frac{n\bar{y}\sigma_0^2 + \mu_0\sigma^2}{n\sigma_0^2 + \sigma^2} \quad \text{and} \quad V(\mu|\boldsymbol{y}) = \tilde{\sigma}^2 = \frac{\sigma^2\,\sigma_0^2}{n\sigma_0^2 + \sigma^2},$$

respectively, since the density of the $N(\tilde{\mu}, \tilde{\sigma}^2)$ is given by

$$f(x) = \frac{1}{\sqrt{2\pi\tilde{\sigma}^2}} \exp\left(-\frac{(x - \tilde{\mu})^2}{2\tilde{\sigma}^2}\right) \propto \exp\left(-\frac{1}{2\tilde{\sigma}^2}(x^2 - 2x\tilde{\mu})\right).$$

The above mentioned posterior mean and variance can also be written as

$$\tilde{\mu} = w\bar{y} + (1 - w)\mu_0 \quad \text{and} \quad \tilde{\sigma}^2 = w\frac{\sigma^2}{n},$$

where $w = n\sigma_0^2/(n\sigma_0^2 + \sigma^2)$. Consequently, the posterior mean is again expressed as a weighted average of the prior and the sample mean. If the prior variance is low (and hence the prior information concerning the parameter μ is strong), then the posterior mean will be equal to the prior mean, while if the prior variance is high (and hence the prior information concerning the parameter μ is low) then the posterior mean will be equal to the sample mean. Similarly, the posterior variance is expressed as a proportion of the standard error of the sample mean. When the prior variance is large, the posterior distribution is equivalent to the distribution of the sample mean.

1.5.4   Inference for the mean and variance of normal data

The logic for normal data y when the variance is unknown is similar to that described in the previous section. In this case we need to estimate the posterior distribution for the parameters μ and σ^2. The conjugate prior is the normal–inverse gamma distribution $\text{NIG}(\mu_0, c, a, b)$ [see, e.g., Bernardo and Smith (1994)] with density

$$f(\mu, \sigma^2) = f(\mu|\sigma^2)f(\sigma^2),$$

where $f(\mu|\sigma^2) = N(\mu_0, c\sigma^2)$ and $f(\sigma^2) = IG(a, b)$ when $IG(a, b)$ is the inverse gamma distribution with parameters a and b. Note that if $X \sim \text{gamma}(a, b)$, then $X^{-1} \sim IG(a, b)$. For this reason, for normal models the precision $\tau = \sigma^{-2}$ is used instead. This notation is also adopted by WinBUGS software. The precision parameter has a plausible interpretation since it measures the precision of the available information concerning the parameter of interest. If the variance is large, then the precision is low and hence the mean cannot be used as a precise estimate for the random variable Y. On the contrary, when the variance is low, the precision is high and hence the mean can be used as a precise summary of Y.

With this approach, the prior for the parameter vector $\theta = (\mu, \tau)^T$ is a normal–gamma distribution, denoted by $\text{NG}(\mu, c, a, b)$, with density

$$f(\mu, \tau) = (2\pi c)^{-1/2}\tau^{1/2}\exp\left(-\frac{\tau}{2c}(\mu - \mu_0)^2\right) \times \frac{b^a}{\Gamma(a)}\tau^{a-1}e^{-b\tau}.$$

The resulting posterior density $f(\mu, \sigma^2|y)$ is a $\text{NIG}(\tilde{\mu}, \tilde{c}, \tilde{a}, b)$ [or equivalently $\text{NG}(\tilde{\mu}, \tilde{c}, \tilde{a}, b)$ for μ and τ] with

$$\tilde{\mu} = w\bar{y} + (1 - w)\mu_0 \text{ with } w = \frac{nc}{1 + nc}$$

$$\tilde{c} = \frac{w}{n}$$

$$\tilde{a} = a + \frac{n}{2}$$

$$\tilde{b} = b + \frac{\text{SS}}{2}$$

since

$$f(\mu, \tau|y) \propto \tau^{(n+1)/2+a-1}\exp\left\{-\frac{\tau}{2}\left(\frac{1 + nc}{c}\right)\left(\mu - \frac{\mu_0 + nc\bar{y}}{1 + nc}\right)^2 - \tau\left(b + \frac{\text{SS}}{2}\right)\right\}$$

$$\propto \tau^{n/2+a+1/2-1}\exp\left\{-\frac{\tau}{2}\left(\frac{n}{w}\right)(\mu - \tilde{\mu})^2 - \tau\left(b + \frac{\text{SS}}{2}\right)\right\},$$

where SS is given by

$$\text{SS} = \sum_{i=1}^{n}y_i^2 + \frac{\mu_0^2}{c} - \frac{(nc\bar{y} + \mu_0)^2}{c(nc + 1)}$$

$$= \sum_{i=1}^{n}(y_i - \bar{y})^2 + \frac{w}{c}(\bar{y} - \mu_0)^2 = (n - 1)s^2 + \frac{w}{c}(\bar{y} - \mu_0)^2,$$

and s^2 is the sample variance of Y: $s^2 = \sum_{i=1}^{n}(y_i - \bar{y})^2/(n - 1)$.

In multidimensional posterior distributions, focus is given in the analysis of the marginal posterior distributions and their corresponding descriptive measures. In the current example, we focus our attention on the marginals

$$
\begin{aligned}
f(\mu|\boldsymbol{y}) &= \int f(\mu,\tau|\boldsymbol{y})d\tau \propto \left[b + \frac{\mathrm{SS}}{2} + \frac{n}{2w}(\mu - \widetilde{\mu})^2\right]^{-(a+(n+1)/2)} \\
&\propto \left[1 + \frac{n/w}{\mathrm{SS}+2b}(\mu - \widetilde{\mu})^2\right]^{-(n+2a+1)/2} \\
&= t\left(\widetilde{\mu},\ \frac{n}{w}\frac{n+2a}{\mathrm{SS}+2b},\ n+2a\right) \\
f(\tau|\boldsymbol{y}) &= \int f(\mu,\tau|\boldsymbol{y})d\mu \propto \tau^{n/2+a-1}\exp\left\{-\tau\left(\frac{\mathrm{SS}}{2}+b\right)\right\} \\
&= \mathrm{gamma}\left(\frac{n}{2}+a,\ \frac{\mathrm{SS}}{2}+b\right),
\end{aligned}
$$

where $t(\mu,\tau,a)$ is the noncentral Student's t distribution with density

$$
f(x|\mu,\tau,a) \propto \left[1 + \frac{\tau}{a}(x-\mu)^2\right]^{-(a+1)/2};\quad x \in \mathbf{R}
$$

[for details, see Bernardo and Smith (1994)]. The posterior mean and variance of (μ,τ) are given by

$$
\begin{aligned}
E(\mu|\boldsymbol{y}) &= \widetilde{\mu} &\text{and}\quad V(\mu|\boldsymbol{y}) &= \frac{w}{n}\frac{\mathrm{SS}+2b}{n+2a-2}\ \text{if}\ n>2-2a \\
E(\tau|\boldsymbol{y}) &= \frac{n/2+a}{\mathrm{SS}/2+b} &\text{and}\quad V(\tau|\boldsymbol{y}) &= \frac{n/2+a}{(\mathrm{SS}/2+b)^2}\ .
\end{aligned}
$$

Inference for the variance σ^2 can be directly based on the posterior distribution

$$
\sigma^2|\boldsymbol{y} \sim IG\left(\frac{n}{2}+a,\ \frac{\mathrm{SS}}{2}+b\right)
$$

with posterior mean and variance

$$
\begin{aligned}
E(\sigma^2|\boldsymbol{y}) &= \frac{\mathrm{SS}/2+b}{n/2+a-1}\ \text{if}\ a>2-2a\ \text{and} \\
V(\sigma^2|\boldsymbol{y}) &= \frac{(\mathrm{SS}/2+b)^2}{(n/2+a-1)^2(n/2+a-2)}\ \text{if}\ a>4-2a\ .
\end{aligned}
$$

1.5.5 Inference for normal regression models

The results of the previous paragraph can easily be extended for the case of the normal regression model. Here we use the conjugate multivariate normal–inverse gamma prior and present the posterior densities directly without further details.

Let us consider a n-dimensional vector of data $\boldsymbol{y}$ and their corresponding random variable Y. Then the normal regression model can be summarized by

$$
\boldsymbol{Y}|\boldsymbol{\mu},\sigma^2 \sim N_n\left(\boldsymbol{X\beta},\sigma^2\boldsymbol{I}_n\right),
$$

where $\boldsymbol{X}$ is the $n \times p$ data matrix including the values of the covariates, $\boldsymbol{\beta}$ is the $p \times 1$ parameter vector specifying the effect of each covariate, and $\boldsymbol{I}_n$ it the $n \times n$ identity matrix.

The multivariate normal–inverse gamma prior distribution is conjugate to the likelihood described above and is given by

$$[\boldsymbol{\beta}, \sigma^2] \sim \mathrm{NIG}(\boldsymbol{\mu}_\beta, \boldsymbol{V}, a, b),$$

which can be expressed as a product of a multivariate normal distribution (for parameter vector $\boldsymbol{\beta}$) and an inverse gamma prior (for σ^2) given by

$$\boldsymbol{\beta}|\sigma^2 \sim N_p(\boldsymbol{\mu}_\beta, \boldsymbol{V}\sigma^2) \ \text{ and } \ \sigma^2 \sim \mathrm{IG}(a, b).$$

The resulting posterior is also a normal–inverse gamma with parameters

$$
\begin{aligned}
\widetilde{\boldsymbol{\beta}} &= \widetilde{\boldsymbol{\Sigma}}\left(\boldsymbol{X}^T\boldsymbol{y} + \boldsymbol{V}^{-1}\boldsymbol{\mu}_\beta\right) \\
\widetilde{\boldsymbol{\Sigma}} &= \left(\boldsymbol{X}^T\boldsymbol{X} + \boldsymbol{V}^{-1}\right)^{-1} \\
\widetilde{a} &= \frac{n}{2} + a \\
\widetilde{b} &= \frac{\mathrm{SS}}{2} + b \\
\mathrm{SS} &= \boldsymbol{y}^T\boldsymbol{y} - \widetilde{\boldsymbol{\beta}}^T\widetilde{\boldsymbol{\Sigma}}^{-1}\widetilde{\boldsymbol{\beta}} + \boldsymbol{\mu}_\beta^T\boldsymbol{V}^{-1}\boldsymbol{\mu}_\beta.
\end{aligned}
$$

The posterior mean $\widetilde{\boldsymbol{\beta}}$ can be written in a form equivalent to that for the weighted average of the prior and the sample mean in the univariate normal case since

$$\widetilde{\boldsymbol{\beta}} = \boldsymbol{W}\hat{\boldsymbol{\beta}} + (\boldsymbol{I}_p - \boldsymbol{W})\boldsymbol{\mu}_\beta \ \text{ with } \ \boldsymbol{W} = \left(\boldsymbol{X}^T\boldsymbol{X} + \boldsymbol{V}^{-1}\right)^{-1}\boldsymbol{X}^T\boldsymbol{X},$$

where $\hat{\boldsymbol{\beta}} = \left(\boldsymbol{X}^T\boldsymbol{X}\right)^{-1}\boldsymbol{X}^T\boldsymbol{y}$ is the maximum-likelihood estimator. Using more algebra we can obtain the elegant expression of Atkinson (1978)

$$\mathrm{SS} = \mathrm{RSS} + \left(\hat{\boldsymbol{\beta}} - \boldsymbol{\mu}_\beta\right)^T\left[\left(\boldsymbol{X}^T\boldsymbol{X}\right)^{-1} + \boldsymbol{V}\right]^{-1}\left(\hat{\boldsymbol{\beta}} - \boldsymbol{\mu}_\beta\right),$$

where $\mathrm{RSS} = \left(\boldsymbol{y} - \boldsymbol{X}\hat{\boldsymbol{\beta}}\right)^T\left(\boldsymbol{y} - \boldsymbol{X}\hat{\boldsymbol{\beta}}\right) = \boldsymbol{y}^T\boldsymbol{y} - \hat{\boldsymbol{\beta}}^T\boldsymbol{X}^T\boldsymbol{X}\hat{\boldsymbol{\beta}}$ is the residual sum of squares in classical regression analysis. This expression is useful because it gives insight concerning the meaning and the role of SS, which is equal to the traditional sum of squares and a measure of distance between the maximum-likelihood estimator (MLE) and the prior mean.

The marginal distributions are similar as in the simpler case. The marginal posterior distribution of $\boldsymbol{\beta}$ is a multivariate Student distribution with parameters $\widetilde{\boldsymbol{\beta}}$, $\widetilde{\boldsymbol{\Sigma}}(\mathrm{SS}+2b)/(n+2a)$ and $n + 2a$, where the multivariate Student distribution (denoted by MSt) is defined as

$$Y \sim \mathrm{MSt}_p(\boldsymbol{\mu}, \boldsymbol{\Sigma}, a) \Leftrightarrow f(y) \propto \left[1 + \frac{1}{a}(\boldsymbol{y} - \boldsymbol{\mu})^T\boldsymbol{\Sigma}^{-1}(\boldsymbol{y} - \boldsymbol{\mu})\right]^{-(a+p)/2}.$$

The marginal posterior distribution of σ^2 is simply an inverse gamma distribution with the parameters $\widetilde{a}$ and $\widetilde{b}$ given above; for more details, see Bernardo and Smith (1994).

1.5.6 Other conjugate prior distributions

In this section we summarize some well-known conjugate distributions. We focus on the popular exponential family, which includes a wide range of well-known distributions and is also used in the generalized linear models setup.

Let us consider the exponential dispersion family with $Y \sim \text{expf}\left(\vartheta, \phi, a(), b(), c()\right)$

$$f(y|\vartheta, \phi) = \exp\left(\frac{y\vartheta - b(\vartheta)}{a(\phi)} + c(y, \phi)\right) \tag{1.3}$$

for some specific functions $a(\phi)$, $b(\vartheta)$, and $c(y, \phi)$, where ϑ and ϕ are location and scale parameters, respectively.

The likelihood function for a sample of size n is given by

$$f(\boldsymbol{y}|\vartheta, \phi) = \exp\left(\frac{n\bar{y}\vartheta - nb(\vartheta)}{a(\phi)} + \sum_{i=1}^{n} c(y_i, \phi)\right),$$

[for details, see McCullagh and Nelder (1989, pp. 28–29)]. Using a prior distribution of type

$$f(\vartheta|\vartheta_0, \tau_0, \phi) \propto \exp\left(\frac{\vartheta\vartheta_0 - \tau_0 b(\vartheta)}{a(\phi)}\right), \tag{1.4}$$

we end up with a posterior

$$f(\vartheta|\boldsymbol{y}, \phi) \propto \exp\left(\frac{(n\bar{y} + \vartheta_0)\vartheta - (n + \tau_0)b(\vartheta)}{a(\phi)} + \sum_{i=1}^{n} c(y_i, \phi)\right).$$

This posterior has the same form as the prior (1.4) with parameters $n\bar{y} + \vartheta_0$ and $n + \tau_0$ (instead of ϑ_0 and τ_0). The conjugate distribution shown above is conditional on the dispersion parameter ϕ assuming that it is known and fixed.

For example, for the normal model, assuming that σ^2 is known, we have

$$\vartheta = \mu, \quad \phi = \sigma^2, \quad a(\phi) = \sigma^2, \quad b(\vartheta) = \frac{\mu^2}{2}, \quad \text{and } c(y, \vartheta) = -\frac{y^2}{2\sigma^2} - \frac{1}{2}\log(2\Pi\sigma^2).$$

Following (1.4), the conjugate prior is given by

$$f(\mu|\sigma^2) \propto \exp\left(\frac{\vartheta\vartheta_0 - \tau_0\vartheta^2/2}{\sigma^2}\right)$$

$$\propto \exp\left(-\frac{\tau_0}{2\sigma^2}(\vartheta^2 - 2\tau_0^{-1}\vartheta_0\,\vartheta)\right) = N(\tau_0^{-1}\vartheta_0, \tau_0^{-1}\sigma^2).$$

This prior is equivalent to the conjugate prior that we used in Section 1.5.3 with $\mu_0 = \vartheta_0/\tau_0$ and $\sigma_0^2 = \sigma^2/\tau_0$. Members of the exponential family are also the gamma, binomial, and Poisson distributions. Table 1.1 summarizes the most important conjugate distributions.

1.5.7 Illustrative examples

In this subsection we use realistic data examples to illustrate the results shown above, which are based on conjugate prior distributions. Examples of data analysis using the Poisson, binomial, and normal distributions are provided in detail.

Table 1.1 Conjugate prior distributions for commonly used distributions

Distribution	Likelihood	Prior distribution	Posterior parameters
Poisson	$Y_i \sim \text{Poisson}(\lambda)$	$\lambda \sim \text{gamma}(a,b)$	$\tilde{a} = n\bar{y} + a$, $\tilde{b} = n+b$
Binomial	$Y_i \sim \text{binomial}(p, N_i)$	$p \sim \text{beta}(a,b)$	$\tilde{a} = \sum_{i=1}^n y_i + a$, $\tilde{b} = \sum_{i=1}^n N_i + b$
Normal (known σ^2)	$Y_i \sim N(\mu, \sigma^2)$	$\mu|\sigma^2 \sim N(\mu_0, \sigma_0^2)$	$\tilde{\mu} = w\bar{y} + (1-w)\mu_0$, $\tilde{\sigma}^2 = w\sigma^2/n$, $w = \sigma_0^2/(\sigma_0^2 + \sigma^2/n)$
Normal	$Y_i \sim N(\mu, \sigma^2)$	$[\mu, \sigma^2] \sim \text{NIG}(\mu_0, c, a, b)$ $\left[N(\mu_0, c\sigma^2) \times \text{IG}(a,b) \right]$	$\tilde{\mu} = w\bar{y} + (1-w)\mu_0$ $\tilde{a} = n/2 + a$ $\tilde{c} = w/n$, $\tilde{b} = \text{SS}/2 + b$ $\text{SS} = (n-1)s^2 + (\bar{y} - \mu_0)^2 w/c$ $w = nc/(1+nc)$
Gamma (ν known)	$Y_i \sim \text{gamma}(\nu, \theta)$	$\theta|\nu \sim \text{gamma}(a,b)$	$\tilde{a} = n\nu + a$, $\tilde{b} = n\bar{y} + b$
Exponential	$Y_i \sim \text{exponential}(\theta)$ $= \text{gamma}(1, \theta)$	$\theta \sim \text{gamma}(a,b)$	$\tilde{a} = n + a$, $\tilde{b} = n\bar{y} + b$
Negative binomial	$Y_i \sim \text{NB}(p, K_i)$	$p \sim \text{beta}(a,b)$	$\tilde{a} = \sum_{i=1}^n K_i + a$, $\tilde{b} = \sum_{i=1}^n y_i + b$
Multinomial	$\mathbf{Y}_i \sim \text{multinomial}(\mathbf{a}, N_i)$	$\mathbf{a} \sim \text{Dirichlet}(\mathbf{a}_0)$	$\tilde{\mathbf{a}} = \sum_{i=1}^n \mathbf{y}_i + \mathbf{a}_0$
Exponential family (ϕ known)	$Y_i \sim$ $\text{expf}\left(\vartheta, \phi, a(), b(), c() \right)$	$f(\vartheta|\phi) \propto$ $\exp\left\{ [\vartheta\vartheta_0 - \tau_0 b(\vartheta)]/a(\phi) \right\}$	$\tilde{\vartheta} = n\bar{y} + \vartheta_0$, $\tilde{\tau} = n + \tau_0$

Example 1.2. Goals scored by the national football team of Greece in Euro 2004 (Poisson data). Let us consider the number of goals scored by a team in association football (soccer). The Poisson distribution is widely used to model such data although a slight overdispersion is observed (Lee, 1997; Karlis and Ntzoufras, 2000; Karlis and Ntzoufras, 2003a). In this example we consider the final scores of the National team of Greece in the Euro 2004 cup competition, where Greece surprisingly won the competition; see Table 1.2. We are interested in estimating the posterior distribution of expected number of goals scored and conceded by Greece as well as the total number of goals scored by both teams in each game played by Greece (the latter is frequently used for betting purposes).

Table 1.2 Games of Greece in the Euro 2004 competition

	Opponent	Goals scored (In favor – against)	Total
1	Portugal	$2-1$	3
2	Spain	$1-1$	2
3	Russia	$1-2$	3
4	France	$1-0$	1
5	Czech Republic	$0-0 \, (1-0)^a$	0 (1)
6	Portugal	$1-0$	1
	Sum	$6-4$	10
	Sample mean $\bar{x}$	$1.00 - 0.67$	1.67

[a] Normal time score = 0–0; score after 15 times of extra time = 1–0.

To model the goals scored and conceded by Greece (denoted by y_i^s and y_i^c, respectively) and the total number of goals ($y_i^t = y_i^s + y_i^c$), we use the Poisson distribution. Hence we write

$$y_i^k \sim \text{Poisson}(\theta^k), \text{ for } k \in \{s, c, t\} \text{ and } i = 1, 2, \ldots, 6,$$

where θ^s and θ^c respectively are the expected number of goals scored and conceded by Greece and θ^t are the expected total number of goals scored in each game.

If we wish to consider a low-information prior, then we can use a gamma$(0.001, 0.001)$ prior with mean equal to one and variance equal to 1000. Following (1.1), the posterior distributions for the mean goals scored by Greece, against Greece, and in total are expressed in the following gamma distributions:

$$\theta^s | \boldsymbol{y} \sim \text{gamma}(6.001, 6.001),$$
$$\theta^c | \boldsymbol{y} \sim \text{gamma}(4.001, 6.001),$$
$$\theta^t | \boldsymbol{y} \sim \text{gamma}(10.001, 6.001).$$

The posterior means and standard deviations (SD) are given by

$$E(\theta^s | \boldsymbol{y}) = \frac{6.001}{6.001} = 1.000, \quad \text{SD}(\theta^s | \boldsymbol{y}) = \frac{6.001}{6.001^2} = 0.166$$
$$E(\theta^c | \boldsymbol{y}) = \frac{4.001}{6.001} = 0.667, \quad \text{SD}(\theta^c | \boldsymbol{y}) = \frac{4.001}{6.001^2} = 0.111$$
$$E(\theta^t | \boldsymbol{y}) = \frac{10.001}{6.001} = 1.667, \quad \text{SD}(\theta^t | \boldsymbol{y}) = \frac{10.001}{6.001^2} = 0.278$$

Graphical representation of the posterior distributions is given in Figure 1.1.

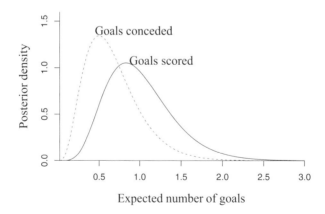

Figure 1.1   Posterior distributions of number of goals scored and conceded by Greece in Euro 2004; low-information prior.

Using an informative prior distribution.   Before the Euro 2004 competition, prior information was available from the group qualifying stage of the competition. Although the quality of the opponents was not as high as in the Euro 2004 competition and all games were played in both "home" and "away" (that is, domestic and foreign) stadiums, we can extract information from these game and specify a prior more informative than the one used above.

From these games, the Greek national team was first in its group scoring 8 goals and accepting only 4 in 8 games. From this information we can construct the posterior distribution after the completion of the qualifying group stage. Hence, according to (1.1), our posterior after considering these games is a gamma$(8, 8)$ for goals scored in favor and a gamma$(4, 8)$ for goals scored against Greece. Although these posterior distributions can be directly used as prior distributions for analysis of the data in the Euro 2004 competition, we propose slightly modifying the prior by increasing the variance in order to reflect additional uncertainty because this information resulted under different conditions. The finally imposed prior assumes the same means (1 and 0.5) as the ones above, but the variance is multiplied by the sample size of the prior data. In this way, the prior approximately contributes information equivalent to one data point to the posterior distribution. Therefore, the finally imposed priors are respectively gamma$(1, 1)$ and gamma$(1, 2)$ for goals scored in favor and against Greece. Following (1.1), the posterior distribution for the goals scored by Greece is now given by $\theta^s|\boldsymbol{y} \sim$ gamma$(7, 7)$, with mean 1 and standard deviation 0.38 goals per game, while the posterior for the goals conceded by Greece is given by $\theta^c|\boldsymbol{y} \sim$ gamma$(5, 8)$, with mean 0.625 and standard deviation 0.078 per game. These posterior distributions are also illustrated in Figure 1.2.

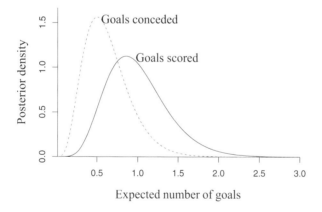

Figure 1.2 Posterior distributions of number of goals scored and conceded by Greece in Euro 2004 including prior information from qualifying group games.

In the analysis above we have excluded the goal scored in the extra time played in the game against the Czech Republic. This can be easily incorporated by considering the factor of time. In this case we can assume that $y_i \sim Poisson(t_i\theta)$, where t_i is equal to one for usual games of 90 minutes of normal play, $t_i = 1 + \frac{15}{90} = 1.167$ for games with 15 minutes of extra time and $t_i = 1 + \frac{30}{90} = 1.33$ for games with 30 minutes of extra time. After the inclusion of extra time goals, the posterior distribution is

$$\theta|\boldsymbol{y} \sim \text{gamma}\left(\sum_{i=1}^{n} t_i y_i + a, \ \sum_{i=1}^{n} t_i + b\right).$$

Using the data of our example, the posterior distributions are slightly changed to

$$\theta^s|\boldsymbol{y} \sim \text{gamma}(7.168, 6.618) \ \text{ and } \ \theta^c|\boldsymbol{y} \sim \text{gamma}(4.001, 6.618)$$

in the noninformative case and in the informative case, to

$$\theta^s | \boldsymbol{y} \sim \text{gamma}(8.168, 7.618) \quad \text{and} \quad \theta^c | \boldsymbol{y} \sim \text{gamma}(5.001, 8.618).$$

In such problems, we might also be interested in inferences about the difference of the means $\theta^s - \theta^c$ as well as making predictions based on the probability of winning a game. The first issue can easily be solved by simply sampling from each gamma distribution, calculating the difference, and estimating the posterior distribution. This is a simple implementation of simulation and Monte Carlo techniques, presented in the following sections. To describe the second issue, we first need to define the predictive distribution, which is presented in Chapter 10.

> **Example 1.3. Estimating the prevalence of a disease (Bernoulli data).** Suppose that we are interested in the estimation of the prevalence of a disease, for example, coronary heart disease, for men aged 30–40. Data from a prospective study of sample size equal to 1250 men have indicated 5 men experiencing at least one incident. We are interested on estimating the prevalence rate of the disease for this specific age group.

Following (1.2), using a beta$(0.01, 0.01)$ low-information prior, we have

$$\pi | \boldsymbol{y} \sim \text{beta}(5.01, 1245.01),$$

with mean equal to 4 incidents over 1000 men and standard deviation equal to 1.89 incidents over 1000 men; see Figure 1.3 for graphical representation of the posterior distribution of prevalence. Alternatively, we may use the uniform $U(0, 1)$ distribution as a noninformative prior. Under this prior setup, the posterior is $\pi | \boldsymbol{y} \sim \text{beta}(6, 1246)$, with posterior mean of the prevalence rate equal to 4.8 incidents per 1000 males and standard deviation equal to 1.95 incidents per 1000 males.

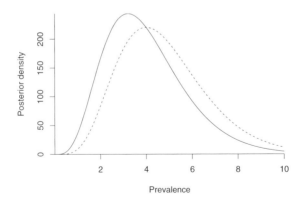

Figure 1.3   Posterior distribution of prevalence rate (per 1000 males) for coronary heart disease: solid line — beta$(0.01, 0.01)$ prior; dashed line — uniform $U(0, 1)$ prior.

A simple credible interval can be given by the 2.5% and 97.5% quantiles of the beta distribution. In our case, the prevalence will be in the interval $(1.3, 8.2)$ with probability 95% for the beta$(0.01, 0.01)$ and $(1.76, 9.3)$ for the uniform prior.

Example 1.4 . Kobe Bryant's field goals in NBA (binomial data). In this example we consider the field goals and the corresponding attempts of a successful NBA player, Kobe Bryant. Data were downloaded from the Yahoo sports page and are presented in Table 1.3.

Having observed the data of Table 1.3, we are interested in comparing the performance of the player in terms of the success probability (or percentage) of field goals across different seasons.

Table 1.3   Kobe Bryant's field goals for seasons 1999–2007

| Season | Games | Field goals | | |
		Successes	Attempts	Percent (%)
1999/00	66	554	1183	46.8
2000/01	68	701	1510	46.4
2001/02	80	749	1597	46.9
2002/03	82	868	1924	45.1
2003/04	65	516	1178	43.8
2004/05	66	573	1324	43.3
2005/06	80	978	2173	45.0
2006/07	42	399	845	47.2
Total	749	5338	11,734	45.5

Source: Yahoo sports.

In Table 1.4, we present the posterior details, including the parameters of the beta distribution, the means and standard deviations, as well as the 95% posterior intervals. In all calculations a beta prior distribution with parameters equal to 0.01 has been used; 95% posterior intervals (based on the 2.5% and 97.5% posterior percentiles) are also depicted using error bars in Figure 1.4 in order to clearly compare the Kobe Bryant's performance across different seasons.

Table 1.4   Posterior summaries for Kobe Bryant's field goals for seasons 1999–2007

| Year | Games | Successes | Attempts | Posterior parameters | | Posterior summaries of field goal success percentage | | | |
				$\tilde{a}$	$\tilde{b}$	Mean	SD^a	$Q_{0.025}$	$Q_{0.975}$
1999/00	66	554	1183	554.01	629.01	46.8	1.5	44.0	49.7
2000/01	68	701	1510	701.01	809.01	46.4	1.3	43.9	48.9
2001/02	80	749	1597	749.01	848.01	46.9	1.2	44.5	49.4
2002/03	82	868	1924	868.01	1056.01	45.1	1.1	42.9	47.3
2003/04	65	516	1178	516.01	662.01	43.8	1.4	41.0	46.6
2004/05	66	573	1324	573.01	751.01	43.3	1.4	40.6	46.0
2005/06	80	978	2173	978.01	1195.01	45.0	1.1	42.9	47.1
2006/07	42	399	845	399.01	446.01	47.2	1.7	43.9	50.6
Total	749	5338	11734	5338.01	6396.01	45.5	0.5	44.6	46.4

^{a} Standard deviation.
Source: Yahoo sports.

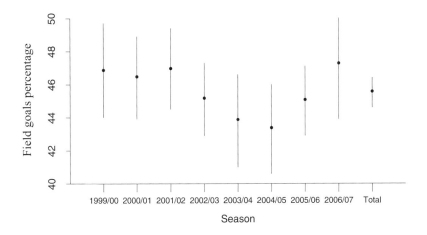

Figure 1.4   95% posterior credible intervals for Kobe Bryant's field goals for seasons 1999–2007; prior =beta$(0.01, 0.01)$.

Updating information over the seasons.   In the analysis above we compared the performance of the player across different seasons. In Bayesian analysis, we may also update our posterior after the end of each season. This will result in improving our confidence for the player's performance (although we do not consider other factors that may be important, such as the age of the player).

In Table 1.5 and Figure 1.5 we can observe how posterior distribution is updated after incorporating sequentially data information of each year. As expected, the final posterior distribution coincides with the corresponding posterior, which combines all seasons' data.

Table 1.5   Summaries of the updated posterior distributions for Kobe Bryant's field goals for seasons 1999–2007

Year	Games	Successes	Attempts	Posterior parameters $\widetilde{a}$	$\widetilde{b}$	Posterior summaries of field goal success percentage Mean	SD^a	$Q_{0.025}$	$Q_{0.975}$
1999/00	66	554	1183	554	629	46.8	1.5	44.0	49.7
2000/01	68	701	1510	1255	1438	46.6	1.0	44.7	48.5
2001/02	80	749	1597	2004	2286	46.7	0.8	45.2	48.2
2002/03	82	868	1924	2872	3342	46.2	0.6	45.0	47.5
2003/04	65	516	1178	3388	4004	45.8	0.6	44.7	47.0
2004/05	66	573	1324	3961	4755	45.4	0.5	44.4	46.5
2005/06	80	978	2173	4939	5950	45.4	0.5	44.4	46.3
2006/07	42	399	845	5338	6396	45.5	0.5	44.6	46.4

^{a}Standard deviation.
Source: Yahoo sports.

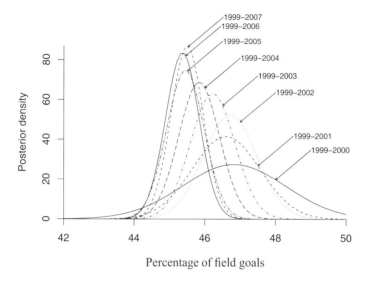

Figure 1.5 Updated posterior densities for Kobe Bryant's percentage of field goals for seasons 1999–2007; prior=beta$(0.01, 0.01)$.

Example 1.5. Body temperature data (normal data). In this example we consider data of Mackowiak et al. (1992), who examined whether the true mean body temperature is 98.6 °F. The same data were reanalyzed and presented by Shoemaker (1996). In the present example, we consider the normal model for inferences about both the mean and the variance of the body temperature.

We adopt the normal model with $y_i \sim N(\mu, \sigma^2)$ and a normal–gamma prior with parameters $\mu = 98.6$, $c = 100$, $a = 0.001$ and $b = 0.001$. This is a low-information prior (since the variance is large) with the mean temperature centered around 98.6 °F which is the widely accepted value.

The sample size, mean, and variance are equal to

$$n = 130, \ \ \overline{y} = 98.24923$$

and

$$s^2 = 0.5375575 \Leftrightarrow \mathrm{SS} = (130 - 1) \times 0.5375575 = 69.34492,$$

respectively, resulting in an NIG posterior distribution for the mean and the variance of temperature with parameters

$$
\begin{aligned}
\widetilde{\mu} &= 0.999923 \times 98.24923 + (1 - 0.999923) \times 98.6 = 98.249 \\
\widetilde{c} &= \frac{w}{n} = \frac{0.999923}{153} = 0.007692 \\
\widetilde{a} &= \frac{150}{2} + 0.001 = 65.001 \\
\widetilde{b} &= \frac{69.34492}{2} + 0.001 = 34.6735 \ .
\end{aligned}
$$

Hence we can summarize the joint posterior density by

$$\mu, \sigma^2 | \boldsymbol{y} \sim \text{NIG}(98.25, \ 0.0077, \ 65.0, \ 34.67).$$

The posterior weight w of the data is equal to $w = (130 \times 100)/(130 \times 100 + 1) = 0.999923$. Since w is very close to one (both sample size and prior variance are large), we deduce that the data likelihood dominates the posterior disrtibution.

The marginal posterior distribution of μ is given by $\mu | \boldsymbol{y} \sim t(98.25, \ 243.725, \ 130.002)$ since

$$\tau = \frac{130}{0.999923} \frac{130.002}{0.02 + 69.345} = 243.7247$$

with mean and standard deviation

$$E(\mu|\boldsymbol{y}) \ = \ 98.25 \ \text{and} \ \text{SD}(\mu|\boldsymbol{y}) = \sqrt{\frac{0.999923}{130} \frac{129 \times 0.5375575 + 0.02}{128.002}} = 0.06456 \ .$$

The marginal posterior distribution of the precision parameter τ is

$$\tau | \boldsymbol{y} \sim \text{gamma}(\widetilde{a} = 65.001, \ \widetilde{b} = 34.67)$$

with mean and variance

$$E(\tau|\boldsymbol{y}) \ = \ \frac{n/2 + a}{\text{SS}/2 + b} = 65.001/34.67346 = 1.875,$$

$$V(\tau|\boldsymbol{y}) \ = \ \frac{n/2 + a}{(\text{SS}/2 + b)^2} = 0.054.$$

The corresponding posterior mean and variance for σ^2 are given by

$$E(\sigma^2|\boldsymbol{y}) \ = \ \frac{\text{SS}/2 + b}{n/2 + a - 1} = \frac{34.67346}{64.001} = 0.542,$$

$$V(\sigma^2|\boldsymbol{y}) \ = \ \frac{(\text{SS}/2 + b)^2}{(n/2 + a - 1)^2(n/2 + a - 2)} = \frac{34.67346^2}{64.001^2 \times 63.001} = 0.00466.$$

Graphical representations of the marginal and the joint posterior densities are given in Figure 1.6.

From the posterior density plot it is obvious that the prior value 98.6 °F is not supported since it corresponds to a value with low posterior density lying at the tails of the distribution (in fact, in our graph is out of the range of the plotted distribution).

Sensitivity analysis. When low-information prior distributions are used, it is common practice to proceed to the implementation of sensitivity analysis with different values of the prior mean or variance.

The sensitivity of the posterior mean over different values of the prior parameter c, which controls the prior variance, is depicted in Figure 1.7. For c we consider $c = 10^k$ for values of $k = -2, -1, 0, 1, 2, 3, 4, 5$. From the graph we clearly see that the posterior mean is quite robust with values ranging from 98.25 to 98.4. Even for relatively low values of c (e.g., for $c = 1$), the estimate is equal to 98.27, which is quite close to the value of 98.25 resulted using large values of c.

Even if we consider as prior mean the extreme value of zero (which is far away from realistic values), the posterior mean is still quite robust; for values of $c \geq 10$ with posterior mean ranging from $98.17 - 98.25$, see Figure 1.8.

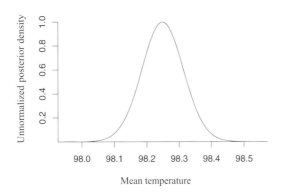

(a) Posterior density for mean body temperature

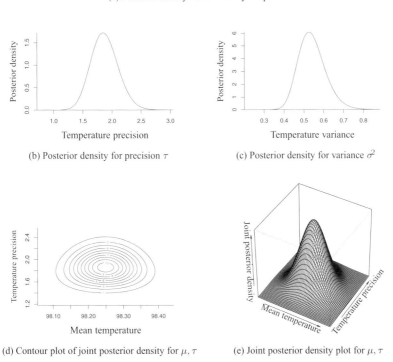

(b) Posterior density for precision τ

(c) Posterior density for variance σ^2

(d) Contour plot of joint posterior density for μ, τ

(e) Joint posterior density plot for μ, τ

Figure 1.6   Posterior plots for parameters of body temperature data

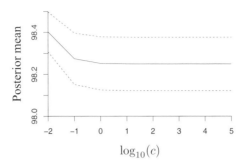

Figure 1.7   Sensitivity plot of posterior mean for different values of $\log_{10}(c)$ with prior mean equal to 98.6; dotted lines represent the 2.5% and 97.5% quantilies of the distribution.

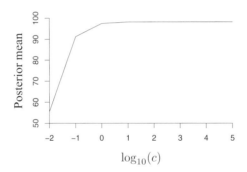

Figure 1.8   Sensitivity plot of posterior mean for different values of $\log_{10}(c)$ with prior mean equal to zero.

The same procedure for $c = 100$ and $\mu \in (98, 99)$ led to a quite robust posterior mean ranging from 98.24921 to 98.24929.

For illustration we have also considered informative priors with values of $c = 10^k$ for $k = -4, -3, -2, -1$ and for $a = b = 0.1$ or $a = b = 1$. In Figures 1.9 and 1.10 we see how the prior becomes increasingly informative for low values of c which controls the marginal prior variance. Also note that even for $c = 0.1$, the posterior is very close to the data (expressed in terms of the distribution of the sample mean $\bar{y}$).

1.6 NONCONJUGATE ANALYSIS

In the example of the body temperature data we have expressed the amount of information as a proportion of the unknown variance of Y. In many occasions we are interested in expressing our prior beliefs in a simpler and more straightforward manner. Usually such prior information is extracted by experts who are not familiar with simple probability notions such as dependence and correlation. Therefore, in the example above, we might need to

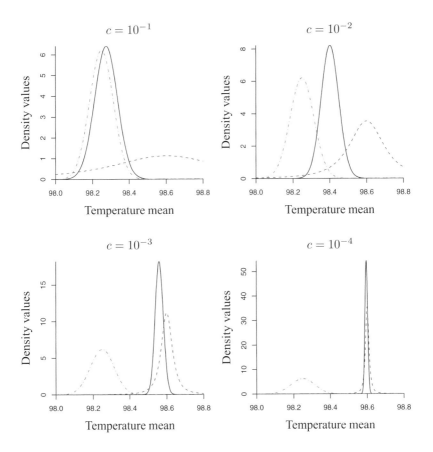

Figure 1.9   Comparison of prior, data, and posterior distribution over different values of c for $a = b = 1$. Lines from left to right: dashed-dotted line $(-\cdot-)$ represents sample distribution of $\overline{y}$; solid line $(—)$ indicates posterior of μ; dashed line $(---)$ denotes prior of μ.

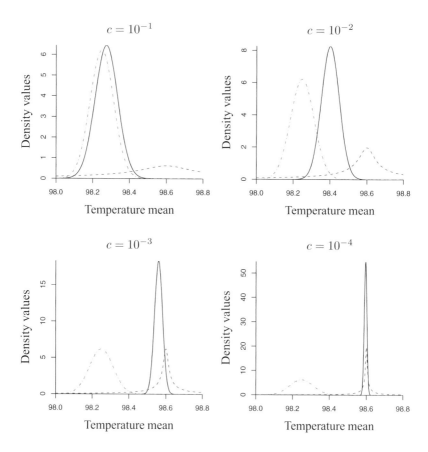

Figure 1.10 Comparison of prior, data, and posterior distribution over different values of c for $a = b = 0.1$. Lines from left to right: dashed-dotted line (— · —) represents sample distribution of $\overline{y}$; solid line (——) indicates posterior of μ; dashed line (- - -) denotes prior of μ.

simplify the prior structure using independent distributions for μ and τ (or equivalently σ^2) and directly specify the prior precision of μ instead of setting it proportional to σ^2. For example, we may consider

$$f(\mu, \sigma^2) = f(\mu)f(\sigma^2) \text{ with } f(\mu) = N(\mu_0, \sigma_0^2) \text{ and } f(\sigma^2) = IG(a, b) .$$

In this case, the resulting posterior distribution is of an unknown form. Consequently, it is difficult to evaluate the posterior summaries and their corresponding marginal densities.

In cases where conjugate priors are considered to be unrealistic or are unavailable, either asymptotic approximations such as Laplace approximation [see, e.g., Tierney and Kadane (1986), Tierney et al. (1989), Erkanli (1994)] or numerical integration techniques [see, e.g., Evans and Swartz (1996)] can be used. Another appealing alternative would be to use simulation-based techniques. These methods generate samples from the posterior distribution and will be used throughout this book since WinBUGS facilitates such methods in order to indirectly estimate the posterior distributions of interest. In the next Chapter, the most important methods for generating samples from the posterior distribution are described and illustrated using simple data examples.

Problems

$P[+|D]$ $P[-|\sim D]$

1.1 Let us consider a medical diagnostic test with 90% sensitivity and 80% specificity for a disease with incidence equal to 1 in 10,000 individuals. Use the Bayes theorem to calculate the probability that an individual has the disease when the test is positive and when the test is negative.

1.2 Let us consider the exponential distribution with density function $f(y|\theta) = \theta e^{-\theta y}$ and an i.i.d. sample $Y_i \sim$ exponential(θ) for $i = 1, \ldots, n$.
   a) Show that a gamma prior distribution is conjugate for θ.
   b) Calculate the posterior mean and variance under this conjugate prior.
   c) For the following data
$$0.4 \ 0.0 \ 0.2 \ 0.1 \ 2.1 \ 0.1 \ 0.9 \ 2.4 \ 0.1 \ 0.2$$
   use the exponential distribution and
   (i) Plot the posterior distribution for gamma prior parameters $a = b = 0.001$.
   (ii) Perform sensitivity analysis for various values of a and b. Produce related plots depicting changes on the posterior mean and variance.

1.3 In the exponential distribution, consider directly the mean parameter $\mu = 1/\theta$.
   a) What is the conjugate prior for μ?
   b) What is the posterior distribution for μ and θ under this setup?
   c) Is the posterior analysis under this approach equivalent to the corresponding one in Problem 1.2?

1.4 For $Y_i \sim$ gamma(ν, θ) assuming that ν is known
   a) Prove that the gamma distribution is a conjugate prior for θ.
   b) Find the posterior mean and variance for θ.
   c) Examine the effect on the posterior density of θ
   (i) Of the known parameter ν
   (ii) Of the sample size n
   (iii) Of the prior parameters

1.5 Let us consider Y_i (for $i = 1, \ldots, n$) be an i.i.d. sample of categorical variables with n categories.

a) Show that a Dirichlet prior is conjugate for the probability of each category.
b) Calculate the posterior mean and variance for the probability of each category.
c) Calculate the posterior correlations between the probabilities of two different categories.

1.6   Let us consider the following 2×2 contingency table:

X: Risk factor	**Y:** disease status 1: present	2: absent	Marginal of X
1: present	211	320	531
2: absent	343	1301	1644
Marginal of Y	554	1621	2175

a) Consider the multinomial distribution and the conjugate Dirichlet distribution to analyze the data tabulated above.
b) Calculate the posterior means and variances for the cell probabilities.
c) Obtain plots of the estimated density function and boxplots of the posterior distributions of the cell probabilities.

1.7   Let us consider the following 2×2 contingency table of the previous problem:
a) Use the Poisson distribution, that is, assume that $n_{ij} \sim \text{Poisson}(\lambda_{ij})$, and the corresponding conjugate prior to analyze the data.
b) Obtain the posterior mean and variance for the expected cell frequencies according to the two approaches and compare them.
c) Produce plots of the posterior densities for the expected cell frequencies λ_{ij}, under the Poisson assumption, and compare them graphically with the corresponding posteriors obtained by the multinomial model in Problem 1.6.

1.8   Let us consider the soccer/football World Cup 2006 final scores for Italy (winner of the cup):

Stage	Opponents	Normal time Score	Other Comments
Group	Italy – Ghana	2–0	
Group	Italy – USA	1–1	
Group	Czech Republic – Italy	0–2	
Round of 16	Italy – Australia	1–0	
Quarter-final	Italy – Ukraine	3–0	
Semifinal	Germany – Italy	0–0	0–2 (a.e.t.)[a]
Final	Italy – France	1–1	1–1 (a.e.t.)[a], 5–3 (pen.)[b]

[a] Score after extra time of 30 minutes.
[b] Score in penalty kicks.

a) Use the Poisson distribution to analyze the data tabulated above. Compare the goals scored and conceded by the Italian national team.
b) Produce similar analysis using the multinomial distribution for the score difference during the normal time of a game (90 minutes).
c) Perform similar analysis for the corresponding results of national team of France which was the finalist team of the tournament (see data in the table below). Com-

pare the results with the corresponding one for the Italian national team. Which team seems to be better according to your (simple) analysis?

Stage	Opponents	Normal time score	Other comments
Group	France – Switzerland	0–0	
Group	France – South Korea	1–1	
Group	Togo – France	0–2	
Round of 16	Spain – France	1–3	
Quarter-final	Brazil – France	0–1	
Semifinal	Portugal – France	0–1	
Final	Italy – France	1–1	1–1(a.e.t.)[a], 5–3 (pen.)[b]

[a] Score after extra time of 30 minutes.
[b] Score in penalty kicks.

1.9 Assume that a friend tosses a coin 10 times and tells you that "heads" appeared less than 4 times.
  a) Calculate the posterior density for the success probability π using a beta(a, a) prior density.
  b) Plot the posterior density for $a = 1$ and $a = 2$.
  c) Calculate the posterior mean of π.
  d) Plot the posterior mean over different values of the prior parameter a and examine the sensitivity of the results.

1.10 Consider the following data:

```
0.671 1.412 -2.119 1.224 -1.168 -0.860 1.936 3.396 4.808 -1.259
0.275 1.820  2.417 2.929  7.020  0.483 6.483 2.966 0.942 -3.398
2.846 3.840  6.640 1.018  2.747  1.857 7.270 2.734 4.325 -1.222
```

  a) Assuming the normal distribution for these data and the corresponding conjugate prior, calculate the posterior mean and variance.
  b) Produce graphical representations of the marginal posterior distributions for μ and σ^2 as well as their corresponding joint posterior distribution.

CHAPTER 2

MARKOV CHAIN MONTE CARLO ALGORITHMS IN BAYESIAN INFERENCE

The present chapter provides a short introduction concerning the use random-number generation (or stochastic simulation) methods in Bayesian inference. The focus is on Markov chain Monte Carlo (MCMC) methods that are widely used in Bayesian inference. The description of the most popular MCMC methods is accompanied by detailed examples implemented in R statistical and programming software. Users not familiar with R can follow the comprehensive pseudocode that describes the corresponding code.

The chapter is divided into three main sections. First the reader is introduced to the basic notions of simulation and Monte Carlo integration. In Section 2.2, Markov chain Monte Carlo methods and their general setup are described. Finally, the most popular methods are described and illustrated using simple examples. The methods are described and illustrated in detail to introduce the reader to the implementation logic of such algorithms. Readers who wish to learn more about the technical and theoretical details of this topic should refer to more specialized books such as those by Gilks et al. (1996), Givens and Hoeting (2005), and Gamerman and Lopes (2006).

2.1 SIMULATION, MONTE CARLO INTEGRATION, AND THEIR IMPLEMENTATION IN BAYESIAN INFERENCE

In quantitative sciences, the problem of evaluation of integrals of the type

$$I = \int_x g(x)dx$$

is often required. Several solutions have been proposed in the literature, including either approximations or computationally intensive methods. One of them is based on generating random samples and then obtaining the integral shown above by its statistical unbiased estimate, the sample mean. Hence let us assume that the density function $f(x)$ of a random variable enables us to easily generate random values. This can be expressed as

$$I = \int_x \left[\frac{g(x)}{f(x)} \right] f(x)dx = \int_x g^*(x)f(x)dx,$$

where $g^*(x) = g(x)/f(x)$. Hence the integral I can be efficiently estimated by

1. Generating $x^{(1)}, x^{(2)}, \ldots, x^{(T)}$ from the target distribution with probability density function (p.d.f.) $f(x)$

2. Calculating the sample mean

$$\hat{I} = \frac{1}{T} \sum_{t=1}^{T} \left[\frac{g(x^{(t)})}{f(x^{(t)})} \right]. \tag{2.1}$$

This concept was known from the early days of the electronic computers and was originally adopted by the research team of Metropolis in Los Alamos (Anderson, 1986; Metropolis and Ulam, 1949).

The main advantage of this approach is its simplicity. Even if integrals are tractable, nowadays it is much easier to generate samples than calculate high-dimensional integrals. The accuracy of sample mean (2.1) as an estimate of the integral I is of order $O(T^{-1/2})$ (Givens and Hoeting, 2005, p. 144), which implies slow convergence toward the true value. Nevertheless, the generated sample size T here can be specified by the researcher. For a suitable large generated sample (say, e.g., $T = 10,000$), this approach is very accurate (Gamerman and Lopes, 2006, p. 96).

The method described above is directly applicable to many problems in Bayesian inference. Hence for every function of the parameter of interest $G(\theta)$, we can calculate the posterior mean and variance by simply

1. Generating a sample $\theta^{(1)}, \theta^{(2)}, \ldots, \theta^{(T)}$ from the posterior distribution $f(\theta|y)$.

2. Calculating the sample mean of $G(\theta)$ by simply calculating the quantity

$$\hat{I} = \frac{1}{T} \sum_{t=1}^{T} G(\theta^{(t)}).$$

Simulation can also be used to estimate and visualize the posterior distribution of $G(\theta)$ itself. This can be done with the assistance of kernel estimates of the values $G(\theta^{(1)}), G(\theta^{(2)}), \ldots, G(\theta^{(T)})$.

The main problem in the above mentioned procedure is how to generate from the posterior density $f(\theta|y)$ which, in most cases, is not straightforward. The posterior distributions resulting from conjugate priors is the simplest possible case in which interest may lie in the computation of posterior summaries of a function $G(\theta)$ of the parameters θ that will not be analytically available. In order to generate random values, a number of approaches are available (see References list at the end of this book), such as the method using the inverse cumulative distribution function, rejection sampling algorithms, and importance sampling methods; for further details, see Carlin and Louis (2000, pp. 129–137), Gelman et al. (1995, chap. 10), and Givens and Hoeting (2005, pp. 145–162).

Example 2.1. A simple example: Risk measures in medical research. In bio-statistics a 2×2 contingency table frequently arises by the cross-tabulation of the disease under investigation (denoted by Y with $Y = 1$ to indicate a disease case and zero otherwise) and an exposure risk factor (denoted by X with value one if the subject is exposed to the risk factor and zero otherwise). In such cases, the re-searcher is interested in estimating certain measures of risk such as the attributable risk (AR), the relative risk (RR), and the odds ratio (OR). To be more specific, the attributable risk (or risk difference) is given by

$$ AR = P(Y = 1|X = 1) - P(Y = 1|X = 0) = \pi_1 - \pi_0, $$

the relative risk is defined as

$$ RR = \frac{P(Y = 1|X = 1)}{P(Y = 1|X = 0)} = \frac{\pi_1}{\pi_0}, $$

and finally the odds ratio is equal to

$$ OR = \frac{P(Y = 1|X = 1)P(Y = 0|X = 0)}{P(Y = 0|X = 1)P(Y = 1|X = 0)} = \frac{\pi_1(1 - \pi_0)}{\pi_0(1 - \pi_1)}. $$

In this problem we have a product binomial likelihood given by

$$ Y_1 \sim \text{binomial}(\pi_0, n_0) \text{ and } Y_0 \sim \text{binomial}(\pi_1, n_1) $$

with the parameters of interest as $\boldsymbol{\theta} = (\pi_0, \pi_1)$. Assuming beta priors for the success probabilities π_0, π_1 with parameters (a_0, b_0) and (a_1, b_1), respectively, we end up with posteriors given by

$$ \pi_0|\boldsymbol{y} \sim \text{beta}(y_0 + a_0, n_0 + b_0) \text{ and } \pi_1|\boldsymbol{y} \sim \text{beta}(y_1 + a_1, n_1 + b_1). $$

The posterior distribution of AR, RR, and OR cannot be obtained in a straightforward man-ner. Derivation of approximate expressions may be based on large sample approximations of π_0 and π_1 and their logarithm as in the classical approach [see, e.g., Woodworth (2004, chap. 7)].

By simulating values directly from the posterior distributions $f(\pi_0|\boldsymbol{y})$ and $f(\pi_1|\boldsymbol{y})$, the estimation of the posterior distribution of the risk measures and their corresponding summaries can be easily obtained without further analytical or approximate calculations.

Let us consider a simple illustrative example where we have traced 25 cases over a sample of 300 subjects exposed to a risk factor (exposed group, $X = 1$) while we have traced 30 cases over 900 subjects in the nonexposed group ($X = 0$). From these givens, $y_1 = 25$, $n_1 = 300$, $y_0 = 30$, and $n_0 = 900$. Hence, for prior parameters $a_0 = b_0 = a_1 = b_1 = 1$, the resulting posteriors are given by

$$ \pi_0|\boldsymbol{y} \sim \text{beta}(26, 301) \text{ and } \pi_1|\boldsymbol{y} \sim \text{beta}(31, 901). $$

We can estimate the posterior distributions of the risk measures AR, RR, and OR using the following simple steps. For $t = 1, \ldots T$

1. Generate $\pi_0^{(t)} \sim \text{beta}(26, 301)$

2. Generate $\pi_1^{(t)} \sim \text{beta}(31, 901)$.

3. Calculate $\text{AR}^{(t)}$, $\text{RR}^{(t)}$ and $\text{OR}^{(t)}$ using the expressions

$$\text{AR}^{(t)} = \pi_1^{(t)} - \pi_0^{(t)}, \quad \text{RR}^{(t)} = \frac{\pi_1^{(t)}}{\pi_0^{(t)}}, \quad \text{and} \quad \text{OR}^{(t)} = \frac{\pi_1^{(t)} \left(1 - \pi_0^{(t)}\right)}{\left(1 - \pi_1^{(t)}\right) \pi_0^{(t)}}.$$

After we completing this generation, we obtain a simulated sample of size T from the posterior distributions of AR, RR, and OR. Posterior summaries can be easily obtained using sample estimates from this sample. Implementation of this simple generation in pseudocode commands as well as in R code is outlined in Table 2.1. Results are summarized in Table 2.2 along with histograms of the generated values, and their corresponding kernel estimates of the posterior densities are depicted in Figure 2.1; for more details concerning kernel estimates, see, for example, Scott (1992). Later in this book, we directly plot the posterior density estimates instead of the corresponding histograms of the simulated values.

Table 2.1 Pseudocode and R code (and results) for generation of AR, RR and OR from corresponding posterior distributions in Example 2.1

Algorithm (pseudocode)	R code
Set up prior parameters	`> a<-1; a0<-a1<-b0<-b1<-a`
Set up data	`> y1<- 25; y0<- 30; n1<-300; n0<-900`
Generate 1000 observations from the posterior of π_0 and π_1	`> p0 <- rbeta(1000, y0+a0, n0+b0)` `> p1 <- rbeta(1000, y1+a1, n1+b1)`
Calculate the corresponding values for risk measures	`> AR <- p1-p0; RR <- p1/p0` `> OR <- p1*(1-p0)/(p0*(1-p1))`
Calculate posterior means	`> mean(AR); mean(RR); mean(OR)`

(results)	`[1] 0.04645852`
	`[1] 2.469779`
	`[1] 2.605217`

Calculate posterior standard deviations	`> sd(AR); sd(RR); sd(OR)`

(results)	`[1] 0.01578586`
	`[1] 0.6198302`
	`[1] 0.6964608`

Calculate posterior credible intervals for AR	`> quantile(AR,c(0.025, 0.975))`
(results)	`2.5% 97.5%` `0.01792363 0.07805458`
Calculate posterior credible intervals for RR	`> quantile(RR,c(0.025, 0.975));`
(results)	`2.5% 97.5%` `1.456754 3.770378`
Calculate posterior credible intervals for OR	`> quantile(OR,c(0.025, 0.975))`
(results)	`2.5% 97.5%` `1.482756 4.100452`

Plot the estimated posterior densities	`> par(mfrow=c(3,1))`
	`> plot(density(AR), main='', xlab='Attributable Risk',`
	`+ ylab='Posterior Density')`
	`> plot(density(RR), main='', xlab='Relative Risk',`
	`+ ylab='Posterior Density')`
	`> plot(density(OR), main='', xlab='Odds Ratio',`
	`+ ylab='Posterior Density')`

Table 2.2   Posterior summaries of risk measures obtained from simulation of 1000 values

Risk measure	Mean	Standard deviation	Percentiles 2.5%–97.5%
Attributable risk (AR)	0.046	0.016	0.018 – 0.078
Relative risk (RR)	2.470	0.620	1.457 – 3.770
Odds ratio (AR)	2.605	0.696	1.483 – 4.100

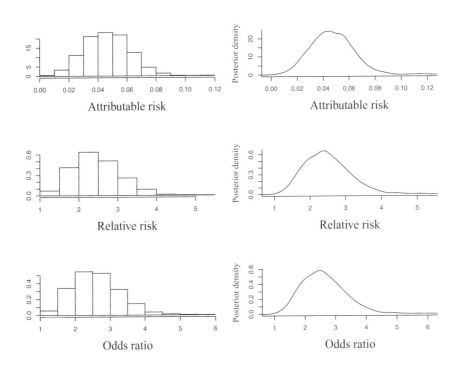

Figure 2.1   Histograms of posterior samples and plots of estimated posterior densities of AR, RR, and OR using 1000 simulated values. [*Note*: Posterior densities have been estimated using kernel density estimates (command density in R); see example in Scott (1992) for details.]

2.2   MARKOV CHAIN MONTE CARLO METHODS

The simulation methods described in the previous section (also called methods of "direct" simulation) cannot be applied in all cases. They refer mostly to unidimensional distributions. Moreover, some of them are focused on the effective computation of specific integrals and cannot be used to obtain samples from any posterior distribution of interest (Givens and Hoeting, 2005, p. 183). Simulation techniques based on Markov chains overcome such problems because of their generality and flexibility. These characteristics, along with the massive development of computing facilities, have made Markov chain Monte Carlo (MCMC) techniques popular since the early 1990s. MCMC methods are not new, as they were introduced into physics in 1953 in a simplified version by Metropolis and his associates (Metropolis et al., 1953). Intermediate landmark publications include the generalization of Metropolis algorithm by Hastings (1970) and development of the Gibbs sampler by Geman and Geman (1984). Nevertheless, it took about 35 years until MCMC methods were rediscovered by Bayesian scientists (Tanner and Wong, 1987; Gelfand et al., 1990; Gelfand and Smith, 1990) and became one of the main computational tools in modern statistical inference.

Markov chain Monte Carlo techniques enabled quantitative researchers to use highly complicated models and estimate the corresponding posterior distributions with accuracy. In this way, MCMC methods have greatly contributed to the development and propagation of Bayesian theory. Extensive details of the use of MCMC methods can be found in Gilks et al. (1996). BUGS (Spiegelhalter et al., 1996*a*) and WinBUGS software (Spiegelhalter et al., 2003*d*) use MCMC techniques to generate samples from posterior distribution of complicated models, providing an effective way to evaluate Bayesian models.

In the following sections, we will provide the elementary notions of MCMC algorithms in order to introduce the reader to the terminology and the logic of MCMC simulation. MCMC techniques are based on the construction of a Markov chain that eventually "converges" to the target distribution (called *stationary* or *equilibrium*) which, in our case, is the posterior distribution $f(\boldsymbol{\theta}|\boldsymbol{y})$. This is the main way to distinguish MCMC algorithms from "direct" simulation methods, which provide samples directly from the target — posterior distribution. Moreover, the MCMC output is a dependent sample since it is generated from a Markov chain, in contrast to the output of "direct" methods, which is an independent sample. Finally, MCMC methods incorporate the notion of an *iterative* procedure (for this reason they are frequently called *iterative methods*) since in every step they produce values depending on the previous one.

2.2.1 The algorithm

A *Markov chain* is a stochastic process $\{\boldsymbol{\theta}^{(1)}, \boldsymbol{\theta}^{(2)}, \dots, \boldsymbol{\theta}^{(T)}\}$ such that

$$f\left(\boldsymbol{\theta}^{(t+1)}\big|\boldsymbol{\theta}^{(t)}, \dots, \boldsymbol{\theta}^{(1)}\right) = f\left(\boldsymbol{\theta}^{(t+1)}\big|\boldsymbol{\theta}^{(t)}\right);$$

that is, the distribution of $\boldsymbol{\theta}$ at sequence $t+1$ given all the preceding $\boldsymbol{\theta}$ values (for times $t, t-1, \dots, 1$) depends only on the value $\boldsymbol{\theta}^{(t)}$ of the previous sequence t. Moreover, $f\left(\boldsymbol{\theta}^{(t+1)}\big|\boldsymbol{\theta}^{(t)}\right)$ is independent of time t. Finally, when the Markov chain is irreducible, aperiodic, and positive-recurrent, as $t \to \infty$ the distribution of $\boldsymbol{\theta}^{(t)}$ converges to its equilibrium distribution, which is independent of the initial values of the chain $\boldsymbol{\theta}^{(0)}$; for details, see Gilks et al. (1996).

In order to generate a sample from $f(\boldsymbol{\theta}|\boldsymbol{y})$, we must construct a Markov chain with two desired properties: (1) $f(\boldsymbol{\theta}^{(t+1)}|\boldsymbol{\theta}^{(t)})$ should be "easy to generate from", and (2) the equilibrium distribution of the selected Markov chain must be the posterior distribution of interest $f(\boldsymbol{\theta}|\boldsymbol{y})$.

Assuming that we have constructed a Markov chain with these requirements, we then

1. Select an initial value $\boldsymbol{\theta}^{(0)}$.

2. Generate T values until the equilibrium distribution is reached.

3. Monitor the converge of the algorithm using convergence diagnostics. If convergence diagnostics fail, we then generate more observations.

4. Cut off the first B observations.

5. Consider $\{ \boldsymbol{\theta}^{(B+1)}, \boldsymbol{\theta}^{(B+2)}, \cdots, \boldsymbol{\theta}^{(T)} \}$ as the sample for the posterior analysis.

6. Plot the posterior distribution (usually focus is on the univariate marginal distributions).

7. Finally, obtain summaries of the posterior distribution (mean, median, standard deviation, quantiles, correlations).

In these steps, we refer to *convergence diagnostics*, which are statistical tests that attempt to identify cases where convergence is not achieved. More details follow in the next section, along with terminology and additional implementation details of the preceding steps.

2.2.2 Terminology and implementation details

In this section we present the basic concepts related to MCMC algorithms. It is divided into four subsections. Section 2.2.2.1 briefly introduces the reader to the basic terminology used in the topic, while Sections 2.2.2.2–2.2.2.4 focus on more specific matters such as the analysis of the MCMC sample, estimation of Monte Carlo variability measures, and convergence of the algorithm.

2.2.2.1 *Definitions and initial terminology.*   In this subsection we present epigrammatically the most important notions of MCMC methodology. We avoided unnecessary long discussion of the topic in order to to introduce readers in a straightforward manner. More detailed description of related topics can be found in more specialized books related to MCMC methodology; see for example in Gilks et al. (1996), Gamerman and Lopes (2006), and Givens and Hoeting (2005).

Equilibrium distribution.   This is called the *stationary* or *target distribution* of the MCMC algorithm. The notion of the equilibrium distribution is related to the Markov chain used to construct the MCMC algorithm. Such chains stabilize to the equilibrium/stationary distribution after a number of time sequences $t > B$. Therefore, in a Markov chain, the distribution of $\boldsymbol{\theta}^{(t)}$ and $\boldsymbol{\theta}^{(t+1)}$ will be identical and equal to the equilibrium/stationary distribution. Equivalently, once it reaches its equilibrium (distribution), an MCMC scheme generates dependent random values from the corresponding stationary distribution (Robert and Casella, 2004, pp. 206–207).

Convergence of the algorithm.   With the term *convergence* of an MCMC algorithm, we refer to situations where the algorithm has reached its equilibrium and generates values from the desired target distribution. Generally it is unclear how much we must run an algorithm to obtain samples from the correct target distributions. Several diagnostic tests have been developed to monitor the convergence of the algorithm; for more details, see Section 2.2.2.4.

Iteration.   *Iteration* refers to a cycle of the algorithm that generates a full set of parameter values from the posterior distribution. It is frequently used to denote an observation of simulated values. Here iteration is denoted by superscript number or the corresponding index t in parentheses. For example, $\boldsymbol{\theta}^{(5)}$ and $\boldsymbol{\theta}^{(t)}$ respectively denote the values of random vector $\boldsymbol{\theta}$ generated at the 5th and tth iterations of the algorithm.

Total number of iterations T.   This refers to the total number of the iterations of the MCMC algorithm .

Initial values of the chain $\theta^{(0)}$.   The starting values used to initialize the chain are simply called initial values. These initial values may influence the posterior summaries if they are far away from the highest posterior probability areas and the sample size of the simulated sample T is insufficient to eliminate its effect. We can mitigate or avoid the influence of the initial values by removing the first iterations of the algorithm or letting the algorithm run for a large number of iterations or obtain different samples with different starting points. Other researchers, in order to ensure that the values will be close to the center of the posterior distribution, select as starting points the posterior mode or the maximum likelihood values if they are easy to obtain. The latter choice often provides poor starting points according to many researchers, including Radford Neal (Kass et al., 1998, p. 96). Other reasonable starting points can be based on the mean or the mode of the prior distribution when informative priors are used. Finally, for problems with multiple modes or "ridges" (modes in highly correlated multivariate densities), using multiple chains with different starting points is highly recommended; for more details, see Brooks (1998) and the fruitful discussion by Kass et al. (1998).

Burnin period.   In the *burnin period* the first B iterations are eliminated from the sample in order to avoid the influence of the initial values. If the generated sample is large enough, the effect of this period on the calculation of posterior summaries is minimal.

Thinning interval or sampling lag.   As has already been mentioned, the final MCMC generated sample is not independent. For this reason, we need to monitor the autocorrelations of the generated values and select a sampling lag $L > 1$ after which the corresponding autocorrelation are low. Then, we can produce an independent sample by keeping the first generated values in every batch of L iterations. Hence, if we consider a lag (or thin interval) of three iterations then we keep the first every three iterations (i.e., we keep observations 1, 4, 7, etc.). This tactic is also followed to save storage space or computational speed in high-dimensional problems.

Iterations kept, T'.   These are the number of the iterations retained after discarding the initial burnin iterations (i.e., $T' = T - B$). If we also consider a sampling lag $L > 1$, then the total number of iterations kept refers to the final independent sample used for posterior analysis.

MCMC output.   This refers to the MCMC generated sample. We often refer to the MCMC output as the sample after removing the initial iterations (produced during the burnin period) and considering the appropriate lag.

Output analysis.   This refers to analysis of the MCMC output sample. It includes both the monitoring procedure of the algorithm's convergence and analysis of the sample used for the description of the posterior distribution and inference about the parameters of interest; for more details, see Sections 2.2.2.2 and 2.2.2.4, which follow.

2.2.2.2 *Describing the target distribution using MCMC output.*   The MCMC output provides us with a random sample of the type

$$\theta^{(1)}, \theta^{(2)}, \dots, \theta^{(t)}, \dots, \theta^{(T')}.$$

From this sample, for any function $G(\boldsymbol{\theta})$ of the parameters of interest $\boldsymbol{\theta}$ we can

1. Obtain a sample of the desired parameter $G(\boldsymbol{\theta})$ by simply considering

$$G\left(\boldsymbol{\theta}^{(1)}\right), G\left(\boldsymbol{\theta}^{(2)}\right), \dots, G\left(\boldsymbol{\theta}^{(t)}\right), \dots, G\left(\boldsymbol{\theta}^{(T')}\right).$$

2. Obtain any posterior summary of $G(\boldsymbol{\theta})$ from the sample using traditional sample estimates. For example, we can estimate the posterior mean by

$$\widehat{E}\left(G(\boldsymbol{\theta})\middle|\boldsymbol{y}\right) = \overline{G(\boldsymbol{\theta})} = \frac{1}{T'}\sum_{t=1}^{T'} G\left(\boldsymbol{\theta}^{(t)}\right)$$

and the posterior standard deviation by

$$\widehat{SD}\left(G(\boldsymbol{\theta})\middle|\boldsymbol{y}\right) = \frac{1}{T'-1}\sum_{t=1}^{T'}\left[G\left(\boldsymbol{\theta}^{(t)}\right) - \widehat{E}\left(G(\boldsymbol{\theta})\middle|\boldsymbol{y}\right)\right]^2.$$

Other measures of interest might be the posterior median or quantiles (e.g., 2.5% and 97.5% percentiles will provide a 95% credible interval). Finally, the posterior mode can be estimated by an MCMC sample by simply tracing the value of $\boldsymbol{\theta}$, which maximizes the posterior. This estimate cannot be considered as a reliable one, and hence it is preferable to directly consider optimization methods if interest lies in estimating the mode of the posterior distribution.

3. Calculate and monitor correlations between parameters.

4. Produce plots of the marginal posterior distributions (histograms, density plots, error bars, boxplots, etc.).

2.2.2.3 Monte Carlo error.
In the analysis of the MCMC output, an important measure that must be reported and monitored is the *Monte Carlo error* (MC error), which measures the variability of each estimate due to the simulation. MC error must be low in order to calculate the parameter of interest with increased precision. It is proportional to the inverse of the generated sample size that can be controlled by the user. Therefore, for a sufficient number of iterations T, the quantity of interest can be estimated with increased precision.

The two most common ways to estimate MC error are: the *batch mean* method and the *window estimator* method. The first one is simple and easy to implement and, for this reason, popular, while the latter is more precise.

In order to calculate the MC error using the batch means method, we simply partition the resulting output sample in K batches (usually $K = 30$ or $K = 50$). Both the number of batches K and the sample size of each batch $\nu = T'/K$ must be sufficiently large in order to enable us to estimate the variance consistently and also eliminate autocorrelations (Carlin and Louis, 2000, p. 172).

To calculate the Monte Carlo error of the posterior mean of $G(\boldsymbol{\theta})$, we first calculate each batch mean $\overline{G(\boldsymbol{\theta})}_b$ by

$$\overline{G(\boldsymbol{\theta})}_b = \frac{1}{\nu}\sum_{t=(b-1)\nu+1}^{b\nu} G\left(\boldsymbol{\theta}^{(t)}\right)$$

for each batch $b = 1, \ldots, K$, and the overall sample mean by

$$\overline{G(\boldsymbol{\theta})} = \frac{1}{T'} \sum_{t=1}^{T'} G(\boldsymbol{\theta}^{(t)}) = \frac{1}{K} \sum_{b=1}^{K} \overline{G(\boldsymbol{\theta})}_b,$$

assuming that we keep $\boldsymbol{\theta}^{(1)}, \ldots, \boldsymbol{\theta}^{(T')}$ observations. Then an estimate of the MC error is simply given by the standard deviation of the batch means estimates $\overline{G(\boldsymbol{\theta})}_b$

$$
\begin{aligned}
\mathrm{MCE}\big[G(\boldsymbol{\theta})\big] &= \widehat{\mathrm{SE}}\left[\overline{G(\boldsymbol{\theta})}\right] = \sqrt{\frac{1}{K}} \widehat{\mathrm{SD}}\left[\overline{G(\boldsymbol{\theta})}_b\right] \\
&= \sqrt{\frac{1}{K(K-1)} \sum_{b=1}^{K} \left(\overline{G(\boldsymbol{\theta})}_b - \overline{G(\boldsymbol{\theta})}\right)^2}.
\end{aligned}
$$

The procedure for calculating the MC error for any other posterior quantity of interest $\hat{U} = U(\boldsymbol{\theta}^{(1)}, \ldots, \boldsymbol{\theta}^{(T')})$ is equivalent. To estimate the corresponding Monte Carlo error, we calculate $\hat{U}_b = U(\boldsymbol{\theta}^{((b-1)\nu+1)}, \ldots, \boldsymbol{\theta}^{(b\nu)})$ from each batch $b = 1, \ldots, K$ and then the MC error by

$$\mathrm{MCE}(\hat{U}) = \sqrt{\frac{1}{K(K-1)} \sum_{b=1}^{K} \left(\hat{U}_b - \hat{U}\right)^2}.$$

The batch mean estimator of the Monte Carlo error is discussed in more detail by Hastings (1970), Geyer (1992), Roberts (1996, p. 50), Carlin and Louis (2000, p. 172), and Givens and Hoeting (2005, p. 208).

The second method (window estimator) is based on the expression of the variance in autocorrelated samples given by Roberts (1996, p. 50)

$$\mathrm{MCE}\big[G(\boldsymbol{\theta})\big] = \frac{\widehat{\mathrm{SD}}\big[G(\boldsymbol{\theta})\big]}{\sqrt{T'}} \sqrt{1 + 2 \sum_{k=1}^{\infty} \hat{\rho}_k\big[G(\boldsymbol{\theta})\big]},$$

where $\hat{\rho}_k\big[G(\boldsymbol{\theta})\big]$ is the estimated autocorrelation of lag k, that is, the correlation between parameters $G(\boldsymbol{\theta}^{(t)})$ and $G(\boldsymbol{\theta}^{(t+k)})$. Thus, it is obvious that for large k the autocorrelations will not be estimated reliably from the sample because of the small number of remaining observations. Moreover, in practice the autocorrelation will be close to zero for a sufficiently large k. For this reason, we identify a window w after which autocorrelations are considerably low [say, < 0.1 (Carlin and Louis, 2000)] and discard $\hat{\rho}_k$ with $k > w$ from the preceding MC error estimate. Hence, this window based modified MC error estimate is given by

$$\mathrm{MCE}\big[G(\boldsymbol{\theta})\big] = \frac{\widehat{\mathrm{SD}}\big[G(\boldsymbol{\theta})\big]}{\sqrt{T'}} \sqrt{1 + 2 \sum_{k=1}^{w} \hat{\rho}_k\big[G(\boldsymbol{\theta})\big]}.$$

The quantity $\mathrm{ESS} = N / \sqrt{1 + 2 \sum_{k=1}^{w} \hat{\rho}_k\big[G(\boldsymbol{\theta})\big]}$ is also referred to as the *effective sample size* (Kass et al., 1998).

Other methods for estimating Monte Carlo error are discussed by Geyer (1992) and Carlin and Louis (2000, pp. 170–172).

2.2.2.4 Convergence of the algorithm.   This term refers to whether the algorithm has reached its equilibrium (target) distribution. If this is true, then the generated sample comes from the correct target distribution. Hence, monitoring the convergence of the algorithm is essential for producing results from the posterior distribution of interest.

There are many ways to monitor convergence. The simplest way is to monitor the MC error (calculated as described above) since small values of this error will indicate that we have calculated the quantity of interest with precision. Monitoring autocorrelations is also very useful since low or high values indicate fast or slow convergence, respectively.

A second way is to monitor the *trace plots*: the plots of the iterations versus the generated values. If all values are within a zone without strong periodicities and (especially) tendencies, then we can assume convergence. An example of trace plots from an MCMC run is provided in Figure 2.2. In the first trace plot, we can clearly see the burnin period (within the gray box), which must be discarded from the final sample. After this period the generated sampled values, are stabilized within a zone. In the second plot, the initial 200 iterations have been discarded to monitor the sampled values which demonstrate much better behavior with small periodicities (up and down periods in the graph). Finally, generated observations of the last trace plot are more convincing in terms of convergence, with all generated values within a parallel zone and no obvious tendencies or periodicities.

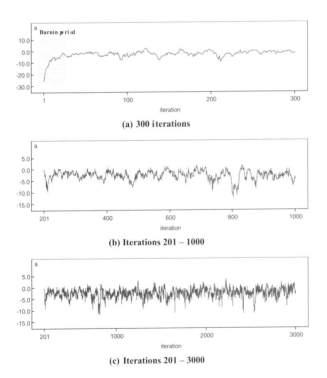

(a) 300 iterations

(b) Iterations 201 – 1000

(c) Iterations 201 – 3000

Figure 2.2   Trace plots (iterations vs. generated values) for (a) 300 iterations, (b) 1000 iterations after discarding the first 200, and (c) 3000 iterations after discarding the first 200.

Another useful plot is produced by depicting the evolution of the ergodic mean of a quantity over the number of iterations. The term *ergodic mean* refers to the mean value until the current iteration. If the ergodic mean stabilizes after some iterations, then this is an indication of the convergence of the algorithm.

Another tactic, which is very efficient in practice, is to run multiple chains with different starting points. When we observe that the lines of different chains mix or cross in trace and/or the ergodic mean plots, then convergence is ensured.

Finally, several statistical tests have been developed and used as convergence diagnostics (Cowles and Carlin, 1996; Brooks and Roberts, 1998). CODA (Best et al., 1996) and BOA (Smith, 2005) software programs have been developed in order to implement such diagnostics to the output of BUGS and WinBUGS software. Note that all convergence diagnostics work like alarms that sound when they detect an unexpected anomaly in the MCMC output. Each diagnostic test is constructed to detect different problems and hence, in most cases, all diagnostics must be applied to ensure that convergence has been reached.

To summarize, experienced users follow simple and fast tactics for monitoring convergence, including plotting autocorrelations, trace plots, and ergodic means. More advanced techniques such as multiple-chain comparisons and convergence diagnostics must be used for high dimensional complicated posterior distributions; see Kass et al. (1998, sec. 2.2) for further suggestions on the subject. We may improve the mixing of the chain, reducing in this way the time to reach convergence, by implementing approaches such as transformation of the parameters of interest or selecting an efficient method appropriate for the specific problem under consideration; for more details, see Gilks and Roberts (1996).

2.3   POPULAR MCMC ALGORITHMS

The two most popular MCMC methods are: the Metropolis–Hastings algorithm (Metropolis et al., 1953; Hastings, 1970) and the Gibbs sampling (Geman and Geman, 1984).

Many variants and extensions of these algorithms have been developed. Although they are based on the principles of the original algorithms, most of these algorithms are more advanced and complicated than the original ones and usually focus on specific problems. Some important more recent developments reported in the MCMC literature are the slice sampler (Higdon, 1998; Damien et al., 1999; Neal, 2003), the reversible jump MCMC (RJMCMC) algorithm (Green, 1995), and perfect sampling (Propp and Wilson, 1996; Møller, 1999).

In the following subsections, we focus on these two most popular methods (Metropolis–Hastings algorithm and the Gibbs sampler). For additional information concerning MCMC-specific methods, see Gilks et al. (1996), Robert and Casella (2004), Givens and Hoeting (2005), and Gamerman and Lopes (2006).

2.3.1   The Metropolis–Hastings algorithm

Metropolis et al. (1953) originally formulated the *Metropolis algorithm*, by introducing the Markov-chain-based simulation methods used in science. Later, Hastings (1970) generalized the original method in what is known as the *Metropolis–Hastings* algorithm. The latter is considered to be the general formulation of all MCMC methods. Green (1995) further generalized the Metropolis–Hastings algorithm by introducing *reversible jump Metropolis–Hastings* algorithms for sampling from parameter spaces with different dimensions.

Let as assume a target distribution $f(x)$ from which we wish to generate a sample of size T. The Metropolis–Hastings algorithm can be described by the following iterative steps; where $x^{(t)}$ is the vector of generated values in t iteration of the algorithm:

1. Set initial values $x^{(0)}$.

2. For $t = 1, \ldots, T$ repeat the following steps

   a. Set $x = x^{(t-1)}$

   b. Generate new candidate values x' from a proposal distribution $q(x \rightarrow x') = q(x'|x)$.

   c. Calculate
   $$\alpha = \min\left(1, \frac{f(x')q(x|x')}{f(x)\,q(x'|x)}\right).$$

   d. Update $x^{(t)} = x'$ with probability α and $x^{(t)} = x = x^{(t-1)}$ with probability $1 - \alpha$

The Metropolis–Hastings algorithm will converge to its equilibrium distribution regardless of whatever proposal distribution q is selected. Nevertheless, in practice, the choice of the proposal is important since poor choices will considerably delay convergence towards the equilibrium distribution.

The algorithm outlined above can be directly implemented in Bayesian framework by substituting x by the parameters of interest θ and the target distribution $f(x)$ by the posterior distribution $f(\theta|y)$. Thus, in Bayesian inference, the algorithm is summarized as follows:

1. Set initial values $\theta^{(0)}$.

2. For $t = 1, \ldots, T$ repeat the following steps

   a. Set $\theta = \theta^{(t-1)}$

   b. Generate new candidate parameter values θ' from a proposal distribution $q(\theta'|\theta)$.

   c. Calculate
   $$\alpha = \min\left(1, \frac{f(\theta'|y)q(\theta|\theta')}{f(\theta|y)\,q(\theta'|\theta)}\right).$$

   d. Update $\theta^{(t)} = \theta'$ with probability α; otherwise set $\theta^{(t)} = \theta$.

An important characteristic of the algorithm is that we do not need to evaluate the normalizing constant $f(y)$ involved in $f(\theta|y)$ since it cancels out in α. Hence the acceptance probability is simplified to

$$\alpha = \min\left(1, \frac{f(y|\theta')f(\theta')q(\theta|\theta')}{f(y|\theta)f(\theta)\,q(\theta'|\theta)}\right).$$

In order to simplify notation, in the following we denote the current state of the chain without any superscripts (e.g., by x or θ).

Special cases of the Metropolis–Hastings algorithm are the *random-walk Metropolis*, the *independence sampler*, the *single-component Metropolis–Hastings*, and the *Gibbs sampler*. These frequently used algorithm adaptations are described below.

2.3.1.1 *Random-walk Metropolis*
In the original Metropolis algorithm (Metropolis et al., 1953), only symmetric proposals of type $q(\theta'|\theta) = q(\theta|\theta')$ were considered. Random-walk Metropolis is a special case with $q(\theta'|\theta) = q(|\theta' - \theta|)$. Both cases result in an acceptance probability that depends only on the posterior (target) distribution

$$\alpha = \min\left(1, \frac{f(\theta'|y)}{f(\theta|y)}\right) = \min\left(1, \frac{f(y|\theta')f(\theta')}{f(y|\theta)f(\theta)}\right).$$

A usual proposal of this type is a multivariate normal $q(\boldsymbol{\theta}'|\boldsymbol{\theta}) \equiv N_d(\boldsymbol{\theta}, \overline{S}_{\boldsymbol{\theta}})$, where d is the dimension of $\boldsymbol{\theta}$. The covariance matrix $\overline{S}_{\boldsymbol{\theta}}$ controls the convergence speed of the algorithm. The values of the variances $\overline{S}_{ii}$ determine how close the proposed and current values will be. Small values of $\overline{S}_{ii}$ will result in high acceptance rates but slow convergence since the algorithm will need a large number of iterations to explore the parameter space. In this case, large autocorrelations will appear in the output analysis. On the other hand, high values of the proposal variances $\overline{S}_{ii}$ will result in low acceptance rates. As a consequence, for a large number of iterations the algorithm will stick with the same values, again resulting in poor exploration of the parameter space and a highly autocorrelated sample. In order to specify the elements of $\overline{S}_{\boldsymbol{\theta}}$, the user has to test the performance of the sampler by using several values until the desired acceptance rate is achieved; see Roberts and Rosenthal (2001) for details. This empirical procedure is called *tuning of the proposal distribution*. The corresponding quantities are called *tuning parameters* of the proposal. In random walk samplers the optimal acceptance rate according to Roberts et al. (1997) and Neal and Roberts (2008) is around 25%, ranging from 0.23 for large dimensions to 0.45 for the univariate case; see also in Roberts and Rosenthal (2001). Here we recommend tuning the variance of the proposal density such that the acceptance rates lie within the interval of 20–40%, which are the values also proposed and used by Spiegelhalter et al. (2003d, p. 6). This range is in concordance with the range of 10–40% also suggested by Roberts and Rosenthal (2001) after observing that "there is little to be gained from fine tuning of acceptance rates."

The correlations involved in the proposal covariance matrix $\overline{S}_{\boldsymbol{\theta}}$ influence the direction of the proposed movement. In order to construct an effective Metropolis algorithm, the correlation matrix of the proposal and the target posterior distribution must be equivalent. A good strategy is to run a small pilot run using diagonal $\overline{S}_{\boldsymbol{\theta}}$ (i.e., independent normal proposals) to roughly estimate the correlation structure of the target posterior distribution and then rerun the algorithm using the corresponding estimated variance–covariance matrix $\widetilde{\Sigma}_{\boldsymbol{\theta}}$. After the posterior variance–covariance matrix has been estimated, Gelman et al. (1995, pp. 334–335) suggest that the proposal variance–covariance matrix be set equal to $\overline{S}_{\boldsymbol{\theta}} = c^2 \widetilde{\Sigma}_{\boldsymbol{\theta}}$ with $c^2 \approx 5.8/d$, where d is the dimension of the parameter vector.

2.3.1.2 *The independence sampler.*

The independence sampler is a Metropolis–Hastings algorithm where the proposal distribution does not depend on the previous state $\boldsymbol{\theta}^{(t-1)}$ of the chain. For example a frequent choice is a multivariate normal distribution of the type $\boldsymbol{\theta}' \sim N_d(\overline{\boldsymbol{\theta}}, \overline{S}_{\boldsymbol{\theta}})$. The parameters (mean and variance) of this proposal can be obtained using approximation methods or any available previous experience or expert information.

The independence sampler is efficient when the proposal $q(\boldsymbol{\theta})$ is a good approximation of the target posterior distribution $f(\boldsymbol{\theta}|\boldsymbol{y})$. Good independent proposal densities can be based on Laplace approximation (Tierney and Kadane, 1986; Tierney et al., 1989; Erkanli, 1994). Thus, a generally successful proposal can be obtained by a multivariate normal distribution with mean equal to the posterior mode $\widetilde{\boldsymbol{\theta}}$ and precision matrix

$$\mathbf{H}(\widetilde{\boldsymbol{\theta}}) = \left(-\frac{\partial^2 \log f(\boldsymbol{\theta}|\boldsymbol{y})}{\partial \theta_i \partial \theta_j} \Big|_{\boldsymbol{\theta}=\widetilde{\boldsymbol{\theta}}} \right),$$

that is, minus the second derivative matrix of the log-posterior density

$$\log f(\boldsymbol{\theta}|\boldsymbol{y}) = \text{constant} + \log f(\boldsymbol{y}|\boldsymbol{\theta}) + \log f(\boldsymbol{\theta})$$

evaluated at the posterior mode $\tilde{\boldsymbol{\theta}}$. Consequently, an efficient proposal is given by

$$q(\boldsymbol{\theta}) = N_d \left(\tilde{\boldsymbol{\theta}}, \; \left[\mathbf{H}(\tilde{\boldsymbol{\theta}}) \right]^{-1} \right).$$

The posterior mode can be evaluated by usual optimization methods. When low information prior is used, then an adequate proposal can be obtained by setting the mean equal to the corresponding maximum-likelihood estimator (MLE) and the precision equal to its observed Fisher information matrix.

The acceptance probability, when proposing a transition from $\boldsymbol{\theta}$ to $\boldsymbol{\theta}'$, is given by

$$\alpha = \min \left(1, \frac{f(\boldsymbol{\theta}'|\boldsymbol{y}) \, q(\boldsymbol{\theta})}{f(\boldsymbol{\theta}|\boldsymbol{y}) \, q(\boldsymbol{\theta}')} \right) = \min \left(1, \frac{f(\boldsymbol{y}|\boldsymbol{\theta}') f(\boldsymbol{\theta}') \, q(\boldsymbol{\theta})}{f(\boldsymbol{y}|\boldsymbol{\theta}) f(\boldsymbol{\theta}) \, q(\boldsymbol{\theta}')} \right),$$

which can be reexpressed as

$$\alpha = \min \left(1, \frac{w(\boldsymbol{\theta}')}{w(\boldsymbol{\theta})} \right),$$

where $w(\boldsymbol{\theta}) = f(\boldsymbol{\theta}|\boldsymbol{y})/q(\boldsymbol{\theta})$ is the ratio between the target and the proposal distribution and is equivalent to the importance weight used in importance sampling. In fact, the two approaches are very close, with the latter giving more weight to points with high weights $w(\boldsymbol{\theta})$ (Brooks, 1998).

In contrast to the random-walk Metropolis, where the optimal acceptance rate is around 0.25, in the independence sample the acceptance rate must be high enough to obtain an efficient algorithm. Although high acceptance rates indicate that the proposal is a sufficient approximation of the target posterior distribution, according to Chib and Greenberg (1995, p. 330), the tails of the proposal density must be fatter than the corresponding ones of the posterior distribution in order to obtain an efficient algorithm; see also Gamerman and Lopes (2006, pp. 199–204) for an interesting discussion concerning the selection of the proposal in this case.

2.3.2 Componentwise Metropolis–Hastings

In componentwise Metropolis–Hastings algorithm, the parameter vector $\boldsymbol{\theta}$ is divided into subvectors that are updated sequentially using Metropolis–Hastings steps. Gibbs sampling, which is discussed in Section 2.3.3, is a special case of this algorithm. The algorithm is also called *Metropolis within Gibbs* (see Section 2.3.4 for details) or single-component Metropolis–Hastings algorithm when univariate components are updated sequentially.

In each step of the single-component Metropolis algorithm, a candidate value θ'_j of the jth component of the vector $\boldsymbol{\theta}$ is proposed by $q_j\left(\theta'_j|\boldsymbol{\theta}^{(t-1)}\right)$. The algorithm can be summarized by the following steps:

1. Set initial values $\boldsymbol{\theta}^{(0)}$.

2. For $t = 1, \dots, T$ repeat the following steps

   a. Set $\boldsymbol{\theta} = \boldsymbol{\theta}^{(t-1)}$.

   b. For $j = 1, \dots, d$

      (1) Generate new candidate parameter values θ'_j for j component of vector $\boldsymbol{\theta}'$ from a proposal distribution $q(\theta'_j|\boldsymbol{\theta})$.

(2) Calculate

$$
\begin{aligned}
\alpha &= \min\left(1, \frac{f(\theta'_j|\boldsymbol{\theta}_{\backslash j}, \boldsymbol{y})q(\theta_j|\theta'_j, \boldsymbol{\theta}_{\backslash j})}{f(\theta_j|\boldsymbol{\theta}_{\backslash j}, \boldsymbol{y})q(\theta'_j|\theta_j, \boldsymbol{\theta}_{\backslash j})}\right) \\
&= \min\left(1, \frac{f(\boldsymbol{y}|\theta'_j, \boldsymbol{\theta}_{\backslash j})f(\theta'_j, \boldsymbol{\theta}_{\backslash j})q(\theta_j|\theta'_j, \boldsymbol{\theta}_{\backslash j})}{f(\boldsymbol{y}|\theta_j, \boldsymbol{\theta}_{\backslash j})f(\theta_j, \boldsymbol{\theta}_{\backslash j})q(\theta'_j|\theta_j, \boldsymbol{\theta}_{\backslash j})}\right),
\end{aligned}
$$

where $\boldsymbol{\theta}_{\backslash j}$ is the vector $\boldsymbol{\theta}$ excluding its jth component θ_j [i.e., $\boldsymbol{\theta}_{\backslash j} = (\theta_1, \theta_2, \ldots, \theta_{j-1}, \theta_{j+1}, \ldots, \theta_d)$].
(3) Update $\theta_j = \theta'_j$ with probability α.

c. Set $\boldsymbol{\theta}^{(t)} = \boldsymbol{\theta}$

One generated observation $\boldsymbol{\theta}^{(t)}$ is obtained after updating all components of the parameter vector. Generally, the sequence of updating the elements of $\boldsymbol{\theta}$ does not influence the convergence of the algorithm. Nevertheless, to ensure randomness, random selection of the updating sequence may be also used; this is called *random scan*.

The advantage of this MCMC scheme is that the sampler is decomposed in several univariate steps in which random-number generation is usually straightforward. On the other hand, convergence of the chain cannot be accelerated by using multivariate proposal densities with appropriate correlation structure. To improve convergence in such cases, sampling can be implemented by parameter blocking. The parameter vector is divided into subvectors with correlated elements, which are called *blocks*, and each of them is updated in separate Metropolis steps. More details concerning blocking can be found in Gilks et al. (1996).

2.3.2.1 Simple examples. Two simple examples illustrating how we can build effective Metropolis–Hastings algorithms are presented in this subsection. In the first example we use a univariate binomial example for Kobe Bryant's field goals and in the second example we consider a simple logistic regression model with two parameters under consideration. Detailed implementation of all methods, comments concerning their performance, and interpretation of the results are provided.

Example 2.2. Univariate example: Posterior distribution of odds and log-odds in binomial data. Consider the Kobe Bryant's success data for 2006-2007 season. As we have already mentioned, posterior distribution of the success probability can be easily obtained analytically using a conjugate prior distribution. Frequently, when binomial data are used, there is focus on the estimation of odds. Its posterior distribution can be easily evaluated either analytically or by using a sample derived from direct sampling.

In this example, for illustration, the likelihood is expressed in terms of the log-odds $\theta = \log[\pi/(1-\pi)]$ instead of the original parameter π; log-odds are a convenient parameter choice since they are defined in the set of real numbers $\mathbb{R}$ and, thus, a normal prior can be adopted.

Estimating the log-odds using a random-walk Metropolis. Let us first rewrite the likelihood as a function of θ:

$$
f(y|\theta) = \binom{N}{y}\left(\frac{e^\theta}{1+e^\theta}\right)^y \left(\frac{1}{1+e^\theta}\right)^{N-y} = \binom{N}{y}\frac{e^{\theta y}}{(1+e^\theta)^N}
$$

$$= \exp\left\{\log\binom{N}{y} + \theta y - N\log(1 + e^\theta)\right\}.$$

Since $\theta \in \mathbb{R}$, it is natural to consider as a low-information prior a normal distribution with zero mean and large variance $\theta \sim N(\mu_\theta, \sigma_\theta^2)$, which results in

$$f(\theta|y) \propto \frac{e^{\theta y}}{(1 + e^\theta)^N} \exp\left[-\frac{1}{2}\left(\frac{\theta - \mu_\theta}{\sigma_\theta}\right)^2\right]. \tag{2.2}$$

The above distribution is univariate and therefore we can easily plot it to obtain a general overview of the posterior distribution. Nevertheless, it is still difficult to summarize the properties of the posterior distribution since evaluation of its summaries are not straightforward. Generating a random sample from (2.2) simplifies the evaluation of such posterior summaries.

To avoid numeric overflow problems, it is highly recommended to use the log-scale in all statistical computations. Using the log-scale in the following, a simple random-walk Metropolis can be summarized by

- Specify initial value $\theta^{(0)}$

- For $t = 1, \ldots, T$

    1. Set $\theta = \theta^{(t-1)}$

    2. Propose a new value θ' from $N(\theta, \bar{s}_\theta^2)$

    3. Calculate $\log\alpha = \min(0, A)$ with A given by

    $$A = \log\frac{f(y|\theta')f(\theta')}{f(y|\theta)f(\theta)} = (\theta' - \theta)\left(y - \frac{\theta' + \theta - 2\mu_\theta}{2\sigma_\theta^2}\right) - N\log\frac{1 + e^{\theta'}}{1 + e^\theta}$$

    4. Set $\theta^{(t)} = \theta'$ with probability α and $\theta^{(t)} = \theta$ with the remaining probability.

Parameter $\bar{s}_\theta^2$ is a tuning parameter that needs to be calibrated such that it achieves an acceptance rate approximately equal to 25%. Table 2.3 provides a detailed description of the algorithm on the left and the corresponding R commands on the right.

The algorithm and the corresponding program can be divided into two main parts: the *preamble*, where all parameters are initialized and defined, and the *main MCMC algorithm*, where the iterative procedure takes place (the for loop part). In the preamble we define the data, and the prior parameters and initialize any vectors that will be used by the algorithm.

In order to identify a well-performed value for $\bar{s}_\theta^2$, the algorithm needs to be run using several values until an acceptance rate close to 0.25 is achieved. For this reason, an additional counter must be added in the program. In order to initialize such counter in R, we add the following command

```
>   acc.prob<-0
```

and then update this counter within the `if` statement

```
>   if( u < loga ) {
+        current.theta<-proposed.theta
+        acc.prob <- acc.prob + 1
+        }
```

Table 2.3   Algorithm and R code for running a random walk Metropolis algorithm for log-odds in binomial data

Algorithm (pseudocode)	R code
Set up data	`> y<-399; N<-845`
Set number of iterations T	`> Iterations<-2500`
Set prior parameters: $\mu_\theta = 0, \sigma_\theta^2 = 10,000$	`> mu.theta<-0; s.theta<-100`
Set proposal parameter: $\bar{s}_\theta$	`> prop.s <- 0.35`
Initialize vector of sampled values $\boldsymbol{\theta}$	`> theta<-numeric(Iterations)`
Set initial current $\theta^{(0)}$	`> current.theta<-0`
For $t = 1, \dots, T$ repeat	`> for (t in 1:Iterations){`
Propose θ' from $N(\mu_\theta, \bar{s}_\theta^2)$	`>   prop.theta<-rnorm( 1, current.theta, prop.s )`
Calculate $\log \alpha =$ log-likelihood(θ')	`>   loga <-( (prop.theta*y - N*log(1+exp(prop.theta)))`
$-$ log-likelihood(θ)	`+     -(current.theta * y -N*log(1+exp(current.theta)))`
$+$ log-prior(θ')	`+     +dnorm(prop.theta, mu.theta, s.theta, log=TRUE)`
$-$ log-prior(θ)	`+     -dnorm(current.theta, mu.theta, s.theta,`
	`log=TRUE) )`
Generate u from $U(0,1)$	`>   u<-runif(1)`
Set $u = \log u$	`>   u<-log(u)`
If $u < \log \alpha$ then set $\theta^{(t)} = \theta'$	`>   if( u < loga ) current.theta<-prop.theta`
else $\theta^{(t)} = \theta$	`>   theta[t]<-current.theta`
	`> }`
End of `for` loop	

substituting line 16 of the code provided in Table 2.3. The object `acc.prob` provides the number of the accepted moves within the MCMC run.

After tuning, $\bar{s}_\theta = 0.35$ was finally adopted, achieving acceptance rates close to the target value of 0.25. Trace, ergodic mean, and autocorrelation function (ACF) plots for chains with $\bar{s}_\theta = 0.1, 0.2, 0.35$ and 0.85 (with corresponding acceptance rates approximately equal to 62%, 38%, 23%, and 11%, respectively) are illustrated in Figure 2.3. For $\bar{s}_\theta = 0.85$, long periods of constant θ are observed in the trace plot. High autocorrelations are observed in chains for $\bar{s}_\theta = 0.1$ and $\bar{s}_\theta = 0.85$. All posterior means appear to have converged to the same value.

Figure 2.4 demonstrates trace plots for chains using starting values $\theta^{(0)} = -3, -1.5, 0, 1, 5$. Trace plots for all observations and for observations after discarding a burnin period of 500 iterations are provided in the left and right columns, respectively, of Figure 2.4. The effect of the initial value is evident in the trace plots of the first and the last chains. This effect is also clear in Figure 2.5, where the evolution of the ergodic means for all iterations (on the left) and for observations after discarding the burnin period (on the right) is depicted.

Monte Carlo error values for the posterior mean using 25 batches of size equal to 80 iterations per batch (after discarding the initial 500 iterations) were found equal to 0.004, 0.0032, 0.00397, and 0.0052 for $\bar{s}_\theta = 0.1, 0.2, 0.35$, and 0.85 respectively. Hence the choice of $\bar{s}_\theta = 0.2$ (with acceptance rate equal to 0.38) is the best among the four proposal variance values used in this example, achieving the lowest sampling variability as expressed by the calculated Monte Carlo errors. The estimated posterior density in contrast to the true one [given by Eq. (2.2)] is portrayed in Figure 2.6.

Having generated the posterior sample for θ, we can easily obtain the corresponding sample for the odds o and the success probability π by equations

$$o^{(t)} = e^{\theta^{(t)}} \text{ and } \pi^{(t)} = \frac{o^{(t)}}{1 + o^{(t)}} = \frac{e^{\theta^{(t)}}}{1 + e^{\theta^{(t)}}} \text{ for } t = 1, \dots, T$$

and the corresponding R commands

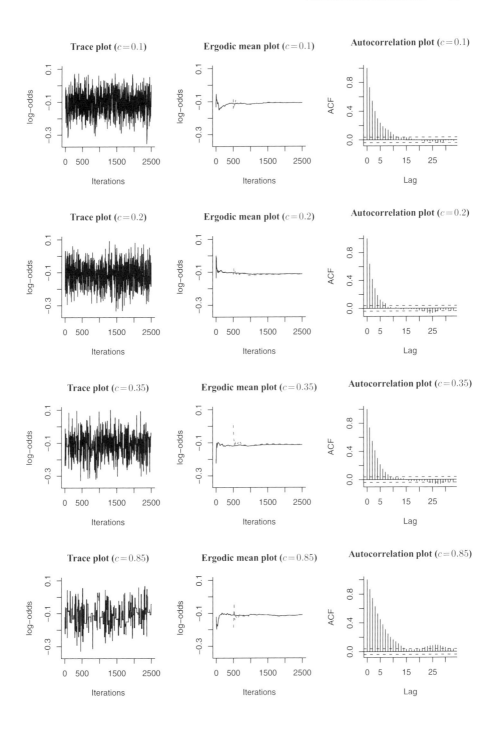

Figure 2.3   Diagnostic plots of Metropolis random-walk algorithm for various proposal parameters $c = \bar{s}_\theta$.

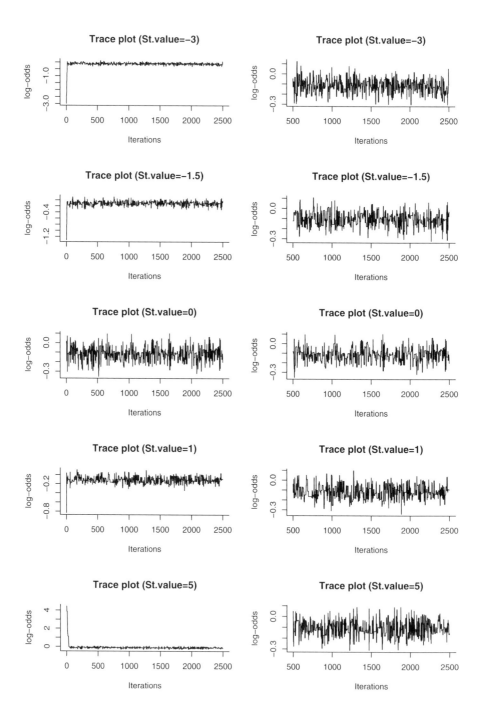

Figure 2.4   Trace plots of Metropolis random-walk algorithm for various starting values $\theta^{(0)}$ (St.=starting).

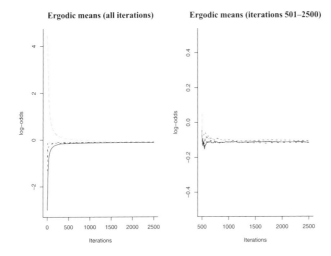

Figure 2.5   Ergodic mean plots of Metropolis random-walk algorithm for various starting values $\theta^{(0)}$.

```
> o<-exp(theta)
> p<-exp(theta)/(1+exp(theta))
```

Density plots of the estimated posterior distributions for the success odds o and probability π are given in Figures 2.7 and 2.8, respectively. Finally, posterior descriptive measures are given in Table 2.4.

Table 2.4   Posterior summaries for Kobe Bryant field goal success rates for season 2006–2007[a]

Posterior summary	Parameter		
	Log-odds (θ)	Odds (o)	Probability (π)
Mean	-0.112	0.896	0.472
Median	-0.109	0.897	0.473
Standard deviation	0.072	0.065	0.018
2.5% percentile	-0.261	0.770	0.435
97.5% percentile	0.026	1.026	0.507

[a]Random-walk, burnin $B = 500$ iterations; iterations kept $T' = 2000$.

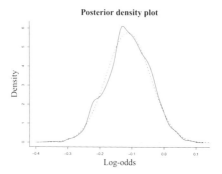

Figure 2.6   Posterior density plots of success log-odds θ of Kobe Bryant's field goals for season 2006–2007; dotted line indicates the true posterior given by Equation (2.2).

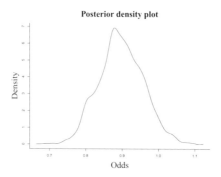

Figure 2.7   Posterior density plots of success odds $o = \pi/(1 - \pi)$ of Kobe Bryant's field goals for season 2006–2007.

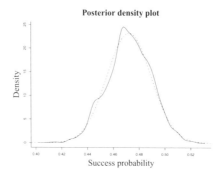

Figure 2.8   Posterior density plots of success probability π of Kobe Bryant's field goals for season 2006–2007; dotted line indicates the true posterior based on the posterior density beta$(399.01, 446.01)$ resulting from conjugate analysis.

Working using the odds. Alternatively, we may work directly using the success odds as the parameter of interest in the model likelihood. In this case the likelihood is expressed by

$$f(y|o) = \exp\left\{\log\binom{N}{y} + y\log o - N\log(1+o)\right\}.$$

Since $o > 0$, we can use either a gamma or a log-normal prior distribution. Here we adopt a log-normal prior distribution that is equivalent to the one used in the log-odds-based analysis since

$$\theta = \log o \sim N(\mu_\theta, \sigma_\theta^2) \Leftrightarrow o \sim \mathrm{LN}(\mu_\theta, \sigma_\theta^2),$$

where $\mathrm{LN}(\mu, \sigma^2)$ denotes the log-normal distribution with parameters μ and σ^2. With this prior, the posterior is given by

$$f(o|y) \propto o^{y-1}(1+o)^{-N}\exp\left[-\frac{1}{2}\left(\frac{\log o - \mu_\theta}{\sigma_\theta}\right)^2\right].$$

Since the parameter of interest o is positive, we cannot use a normal proposal distribution directly. We can still use a normal random walk on $\theta = \log o$, which results to a multiplicative version of the random-walk algorithm. The iterative step can now be written as

- For $t = 1, \ldots, T$

    1. Set $o = o^{(t-1)}$
    2. Propose a new value θ' from $N(\log o, \bar{s}_\theta^2)$. Set $o' = e^{\theta'}$.
    3. Calculate $\log\alpha = \min(0, A)$ with A given by

$$A = \log\frac{f(y|o')f(o')}{f(y|o)f(o)} \times \frac{o'}{o} = \left(\log\frac{o'}{o}\right)\left(y - \frac{\log(o'\,o) - 2\mu_\theta}{2\sigma_\theta^2}\right) - N\log\frac{1+o'}{1+o}$$

    4. Set $o^{(t)} = o'$ with probability α and $o^{(t)} = o$ with probability $1-\alpha$.

Variations of this algorithm can be constructed using any distribution defined on the set of the positive real numbers with mean equal to the parameter value of the previous state of the chain. Using the logic described above, a gamma proposal of type

$$o' \sim \mathrm{gamma}\left(\frac{o}{b}, \frac{1}{b}\right)$$

with mean equal to the previous value o of the chain and variance equal to $b\,o$ can be adopted. Parameter b is a tuning parameter of the proposal variance that must be calibrated to achieve appropriate acceptance rates. With this approach the acceptance probability is given by $\log\alpha = \min(0, A)$ with

$$A = \log\frac{f(y|o')f(o')}{f(y|o)f(o)} + \log\frac{f_\Gamma(o; o/b, 1/b)}{f_\Gamma(o'; o/b, 1/b)}$$

where $f_\Gamma(x; a, b)$ is the probability density function of the gamma(a, b) distribution.

An independence sampler for the log-odds for simple binomial data. An efficient independence Metropolis–Hastings sampler for θ can be constructed by considering the posterior mode as the mean of the proposal density. The posterior mode can be obtained by solving the equation $\partial \log f(\theta|y)/\partial \theta = 0$. In this example, the first derivative of the log–posterior density is given by

$$\frac{\partial \log f(\theta|y)}{\partial \theta} = y - N \frac{e^{\theta}}{1 + e^{\theta}} - \frac{\theta - \mu_{\theta}}{\sigma_{\theta}^2}, \tag{2.3}$$

for which an iterative numerical method can be used to obtain the posterior mode. In the case of a low-information prior distribution, the contribution of the prior part, appearing on the right side of (2.3), to the posterior will be negligible. Thus, the MLE $\hat{\theta} = \log[y/(N-y)]$ can be directly considered as the mean $\bar{\mu}_{\theta}$ of the proposal distribution. The proposal variance $\bar{s}_{\theta}^2$ can be set equal to the posterior variance $\tilde{\sigma}_{\theta}^2$ approximated by

$$
\begin{aligned}
\tilde{\sigma}_{\theta}^2 &\approx \left[-\frac{\partial^2 \log f(y|\theta)}{\partial \theta^2} \Big|_{\theta=\tilde{\theta}} \right]^{-1} = \left[N \frac{y}{N} \frac{N-y}{N} + \frac{1}{\sigma_{\theta}^2} \right]^{-1} = \left[\frac{y(N-y)}{N} + \frac{1}{\sigma_{\theta}^2} \right]^{-1} \\
&= \left[\left(\frac{1}{y} + \frac{1}{N-y} \right)^{-1} + \frac{1}{\sigma_{\theta}^2} \right]^{-1} = \underbrace{\left(\frac{1}{y} + \frac{1}{N-y} \right)}_{\text{Var}(\hat{\theta})} \left[1 + \frac{\left(\frac{1}{y} + \frac{1}{N-y} \right)}{\sigma_{\theta}^2} \right]^{-1}
\end{aligned} \tag{2.4}
$$

where $\text{Var}(\hat{\theta}) = [y^{-1} + (N-y)^{-1}]$ is the approximate variance of the maximum likelihood estimator.

The iterative step of the algorithm can be described as follows:

- For $t = 1, \dots, T$

    1. Set $\theta = \theta^{(t-1)}$.
    2. Propose a new value θ' from $N(\bar{\mu}_{\theta}, \bar{s}^2)$.
    3. Calculate $\log \alpha = \min(0, A)$ with A given by

$$
\begin{aligned}
A &= \log \frac{f(y|\theta')f(\theta')}{f(y|\theta)f(\theta)} + \log \frac{f_N(\theta; \bar{\mu}_{\theta}, \bar{s}_{\theta}^2)}{f_N(\theta'; \bar{\mu}_{\theta}, \bar{s}_{\theta}^2)} \\
&= (\theta' - \theta) \left(y - \frac{\theta' + \theta - 2\mu_{\theta}}{2\sigma_{\theta}^2} + \frac{\theta' + \theta - 2\bar{\mu}_{\theta}}{2\bar{s}_{\theta}^2} \right) - N \log \frac{1 + e^{\theta'}}{1 + e^{\theta}},
\end{aligned}
$$

    where $f_N(x; \mu, \sigma^2)$ is the probability density function of a $N(\mu, \sigma^2)$ evaluated at point x.
    4. Set $\theta^{(t)} = \theta'$ with probability α and $\theta^{(t)} = \theta$ with probability $1 - \alpha$.

Details of the algorithm are given on the left column, with the corresponding R commands in the right column of Table 2.5.

The proposed scheme is very effective with acceptance probability higher than 90%. As we have already mentioned, in the independence sampler we wish to obtain high acceptance rates in contrast to the random-walk approach, where the optimal acceptance rates are around 25%. Plots of the output are provided in Figure 2.9 and posterior summaries in the last column of Table 2.6. In the sampling scheme presented above we have selected as starting point the value of zero. An obvious "good" choice is the posterior mode $\tilde{\theta}$. If we initiate

Table 2.5   Algorithm and R code for running an independence Metropolis–Hastings algorithm for log-odds in binomial data

Algorithm (pseudocode)	R code
Set up data	`> y<-399; N<-845`
Set number of iterations T	`> Iterations<-1500`
Set prior parameters: $\mu_\theta = 0$, $\sigma_\theta^2 = 10,000$	`> mu.theta<-0; s.theta<-100`
Set proposal parameters	`> prop.mu <- log( y/(N-y) )`
	`> mle.var <- 1/y+1/(N-y); w<-1/(1+mle.var/s.theta^2)`
	`> prop.s <- sqrt( mle.var *w )`
Initialize vector of sampled values $\boldsymbol\theta$	`> theta <- numeric(Iterations)`
Initialize acceptance probability counter	`> acc.prob <- 0`
Set initial current $\theta^{(0)}$	`> current.theta<-0`
For $t = 1, \ldots, T$ repeat	`> for (t in 1:Iterations){`
Propose θ' from $N(\mu_\theta, \bar{s}_\theta^2)$	`>    prop.theta <- rnorm( 1, prop.mu, prop.s )`
Calculate $\log\alpha$ =log-likelihood(θ')	`>    loga <-( (prop.theta*y - N*log(1+exp(prop.theta)))`
− log-likelihood(θ)	`+      -(current.theta * y -N*log(1+exp(current.theta)))`
+ log-prior(θ')	`+      +dnorm(prop.theta, mu.theta, s.theta, log=TRUE)`
− log-prior(θ)	`+      -dnorm(current.theta, mu.theta, s.theta,`
	`log=TRUE)`
+ log-proposal($\theta; \bar{\mu}_\theta, \bar{s}_\theta^2$)	`+      +dnorm(current.theta, prop.mu, prop.s, log=TRUE)`
− log-proposal($\theta'; \bar{\mu}_\theta, \bar{s}_\theta^2$)	`+      -dnorm(prop.theta, prop.mu, prop.s, log=TRUE) )`
Generate u from $U(0, 1)$	`>    u<-runif(1)`
Set $u = \log u$	`>    u<-log(u)`
If $u < \log\alpha$ then set $\theta^{(t)} = \theta'$	`>    if( u < loga ) {`
and update acceptance probability	`>      current.theta<-prop.theta`
counter	`>      acc.prob <- acc.prob+1`
	`>    }`
else $\theta^{(t)} = \theta$	`>    theta[t]<-current.theta`
End of for loop	`> }`

the chain from remote values [e.g., $\theta^{(0)} = 5$], you will observe that the chain 'sticks' in this value. This can be avoided by considering a larger proposal variance of type $\bar{s}_\theta^2 = c_\theta^2 \tilde{\sigma}_\theta$, where c_θ is a tuning parameter and $\tilde{\sigma}_\theta^2$ is approximated using (2.4). This parameter must be specified to allow the chain to move even for remote values. For $c_\theta > 1$, we specify a proposal with larger variance and fatter tails than the target distribution as recommended by Chib and Greenberg (1995).

Table 2.6   Posterior summaries for Kobe Bryant's field goal success rates for season 2006–2007[a]

Posterior	Parameter		
summary	Log-odds (θ)	Odds (o)	Probability (π)
Mean	-0.111	0.897	0.472
Median	-0.110	0.895	0.472
Standard deviation	0.068	0.061	0.017
2.5% percentile	-0.247	0.781	0.438
97.5% percentile	0.021	1.022	0.505

[a] Independence sampler, burnin $B = 500$ iterations; iterations kept $T' = 1000$.

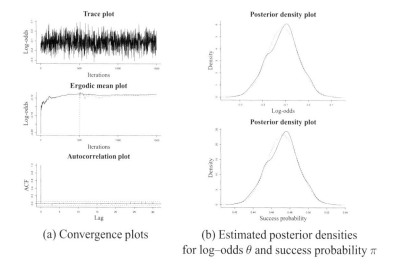

(a) Convergence plots   (b) Estimated posterior densities
for log–odds θ and success probability π

Figure 2.9   Posterior plots for Kobe Bryant's field goal success rates for season 2006–2007 (independence sampler, burnin $B = 500$ iterations; iterations kept $T' = 1000$).

Example 2.3. Metropolis–Hastings algorithms for the simple binary logistic regression: senility symptoms data. Let us consider the data of Agresti (1990, pp. 122–123), in which 54 elderly people completed a subtest of the Wechsler Adult Intelligence Scale (WAIS) resulting to a discrete score with range from 0 to 20. The aim of this study was to identify people with senility symptoms (binary variable) using the WAIS score. Interest also lies in calculating WAIS scores that correspond to increased probability of senility symptoms (i.e., with $\pi > 0.5$). The data of this example can be found in the book's Website and are reproduced with the permission of John Wiley and Sons, Inc.

In order to identify the effect of WAIS (x_i) on the senility symptoms (y_i), we use a simple logistic regression model

$$Y_i \sim \text{binomial}(\pi_i, N_i = 1), \quad \log \frac{\pi_i}{1 - \pi_i} = \beta_0 + \beta_1 x_i \text{ for } i = 1, \ldots, 54.$$

The likelihood of this model is given by

$$
\begin{aligned}
f(\boldsymbol{y}|\beta_0, \beta_1) &= \prod_{i=1}^{n} \left(\frac{e^{\beta_0+\beta_1 x_i}}{1 + e^{\beta_0+\beta_1 x_i}} \right)^{y_i} \left(\frac{1}{1 + e^{\beta_0+\beta_1 x_i}} \right)^{1-y_i} \\
&= \exp \left\{ n\bar{y}\beta_0 + \beta_1 \sum_{i=1}^{n} x_i y_i - \sum_{i=1}^{n} \log \left(1 + e^{\beta_0+\beta_1 x_i} \right) \right\}.
\end{aligned}
$$

A normal prior is frequently used for the parameters of the logistic regression model. Here we consider independent normal prior distributions with zero mean and large variance to express prior ignorance. Hence we consider

$$\beta_j \sim N(\mu_{\beta_j}, \sigma^2_{\beta_j}) \text{ for } j = 0, 1$$

using prior mean $\mu_{\beta_j} = 0$ and large prior standard deviation $\sigma_{\beta_j} = 100$ to express our prior ignorance. This setup results in the posterior density

$$
\begin{aligned}
f(\beta_0, \beta_1 | \boldsymbol{y}) \quad &\propto \quad f(\boldsymbol{y}|\beta_0, \beta_1) f(\beta_0, \beta_1) \\
&\propto \quad \exp \left\{ n\bar{y}\beta_0 + \beta_1 \sum_{i=1}^{n} x_i y_i - \sum_{i=1}^{n} \log\left(1 + e^{\beta_0 + \beta_1 x_i}\right) \right. \\
&\qquad \left. -\frac{1}{2}\left(\frac{\beta_0 - \mu_{\beta_0}}{\sigma_{\beta_0}}\right)^2 - \frac{1}{2}\left(\frac{\beta_1 - \mu_{\beta_1}}{\sigma_{\beta_1}}\right)^2 \right\}.
\end{aligned}
$$

A Metropolis–Hastings algorithm can be built using either a multivariate normal proposal with mean the previous step (random-walk) or with mean the posterior mode (independence sampler).

Bivariate random-walk approach: Independent normals proposal.   For this example, we firstly consider a simple proposal of type

$$
\boldsymbol{\beta}' \quad \sim \quad N_2\left(\boldsymbol{\beta}, \ \mathrm{diag}(\bar{s}_{\beta_0}^2, \bar{s}_{\beta_1}^2) \right) \Leftrightarrow
$$

$$
\begin{pmatrix} \beta_0' \\ \beta_1' \end{pmatrix} \quad \sim \quad N_2\left(\begin{pmatrix} \beta_0 \\ \beta_1 \end{pmatrix}, \ \begin{pmatrix} \bar{s}_{\beta_0}^2 & 0 \\ 0 & \bar{s}_{\beta_1}^2 \end{pmatrix} \right).
$$

The iterative step of the algorithm can be summarized by the following steps

- For $t = 1, \ldots, T$

  1. Set $\boldsymbol{\beta} = (\beta_0^{(t-1)}, \beta_1^{(t-1)})^T$.

  2. Propose new values $\boldsymbol{\beta}' = (\beta_0', \beta_1')^T$ from $N(\beta_0, \bar{s}_{\beta_0}^2)$ and $N(\beta_1, \bar{s}_{\beta_1}^2)$, respectively.

  3. Calculate $\log \alpha = \min(0, A)$ with A given by

  $$
  A \quad = \quad \log \frac{f(\boldsymbol{y}|\beta_0', \beta_1') f(\beta_0', \beta_1')}{f(\boldsymbol{y}|\beta_0, \beta_1) f(\beta_0, \beta_1)}.
  $$

  4. Update $\boldsymbol{\beta}^{(t)} = \boldsymbol{\beta}'$ with probability α or keep the same values with probability $1 - \alpha$.

Table 2.7 demonstrates in more detail the algorithm in parallel to the appropriate R code.

Table 2.7   Algorithm and R code for running a random-walk Metropolis algorithm with independent normal proposals for logistic regression model in Example 2.3

Algorithm (pseudocode)	R code
Set up data	`> wais<-read.table('wais.dat',header=TRUE)` `> y<-wais$senility; x<-wais$wais`
Set number of iterations T	`> Iterations<-2500`
Set prior parameters: $\mu_{\beta_0} = \mu_{\beta_1} = 0,$ $\sigma_{\beta_0} = \sigma_{\beta_1} = 100$	`> mu.beta<-c(0,0); s.beta<-c(100,100)`
Set proposal parameters	`> prop.s<-c(0.2,0.2)`
Initialize vector of sampled values β	`> beta <- matrix(nrow=Iterations, ncol=2)`
Initialize acceptance probability counter	`> acc.prob <- 0`
Set initial current $\beta^{(0)}$	`> current.beta<-c(0,0)`
For $t = 1, \ldots, T$ repeat	`> for (t in 1:Iterations){`
Propose β' from $N(\beta_j, \bar{s}^2_{\beta_j}); j = 0, 1$	`>   prop.beta<- rnorm( 2, current.beta, prop.s )`
Calculate the linear predictor η for current and proposal parameters	`>   cur.eta<-current.beta[1]+current.beta[2]*x` `>   prop.eta<-prop.beta[1]+prop.beta[2]*x`

Calculate $\log \alpha$ =log-likelihood(y; β') $-$ log-likelihood(y; β) $+$ log-prior(β') $-$ log-prior(β)	`>   loga <-(sum( y*prop.eta - log( 1+exp(prop.eta) ) )`
	`+    -sum( y*cur.eta - log( 1+exp(cur.eta ) ) )`
	`+    +sum(dnorm(prop.beta, mu.beta,s.beta,log=TRUE))`
	`+    -sum(dnorm(current.beta,mu.beta,s.beta,log=TRUE)))`

Generate u from $U(0, 1)$	`>   u<-runif(1)`
Set $u = \log u$	`>   u<-log(u)`

If $u < \log \alpha$ then set $\beta^{(t)} = \beta'$ and update acceptance probability counter	`>   if( u < loga ) {`
	`>     current.beta<-prop.beta`
	`>     acc.prob <- acc.prob+1`

else $\beta^{(t)} = \beta$	`>     }` `>   beta[t,]<-current.beta`
End of for loop	`> }`

COMPUTATIONAL NOTE

Frequently in binomial regression models, we face problems in calculation of the likelihood when success probabilities are close to one. In such cases, in $\log(1 + e^{\eta_i})$ for large $\eta_i = \beta_0 + \beta_1 x_i$, the exponential term e^{η_i} becomes higher than the precision limit of the software used (e.g., in R, when $\eta > 709$ then e^η is set equal to infinity). In order to avoid this problem, we set

$$\log(1 + e^\eta) \approx \eta \text{ for } \eta > \text{precision}$$

where precision is the computational threshold for the platform used. The corresponding R code is given by

```
precision<-700
logq<-log( 1+exp(eta) ); logq[eta>precision]<-eta[eta>precision]
loglike<-sum( y*eta - N*logq )
```

In our case, tuning parameters are set equal to $\bar{s}_{\beta_j} = 0.2$ for both $j = 0$ and $j = 1$ achieving acceptance rates of $\sim 20\%$. The algorithm was initially run for 2500 iterations. Diagnostic plots in Figure 2.10i indicate high autocorrelation and lack of convergence since ergodic means have not stabilized. For this reason, the number of iterations was increased to 55,000 to ensure convergence excluding the initial 8000 iterations as burnin period. In order to eliminate high autocorrelations, we consider a thinning interval equal to 650 iterations (see Figure 2.10ii).

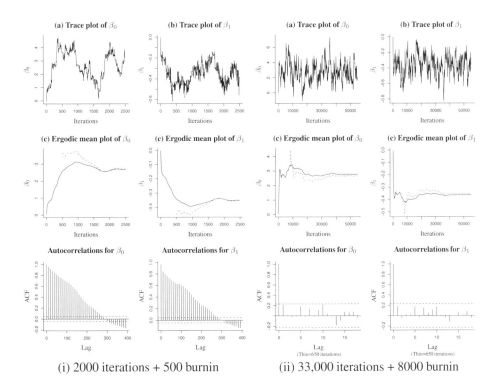

(i) 2000 iterations + 500 burnin   (ii) 33,000 iterations + 8000 burnin

Figure 2.10   MCMC diagnostic plots for logistic regression parameters β_0 and β_1 of Example 2.3 using a random-walk algorithm with independent normal proposals ($\overline{s}_{\beta_0} = \overline{s}_{\beta_1} = 0.2$).

To visualize the evolution of the chain, the bivariate posterior contour is plotted along with the MCMC moves and the simulated values. In Figure 2.11, the moves of the chain can be monitored for the initial 100, 1000, and 10,000 iterations (with appropriate lags). A scatterplot of the generated sample superimposed on the posterior contour is also provided in Figure 2.11d. Although from the latter figure we can conclude that the algorithm has reached convergence, from plots 2.11a–c it is evident that the exploration of the posterior distribution is slow. In the initial 100 iterations (Figure 2.11a), only a small portion of the posterior has been explored. Even for 1000 iterations (Figure 2.11b) the chain has not produced values from the whole space.

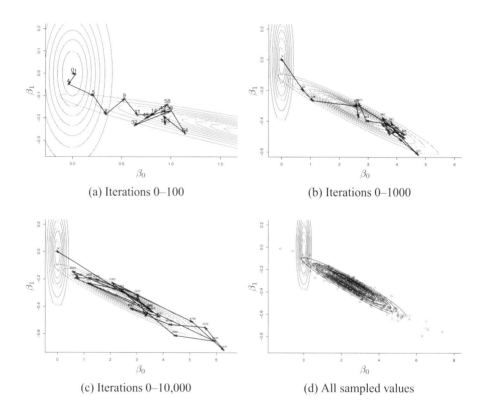

(a) Iterations 0–100

(b) Iterations 0–1000

(c) Iterations 0–10,000

(d) All sampled values

Figure 2.11   MCMC moves and generated values for logistic regression parameters of Example 2.3 using a random walk algorithm with independent normal proposals; the contour on upper left of each graph refers to the proposal distribution at the initial $(0,0)$ step.

Extending the bivariate random-walk approach: Bivariate normal proposal. From Figure 2.11, we observe high negative correlation between β_0 and β_1 with estimated value (from the previous MCMC scheme) equal to $r = -0.964$. This is the reason for the slow convergence of the previous algorithm. Using an independent normal proposal in a highly correlated space results in an inefficient MCMC scheme with slow convergence even in simple cases such as the one illustrated here.

The rate of convergence of the algorithm can be considerably increased by using a multivariate normal proposal distribution. The correlation of the proposal distribution must be similar to the posterior one to allow for proposed moves in the same direction. To solve this problem, the Fisher information matrix can be facilitated in order to construct an efficient algorithm which has similar logic as the random-walk algorithm. Thus, an efficient proposal distribution is given by

$$\boldsymbol{\beta}' \sim N_2\left(\boldsymbol{\beta},\; c_\beta^2 \left[\mathbf{H}(\widetilde{\boldsymbol{\beta}})\right]^{-1}\right),$$

where c_β is a tuning parameter, specified such that the desired acceptance rate is achieved.

In the logistic regression case with p regressors for binomial data with $Y_i \sim \text{binomial}(\pi_i, N_i)$, the log-likelihood is written by

$$\ell(\boldsymbol{\beta}) = \sum_{i=1}^{n} \log\binom{N_i}{y_i} + \left\{\sum_{i=1}^{n}\left(\sum_{j=0}^{p} x_{ij}\beta_j\right) y_i\right\} - \sum_{i=1}^{n} N_i \log\left\{1 + \exp\left(\sum_{j=0}^{p} x_{ij}\beta_j\right)\right\}.$$

The first order partial derivatives are given by

$$\frac{\partial \ell(\boldsymbol{\beta})}{\partial \beta_k} = \sum_{i=1}^{n} x_{ik} y_i - \sum_{i=1}^{n} N_i x_{ik}\left\{\frac{\exp\left(\sum_{j=0}^{p} x_{ij}\beta_j\right)}{1 + \exp\left(\sum_{j=0}^{p} x_{ij}\beta_j\right)}\right\}$$

and the second ones by

$$\frac{\partial^2 \ell(\boldsymbol{\beta})}{\partial \beta_l \partial \beta_k} = -\sum_{i=1}^{n} N_i x_{ik} x_{il} \frac{\exp\left(\sum_{j=0}^{p} x_{ij}\beta_j\right)}{\left[1 + \exp\left(\sum_{j=0}^{p} x_{ij}\beta_j\right)\right]^2}.$$

Hence in the simple logistic regression we can write

$$\mathbf{H}(\boldsymbol{\beta}) = \begin{pmatrix} \sum_{i=1}^{n} N_i \dfrac{\exp\left(\beta_0 + \beta_1 x_i\right)}{\left[1 + \exp\left(\beta_0 + \beta_1 x_i\right)\right]^2} & \sum_{i=1}^{n} x_i N_i \dfrac{\exp\left(\beta_0 + \beta_1 x_i\right)}{\left[1 + \exp\left(\beta_0 + \beta_1 x_i\right)\right]^2} \\ \sum_{i=1}^{n} x_i N_i \dfrac{\exp\left(\beta_0 + \beta_1 x_i\right)}{\left[1 + \exp\left(\beta_0 + \beta_1 x_i\right)\right]^2} & \sum_{i=1}^{n} x_i^2 N_i \dfrac{\exp\left(\beta_0 + \beta_1 x_i\right)}{\left[1 + \exp\left(\beta_0 + \beta_1 x_i\right)\right]^2} \end{pmatrix} + \boldsymbol{\Sigma}_\beta^{-1},$$

where $\boldsymbol{\Sigma}_\beta$ is the prior variance covariance matrix of $\boldsymbol{\beta}$ assuming a multivariate normal prior distribution. The matrix $\mathbf{H}(\boldsymbol{\beta})$ can be expressed in the form

$$\mathbf{H}(\boldsymbol{\beta}) = \boldsymbol{X}^T \text{diag}(h_i)\boldsymbol{X} + \boldsymbol{\Sigma}_\beta^{-1},$$

where $h_i = N_i \exp\left(\sum_{j=0}^{p} x_{ij}\beta_j\right)\left[1 + \exp\left(\sum_{j=0}^{p} x_{ij}\beta_j\right)\right]^{-2}$, $\boldsymbol{X}$ is the $n \times p$ data or design matrix of the model, and h_i is related to the dispersion parameter of the model's distributional structure (McCullagh and Nelder, 1989). In the simple regression case $\boldsymbol{X} = (\mathbf{1}_n \; \boldsymbol{x})$, where $\mathbf{1}_n$ is a n-dimensional vector of ones and $\boldsymbol{x}$ is a vector with elements x_i equal to the observed values of WAIS.

In the expression above, calculation of the posterior mode $\widetilde{\beta}$ via an optimization method is required. If we wish to avoid this optimization step, we may substitute components h_i by some naive estimates originating from the response data y_i. In the logistic regression case these components are equal to $h_i = N_i \pi_i (1 - \pi_i)$. A simple approach is to set probabilities equal to values directly estimated from the original data. If the model is well fitted, then the predicted values will be close to the sample estimates. This approach can be easily adopted for binomial but not for Bernoulli (zero–one) data, since, in the latter case, all estimated π values will be either one or zero, resulting to a poor approximation of the probabilities π_i needed in h_i. An effective alternative is to use the current values of β to estimate structure of the variance covariance matrix. Using this approach, the proposal distribution is given by

$$\beta' \sim N_2 \left(\beta, c_\beta^2 \left[\boldsymbol{X}^T \text{diag}\left(\frac{N_i \exp\left(\sum_{j=0}^{p} x_{ij}\beta_j\right)}{\left[1 + \exp\left(\sum_{j=0}^{p} x_{ij}\beta_j\right)\right]^2}\right) \boldsymbol{X} + \Sigma_\beta^{-1}\right]^{-1}\right).$$

Note that this proposal is not symmetric since the variance–covariance matrix when moving from $\beta \to \beta'$ is different from the corresponding one in the inverse move $\beta' \to \beta$. The iterative step of the algorithm can be described by the following steps:

- For $t = 1, \ldots, T$

    1. Set $\beta = (\beta_0^{(t-1)}, \beta_1^{(t-1)})^T$.

    2. Calculate

$$\boldsymbol{h}(\beta) = \text{diag}\left(\frac{N_i \exp\left(\sum_{j=0}^{p} x_{ij}\beta_j\right)}{\left[1 + \exp\left(\sum_{j=0}^{p} x_{ij}\beta_j\right)\right]^{-2}}\right),$$

$$\boldsymbol{H}(\beta) = \boldsymbol{X}^T \boldsymbol{h}(\beta)\boldsymbol{X} + \Sigma_\beta^{-1} \text{ and } \boldsymbol{S}_\beta = c_\beta^2 [\boldsymbol{H}(\beta)]^{-1}.$$

    3. Propose a new value $\beta' = (\beta_0', \beta_1')^T$ from $N_2(\mu_\beta, \boldsymbol{S}_\beta)$.

    4. Calculate $\boldsymbol{H}(\beta') = \boldsymbol{X}^T \boldsymbol{h}(\beta')\boldsymbol{X} + \Sigma_\beta^{-1}$ and $\boldsymbol{S}_{\beta'} = c_\beta^2 [\boldsymbol{H}(\beta')]^{-1}$.

    5. Calculate $\log \alpha = \min(0, A)$ with A given by

$$\begin{aligned} A &= \log \frac{f(\boldsymbol{y}|\beta_0', \beta_1')f(\beta_0', \beta_1')}{f(\boldsymbol{y}|\beta_0, \beta_1)f(\beta_0, \beta_1)} + \log \frac{f_N(\beta|\beta', \boldsymbol{S}_{\beta'})}{f_N(\beta'|\beta, \boldsymbol{S}_\beta)} \\ &= \log \frac{f(\boldsymbol{y}|\beta_0', \beta_1')f(\beta_0', \beta_1')}{f(\boldsymbol{y}|\beta_0, \beta_1)f(\beta_0, \beta_1)} + \frac{1}{2}\log \frac{\left|\boldsymbol{S}_{\beta'}^{-1}\right|}{\left|\boldsymbol{S}_\beta^{-1}\right|} \\ &\quad - \frac{1}{2}(\beta - \beta')^T \left[\boldsymbol{S}_{\beta'}^{-1} - \boldsymbol{S}_\beta^{-1}\right](\beta - \beta'). \end{aligned}$$

6. Update $\boldsymbol{\beta}^{(t)} = \boldsymbol{\beta}'$ with probability α or keep the same values for $\boldsymbol{\beta}^{(t)}$ with probability $1 - \alpha$.

Table 2.8 demonstrates the above algorithm in parallel to the appropriate R code.

Table 2.8   Algorithm and R code for the Metropolis–Hastings algorithm with dependent normal proposal distributions for logistic regression model in Example 2.3 (The MASS package must be loaded in order to be able to use the `mvrnorm` function)

Algorithm (pseudocode)	R code
Set up data	`> wais<-read.table('wais.dat',header=TRUE)`
	`> y<-wais$senility; x<-wais$wais`
Set up matrix $\boldsymbol{X}$	`> n<-length(y); X<-cbind( rep(1,n), x )`
Set number of iterations T	`> Iterations<-2500`
Set prior parameters: $\mu_{\beta_0} = \mu_{\beta_1} = 0$,	`> mu.beta<-c(0,0); s.beta<-c(100,100)`
$\qquad\qquad\sigma_{\beta_0} = \sigma_{\beta_1} = 100$	
Set proposal parameters	`> c.beta<- 1.75`
Initialize vector of sampled values $\boldsymbol{\beta}$	`> beta <- matrix(nrow=Iterations, ncol=2)`
Initialize acceptance probability counter	`> acc.prob <- 0`
Set initial current $\beta^{(0)}$	`> current.beta<-c(0,0)`
For $t = 0, \ldots, T - 1$ repeat	`> for (t in 1:Iterations){`
Calculate log-likelihood and precision for $\boldsymbol{\beta}$	`>   cur<-calculate.loglike( current.beta, x, y)`
Propose $\boldsymbol{\beta}'$ from bivariate normal	`>   cur.T <- (1/c.beta^2)*(cur$H+diag(1/s.beta^2))`
Calculate log-likelihood and precision for $\boldsymbol{\beta}'$	`>   prop.beta<- mvrnorm(1,current.beta,solve(cur.T))`
Calculate $\log \alpha$ =log-likelihood$(\boldsymbol{y}; \boldsymbol{\beta}')$	`>   prop<-calculate.loglike( prop.beta, x, y)`
$\quad$ − log-likelihood$(\boldsymbol{y}; \boldsymbol{\beta})$	`>   prop.T <-(1/c.beta^2)*(prop$H+diag(1/s.beta^2))`
$\quad$ + log-prior$(\boldsymbol{\beta}')$	`>   loga <-( prop$loglike`
$\quad$ − log-prior$(\boldsymbol{\beta})$	`>    - cur$loglike`
$\quad$ + log-proposal$(\boldsymbol{\beta}' \to \boldsymbol{\beta})$	`+   +sum(dnorm(prop.beta, mu.beta,s.beta,log=TRUE))`
	`+   -sum(dnorm(current.beta,mu.beta,s.beta,log=TRUE))`
	`+   +as.numeric( 0.5*log( det(cur.T) )`
$\quad$ − log-proposal$(\boldsymbol{\beta} \to \boldsymbol{\beta}')$	`-0.5*t(current.beta-prop.beta) %*% prop.T %*%`
	`(current.beta-prop.beta)`
	`+   -0.5*log( det(cur.T) ) +0.5*t(prop.beta-current.beta)`
	`%*% cur.T %*% (prop.beta-current.beta ) ))`
Generate u from $U(0, 1)$	`>   u<-runif(1)`
Set $u = \log u$	`>   u<-log(u)`
If $u < \log \alpha$ then	`>   if( u < loga ) {`
$\quad$ set $\boldsymbol{\beta}^{(t)} = \boldsymbol{\beta}'$	`>     current.beta<-prop.beta`
$\quad$ and update acceptance probability counter	`>     acc.prob <- acc.prob+1`
	`>   }`
else $\boldsymbol{\beta}^{(t)} = \boldsymbol{\beta}$	`>   beta[t,]<-current.beta`
End of for loop	`> }`
Set up a function to calculate log-likelihood and	`># ------- FUNCTION calculate.loglike ---------------`
precision	`># input:  beta, x, y`
	`># output:  loglike and precision`
	`> calculate.loglike<-function( b, x, y){`
	`>   n<-length(x); X<-cbind( rep(1,n), x )`
Set up precision limit	`>   precision<-700`
Calculate linear predictor η_i for $\boldsymbol{\beta}$	`>   eta<-b[1]+b[2]*x`
Calculate $\log(1 + e^{\eta_i})$	`>   logq <- log(1+exp(eta))`
If $\eta_i >$ precision, set $\log(1 + e^{\eta_i}) = \eta_i$	`>   logq[eta>precision]<-eta[eta>precision]`
Calculate log-likelihood	`>   loglike<- sum( y*eta - logq )`
Truncate η_i in order to calculate h_i	`>   eta[eta>precision]<-precision`
Calculate h_i	`>   h <- 1/((1+exp(-eta))*(1+exp(eta)))`
Calculate matrix $\boldsymbol{H}$	`>   H <- t(x) %*% diag( h ) %*% x`
Return the log-likelihood and $\boldsymbol{H}$	`>   return(list( loglike=loglike, H=T ))`
	`> }`

In our example we consider binary (Bernoulli) data; hence $N_i = 1$ for all $i = 1, \ldots, n$. The scheme shown above was used to simulate samples of 1500 and 11000 iterations. Parameter c_β was set equal to 2.5 achieving an acceptance rate around 20%. The diagnostic plots (in Figure 2.12) demonstrate much better behavior than do the algorithm of the previous paragraph. In Figure 2.12i we observe that the autocorrelation lag is about 30 iterations, while in Figure 2.12ii (chain with 11,000 iterations and burnin period of 1000 iterations), the ergodic means have been stabilized. Moreover, the contour plots in Figure 2.13 demonstrate that the chain is highly mobile and even in the initial 100 iterations (Figure 2.13a) the algorithm has explored an important portion of the posterior distribution.

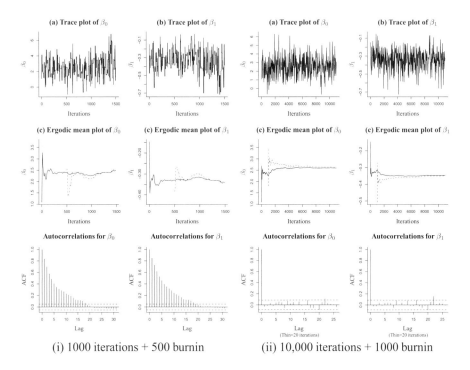

(i) 1000 iterations + 500 burnin   (ii) 10,000 iterations + 1000 burnin

Figure 2.12   MCMC diagnostic plots for logistic regression parameters β_0 and β_1 of Example 2.3 using a bivariate normal proposal distribution.

Single-component random-walk approach.   In practice, MCMC schemes are adopted in which one parameter at each time is updated iteratively. This procedure is called the *single-component Metropolis–Hastings algorithm* (or *Metropolis within Gibbs* or *componentwise Metropolis–Hastings algorithm*) and is simpler to design and easier in terms of the specification of the tuning parameters since in each step we deal with univariate distributions. This iterative procedure is summarized as follows:

- For $t = 1, \dots, T$

    1. Set $\boldsymbol{\beta} = (\beta_0^{(t-1)}, \beta_1^{(t-1)})^T$.
    2. Propose a new value β_0' from $N(\beta_0, \overline{s}_{\beta_0}^2)$.
    3. Set $\boldsymbol{\beta}' = (\beta_0', \beta_1^{(t-1)})^T$.
    4. Calculate $\log \alpha = \min(0, A)$ with A given by

$$A = \log \frac{f(\boldsymbol{y}|\beta_0', \beta_1)f(\beta_0', \beta_1)}{f(\boldsymbol{y}|\beta_0, \beta_1)f(\beta_0, \beta_1)}.$$

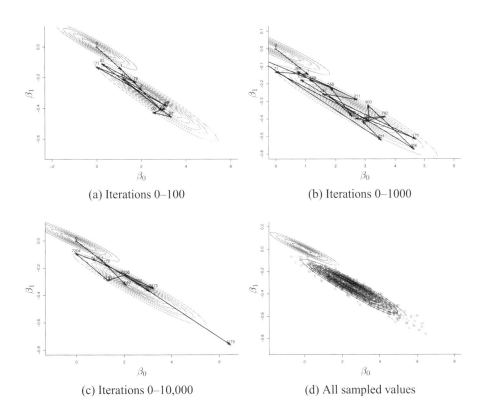

(a) Iterations 0–100

(b) Iterations 0–1000

(c) Iterations 0–10,000

(d) All sampled values

Figure 2.13   MCMC moves and generated values for logistic regression parameters of Example 2.3 using a bivariate normal proposal; the contour on upper left of each graph refers to the proposal distribution at the initial $(0, 0)$ step.

5. Update $\boldsymbol{\beta} = \boldsymbol{\beta}'$ with probability α or keep the same values for $\boldsymbol{\beta}$ with probability $1 - \alpha$.

6. Propose a new value β_1' from $N(\beta_1, \bar{s}_{\beta_1}^2)$.

7. Set $\boldsymbol{\beta}' = (\beta_0, \beta_1')^T$.

8. Calculate $\log \alpha = \min(0, A)$ with A given by

$$A = \log \frac{f(\boldsymbol{y}|\beta_0, \beta_1')f(\beta_0, \beta_1')}{f(\boldsymbol{y}|\beta_0, \beta_1)f(\beta_0, \beta_1)}.$$

9. Update $\boldsymbol{\beta} = \boldsymbol{\beta}'$ with probability α or keep the same values for $\boldsymbol{\beta}$ with probability $1 - \alpha$.

10. Set $\boldsymbol{\beta}^{(t)} = \boldsymbol{\beta}$

Table 2.9 demonstrates in more detail the algorithm in parallel to the appropriate R code as implemented for the logistic regression example using the senility symptoms data.

Table 2.9 Algorithm and R code for a single-component random-walk Metropolis algorithm for logistic regression parameters of Example 2.3

Algorithm (pseudocode)	R code
Set up data	`> wais<-read.table('wais.dat',header=TRUE)`
	`> y<-wais$senility; x<-wais$wais`
Set number of iterations T	`> Iterations<-3500`
Set prior parameters: $\mu_{\beta_0} = \mu_{\beta_1} = 0$, $\sigma_{\beta_0} = \sigma_{\beta_1} = 100$	`> mu.beta<-c(0,0); s.beta<-c(100,100)`
Set proposal parameters	`> prop.s<-c(1.5,0.15)`
Initialize vector of sampled values $\boldsymbol{\beta}$	`> beta <- matrix(nrow=Iterations, ncol=2)`
Initialize acceptance probability counter	`> acc.prob <- c(0,0)`
Set initial current $\beta^{(0)}$	`> current.beta<-c(0,0)`
For $t = 1, \ldots, T$ repeat	`> for (t in 1:Iterations){`
Set $\boldsymbol{\beta}' = (\beta_0, \beta_1)$	`>    prop.beta<- current.beta`
Propose β_0' from $N(\beta_0, \bar{s}_{\beta_0}^2)$	`>    prop.beta[1]<- rnorm( 1, current.beta[1],`
	`prop.s[1] )`
Calculate the linear predictor η for current and proposal parameters	`>    cur.eta <-current.beta[1]+current.beta[2]*x`
	`>    prop.eta<-prop.beta[1] +prop.beta[2] *x`
Calculate $\log \alpha$ =log-likelihood($\boldsymbol{y}; \boldsymbol{\beta}'$)	`>    loga <-(sum( y*prop.eta - log( 1+exp(prop.eta) ) )`
$\quad$ − log-likelihood($\boldsymbol{y}; \boldsymbol{\beta}$)	`+     -sum( y*cur.eta - log( 1+exp(cur.eta ) ) )`
$\quad$ + log-prior($\boldsymbol{\beta}'$)	`+     +sum(dnorm(prop.beta, mu.beta,s.beta,log=TRUE))`
$\quad$ − log-prior($\boldsymbol{\beta}$)	`+     -sum(dnorm(current.beta,mu.beta,s.beta,log=TRUE)))`
Generate u from $U(0, 1)$	`>    u<-runif(1)`
Set $u = \log u$	`>    u<-log(u)`
If $u < \log \alpha$ then	`>    if( u < loga ) {`
$\quad$ set $\boldsymbol{\beta} = \boldsymbol{\beta}'$	`>      current.beta<-prop.beta`
$\quad$ and update acceptance probability counter	`>      acc.prob[1] <- acc.prob[1]+1`
	`>    }`
Set $\boldsymbol{\beta}' = (\beta_0, \beta_1)$	`>    prop.beta<- current.beta`
Propose β_1' from $N(\beta_1, \bar{s}_{\beta_1}^2)$	`>    prop.beta[2]<- rnorm( 1, current.beta[2],`
	`prop.s[2] )`
Calculate linear predictor η for current and proposal parameters	`>    cur.eta <-current.beta[1]+current.beta[2]*x`
	`>    prop.eta<-prop.beta[1] +prop.beta[2] *x`
Calculate $\log \alpha$ =log-likelihood($\boldsymbol{y}; \boldsymbol{\beta}'$)	`>    loga <-(sum( y*prop.eta - log( 1+exp(prop.eta) ) )`
$\quad$ − log-likelihood($\boldsymbol{y}; \boldsymbol{\beta}$)	`+     -sum( y*cur.eta - log( 1+exp(cur.eta ) ) )`
$\quad$ + log-prior($\boldsymbol{\beta}'$)	`+     +sum(dnorm(prop.beta, mu.beta,s.beta,log=TRUE))`
$\quad$ − log-prior($\boldsymbol{\beta}$)	`+     -sum(dnorm(current.beta,mu.beta,s.beta,log=TRUE)))`
Generate u from $U(0, 1)$	`>    u<-runif(1)`
Set $u = \log u$	`>    u<-log(u)`
If $u < \log \alpha$ then	`>    if( u < loga ) {`
$\quad$ set $\boldsymbol{\beta} = \boldsymbol{\beta}'$	`>      current.beta<-prop.beta`
$\quad$ and update acceptance probability counter	`>      acc.prob[2] <- acc.prob[2]+1`
	`>    }`
Set $\boldsymbol{\beta}^{(t)} = \boldsymbol{\beta}$	`>    beta[t,]<-current.beta`
End of for loop	`> }`

This algorithm will move more frequently than the corresponding one, which updates all parameters in a single step since it also allows us to change only specific parameters in each step. This is beneficial when high correlations between parameters exist, and we wish to avoid using multivariate proposals.

Initially, a sample of 1500 iterations was generated excluding the first 500 observations as a burnin period using $\bar{s}_{\beta_0} = 1.75$ and $\bar{s}_{\beta_1} = 0.2$, which result in acceptance rates approximately equal to 20%.

As we can see from Figure 2.14i, the behavior of the chain was improved in comparison to the initial one-block random-walk algorithm. Nevertheless, it is still slightly worse than the second approach, which used a multivariate normal proposal distribution with variance–covariance matrix equivalent to the posterior one. According to Figure 2.14ii, the algorithm has reached convergence. A thinning interval equal to 90 iterations is required to obtain an independent sample.

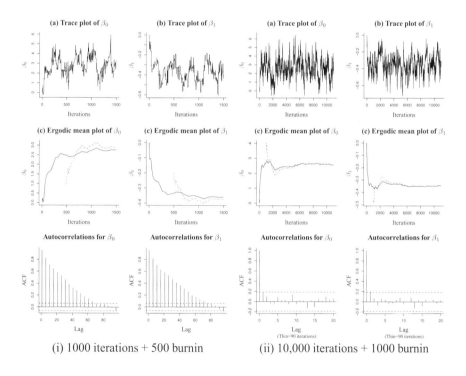

(i) 1000 iterations + 500 burnin   (ii) 10,000 iterations + 1000 burnin

Figure 2.14   MCMC diagnostic plots for logistic regression parameters β_0 and β_1 of Example 2.3 using a single-component random-walk algorithm.

Moves of the algorithm in the bivariate space are depicted in Figure 2.15. In the first plot (iterations 1–30), both vertical and horizontal moves are visible in addition to the usual diagonal moves appearing in all previously illustrated algorithms. Such moves correspond to updating only one of the two parameters in the corresponding iteration. This depiction is typical of how single-component Metropolis–Hastings algorithms explore the parameter space.

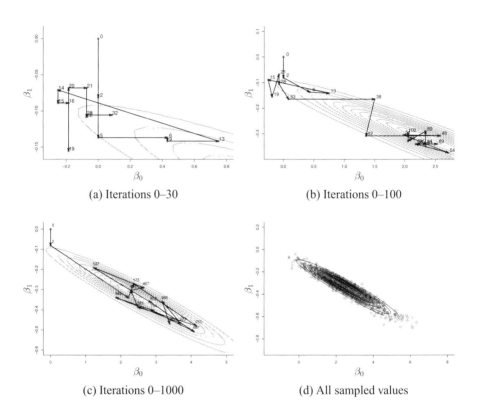

(a) Iterations 0–30   (b) Iterations 0–100

(c) Iterations 0–1000   (d) All sampled values

Figure 2.15   MCMC moves and generated values for logistic regression parameters of Example 2.3 using a single-component random-walk algorithm; the contour on upper left of each graph refers to the proposal distribution at the initial $(0, 0)$ step.

Comparison of the three algorithms. Table 2.10 provides posterior summaries from the samples generated using the three algorithms described above. Additional details for the MCMC algorithms are given in Table 2.11. All approaches have reached convergence, and their posterior means and standard deviations are similar. Some discrepancies exist in the posterior quantiles, indicating that additional iterations might be required in order to estimate them more precisely. Monte Carlo errors, for the posterior means are also provided. For the second MCMC scheme, MC errors are considerably lower than for the other two algorithms. The first and the third MCMC schemes have similar MC errors but the length of the first chain is about 5 times as high as the length of the latter (47,000 vs. 10,000 iterations kept).

Table 2.10 Posterior summaries of MCMC outputs for Example 2.3

Parameter	Algorithm[a]	Mean	MC error	SD[b]	2.5%	97.5%
β_0	RW with independent normal proposals	2.658	0.144	1.322	0.192	5.319
	MH with bivariate normal proposal	2.624	0.036	1.217	0.437	5.226
	Single-component RW	2.563	0.120	1.208	0.205	5.171
β_1	RW with independent normal proposals	-0.353	0.013	0.127	-0.617	-0.121
	MH with bivariate normal proposals	-0.350	0.004	0.117	-0.607	-0.146
	Single Component RW	-0.343	0.012	0.116	-0.601	-0.127

[a] RW=random-walk Metropolis algorithm; MH=Metropolis–Hastings algorithm.
[b] Standard deviation.

Table 2.11 MCMC details for algorithms used in Example 2.3

Algorithm[a]	Burnin (B)	Iterations kept (T')	Lag	Proposal parameters	Acceptance rate
RW with independent normal proposals	8000	47,000	650	$\bar{s}_{\beta_j} = 0.20$	0.208
MH with bivariate normal proposal	1000	10,000	20	$c_\beta = 2.50$	0.196
Single-component RW	1000	10,000	90	$\bar{s}_{\beta_0} = 1.75$	0.245[b]
				$\bar{s}_{\beta_1} = 0.20$	0.212[b]

[a] RW=random-walk Metropolis algorithm; MH=Metropolis–Hastings algorithm.
[b] Overall acceptance rate=0.398 (probability of updating at least one parameter).

Finally, differences, in terms of convergence, between the three algorithms are depicted in Figure 2.16 via the evolution of the ergodic means of β_0 and β_1. From these plots, it is clear that the means of the simple independent normal random-walk algorithm (solid line) converge to the correct value slower than the corresponding means for the other two methods.

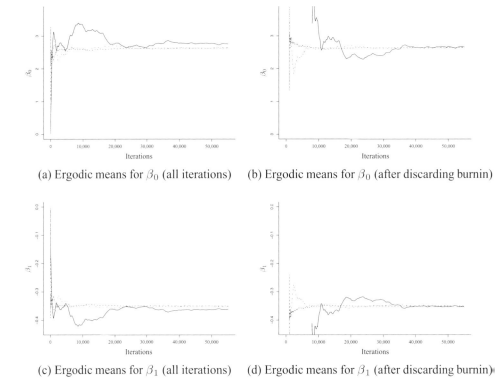

   (a) Ergodic means for β_0 (all iterations)   (b) Ergodic means for β_0 (after discarding burnin)

   (c) Ergodic means for β_1 (all iterations)   (d) Ergodic means for β_1 (after discarding burnin)

Figure 2.16   Ergodic mean plots for each algorithm for logistic regression parameters of Example 2.3.

Estimating WAIS with $\pi > 0.5$.   In order to identify which scores of WAIS result in increased probability of senility symptoms, we can follow either of two approaches. The first approach is to estimate all π values under all (or some) possible values of WAIS and trace which posterior distributions are clearly over the threshold value of 0.5.

Using this first approach for each set of sampled values $\boldsymbol{\beta}^{(t)}$, we can calculate

$$\pi^{(t)}(x) = \frac{\exp\left(\beta_0^{(t)} + \beta_1^{(t)}x\right)}{1 + \exp\left(\beta_0^{(t)} + \beta_1^{(t)}x\right)} \quad \text{for } x = 0, 1, 2, \ldots, 20$$

and hence obtain a sample for each probability and estimate the posterior 95% or 99% credible interval for each possible score. These details as well as the probability $P(\pi(x) > 0.5)$ are provided in Table 2.12. If inference is based on the 95% credible intervals, then values of WAIS lower than 10 (WAIS<10) indicate which subjects demonstrate high probability of having senility symptoms.

Table 2.12 Posterior summaries for probability of senility symptoms for all possible WAIS values in Example 2.3

WAIS		Percentiles		
score (x)	Mean	2.5%	97.5%	$P(\pi(x) > 0.5)$
0	0.894	0.608	0.995	0.993
1	0.869	0.566	0.990	0.991
2	0.837	0.529	0.983	0.984
3	0.797	0.491	0.969	0.972
4	0.748	0.452	0.945	0.950
5	0.688	0.411	0.907	0.912
6	0.619	0.371	0.848	0.807
7	0.541	0.328	0.763	0.630
8	0.459	0.270	0.659	0.342
9	0.376	0.222	0.548	0.075
10	0.300	0.173	0.446	0.005
11	0.234	0.123	0.369	0.000
12	0.180	0.077	0.304	0.000
13	0.138	0.047	0.254	0.000
14	0.105	0.029	0.218	0.000
15	0.080	0.016	0.187	0.000
16	0.061	0.009	0.158	0.000
17	0.047	0.005	0.136	0.000
18	0.037	0.003	0.117	0.000
19	0.028	0.002	0.100	0.000
20	0.022	0.001	0.088	0.000

Alternatively, the logistic regression equation may be used to obtain an expression for the threshold probability value of $\pi = 0.5$, which results in zero logit. Thus, denoting by WAIS* the value that corresponds to probability of having senility symptoms equal to 0.5, we obtain the expression $\beta_0 + \beta_1 \text{WAIS}^* = 0 \Leftrightarrow \text{WAIS}^* = -\beta_0/\beta_1$. We obtain a posterior sample for WAIS* by simply setting $\text{WAIS}^{*(t)} = -\beta_0^{(t)}/\beta_1^{(t)}$, for $t = 1, 2, \ldots, T$. This approach is easier to implement and can be used even when variable x is continuous.

Implementing this transformation to the output of the second MCMC scheme, we obtain a posterior sample of WAIS* with posterior mean equal to 7.1 and 95th and 99th percentiles equal to 9.2 and 9.8, respectively. Hence, using these quantiles, we conclude that the value of WAIS equal to 10 (WAIS* = 10) safely discriminates high and low-risk subjects (using WAIS< 10 and WAIS≥ 10, respectively). This threshold value is the same as the corresponding one attained by the first approach for the calculation of WAIS *.

2.3.3 The Gibbs sampler

The Gibbs sampler was introduced by Geman and Geman (1984). It is a special case of single-component Metropolis–Hastings algorithm using as proposal density $q\left(\boldsymbol{\theta}'|\boldsymbol{\theta}^{(t)}\right)$ the full conditional posterior distribution $f(\theta_j|\boldsymbol{\theta}_{\setminus j}, \boldsymbol{y})$, where $\boldsymbol{\theta}_{\setminus j} = (\theta_1, \ldots, \theta_{j-1}, \theta_{j+1}, \ldots, \theta_d)^T$.

Such proposal distributions result in acceptance probability $\alpha = 1$, and therefore the proposed move is accepted in all iterations. Although Gibbs sampling is a special case of Metropolis–Hasting algorithm, it is usually cited as a separate simulation technique because of its popularity and convenience. One advantage of the Gibbs sampler is that, in each step, random values must be generated from unidimensional distributions for which a wide variety of computational tools exists (Gilks, 1996). Frequently, these conditional distributions have a known form and, thus, random numbers can be easily simulated using standard functions in statistical and computing software. Gibbs sampling is always moving to new values and, most importantly, does not require specification of proposal distributions. On the other hand, it can be ineffective when the parameter space is complicated or the parameters are highly correlated.

The algorithm can be summarized by the following steps:

1. Set initial values $\boldsymbol{\theta}^{(0)}$.

2. For $t = 1, \ldots, T$ repeat the following steps

     a. Set $\boldsymbol{\theta} = \boldsymbol{\theta}^{(t-1)}$

     b. For $j = 1, \ldots, d$, update θ_j from $\theta_j \sim f\left(\theta_j \mid \boldsymbol{\theta}_{\backslash j}, \boldsymbol{y}\right)$.

     c. Set $\boldsymbol{\theta}^{(t)} = \boldsymbol{\theta}$ and save it as the generated set of values at $t + 1$ iteration of the algorithm.

Hence, given a particular state of the chain $\boldsymbol{\theta}^{(t)}$, we generate the new parameter values by

$$
\begin{aligned}
\theta_1^{(t)} &\quad \text{from} \quad f(\theta_1 | \theta_2^{(t-1)}, \theta_3^{(t-1)}, \ldots, \theta_p^{(t-1)}, \boldsymbol{y}), \\
\theta_2^{(t)} &\quad \text{from} \quad f(\theta_2 | \theta_1^{(t)}, \theta_3^{(t-1)}, \ldots, \theta_p^{(t-1)}, \boldsymbol{y}), \\
\theta_3^{(t)} &\quad \text{from} \quad f(\theta_3 | \theta_1^{(t)}, \theta_2^{(t)}, \theta_4^{(t-1)}, \ldots, \theta_p^{(t-1)}, \boldsymbol{y}), \\
&\quad \vdots \\
\theta_j^{(t)} &\quad \text{from} \quad f(\theta_j | \theta_1^{(t)}, \theta_2^{(t)}, \ldots, \theta_{j-1}^{(t)}, \theta_{j+1}^{(t-1)}, \ldots, \theta_p^{(t-1)}, \boldsymbol{y}), \\
&\quad \vdots \\
\theta_p^{(t)} &\quad \text{from} \quad f(\theta_p | \theta_1^{(t)}, \theta_2^{(t)}, \ldots, \theta_{p-1}^{(t)}, \boldsymbol{y}).
\end{aligned}
$$

Generating values from $f(\theta_j | \boldsymbol{\theta}_{\backslash j}, \boldsymbol{y}) = f(\theta_j | \theta_1^{(t)}, \ldots, \theta_{j-1}^{(t)}, \theta_{j+1}^{(t-1)}, \ldots, \theta_p^{(t-1)}, \boldsymbol{y})$ is relatively easy since it is a univariate distribution and can be written as $f(\theta_j | \boldsymbol{\theta}_{\backslash j}, \boldsymbol{y}) \propto f(\boldsymbol{\theta} | \boldsymbol{y})$, where all the variables except θ_j are held constant at their given values. More detailed description of the Gibbs sampler is given by Casella and George (1992) and Smith and Roberts (1993), while early applications of the Gibbs sampling are provided by Gelfand and Smith (1990) and Gelfand et al. (1990).

2.3.3.1   *A simple example using the Gibbs sampler*

Example 2.4. Body temperature data revisited: Analysis with nonconjugate prior. Let us consider the data of example 1.5 but now with the following prior distribution

$$
\mu \sim N(\mu_0, \sigma_0^2) \text{ and } \sigma^2 \sim IG(a_0, b_0)
$$

instead of the conjugate prior distribution presented in Section 1.5.7.

In order to construct a Gibbs sampler for this model, we need to calculate the conditional distributions $f(\mu|\sigma^2, \boldsymbol{y})$ and $f(\sigma^2|\mu, \boldsymbol{y})$ and sample sequentially from these two distributions. After some calculations, we obtain

$$\mu|\sigma^2, \boldsymbol{y} \sim N\left(w\bar{y} + (1-w)\mu_0, \ w\frac{\sigma^2}{n} \right), \text{ where } w = \frac{\sigma_0^2}{\sigma^2/n + \sigma_0^2}$$

and

$$\sigma^2|\mu, \boldsymbol{y} \sim \text{IG}\left(a_0 + \frac{n}{2}, \ b_0 + \frac{1}{2}\sum_{i=1}^{n}(y_i - \mu)^2 \right).$$

Using these results, the Gibbs sampler is summarized as follows:

- For $t = 1, \ldots, T$

    1. Set $\mu = \mu^{(t-1)}$, $\sigma = \sigma^{(t-1)}$ and $\boldsymbol{\theta} = (\mu, \sigma^2)^T$.
    2. Calculate $w = \sigma_0^2/(\sigma^2/n + \sigma_0^2)$, $m = w\bar{y} + (1-w)\mu_0$, and $s^2 = w\sigma^2/n$.
    3. Generate μ from $N(m, s^2)$
    4. Set $\mu^{(t)} = \mu$
    5. Calculate $a = a_0 + n/2$ and $b = b_0 + \frac{1}{2}\sum_{i=1}^{n}(y_i - \mu)^2$.
    6. Generate τ from $G(a, b)$
    7. Set $\sigma^2 = 1/\tau$ and $\sigma^{(t)} = \sigma$.

Table 2.13 demonstrates in more detail the algorithm in parallel to the appropriate R code for the simple regression model of the body temperature data.

Table 2.13   Algorithm and R code for Gibbs sampler of Example 2.4

Algorithm (pseudocode)	R code
Set up data	`> y<-normtemp$temp; bary<-mean(y); n<-length(y)`
Set number of iterations T	`> Iterations<-3500`
Set prior parameters: $\mu_0 = 0$, $\sigma_0 = 100$, $a_0 = b_0 = 0.001$	`> mu0<-0; s0<-100; a0<-0.001; b0<-0.001`
Initialize vectors of sampled values μ, σ	`> theta <- matrix(nrow=Iterations, ncol=2)`
Set initial current $\mu^{(0)}$ and $\sigma^{(0)}$	`> cur.mu<-0; cur.tau<-1; cur.s<-sqrt(1/cur.tau)`
For $t = 1, \ldots, T$ repeat	`> for (t in 1:Iterations){`
Calculate $w = \sigma_0^2/(\sigma^2/n + \sigma_0^2)$	`> w<- s0^2/( cur.s^2/n+ s0^2 )`
Calculate $m = w\bar{y} + (1-w)\mu_0$	`> m <- w*bary + (1-w)*mu0`
Calculate $s = \sqrt{w\sigma^2/n}$	`> s <- sqrt( w * cur.s^2/n )`
Generate $\mu \sim N(m, s^2)$	`> cur.mu <- rnorm( 1, m, s )`
Calculate $a = a_0 + n/2$	`> a <- a0 + 0.5*n`
Calculate $b = b_0 + \frac{1}{2}\sum_{i=1}^{n}(y_i - \mu)^2$	`> b <- b0 + 0.5 * sum( (y-cur.mu)^2 )`
Generate $\tau \sim G(a, b)$	`> cur.tau <- rgamma( 1, a, b )`
Set $\sigma = \sqrt{1/\tau}$	`> cur.s <- sqrt(1/cur.tau)`
Set $\mu^{(t)} = \mu$ and $\sigma^{(t)} = \sigma$	`> theta[t,]<-c( cur.mu, cur.s )`
End of for loop	`> }`

In Figure 2.17, trace, ergodic means, and autocorrelation plots for both μ and σ are provided. All convergence plots indicate that the algorithm has reached convergence since trace plots do not present irregularities and ergodic means have been stabilized. Moreover, only the first autocorrelation of the sample is high, indicating a well performed sampler.

From the initial 10 moves of the Gibbs sampler (depicted in Figure 2.18 along with the intermediate steps between the updates of μ and σ), we observe that the chain is highly mobile, moving directly to the center of the target posterior distribution. The finally generated sample is provided in the scatterplot of Figure 2.19.

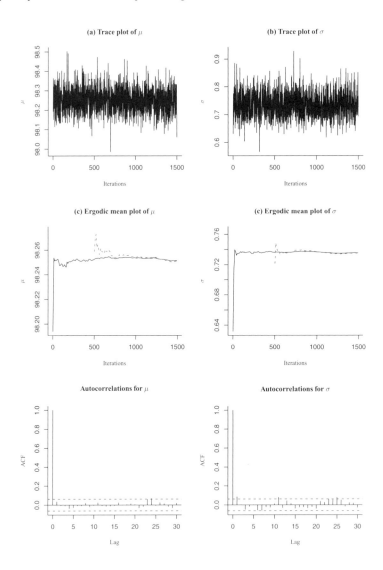

Figure 2.17 MCMC diagnostic plots for normal model using Gibbs sampler for Example 2.4.

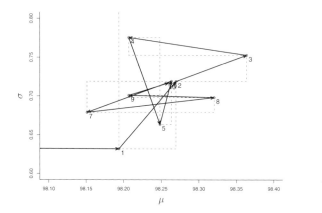

Figure 2.18   Initial 10 moves of Gibbs sampler for Example 2.4; dashed lines indicate intermediate steps.

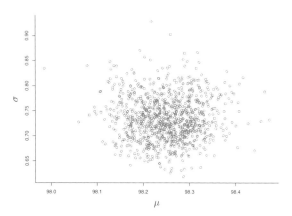

Figure 2.19   Scatterplot of generated sample using Gibbs sampler in Example 2.4.

2.3.4   Metropolis within Gibbs

Nowadays, the wide range of available computational tools for generating random values from univariate distributions allow us to implement the Gibbs sampler in a variety of cases, even when the resulting conditional posterior distribution is cumbersome. Nevertheless, on some occasions, it is convenient to use simple Metropolis–Hastings steps to generate from these univariate conditional posterior distributions. This approach is called the *Metropolis within Gibbs algorithm* and it is simply a componentwise Metropolis–Hastings algorithm in which some components of the parameter vector are directly generated from the corresponding full conditional posterior distributions. This combination of Metropolis and Gibbs steps is frequently used in practice. In this way, the user can easily incorporate blocking, take advantage of specific easy-to-generate full conditionals, and generally build an efficient and flexible MCMC algorithm.

2.3.5   The slice Gibbs sampler

The slice sampler is essentially based on Gibbs sampling. It is used mainly when the full conditional posterior distributions do not have a convenient form. This method augments the parameter space by adding a set of convenient random variables (called *auxiliary variables*) that retain the marginal posterior distribution of interest unchanged but convert all conditionals to distributions of standard form. In this way a simple Gibbs sampler is directly applicable. The method is also referred as the *auxiliary variables method* and was introduced in statistical physics in the late 1990s. It was further developed in the context of statistical science by Higdon (1998), Damien et al. (1999), and Neal (2003).

The idea can be summarized as follows. Let us consider a target distribution $g(x)$ that is difficult to generate from. We introduce a new variable u with conditional $f(u|x)$. Then the joint distribution distribution can be written as

$$f(u, x) = f(u|x)g(x),$$

while the marginal distribution $f(x)$ is equal to the original target distribution $g(x)$ since

$$f(x) = \int f(u, x)du = \int f(u|x)g(x)du = g(x) \ .$$

In this way we can set up the following Gibbs sampler, which generates values from the joint distribution $f(u, x)$ and the corresponding marginals $f(u)$ and $f(x) = g(x)$:

1. Generate $u \sim f(u|x)$

2. Generate $x \sim f(u|x)g(x)$

Since $f(u|x)$ participates in both Gibbs steps above, it must be specified in such way that both $f(u|x)$ and $f(u|x)g(x)$ are convenient in terms of simulation. A usual choice for $f(u|x)$ is the uniform distribution $U\left(0, g(x)\right)$ since

$$f(u, x) = \frac{1}{g(x)}g(x)I\left(0 < u < g(x)\right) = I\left(0 < u < g(x)\right)$$

and

$$f(x) = \int I\left(0 < u < g(x)\right)du = \int_0^{g(x)} du = [u]_0^{g(x)} = g(x),$$

where $I(x)$ is the indicator function taking value equal to one if x is true and zero otherwise. Then the Gibbs sampling is summarized by the following steps:

1. Generate $u^{(t)} \sim U\left(0, g\left(x^{(t-1)}\right)\right)$

2. Generate $x^{(t)} \sim U\left(x : 0 \le u^{(t)} \le g(x)\right)$.

When we focus on the Bayesian context, it is usual practice to facilitate $\boldsymbol{u} = (u_1, \ldots, u_n)$ auxiliary variables coming from uniform distribution defined within the interval from zero to the likelihood ordinate $f(y_i|\boldsymbol{\theta})$. Thus the joint distribution will be given by

$$f(\boldsymbol{\theta}, \boldsymbol{u}|\boldsymbol{y}) \propto \left\{\prod_{i=1}^{n} I\left(0 \le u_i \le f(y_i|\boldsymbol{\theta})\right)\right\} f(\boldsymbol{\theta}),$$

resulting in a Gibbs sampler of type

1. Set $\boldsymbol{\theta} = \boldsymbol{\theta}^{(t-1)}$.

2. For $i = 1, \ldots, n$, generate $u_i^{(t)} \sim U\left(0, f(y_i|\boldsymbol{\theta})\right)$.

3. For $j = 1, \ldots, d$, update $\theta_j \sim f(\theta_j) \prod_{i=1}^{n} I\left(0 \le u_i^{(t)} \le f(y_i|\boldsymbol{\theta})\right)$.

4. Set $\boldsymbol{\theta}^{(t)} = \boldsymbol{\theta}$.

Clearly, in 2, we generate values for the parameters from the corresponding prior distribution truncated to satisfy the condition $f(y_i|\boldsymbol{\theta}) \ge u$ implied by the auxiliary variables. According to Damien et al. (1999), the sampling scheme described above can be easily implemented in a wide variety of popular statistical models, including generalized linear models.

Although, by using the slice sampler, we avoid the specification of the proposal densities embedded in Metropolis–Hastings algorithms, we still need to find a convenient augmentation scheme. The main advantage of the algorithm is that, after finding the appropriate augmentation scheme, the method is directly applicable in all sets of data without computational difficulties. On the other hand, the resulting chain is usually highly autocorrelated. This is due to the fact that the sampling space is greatly extended which considerably delays convergence of the MCMC algorithm (Neal, 2003). Details of the slice sampler can also be found in Carlin and Louis (2000, pp. 167–170), Givens and Hoeting (2005, pp. 219–223), and Robert and Casella (2004, chap. 8).

2.3.6   A simple example using the slice sampler

Example 2.5. Slice sampler for binary logistic regression: Senility symptoms data revisited. Let us now consider the senility symptoms data presented in Example 2.3. We will now generate the posterior distribution using a simple slice sampler [for more details, see Damien et al. (1999)].

In order to build the slice sampler, we use u_i such that

$$f(\boldsymbol{u}, \beta_0, \beta_1|\boldsymbol{y}) \propto \prod_{i=1}^{n} I\left(u_i \le \frac{e^{\beta_0 y_i + \beta_1 x_i y_i}}{1 + e^{\beta_0 + \beta_1 x_i}}\right) \exp\left(-\frac{(\beta_0 - \mu_{\beta_0})^2}{2\sigma_{\beta_0}^2} - \frac{(\beta_1 - \mu_{\beta_1})^2}{2\sigma_{\beta_1}^2}\right).$$

This posterior has the "correct" marginal $f(\beta_0, \beta_1 | \boldsymbol{y})$ since

$$
f(\beta_0, \beta_1 | \boldsymbol{y}) = \int_0^{U_1} \int_0^{U_2} \cdots \int_0^{U_n} f(\boldsymbol{u}, \beta_0, \beta_1 | \boldsymbol{y}) d\boldsymbol{u}
$$

$$
\text{with} \quad U_i = \frac{\exp(\beta_0 y_i + \beta_1 x_i y_i)}{1 + \exp(\beta_0 + \beta_i x_i)} \quad \text{for } i = 1, 2, \ldots, n
$$

$$
\propto \left\{ \prod_{i=1}^{n} \int_0^{U_i} du_i \right\} \exp\left(-\frac{(\beta_0 - \mu_{\beta_0})^2}{2\sigma_{\beta_0}^2} - \frac{(\beta_1 - \mu_{\beta_1})^2}{2\sigma_{\beta_1}^2} \right)
$$

$$
\propto \left\{ \prod_{i=1}^{n} \frac{\exp(\beta_0 y_i + \beta_1 x_i y_i)}{1 + \exp(\beta_0 + \beta_1 x_i)} \right\} \exp\left(-\frac{(\beta_0 - \mu_{\beta_0})^2}{2\sigma_{\beta_0}^2} - \frac{(\beta_1 - \mu_{\beta_1})^2}{2\sigma_{\beta_1}^2} \right).
$$

Therefore we sample u_i from

$$
u_i | \boldsymbol{u}_{\backslash i}, \beta_0, \beta_1, \boldsymbol{y} \sim U\left(0, \frac{e^{\beta_0 y_i + \beta_1 x_i y_i}}{1 + e^{\beta_0 + \beta_i x_i}} \right),
$$

β_0 from

$$
\beta_0 | \boldsymbol{u}, \beta_1, \boldsymbol{y} \sim N\left(\mu_{\beta_0}, \sigma_{\beta_0}^2\right) \prod_{i=1}^{n} I\left(u_i \le \frac{e^{\beta_0 y_i + \beta_1 x_i y_i}}{1 + e^{\beta_0 + \beta_1 x_i}} \right),
$$

and β_1 from

$$
\beta_1 | \boldsymbol{u}, \beta_0, \boldsymbol{y} \sim N\left(\mu_{\beta_1}, \sigma_{\beta_1}^2\right) \prod_{i=1}^{n} I\left(u_i \le \frac{e^{\beta_0 y_i + \beta_1 x_i y_i}}{1 + e^{\beta_0 + \beta_1 x_i}} \right).
$$

The conditional posteriors described above are truncated normal distributions. The range of truncation is defined by the inequalities $u_i \le \exp(\beta_0 y_i + \beta_1 x_i y_i)/[1 + \exp(\beta_0 + \beta_1 x_i)]$ which can be written as

$$
\text{For } y_i = 1 \quad \Rightarrow \quad u_i \le \frac{e^{\beta_0 + \beta_1 x_i}}{1 + e^{\beta_0 + \beta_1 x_i}} \quad \Rightarrow \quad \beta_0 + \beta_1 x_i \ge \log \frac{u_i}{1 - u_i}
$$

$$
\text{For } y_i = 0 \quad \Rightarrow \quad u_i \le \frac{1}{1 + e^{\beta_0 + \beta_1 x_i}} \quad \Rightarrow \quad \beta_0 + \beta_1 x_i \le \log \frac{1 - u_i}{u_i}
$$

resulting in

$$
\max_{i:y_i=1} \left(\log \frac{u_i}{1 - u_i} \right) \le \beta_0 + \beta_1 x_i \le \min_{i:y_i=0} \left(\log \frac{1 - u_i}{u_i} \right).
$$

By solving these inequalities with respect to β_0 and β_1, we obtain the truncation intervals for these parameters

$$
l_0 = \max_{i:y_i=1} \left(\log \frac{u_i}{1 - u_i} - \beta_1 x_i \right) \le \beta_0 \le u_0 = \min_{i:y_i=0} \left(\log \frac{1 - u_i}{u_i} - \beta_1 x_i \right)
$$

and

$$
l_1 = \max_{i:y_i=1} \left(x_i^{-1} \left\{ \log \frac{u_i}{1 - u_i} - \beta_0 \right\} \right) \le \beta_1 \le u_1 = \min_{i:y_i=0} \left(x_i^{-1} \left\{ \log \frac{1 - u_i}{u_i} - \beta_0 \right\} \right),
$$

since $x_i > 0$ in our example. Parameters β_0 and β_1 are finally generated from $N(\mu_{\beta_0}, \sigma_{\beta_0}^2)I(l_0, u_0)$ and $N(\mu_{\beta_1}, \sigma_{\beta_1}^2)I(l_1, u_1)$ respectively.

COMPUTATIONAL NOTE (DIC for mixture models WinBUGS)

The conditional probability $P(X \le x | a \le X \le b)$ is given by

$$F_{[a,b]}^T(x) = P(X \le x | a \le X \le b) = \frac{P(a \le X \le x)}{P(a \le X \le b)} = \frac{F(x) - F(a)}{F(b) - F(a)}$$

for $x \in [a, b]$, where $F(x)$ is the cumulative distribution function of X and $F_{[a,b]}^T(x)$ is the corresponding function for X truncated to the interval $[a, b]$. Using the inversion method, we sample $u \sim U(0, 1)$ and set $u = F_{[a,b]}^T(x)$. Then, by solving the equation above, we find that the generated value x is given by

$$x = F^{-1}\left(F(a) + u\{F(b) - F(a)\} \right).$$

Details of the implementation of the above mentioned slice sampler and the corresponding R code are given in Table 2.14

Table 2.14   Algorithm and R code the slice sampler used in Example 2.5

Algorithm (pseudocode)	R code

Set up data	```> wais<-read.table('wais.dat',header=TRUE)```
	```> y<-wais$senility; x<-wais$wais; n<-length(y)```
	```> positive<- y==1```

Set number of iterations T Set prior parameters: $\mu_{\beta_0} = \mu_{\beta_1} = 0,$ $\sigma_{\beta_0} = \sigma_{\beta_1} = 100$	```> Iterations<-2500``` ```> mu.beta<-c(0,0); s.beta<-c(100,100)```

Initialize vector of sampled values β
Initialize acceptance probability counter
Set initial current $\beta^{(0)}, u$
For $t = 1, \dots, T$ repeat
	```> beta <- matrix(nrow=Iterations, ncol=2)```
	```> acc.prob <- 0```
	```> current.beta<-c(0,0); u<-numeric(n)```
	```> for (t in 1:Iterations){```

Calculate $\eta = \beta_0 + \beta_1 x$
Calculate $U = [U_i = e^{y_i \eta_i}/(1 + e^{\eta_i})]$
Generate $u = [u_i \sim U(0, U_i)]$
Calculate $lu = \{lu_i = \log[u/(1-u)]\}$
Calculate $l_0 = \max(lu_i - \beta_1 x_i)$ for all i
    with $y_i = 1$
Calculate $u_0 = \max(-lu_i - \beta_1 x_i)$ for
    all i with $y_i = 0$
Generate $\beta_0 \sim N(\mu_{\beta_0}, \sigma_{\beta_0}^2)I(l_0, u_0)$
	```> eta<-current.beta[1]+current.beta[2]*x```
	```> U<-exp(y*eta)/(1+exp(eta))```
	```> u<-runif(n, rep(0,n), U)```
	```> logitu<-log(u/(1-u))```
	```> logitu1<- logitu[positive]```
	```> l0<- max(logitu1 - current.beta[2]*x[positive])```
	```> logitu2<- -logitu[!positive]```
	```> u0<- min(logitu2 - current.beta[2]*x[!positive])```
	```> unif.random<-runif(1,0,1)```
	```> fa<- pnorm(l0, mu.beta[1], s.beta[1])```
	```> fb<- pnorm(u0, mu.beta[1], s.beta[1])```
	```> current.beta[1] <- qnorm(fa +```
	```unif.random*(fb-fa), mu.beta[1], s.beta[1])```

Calculate $l_1 = \max[(lu_i - \beta_0)/x_i]$ for all
    i with $y_i = 1$
Calculate $u_1 = \max[(-lu_i - \beta_0)/x_i]$ for
    all i with $y_i = 0$
Generate $\beta_1 \sim N(\mu_{\beta_1}, \sigma_{\beta_1}^2)I(l_1, u_1)$
	```> l1<- max((logitu1-current.beta[1])/x[positive])```
	```> u1<- min((logitu2-current.beta[1])/x[!positive])```
	```> unif.random<-runif(1,0,1)```
	```> fa<- pnorm(l1, mu.beta[2], s.beta[2])```
	```> fb<- pnorm(u1, mu.beta[2], s.beta[2])```
	```> current.beta[2] <- qnorm(fa +```
	```unif.random*(fb-fa), mu.beta[2], s.beta[2])```

Set $\beta^{(t)} = \beta$ End of for loop	```> beta[t,]<-current.beta``` ```> }```

Using the slice sampler described above, we have generated a total of 55,000 iterations and discarded the initial 15,000 iterations as a burnin period. From the diagnostic plots (Figure 2.20), we conclude that the ergodic means have been stabilized. Estimated Monte Carlo errors for the posterior means of β_0 and β_1 are equal to 0.110 and 0.0105, respectively. As expected, high autocorrelations are present, indicating slow convergence. This is typical when using slice samplers since the parameter space is expanded. Using the above mentioned slice sampler in this example, we generate $n + 2$ parameters, since one u_i is added for each observation i, instead of two parameters (β_0 and β_1) used in the original

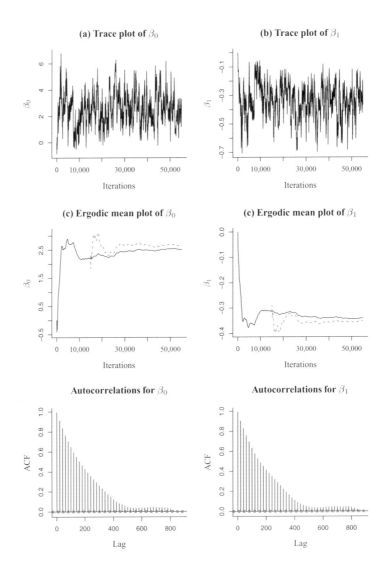

Figure 2.20   Diagnostic plots for slice sampler of logistic regression parameters of senility symptoms data (Example 2.5).

logistic regression model. The posterior means of β_0 and β_1 were found equal to 2.62 and -0.35, respectively, with standard deviations 1.25 and 0.119 and 95% credible intervals $(0.42, 5.28)$ and $(-0.61, -0.15)$, respectively. All these estimated values are close to the ones estimated using the Metropolis–Hastings schemes (see Table 2.10).

2.4 SUMMARY AND CLOSING REMARKS

In this chapter, Markov chain Monte Carlo algorithms are presented. Random-walk Metropolis, independent Metropolis–Hastings algorithms, the Gibbs sampler, and the slice sampler are described in detail accompanied with elaborate illustrations. The aim of this chapter is to introduce users to the logic of MCMC methods and enable them to understand how WinBUGS works and motivate them to construct their own MCMC algorithms (and codes).

The basic MCMC algorithms described above were developed, improved, and massively implemented in Bayesian inference during 1990–2000. During this decade, such methods were considered as standard and more complicated algorithms, implemented in specific problems, and are under further development and assessment.

Two interesting areas of active research regarding MCMC algorithms are (1) the development of varying dimension algorithms for model and variable selection methods (Green, 1995; Sisson, 2005) and (2) the construction of efficient and "automatic" algorithms (in the sense that tuning parameters can be specified using mathematical arguments).

In the following chapters we focus on the presentation and the technical details of Win-BUGS using the methods described here.

Problems

2.1 Consider the contingency table data of Problem 1.6 and the model of independence between X and Y i.e. $\pi_{ij} = \pi_i^X \pi_j^Y$. Assuming a multinomial distribution for the cell frequencies y_{ij} with probabilities π_{ij}, and using beta priors for π_i^X and π_j^Y:

    a) Estimate the posterior distributions for π_i^X and π_j^Y.

    b) Estimate the posterior distributions for π_{ij} using direct sampling.

    c) Compare the posterior distributions of π_{ij} under the models, assuming and not assuming independence between X and Y.

2.2 For the World Cup 2006 soccer data of Italy and France provided in Problem 1.8, using the help of direct sampling, calculate the posterior probabilities

    a) Of winning a game

    b) Of playing overtime (i.e., in case of a draw)

    c) Of winning a game during the overtime.

2.3 Consider the data of Problem 1.10 and the normal distribution with $Y_i \sim N(\mu, \sigma^2)$. Use a prior of the types $\mu \sim N(0, \sigma_\mu^2)$ and $\sigma^2 \sim \mathrm{IG}(a, b)$. Construct the following MCMC schemes, and generate 5000 observations from the posterior distribution.

    a) Use a simple Gibbs scheme (all conditional posterior distributions are of known form).

    b) Use an independence Metropolis scheme to jointly sample μ and $\log \sigma^2$.

    c) Use a bivariate random-walk scheme to jointly sample μ and $\log \sigma^2$.

2.4   For the data of Problem 1.10, assume the Student's t distribution for Y_i. Construct the following MCMC schemes to generate 5000 observations from the posterior distribution.

   a) Use an independence Metropolis scheme to jointly sample μ and $\log \sigma^2$.

   b) Use a bivariate random-walk scheme to jointly sample μ and $\log \sigma^2$.

   c) Use the adaptive rejection sampling of Gilks and Wild (1992) to implement Gibbs sampling (software for adaptive rejection sampling is available on the Web).

2.5   Use the Weibull distribution to model the following survival times (in days):
$$16\ 6\ 7\ 11\ 4\ 9\ 4\ 5\ 17\ 3\ .$$

   a) Construct a Metropolis within Gibbs algorithm to estimate the posterior distributions of the parameters of the Weibull distribution.

   b) Estimate the mean and median survival time under this model.

   c) Do you think that the exponential distribution is more appropriate for the data above mentioned .

2.6   Use the gamma distribution to model the survival times (in days) of Problem 2.5.

   a) Construct a Metropolis within Gibbs algorithm to estimate the posterior distributions of the parameters of the Weibull distribution.

   b) Estimate the mean survival time under this model.

   c) Compare the results with the corresponding ones from the Weibull and the exponential distributions.

2.7   Consider the pumps data in Spiegelhalter et al. (2003a). Construct MCMC algorithms to estimate the posterior distributions for the Poisson models $y_i \sim \text{Poisson}(\lambda_i)$ with

   a) Model 1: $\log \lambda_i = \alpha + \log(t_i)$

   b) Model 2: $\log \lambda_i = \alpha + \beta \log(t_i)$

   c) Model 3: $\lambda_i = \theta_i t_i$ and $\theta_i \sim \text{gamma}(a, b)$.

2.8   For the output obtained from any MCMC scheme of Problems 2.3–2.7

   a) Monitor the convergence graphically (see Section 2.2.2.4).

   b) Estimate Monte Carlo errors for all parameters (see Section 2.2.2.3).

   c) Download CODA and/or BOA packages for R and implement the available convergence diagnostics.

CHAPTER 3

WinBUGS SOFTWARE: INTRODUCTION, SETUP, AND BASIC ANALYSIS

3.1 INTRODUCTION AND HISTORICAL BACKGROUND

WinBUGS is a programming language based software that is used to generate a random sample from the posterior distribution of the parameters of a Bayesian model. The user only has to specify the data, the structure of the model under consideration, and some initial values for the model parameters. BUGS, the ancestor of WinBUGS , became popular during the 1990s, and its last version (v0.6 for DOS and UNIX operating systems) is still available via the BUGS project Website.[1] The acronym BUGS stands for the initials of the phrase "*B*ayesian inference *U*sing *G*ibbs *S*ampling".

The BUGS project was initiated in 1989 by the MRC Biostatistics Unit. The last versions (v0.5 and v0.6) of the "classic" BUGS were available via the project's Website in 1996 and 1997, respectively. The first experimental version of BUGS for windows, WinBUGS , was presented in 1997, while the current version (v1.4.3) was developed jointly with the Imperial College School of Medicine at St. Mary's, London. The project currently includes the development of OpenBUGS at the University of Helsinki in Finland, which is an open source experimental version of WinBUGS . In the project's site[2] a wide variety of add-in software, utilities, related papers, and course material is available. Some examples of related add-in software and WinBUGS expansions include PKBUGS for pharmacokinetic modeling, GeoBUGS for spatial modeling, and the WinBUGS jump interface for model

[1]http://www.mrc-bsu.cam.ac.uk/bugs/classic/contents.shtml.
[2]http://www.mrc-bsu.cam.ac.uk/bugs/

and variable selection. Also independent contributors have provided plug-in software and useful utilities to further assist WinBUGS users. WinBUGS can now be run even from other software packages, such as R, Matlab, and Excel.

The original aim of the WinBUGS project was to develop software for producing MCMC samples from the posterior distribution of the parameters of a desired model. Such model can be specified in WinBUGS using a relatively simple code (for users who are familiar with other languages) that is similar to the popular S language used by Splus and R statistical programs. Users not familiar with programming can specify the model structure by drawing its directed graphical structure in the DOODLE interface of WinBUGS. This tool automatically generates the corresponding model code.

The wide range of models that can be implemented by WinBUGS is one of the main factors that has considerably increased its popularity. Moreover, it is free of charge with high quality support by the research team of the project. Documentation includes a detailed accompanying manual and the three volumes of examples that cover a wide range of applications. WinBUGS users are only asked to register in the project's Website in order to receive by email a "key" file that (currently) must be renewed every 6 months.

OpenBugs is an open source version of WinBUGS. WinBUGS 1.4.3 is recommended by the BUGS team for standard use since OpenBUGS is still in experimental phase. Nevertheless, the corresponding compiled versions of OpenBUGS for windows and LINUX are named WinBUGS versions 2.2 and LinBUGS, respectively. In this book we focus on the WinBUGS version 1.4.3.

Finally, to close this short introduction, all WinBUGS users may have noticed the "health" warnings that always appear in the first page of BUGS and WinBUGS manuals:

Beware — Gibbs sampling can be dangerous!

This phrase resembles and reminds us of the health warnings that appear in all cigarette packets. Experienced MCMC users will consent with this comment for two reasons: (1) running a Gibbs sampler (and generally an MCMC algorithm) is usually a laborious procedure that requires careful computation of the conditional posterior densities and coding of the corresponding algorithm using a programming language, and (2) WinBUGS users must be familiar with the basic notions of Bayesian statistical theory and computation (focusing on MCMC); otherwise they may not specify the model correctly, or may interpret the posterior results incorrectly, or might not obtain results from a converged chain or even stick with the first computation problem of the algorithm (e.g., with a simple overflow problem).

3.2 THE WinBUGS ENVIRONMENT

3.2.1 Downloading and installing WinBUGS

During the writing of this book, the latest version of WinBUGS (version 1.4.3) was freely available via the BUGS project Website.[3] In order to install it, you need to download the WinBUGS installation file and the additional cumulative patches (if available)

After completing installation, the following license agreement appears when you open WinBUGS:

[3]http://www.mrc-bsu.cam.ac.uk/bugs/.

If the procedure is successfully completed, after restarting WinBUGS and retrieving the online help of the program (using the F1 key), the screen similar to the following must appear, with the new updated version appearing as highlighted below:

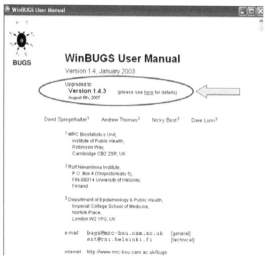

3.2.2   A short description of the menus

After completing installation, the following menu bar, which is identical in all Microsoft windows programs, appears in WinBUGS :

They can categorized as follows according to their functionality: basic operations (File, Window and Help), file editing operations (Tools, Edit, and Text), MCMC functions (Info, Model, Inference, and Options), and specialized operations (Doodle, and Map).

File, Window, and Help menus may be included in the first category of WinBUGS procedures since all of them refer to functions common to every software. The File menu provides the basic file handling capabilities. Using this menu, the user can open, save, print, and perform other related file management operations. The Window menu controls opened windows within WinBUGS. The main operations are the Tile Horizontal/Vertical and Arrange Icons commands usually available in all Windows programs. Using the Help menu, the user can read details concerning the license and the package version as well as browse the manual and the two volumes of examples that are available in the current version of WinBUGS.

File editing operations are performed using menus Tools, Edit, Attributes, and Text. All files edited in WinBUGS can be saved as *compound document* files that have an odc suffix. Text as well as graphics and plots produced by the WinBUGS output analysis can be saved in such files. The Tools menu provides commands for managing and editing compound document files, including encode/decode commands used to run software upgrades and license keys. The Edit menu is equivalent to the corresponding menu included in most Windows packages. It includes basic file editing commands such as cut, copy and paste. Fonts size, type, and color can be controlled via the Attributes menu. The Text menu provides useful tools for the manipulation of text in a compound document. It includes find/search basic operations and other text editing commands.

In the third category of menus, MCMC-related menus Info, Model, Inference and Options can be included. The Info menu provides information concerning the status of the current model and its components. It can be used to open and clear a log file or window. By log file or window, we refer to the location where results and messages produced by WinBUGS are printed. By default, WinBUGS opens an output window that can be saved as an odc file. Moreover, using the Node Info, the user can monitor the current values of any component of the model (random or constant). Finally, using the components command, we can monitor how many WinBUGS modules are currently active.

The Model menu is the most important menu of Win-BUGS since it checks the syntax of the model code, runs the MCMC algorithm, and obtains the posterior sample. In more detail, it is used to

- Specify the model (compile model code, load data and initial values) via the Specification tool

- Update the chain (Update)

- Monitor the acceptance rate of the Metropolis–Hastings algorithm (Monitor Met)

- Save the current values of the variables of interest (Save State)

- Specify the random seed number used to initialize the random number generation procedure (Seed)

- Run a script code (Script) that generates a list of WinBUGS commands without using the corresponding menus (see Appendix B for more details)

The Inference menu is also valuable since, with its available set of operations, we can

- Monitor the MCMC output

- Check the convergence

- Obtain posterior summaries and plots

Generally, with this menu we can infer for the posterior distributions of the parameters of interest using the generated sample. The corresponding set of inference operations was limited in classic BUGS and most of the output analysis was performed using other statistical programs.

The menu item `Samples` is the most frequently used tool. With this tool, the user can perform a variety of posterior output analyses producing, among others, summaries and graphical representations of the posterior distribution and the corresponding MCMC sample. Further analysis can be obtained using the remaining items of the `Inference` menu, including

- Comparison plots (using `Compare`)

- The posterior correlation matrix estimates (via the `Correlations` item)

- Summary measures of the posterior distributions of interest (`Summary`)

- Evaluation of the posterior distribution of the ordering/ranks of one vector (`Rank`)

- Calculation of the deviance information criterion (DIC) of Spiegelhalter et al. (2002) for checking the goodness of fit (`DIC` item)

The `Options` menu has three items:

- `Output options`, which controls the location where the output will printed (screen or log file) as well as the number of digits for the numeric output

- `Blocking options`, which specifies whether the updating method of Gamerman (1997) for the simulation of blocks of fixed-effect parameters will be used

- `Update options`, which selects the simulation method for each component

Finally, two menus can be referred as specialized ones: the `Doodle` and `Map` menus.

- The `Doodle` menu is used for constructing the model by drawing directed acyclic graphs (DAG). Specification of the model is produced graphically. This tool is convenient and friendly for practitioners who are not familiar with programming but they can alternatively conceive the structure of the model in terms of graphical paths and links. Additional details and a simple example are provided in Appendix A. This graphical interface was not included in the early DOS and LINUX versions of BUGS. It can be considered as an advanced, specialized function since the user can work in WinBUGS without using this module.

- The `Map` menu refers to the GeoBUGS facilities used for spatial modeling. GeoBUGS was been initially developed as an add-in upgrade option but was embodied in WinBUGS version 1.4. GeoBUGS provides convenient tools for the graphical representation, on actual maps, of geographic indices estimated using a Bayesian model.

3.3   PRELIMINARIES ON USING WinBUGS

WinBUGS code and output (including graphs and tables) can be stored in a compound document with the suffix odc. The user can complete all the basic word processing operations using the Tools, Edit, Attributes and Text menus of WinBUGS.

In order to start building a model, you need to open a new compound document using the path File>New.

3.3.1   Code structure and type of parameters/nodes

In order to specify a model, using WinBUGS code, we need to enclose all commands with the model statement. Hence, we write

```
model {
          . . . . . . . . . . . . . . . . . . . . . . . . .
}
```

The model's parameters/variables are grouped into three categories:

1. *Constants*: fixed values usually specified in the data section, which is described below. Explanatory variables, which are prefixed by the study design, are also set as constant parameters.

2. *Stochastic* or *random* components of the model: the random variables of the model that are characterized by a distribution. Both the model parameters θ and the response data (or variables) y_i are stochastic components since they are described by the prior $f(\theta)$ and the response distribution $f(y_i|\theta)$, respectively.

3. *Logical* components of the model: variables specified by a basic mathematical expression (called *logical* in WinBUGS terminology). They are simple functions or transformations of other model parameters. In this book we will refer to them as *deterministic* components. Deterministic components can be either random (hence be randomly distributed) or fixed constants depending on the expression or function used to specify them.

Model parameters (or variables or components) are also called *nodes* from the corresponding terminology in graphical models. In fact, this term is used in the WinBUGS manual since the project's research team facilitates directed acyclic graphs (DAGs) to describe all models; for more details, see Appendix A.

The stochastic nodes are specified using the following syntax

```
Variable ~ distribution( parameter1, parameter2, ... )
```

The tilde sign ~ is used to specify that a variable "follows" a distribution. As Variable, any conventional name can be adopted provided that it is not constrained by WinBUGS for other use. As distribution, a command from the list of the random distributions available by WinBUGS (described in Tables 3.1 and 3.2) can be selected. Most distribution command names are similar to the corresponding ones used in R and Splus packages. For example, the expression $X \sim N(\mu, \sigma^2 = 1/\tau)$ in WinBUGS language is written as

```
X ~ dnorm( mu, tau )
```

where tau is the precision parameter. Note that in WinBUGS, the normal distribution is defined using its precision parameter, not its variance or standard deviation.

A deterministic node can be specified by setting it equal to a mathematical expression using the assignment sign <-. Thus <- stands for the equality sign (=) used in mathematical equations. All such computations are hereafter called as node assignments or simply assignments. For example, we may calculate $y = x + z/3 + 1/w$ using the syntax

```
y <- x + z/3 + 1/w
```

In the case of the normal distribution, we may define by

```
s2 <- 1/tau
```

the variance of the distribution given by $\sigma^2 = 1/\tau$. If tau is a constant node then s2 will be also a constant node. If tau is a stochastic component [i.e., with a prior distribution $f(\tau)$ attached to it], then s2 will be also a random variable. In this case WinBUGS will produce a sample from its posterior distribution $f(\sigma^2|y)$. In order to facilitate the specification of deterministic nodes, a variety of function commands are available; for details, see Section 3.4.1 and the related summary in Table 3.3.

Generally a simple model of the type

```
model {
    Variable ~ distribution (parameter.value1 , parameter.value2 , ...)
}
```

produces random numbers from the distribution used. For example, the following simple code will generate random values for X from the standardized normal distribution:

```
model {
        X~dnorm(0.0, 1.0)
}
```

Note that each node/variable can be only defined once within the WinBUGS model code. Exception is made only for transformations of response variables [see relevant section in Spiegelhalter et al. (2003d)].

Comments can be included in the model code using the character #. Text after this character is ignored by WinBUGS. Comments are useful for reminding to the author or describing to other users the structure of the model.

3.3.2   Scalar, vector, matrix, and array nodes

In the previous section, the basic notions of WinBUGS were described, including the definition of a node and internode differentiation according to type. In this section, we focus on the dimension of a node. The simplest, in terms of computation, node is unidimensional and it is called a *scalar node* or simply a *node*. In WinBUGS, we may also define vector, matrix, and array nodes.

A vector v of length n and elements v_i (for $i = 1, 2, \ldots, n$) is denoted in WinBUGS by v[] and elements v[i], respectively. Similarly, a matrix M of dimension $I \times J$ with elements M_{ij} is denoted by M[,] and M[i,j], respectively. Finally, arrays are high-dimensional arrays and are denoted using the same logic as above. Each array element is denoted by its name followed by indices of each dimension separated by commas included in square brackets. Hence a four-dimensional array A with elements A_{ijkl} is denoted in WinBUGS by A[,,,] and A[i,j,k,l], respectively. R/Splus users are familiar with this notation. Parts of vectors, matrices, or arrays can be extracted using the same syntax as in R/Splus. All elements of a dimension can be extracted by omitting indices in the corresponding index space of the node. Thus, by M[i,] and M[,j] we extract the i row and j column vectors, respectively, of matrix M. Finally, using the expression n:m (where

Table 3.1 Univariate distributions available in WinBUGS

Distribution name	WinBUGS syntax	Probability or density function $f(x)$	Mean	Variance
		Discrete distributions		
(1) Bernoulli	x ~ dbern(p)	$p^x(1-p)^{1-x}$	p	$p(1-p)$
(2) Binomial	x ~ dbin(p, n)	$n!p^x(1-p)^{n-x}/[x!(n-x)!]$	np	$np(1-p)$
(3) Categorical	x ~ dcat(p[])	p_x	$\sum_{x=1}^{K} xp_x$	$\sum_{x=1}^{K}[x - E(x)]^2 p_x$
(4) Negative binomial	x ~ dnegbin(p, r)	$(x+r-1)!p^r(1-p)^x/[x!(r-1)!]$	$r(1-p)/p$	$r(1-p)/p^2$
(5) Poisson	x ~ dpois(lambda)	$\exp(-\lambda)\lambda^x/x!$	λ	λ
		Continuous distributions		
(6) Beta	x ~ dbeta(a, b)	$\Gamma(a+b)x^{a-1}(1-x)^{b-1}/[\Gamma(a)\Gamma(b)]$	$a/(a+b)$	$ab/[(a+b)^2(a+b+1)]$
(7) Chi-squared	x ~ dchisqr(k)	See gamma$(k/2, \frac{1}{2})$	k	$2k$
(8) Double exponential	x ~ ddexp(mu, tau)	$\frac{1}{2}\tau\exp(-\tau|x-\mu|)$	μ	$\sqrt{2}/\tau$
(9) Exponential	x ~ dexp(lambda)	$\lambda e^{-\lambda x}$	$1/\lambda$	$1/\lambda^2$
(10) Gamma	x ~ dgamma(a, b)	$b^a x^{a-1} e^{-bx}/\Gamma(a)$	a/b	a/b^2
(11) Generalized gamma	x ~ gen.gamma(a, b, r)	$rb(bx)^{ra-1}\exp[-(bx)^r]/\Gamma(a)$	$\Gamma(a+1/r)/[b\Gamma(a)]$	$[\Gamma(a+2/r^2)\Gamma(a) - \Gamma(a+1/r)^2]/[\Gamma(a)]^2$
(12) Log-normal	x ~ dlnorm(mu, tau)	$\sqrt{\tau/(2\pi)}\,x^{-1}\exp\left[-\tau/2(\log x-\mu)^2\right]$	$e^{\mu+1/(2\tau)}$	$(e^{1/\tau}-1)e^{2\mu+1/\tau}$
(13) Logistic	x ~ dlogis(mu, tau)	$\tau e^{\tau(x-\mu)}\left[1+e^{\tau(x-\mu)}\right]^{-2}$	μ	$\pi^2/[3\tau^2]$
(14) Normal	x ~ dnorm(mu, tau)	$\sqrt{\tau/(2\pi)}\exp[-\tau(x-\mu)^2/2]$	μ	$1/\tau$
(15) Pareto	x ~ dpar(a, c)	$ac^a x^{-a-1}$	$ab/(a-1)$	$ab^2/[(a-1)^2(a-2)]$
(16) Student's t	x ~ dt(mu, tau, v)	$\Gamma[(v+1)/2]\sqrt{\tau/(2\pi)}\left[\Gamma(v/2)\right]^{-1}\times\left[1+\tau v^{-1}(x-\mu)^2\right]^{-(v+1)/2}$	μ	$v\tau^{-1}/(v-2)$
(17) Uniform	x ~ dunif(a, b)	$1/(b-a)$	$a/(a+b)$	$\frac{1}{12}(b-a)^2$
(18) Weibull	x ~ dweib(v, lambda)	$v\lambda x^{v-1}\exp(-\lambda x^v)$	$\lambda^{-1/v}\Gamma\left(1+v^{-1}\right)$	$\left[\Gamma\left(1+2v^{-1}\right)-\Gamma\left(1+v^{-1}\right)^2\right]\lambda^{-2/v}$

Key: $p \in (0, 1)$; **(1)** $x = 0, 1$; **(2)** $x = 0, 1, 2, \ldots, n$, $p \in (0, 1)$; **(3)** $p[\,]$ is a vector of dimension K, $\sum_{x=1}^{K} p_x = p_x \in (0, 1)$, $\sum_{x=1}^{K} p_x = 1$; **(4)** $x = 0, 1, 2, \ldots$; $r = 1, 2, \ldots$; **(5)** $x = 0, 1, 2, \ldots$; $\lambda > 0$; **(6)** $x \in (0, 1)$, $a, b > 0$; **(8)** $x \in \mathbf{R}$, $\mu \in \mathbf{R}$, $\tau > 0$; **(9)** $x > 0$, $\lambda > 0$; **(10, 11)** $x > 0$, $a, b, r > 0$; **(12)** $x > 0$, $\mu \in \mathbf{R}$, $\tau > 0$; **(13, 14, 16)** $x \in \mathbf{R}$, $\mu \in \mathbf{R}$, $\tau, v > 0$; **(15)** $x > c$, $a, c > 0$; **(17)** $x \in (a, b)$, $a, b \in \mathbf{R}$, $a < b$; **(18)** $x > 0$, $v, \lambda > 0$.

Table 3.2 Multivariate distributions available in WinBUGS

Distribution name	WinBUGS syntax	Probability or density function $f(x)$	Mean	Variance/covariance
Discrete distributions				
(19) Multinomial	x[1:K] ~ dmulti(p[], N)	$N!\left(\prod_{i=1}^{K} x_i!\right)^{-1}\prod_{i=1}^{K} p_i^{x_i}$	$E(X_i) = Np_i$	$V(X_i) = Np_i(1-p_i)$ $\mathrm{Cov}(X_i, X_j) = -Np_ip_j$
Continuous distributions				
(20) Dirichlet	x[1:K] ~ ddirch(a[])	$\Gamma(a)\left[\prod_{i=1}^{K}\Gamma(a_i)\right]^{-1}\prod_{i=1}^{K} x_i^{a_i-1}$	$E(X_i) = a_i/a$	$V(X_i) = a_i(a-a_i)/[a^2(a+1)]$ $\mathrm{Cov}(X_i, X_j) = -a_ia_j/[a^2(a+1)]$
(21) Multivariate normal	x[1:K] ~ dmnorm(mu[], T[,])	$(2\pi)^{-K/2}|T|^{1/2}\exp\left[-\frac{1}{2}(x-\mu)^T T(x-\mu)\right]$	$E(X) = \mu$	$V(X) = T^{-1}$
(22) Multivariate Student's t	x[1:K] ~ dmt(mu[], T[,], v)	$(v\pi)^{-K/2}\Gamma[(v+K)/2]|T|^{1/2}/\Gamma(v/2)$ $\times\left[1+v^{-1}(x-\mu)^T T(x-\mu)\right]^{-(v+K)/2}$	$E(X) = \mu$	$V(X) = v(v-2)^{-1}T^{-1}$
(23) Wishart	x[1:K,1:K] ~ dwish(R[,], v)	$|R|^{v/2}|x|^{(v-K-1)/2}\exp\left[-\frac{1}{2}\mathrm{Tr}(Rx)\right]$	$E(X_{ij}) = vA_{ij}$	$\mathrm{Cov}(X_{ij}, X_{km}) = v(A_{ik}A_{jm} + A_{im}A_{jk})$

Key: **(19)** $x[\,]$ and $p[\,]$ are vectors of dimension K with elements $x[i] = x_i = 0, 1, 2, \ldots$, and $p[i] = p_i \in (0, 1)$ with $\sum_{i=1}^{K} x_i = N$ and $\sum_{i=1}^{K} p_i = 1$; **(20)** $x[\,]$ are vectors of dimension K with elements $x[i] = x_i \in (0, 1)$ and $a[i] = a_i > 0$ with $\sum_{i=1}^{K} x_i = 1$ and $a = \sum_{i=1}^{K} a_i$; **(21, 22)** $x[\,]$ and $mu[\,]$ (x and μ) are vectors of dimension K with elements $x[i] = x_i \in (0, 1)$ and $\mu[i] = \mu_i \in \mathbf{R}$, $T[\,]$ (and T) is a $K \times K$ symmetric precision matrix with elements $T[i, j] = \tau_{ij} > 0$; $v > 0$; **(23)** $x[\,]$ and $R[\cdot,]$ (x and R) are $K \times K$ positive-definite (symmetric) matrices with elements $x[i, j] = x_{ij}$ and $R[i, j] = R_{ij}$; A_{ij} are the element of the matrix $A = R^{-1}$; $v > 0$. Also μ_i: mean of X_i; $V(X_i)$, $\mathrm{Cov}(X_i, X_i)$: variance of X_i; $\mathrm{Cov}(X_i, X_j)$: covariance between X_i and X_j.

n, m are positive integer numbers), we can extract $n, n + 1, \ldots, m$ components of a vector (or array). For example, from a vector x we may extract elements 5–10 using the syntax v[5:10]. Equivalently, using the syntax M[3:6,] and M[,10:11], all elements of rows $3 - 6$, and all elements of columns $10 - 11$ (respectively) of matrix M can be extracted. Similar syntax may be used for higher dimensional arrays. For example, A[1:2, 3:4, 5:6] extracts all components A_{ijk} of the three dimensional array A with $i = 1, 2, j = 3, 4$, and $k = 5, 6$, resulting in a new $2 \times 2 \times 2$ array.

This nomenclature can be summarized by the following notation concerning vector nodes:

1. v[]: all elements of vector v

2. v[i]: v_i; i-th element of v

3. v[n:m]: $v_n, v_{n+1}, \ldots, v_m$; elements $n, n + 1, \ldots, m$ of vector v

The syntax used for matrices and their elements is summarized as follows:

1. M[,]: all elements of matrix M

2. M[i,j]: M_{ij}; element of ith row and jth column of matrix M

3. M[i,]: $M_{i1}, M_{i2}, \ldots, M_{iJ}$; elements of i-th row of matrix M

4. M[,j]: $M_{1j}, M_{2j}, \ldots, M_{Ij}$; elements of j-th column of matrix M

5. M[n:m,]: elements of $n, n + 1, \ldots, m$ rows of matrix M

6. M[,n:m]: elements of $n, n + 1, \ldots, m$ columns of matrix M

7. M[n:m, j]: elements of $n, n + 1, \ldots, m$ rows of j column of matrix M

8. M[i ,n:m]: elements of $n, n + 1, \ldots, m$ columns of i row of matrix M

9. M[n:m,k:l]: elements in $n, n+1, \ldots, m$ rows and $k, k+1, \ldots, l$ columns of matrix M

Finally, some indicative examples for three dimensional arrays of dimension $I \times J \times K$ are summarized as follows:

1. A[,,]: all elements of array A.

2. A[i,j,k]: A_{ijk} element of array A.

3. A[i,,], A[,j,], A[,,k]: elements with first, second, and third dimension equal to i, j, and k, respectively. The result of this syntax is a matrix.

4. A[i,j,]: elements with first and second dimensions equal to i and j. The result is a vector. The results for A[i,,k] and A[,j,k] are equivalent to these by obtaining all elements of the second and first dimension, respectively, with specific values for the remaining dimensions.

5. A[n:m,,]: all elements of A having the first dimension equal to $n, n + 1, \ldots, m$. The result is also an array. The result of the commands A[,n:m,] and A[,,n:m] is similar for the corresponding elements of the second and the third dimension, respectively.

6. A[n1:m1,n2:m2,n3:m3]: elements A_{ijk} with $i = n_1, n_1+1, \ldots, m_1, j = n_2, n_2 + 1, \ldots m_2$ and $k = n_3, n_3 + 1, \ldots, m_3$. The result is also an array.

Within brackets, calculations using the basic operations (+, -, *, and /) are allowed. Division should be avoided unless it is ensured that the result is a positive integer number. Nested indexing of the type x[y[i]] can also be used. This is very convenient when categorical data of i observation (group membership) are stored in vector y. For example, in one-way ANOVA models we want to specify that the mean of the ith subject is equal to $\mu_i = a + b_{g_i}$, where g_i is the group in which the i subject belongs. This can be effectively written in WinBUGS as

```
mu[i] <-  a+ b[ g[i] ]
```

The dimension of a node is defined automatically by the maximum indices used for a node within the WinBUGS model code.

Finally, note that computations directly on vectors, matrices, or arrays are not available in WinBUGS. Hence most of the calculations are performed separately for each elements. Details on this subject are provided in Section 3.4.2.

3.4   BUILDING BAYESIAN MODELS IN WinBUGS

In this section, we focus on specification of a Bayesian model. A full description and summary of the commands are provided, followed by details concerning calculations related to vectors, matrices, and arrays. Definition of a Bayesian model, including the prior and the likelihood specification, is also provided. The section concludes with specification of the data and the initial values.

3.4.1   Function description

As we have already mentioned, a set of functions is available within WinBUGS syntax. A summary of the functions is given in Table 3.3, and the most frequently used commands are described below.

Simple arithmetic functions.   Various simple arithmetic functions are available in Win-BUGS, including

- The absolute value (abs)
- The sine and cosine functions (sin, cos)
- The exponent and the natural logarithm (exp, log)
- The logarithm of the factorial of an integer number (logfact)
- The logarithm of the gamma function (loggam)
- The square root value (sqrt)

All these functions require as an argument a single scalar node. In order to set y equal to a function of x, the following syntax can be used:

```
y <- one.parameter.function(x)
```

Table 3.3 Functions available in WinBUGS

WinBUGS Syntax	Function	Description
1. abs(x)	$|x|$	Absolute value
2. cloglog(x)	$\log(-\log(1-x))$	Complementary log–log function
3. cos(x)	$\cos(x)$	Cosine function
4. cut(x)		Posterior of x is not updated by the likelihood
5. equals(x1, x2)	$f(x_1, x_2) = 1$ when $x_1 = x_2$ $= 0$ otherwise	Binary indicator function for equal nodes
6. exp(x)	e^x	Exponent value
7. inprod(v1[], v2[])	$\sum_i v_{1i} v_{2i}$	Inner product of two vectors
8. interp.lin(x, v1[], v2[])	$v_{2i} + (v_{2,i+1} - v_{2i})$ $\times (x - v_{1i})/(v_{1,i+1} - v_{1i})$	Interpolation line
8. inverse(M[,])	A^{-1}	Inverse of a symmetric positive-definite matrix
9. log(x)	$\log(x)$	Logarithm (ln)
10. logdet(M[,])	$\log|A|$	Logarithm of the determinant of a symmetric positive-definite matrix
11. logfact(k)	$\log(k!)$	Log factorial function of an integer
12. loggam(x)	$\log(\Gamma(x))$	Log gamma function
13. logit(x)	$\log \frac{x}{1-x}$	Logit function
14. max(x1, x2)	$\max(x_1, x_2)$	Maximum of two values
15. mean(v[])	$\bar{v} = \sum_{i=1}^{n} v_i/n$, where n is the length of vector v	Sample mean
16. min(x1, x2)	$\min(x_1, x_2)$	Minimum of two values
17. phi(x)	$P(X \leq x)$, $X \sim N(0,1)$	CDF of standardized normal
18. pow(x, z)	x^z	Power function
19. sin(x)	$\sin(x)$	Sine function
20. sqrt(x)	$\sqrt{x}$	Square root
21. rank(v[], k)	$\sum_i I(v_i \leq v_k)$, where $I(z) = 1$ if z true and 0 otherwise	Rank of s component of a vector
22. ranked(v[], k)	$v_i : \sum_s I(v_s \leq v_i) = k$	Element of a vector with rank s
23. round(x)		Round to the closest integer
24. sd(v[])	$\sqrt{\sum_{i=1}^{n}(v_i - \bar{v})^2/(n-1)}$	Sample standard deviation
25. step(x)	$f(x) = 1$ when $x \geq 0$; 0 otherwise	Binary indicator function of positive nodes
26. sum(v[])	$\sum_i v_i$	Sum of a vector's components
27. trunc(x)		Truncation to the closest smaller than x integer

Key: x, z = single real value or logical or mathematical expression; k = single integer value; v = vector; M = matrix.

For example, if $y = |x|$, then in WinBUGS we write

```
y <- abs(x)
```

Furthermore, commands `round` and `trunc` are used to obtain the closest and the lower closest integer values, respectively.

Arithmetic functions with two parameters include the maximum and the minimum between two values (`max`, `min`) and the power function (`pow`). If we wish to compare two values x_1, x_2 and keep (in another variable y) the maximum one, then the following syntax must be used

```
y <- max(x1, x2)
```

while for the computation of $y = x^z$ we write

```
y <- pow(x, z)
```

Statistical functions. Within WinBUGS, simple statistical functions can be calculated. The sample mean (`mean`), the sample standard deviation (`sd`), and the sum (`sum`) of a vector are available. The syntax of these commands is the same as in the one-parameter arithmetic functions. However, their argument must be a vector node v (denoted by $v[]$).

Two further commands (`rank`, `ranked`) are related to the ranking of the elements of a vector. If we wish to calculate the rank of the kth element of vector v, then we define

```
y <- rank(v[], k)
```

while with the syntax

```
y <- ranked(v[], k)
```

we obtain the element of v with rank equal to k. Note that the minimum and the maximum value of a vector v of length n can be calculated using the functions given above by writing

```
miny <- ranked(v[], 1)
maxy <- ranked(v[], n)
```

respectively. Similarly, the median value can be calculated using the syntax

```
mediany <- ranked(v[], (n+1)/2 )
```

when n is an odd number and using the syntax

```
mediany <- 0.5*(ranked(v[], n/2)+ranked(v[], n/2+1))
```

when n is an even number.

Command `phi` is used to calculate the cumulative distribution function (CDF) of a standardized normal variate. Hence the syntax

```
y <- phi(x)
```

will calculate the value

$$ y = \int_{-\infty}^{x} \frac{1}{\sqrt{2\pi}} \exp\left(-\frac{1}{2}z^2\right) dz $$

of a scalar node x. If we wish to calculate the CDF of $N(\mu, \sigma^2)$ evaluated at x, then we use the syntax

```
s <- sqrt(s2)
y <- phi( (x - mu)/s )
```

Finally, the command `inprod` calculates the inner product of two vectors. Hence the syntax

```
y <- inprod ( v[], w[] )
```

calculates the expression

$$y = \sum_{i=1}^{n} v_i w_i,$$

where v and w are vectors of length n with elements v_i and w_i; $i = 1, 2, \ldots, n$. This function is not actually statistical but a simple vector function. It is included in this section because it is frequently used for calculation of the sample covariance between two vectors v and w, given by the syntax

```
covariance <- (inprod ( v[], w[] ) - n * mean(v[])*mean(w[]))/(n-1)
```

Binary indicators. Two binary indicator functions are currently available in WinBUGS. Command equals compares two values resulting one if these two values are equal and zero otherwise. Thus, we have

$$\texttt{y<-equals(x, z)} \Rightarrow y = \begin{cases} 0 & \text{if } x \neq z \\ 1 & \text{if } x = z \end{cases}.$$

Command step checks whether a node is zero or positive, resulting in value one or zero if this statement is true or false, respectively. Hence, we have

$$\texttt{y<-step(x)} \Rightarrow y = \begin{cases} 0 & \text{if } x < 0 \\ 1 & \text{if } x \geq 0 \end{cases}.$$

These commands are used to evaluate "if" equalities and inequalities. For example, we may check whether inequality $x > z$ is true by writing

```
y <- step( x-z ) - equals( x, z)
```

Both these commands may also be used to indirectly implement "if statements", which are frequently encountered in programming languages. They can also be used to facilitate vector computations or for distribution truncations. For example, if we wish to calculate the variable $y = x$ if $x \geq a$, $y = z$ if $x < -a$ and $y = 3$, otherwise then we can write

```
y <- x*step(x-a) + z*(1-step(x+a)) + 3* step(x+a)*(1-step(x-a))
```

This syntax calculates the desired quantity since for $x \geq a$ we find that step(x-a)=1 and step(x+a)=1, resulting in

$$y = x \times 1 + z \times (1-1) + 3 \times 1 \times (1-1) = x.$$

For $-a \leq x < a$, we obtain step(x-a)=0 and step(x+a)=1, resulting in

$$y = x \times 0 + z \times (1-1) + 3 \times 1 \times (1-0) = 3$$

and for $x < a$, we have step(x-a)=0 and step(x+a)=0, resulting in

$$y = x \times 0 + z \times (1-0) + 3 \times 0 \times (1-0) = z.$$

Link functions. Commands `cloglog`, `logit`, `probit`, and `log` can be used on the left side of assignment. They are used to specify the corresponding link functions used in generalized linear models; see Chapter 7 for more details.

Mathematical expressions for each link function are provided in Table 3.3. The `probit` function is the inverse function of the CDF of the standardized normal variate. Thus, syntax

```
probit(y) <- x
```

is equivalent to

```
y <- phi(x)
```

Matrix functions. We can directly calculate the inverse and the logarithm of the determinant of a symmetric positive-definite matrix using the syntax

```
M2[1:K,1:K] <- inverse(M1[,])
y <- logdet(M1[,])
```

respectively, where M1 and M2 are both symmetric positive-definite matrix nodes of dimension $K \times K$ and y is a scalar node. The resulting matrix, which appears in the left part of the assignment for the inverse matrix calculation, must be defined, including its dimension indices within square brackets (i.e., the matrix name must be followed by [1:K,1:K]).

The cut function. This command is used when we do not wish the posterior of each parameter to be updated from the likelihood. It is recommended only for advanced users; for more details, see Spiegelhalter et al. (2003*d*).

3.4.2   Using the `for` syntax and array, matrix, and vector calculations

Direct calculations between vectors, matrices, and arrays are not available in WinBUGS except for specific functions described above. For example, the calculation of $x^T y$, where both x and y are vectors of the same length, is given by

```
z <- inprod( x[], y[] )
```

Other functions concerning vectors are the `sum`, `mean`, `sd`, `rank`, and `ranked`. Also, the commands `inverse` and `logdet` can be directly applied to symmetric positive-definite matrices.

Since functions for multidimensional nodes are limited, most of their calculations are completed separately for each element. For this reason, a syntax repeating similar expressions for each element of a multidimensional node is required. Such repetition is performed in WinBUGS using the `for` syntax:

```
for (i in I1:I2){
... ... ...
}
```

which repeats all commands within curly brackets $I_2 - I_1 + 1$ times, where i is an index taking all integer values from I_1 to I_2 sequentially in each repetition. In terms of programming terminology, such syntax is called a "loop." WinBUGS loop syntax is the same as in R/Splus but here is used to specify nodes (stochastically using ~ or deterministically using <-). For example, we can calculate the sum of two vectors of length n by

```
for (i in 1:n){
        z[i] <- x[i] + y[i]
}
```

or conclude that all elements x_i of a vector $\boldsymbol{x}$ follow independent normal distributions with
mean μ_i and common precision τ by

```
for (i in 1:n){
                x[i] ~ dnorm( mu[i], tau )
}
```

In order to perform similar actions for matrices (and arrays), we need to use "nested"
loops. Hence we can produce the sum of two matrices using the following syntax

```
for (i in 1:I){
      for (j in 1:J){
                C[i,j] <- A[i,j] + B[i,j]
      }
}
```

where I and J are the rows and columns of the matrices $\boldsymbol{A}$, $\boldsymbol{B}$, and $\boldsymbol{C}$. The multiplication
of two matrices $\boldsymbol{A}$ and $\boldsymbol{B}$ of dimensions $I \times K$ and $K \times J$, respectively, is slightly more
complicated. It can be accomplished by using the inprod function:

```
for (i in 1:I){
      for (j in 1:J){
                C[i,j] <- inprod(A[i,],B[,j])
      }
}
```

The range limits I1 and I2 of the index values must be defined as constant nodes in the
data part or the odc file; see Section 3.4.6 for details concerning data specification.

3.4.3   Use of parentheses, brackets and curly braces in WinBUGS

Let us summarize the use of each type of brackets:

1. Parentheses () are used

    - In mathematical expressions or computations
    - In functions surrounding their parameters/arguments [e.g., log(x)]
    - In for loops to declare the values of the index

2. Square brackets [] are used to specify the elements of a vector or array.

3. Curly brackets {} are used to declare the beginning and the end of the model and
    for statements.

3.4.4   Differences between WinBUGS and R/Splus syntax

At this point we summarize the main differences between the R/Splus and WinBUGS syntax.
This is very useful for users of R/Splus, especially for those who they wish to use
R/Splus for data manipulation and then import them in WinBUGS .

First, the order of the commands is not important in WinBUGS since the model is compiled altogether (simultaneously), unlike R/Splus, where the commands are executed
sequentially.

A further difference is that in WinBUGS we do not need to define a vector or array before
using it. Its dimension is set equal to the maximum of indices used for the corresponding
node within the code.

Mathematical expressions are implemented only in unidimensional scalar nodes in Win-BUGS.

Finally, in WinBUGS, an entire vector is denoted by its name followed by square brackets (e.g., x[]) while an array is denoted by its name followed by brackets, including commas to declare the number of dimensions (e.g., x[,,] indicates a three-dimensional array).

3.4.5 Model specification in WinBUGS

Likelihood specification. Let us assume a response variable Y with n observed values stored in a vector $\boldsymbol{y}$ with elements y_i. The stochastic part of the model can be written as

$$Y \sim \text{Distribution}(\boldsymbol{\vartheta})$$

where $\boldsymbol{\vartheta}$ is the parameter vector of assumed distribution. The parameter vector is "linked" with some explanatory variables $X_1, X_2, \ldots, X_p$ using a link function h

$$\boldsymbol{\vartheta} = h(\boldsymbol{\theta}, X_1, X_2, \ldots, X_p),$$

where $\boldsymbol{\theta}$ is a constrained set of parameters used to specify the link function and the final structure of the model. The vector $\boldsymbol{\vartheta}$ is the actual set of parameters to be estimated. Moreover, each subject specific for observed values of $x_{1i}, x_{2i}, \ldots, x_{pi}$ will define a different set of parameters $\boldsymbol{\theta}$ for each subject i given by

$$\boldsymbol{\vartheta}_{(i)} = h(\boldsymbol{\theta}, x_{1i}, x_{2i}, \ldots, x_{pi}) \, .$$

In generalized linear models this function associates (or links) the parameters of the assumed distribution with a linear combination of the explanatory variables. The likelihood of the model is given by

$$f(\boldsymbol{y}|\boldsymbol{\theta}) = \prod_{i=1}^{N} f\left(y_i|\boldsymbol{\vartheta}_{(i)} = h(\boldsymbol{\theta}, x_{1i}, X_{2i}, \ldots, x_{pi})\right) \, .$$

The corresponding WinBUGS syntax is given by

```
for (i in 1:n){
    y[i] ~ distribution.name( parameter1[i], parameter2[i], .... )
    parameter1[i] <-  [function of theta and X's]
    parameter2[i] <-  [function of theta and X's]
    ....
}
```

where distribution.name is one of the prespecified WinBUGS distributions given in Tables 3.1 and 3.2 while parameter1[i], parameter2[i], ... refers to the elements of $\boldsymbol{\vartheta}_{(i)}$.

Let us, for example, consider the case of a simple regression model with

$$Y_i \sim N(\mu_i, \sigma^2) \;\; \text{with} \;\; \mu_i = \alpha + \beta x_i \, .$$

In this example we have $\boldsymbol{\vartheta}_{(i)} = (\mu_i, \tau)$ (where $\tau = \sigma^{-2}$), $\boldsymbol{\theta} = (\alpha, \beta, \sigma^2)$, and h as the function

$$\begin{pmatrix} \mu \\ \tau \end{pmatrix} = h(\boldsymbol{\theta}, x) = \begin{pmatrix} \alpha + \beta x \\ \sigma^{-2} \end{pmatrix} \, .$$

The model described above is defined using the syntax

```
for (i in 1:n){
    y[i]     ~ dnorm( mu[i], tau[i] )
    mu[i]   <- alpha + beta * x[i]
    tau[i] <- 1/sigma2
}
```

Prior specification. To complete the Bayesian model specification, we further need to specify the prior distribution of the model parameters θ. Thus, we complete the specification by writing

```
theta1 ~ distribution.name( .... )
theta2 ~ distribution.name( .... )
....
```

If the parameters θ are stored in a single vector (of length K) and all elements follow a prior distribution of the same type, then a for loop can be used to specify them using the syntax

```
for (j in 1:K){
    theta[j] ~ distribution.name( second.level.parameters )
}
```

Hierarchically structured models can also be specified by allowing the prior parameters of θ to follow a prior distribution. Such specification allows us to define correlations or specify a wide variety of mixture distributions; see in Chapter 9 for more details.

3.4.6 Data and initial value specification

Data can be imported in WinBUGS using two different formats: the *rectangular* and the *list* formats.

3.4.6.1 Rectangular data format. The first data format is simple and similar to the text files used by most statistical packages. It can be used to specify a series of variables — vectors of the same length or a matrix (or array). Concerning simple vectors (for example variables denoted by y, x1, x2, x3), we only need to specify the names followed by square brackets in the first line and the values of each observation in each line, separated by empty spaces. *Be careful: The data must conclude with the command* END *followed by at least one blank line*. This detail has been added in version 1.4 and was not required in earlier versions of the program. Hence, an example is given by the following

```
y[] x1[] x2[] x3[]
10   20   23    12
11   23   11    97
...  ...  ...   ...
44   25   33    12
END
```

← **Blank line**

Similarly, a matrix Y with four columns (i.e., with dimension $n \times 4$) is specified in the rectangular data format by

```
Y[,1] Y[,2] Y[,3] Y[,4]
 10    20    23    12
 11    23    11    97
 ...   ...   ...   ...
 44    25    33    12
END
```

← **Blank line**

A higher-dimensional array is defined similarly by specifying all vectors resulting from leaving the first index empty. Therefore for a $4 \times 2 \times 2$ array A with elements $A_{ijk} = 100i + 10j + k$ is defined by

```
A[,1,1] A[,1,2] A[,2,1] A[,2,2]
111 112 121 122
211 212 221 222
311 312 321 322
411 412 421 422
END
```

← **Blank line**

3.4.6.2 *List data format.*
The list data format is similar to the list objects used in R/Splus. This format can be used to specify single constant numbers (scalar nodes), vectors, matrices, and arrays. The syntax of this data format always begins with the command list followed by parentheses without any separating space. Within the parentheses we specify each variable separated by commas. We can define

1. **Scalars/single numbers**: using the syntax

    ```
    scalar.name = scalar.value
    ```

2. **Vectors**: using the syntax

    ```
    vector.name = c( value1, value2, ..., value-n )
    ```

3. **Matrices**: using the syntax

    ```
    matrix.name = structure(
                  .Data = c( value1, value2, ..., value-k),
                  .Dim  = c( row.number, column.number )
                  )
    ```

4. **Arrays**: using the same syntax as in matrices but the .Dim argument will have at least three values specifying the length of each corresponding dimension.

Matrix elements defined in .Data argument are arranged by row, which means that the elements of the first row are specified first, followed by the elements of the second row, and so on. Note that although this syntax is similar to the corresponding one used in R/Splus, data in these programs are arranged by column (defining initially the elements of the first column followed by those elements of the second column, etc.). Therefore, R/Splus users must be cautious when defining data using this syntax in WinBUGS. For example, syntax

```
A = structure(.Data=c( 1,2,3,4,5,6,7,8,9,10,11,12),
              .Dim = c( 3,4 )
              )
```

defines matrix A as

```
      [,1] [,2] [,3] [,4]
[1,]    1    4    7   10
[2,]    2    5    8   11
[3,]    3    6    9   12
```

while in WinBUGS, it defines matrix A as follows:

```
        [,1] [,2] [,3] [,4]
[1,]      1    2    3    4
[2,]      5    6    7    8
[3,]      9   10   11   12
```

For higher-dimensional arrays, the data in WinBUGS are organized such that the last index of the array changes faster, followed by dimensions from last to first (and indices from left to right). Hence, for a three-way table of dimension $5 \times 3 \times 2$, we first define elements $(x_{11j}; j = 1, 2)$, then elements $(x_{12j}; j = 1, 2)$, $(x_{13j}; j = 1, 2)$, followed by elements $(x_{21j}; j = 1, 2)$, $(x_{22j}; j = 1, 2)$, $(x_{23j}; j = 1, 2)$, and so on. On the other hand, in R/Splus the first index changes faster, followed by indices from right to left. Hence we first specify x_{i11} for all $i = 1, 2, 3, 4, 5$ followed by x_{i21}, x_{i31} x_{i12}, x_{i22}, x_{i32}. Hence, syntax

```
A = structure (.Data=c(1,  2,  3,  4,  5,  6,  7,  8,  9,  10,
                       11, 12, 13, 14, 15, 16, 17, 18, 19, 20,
                       21, 22, 23, 24, 25, 26, 27, 28, 29, 30),
                  .Dim = c( 5,3,2 )
                  )
```

produces, in R/Splus, array A with the following structure

```
, , 1

        [,1] [,2] [,3]
[1,]      1    6   11
[2,]      2    7   12
[3,]      3    8   13
[4,]      4    9   14
[5,]      5   10   15

, , 2

        [,1] [,2] [,3]
[1,]     16   21   26
[2,]     17   22   27
[3,]     18   23   28
[4,]     19   24   29
[5,]     20   25   30
```

while in WinBUGS, it produces array A given by

```
, , 1
        [,1] [,2] [,3]
[1,]      1    3    5
[2,]      7    9   11
[3,]     13   15   17
[4,]     19   21   23
[5,]     25   27   29

, , 2
        [,1] [,2] [,3]
[1,]      2    4    6
[2,]      8   10   12
[3,]     14   16   18
[4,]     20   22   24
[5,]     26   28   30
```

3.4.6.3 *Importing data from* R/Splus.   It is convenient to use R/Splus to specify data and then extract them in a format compatible to the WinBUGS list format. In order to do that, we create a list object in R/Splus with all the desired components. Unfortunately, because of minor differences between WinBUGS and R/Splus syntax, some modifications are needed concerning matrices and arrays.

For arrays we propose the following strategy. Let us assume, for example, that the following matrix x is available in R/Splus

```
      [,1] [,2] [,3]
[1,]    1    2    3
[2,]    4    5    6
```

which can be defined using the command

```
x<-matrix(c(1,2,3,4,5,6),2,3,byrow=TRUE).
```

In R, commands

```
# define a list with object x equal to the transpose x
x2<-list( x=t(x) )
# exctract syntax compatible to R/Splus
dput(x2)
```

will print on the command window a syntax that is close to WinBUGS syntax. Alternatively, syntax dput(x, 'filename.txt') can be used to save the syntax to a file instead of being printed it at the command window of R/Splus.

In more detail, in R (version 2.5), the preceding command produces

```
structure(list(x = structure(c(1, 2, 3, 4, 5, 6),
              .Dim = c(3, 2))), .Names = "x")
```

We can modify this syntax as follows:

1. Remove "structure(" from the beginning and "), .Names = "x")" from the end of the syntax.

2. Add ".Data=" between "x = structure(" and "c(".

3. Reverse the dimension order. Hence substitute ".Dim = c(3, 2)" with ".Dim = c(2,3)".

This procedure finally results in the following WinBUGS list data format:

```
list( x = structure( .Data= c(1, 2, 3, 4, 5, 6),.Dim = c(2, 3) ).
```

The corresponding result of dput command in Splus (version 6.1) is

```
list("x" = matrix(c(1, 2, 3, 4, 5, 6), nrow = 3, ncol = 2) )
```

In this syntax we make the following changes

1. Replace "matrix(" with "structure(.Data=".

2. Remove quotes from the matrix name — replace ""x"" with "x".

3. Reverse the dimension order and use the ".Dim" expression; substitute "nrow = 3, ncol = 2" by ".Dim = c(2,3)".

The result is the same syntax as above:

```
list( x = structure( .Data= c(1, 2, 3, 4, 5, 6), .Dim = c(2, 3) ).
```

For higher-dimensional arrays, the data should be stored in reverse order of the indices. Then, using the command dput, we extract the code as above and proceed by making the appropriate changes. The most important part is to reverse the order of the dimension argument. For example, see the model specification section of the WinBUGS manual (Spiegelhalter et al., 2003*d*) .

If an array is already stored in R/Splus, then we need to use nested for loops to reverse the order of the dimensions. Hence, for a three-dimensional array we can use the syntax

```
dimx<-dim(x)
y<-array(dim=rev(dimx))
for (i in 1:dimx[1]){
    for (j in 1:dimx[2]){
        for (k in 1:dimx[3]){
            y[k,j,i]<-x[i,j,k]
}}}
```

For higher dimensions, additional nested loops are needed.

3.4.6.4 *A simple example of data specification.* Let us assume the following small dataset

y	x1	x2	gender	age
12	2	0.3	1	20
23	5	0.2	2	21
54	9	0.9	1	23
32	11	2.1	2	20

We further need to specify the sample size $n = 4$ and the number of variables $p = 5$. This dataset can be defined using the following list syntax:

```
list( n=4, p=5, y = c(12,23,54,32), x1 = c(2,5,9,11),
                x2 = c(0.3,0.2,0.9,1.1), gender = c(1,2,1,2),
                age = c(20,21,23,20) )
```

while the same data in a matrix form will be specified using the following syntax:

```
list( n=4, p=5,
        datamatrix=structure(
                .Data=c( 12,  2, 0.3,  1,  20, 23,   5, 0.2, 2, 21,
                         54,  9, 0.9,  1,  23, 32,  11, 2.1, 2, 20),
                .Dim=c(4,5)  )
```

The data can be viewed within WinBUGS by the menu path Info>Node Info after compiling the model (see Section 3.5) and are given using the following sequence:

```
datamatrix[1,1]        12.0
datamatrix[1,2]         2.0
datamatrix[1,3]         0.3
datamatrix[1,4]         1.0
datamatrix[1,5]        20.0
datamatrix[2,1]        23.0
datamatrix[2,2]         5.0
datamatrix[2,3]         0.2
datamatrix[2,4]         2.0
datamatrix[2,5]        21.0
datamatrix[3,1]        54.0
datamatrix[3,2]         9.0
datamatrix[3,3]         0.9
datamatrix[3,4]         1.0
datamatrix[3,5]        23.0
datamatrix[4,1]        32.0
datamatrix[4,2]        11.0
```

```
datamatrix[4,3]        2.1
datamatrix[4,4]        2.0
datamatrix[4,5]       20.0
```

3.4.6.5 *A simple example using arrays.* Let us assume that we have the following array

```
, , 1
       [,1] [,2] [,3] [,4]
[1,]     1    4    7   10
[2,]     2    5    8   11
[3,]     3    6    9   12
, , 2
       [,1] [,2] [,3] [,4]
[1,]    13   16   19   22
[2,]    14   17   20   23
[3,]    15   18   21   24
```

and we wish to insert it in WinBUGS including a vector of data with the dimensions of the array and the total number of cells and frequencies. Then, we need to write

```
list(dimx=c(3,4,2), cells=24, n=344,
      x=structure(.Data=c( 1, 13,   4, 16,   7, 19, 10, 22,
                          13, 14, 16, 17, 19, 20, 22, 23,
                           3, 15,   6, 18,   9, 21, 12, 24),
                         .Dim = c(3, 4, 2))                    )
```

In order to ensure that you have correctly specified the data sequence, we propose inserting the table in R/Splus, using the for syntax to reverse the dimension index sequence, and then exporting the data using the command dput as described in the previous subsection. The data imported will appear in the following sequence:

```
x[1,1,1]        1.0
x[1,1,2]       13.0
x[1,2,1]        4.0
x[1,2,2]       16.0
x[1,3,1]        7.0
x[1,3,2]       19.0
x[1,4,1]       10.0
x[1,4,2]       22.0
x[2,1,1]       13.0
x[2,1,2]       14.0
x[2,2,1]       16.0
x[2,2,2]       17.0
x[2,3,1]       19.0
x[2,3,2]       20.0
x[2,4,1]       22.0
x[2,4,2]       23.0
x[3,1,1]        3.0
x[3,1,2]       15.0
x[3,2,1]        6.0
x[3,2,2]       18.0
x[3,3,1]        9.0
x[3,3,2]       21.0
x[3,4,1]       12.0
x[3,4,2]       24.0
```

3.4.6.6 *Mixed and multiple data definition.* In some cases, it is convenient to use a combination of both types of data formats. In this way, we avoid difficulties in the

specification of matrix or arrays elements. Thus, some data can be loaded in a rectangular format and others in a list format. When using a rectangular format, we frequently need to specify some additional data (usually constants such as the data sample size) in a separate list format. In the case of multiple datasets, we need to load each set separately (one at a time) by pressing the `load` button (in `model specification tool`) multiple times; for details, see Section 3.5.

For example, in the first data example above, we can define variables y, x_1, x_2, gender, and age using the simple rectangular format

```
y[]  x1[]  x2[]  gender[]  age[]
 12   2    0.3      1        20
 23   5    0.2      2        21
 54   9    0.9      1        23
 32  11    2.1      2        20
END
```
                                                            ← **Blank line**

and the constants n and p using the following list format

```
list( n=4, p=5)
```

This is similar to the syntax if we wish to load all data in a single array written in a rectangular data format. The only difference is that the first line (with the variable names) must be replaced by

```
datamatrix[,1] datamatrix[,2] datamatrix[,3] datamatrix[,4] datamatrix
   [,5]
```

Alternatively, if we wish to specify the first column as the response with the name y and the rest of the variables in one matrix, then we write

```
y[] datamatrix[,1] datamatrix[,2] datamatrix[,3] datamatrix[,4]
```

For the second case of the $3 \times 4 \times 2$ matrix we can define the array by

```
x[,1,1] x[,2,1] x[,3,1] x[,4,1] x[,1,2] x[,2,2] x[,3,2] x[,4,2]
   1       4       7     10 13    16      19      22
   2       5       8     11 14    17      20      23
   3       6       9     12 15    18      21      24
END
```
                                                            ← **Blank line**

and the remainder using the following list format:

```
list(dimx=c(3,4,2), cells=24, n=24 )
```

For larger arrays it is convenient to specify each matrix defined by the combinations of the other dimensions in separate rectangular formats. For example, in the $3 \times 4 \times 2$ array above we can write

```
x[,1,2] x[,2,2] x[,3,2] x[,4,2]
  13      16      19      22
  14      17      20      23
  15      18      21      24
END
```
                                                            ← **Blank line**

```
x[,1,1] x[,2,1] x[,3,1] x[,4,1]
   1       4       7      10
   2       5       8      11
   3       6       9      12
END
```
                                                            ← **Blank line**

In the latter, the user must specify the data multiple times (one for each data definition used). Note that the rectangular data format with the highest indics (here x[,4,2]) must be read first. In this way WinBUGS will create a matrix of the correct dimension (here $3 \times 4 \times 2$) and then try to fill it in.

3.4.6.7 *Initial values.* Initial values are used to initiate the MCMC sampler. Their format is the same as the list data format. Initial values must be provided for all *stochastic* nodes except for the response data/variables. The user does not need to specify all initial values but can generate all or portion of them. Users must be careful in the case of using randomly generated initial values since they may face problems when certain parameters are initialized using inappropriate values resulting in numerical problems or slow convergence of the algorithm.

3.4.6.8 *Other details.* Missing values are represented using the conventional name NA, which stands for a nonavailable value.

All loaded data must be used in the model specification. For users who do not wish to use specific variables in the current model, we recommend defining a simple model for these variables that will not be connected (and therefore it will not affect) the main model under consideration.

Data are usually stored in the same odc file as the main model code. To conserve space, we can hide or suppress data using a WinBUGS "fold" following the menu path

Tools > Create Fold

Items written between the arrows of a fold can be hidden by clicking on these arrows. This option assists the user to keep the odc file small and easy to read.

3.4.7 An example of a complete model specification

Let us assume that we have a sample of 10 values from the normal distribution and we wish to estimate the posterior distribution of the mean and the variance. The parameter vector $\theta = (\mu, \sigma^2)$. Moreover, we use a normal prior distribution with mean zero and variance equal to 100 for μ, and the inverse gamma distribution for σ^2 with both parameters equal to $\frac{1}{100}$. The latter induces a gamma prior distribution for the precision parameter used in WinBUGS with mean equal to one and variance equal to 100. Both these priors can be regarded as low-information priors since their variance is large. The full model code is given by

```
model{
    # likelihood
    for (i in 1:n){ y[i] ~ dnorm( mu, tau ) }
    mu ~ dnorm( 0, 0.01 )      # prior for mu
    tau~ dgamma( 0.01, 0.01 ) # prior for tau
    #
    # deterministic definition of variance
    sigma.squared <-1/tau
    # deterministic definition of st.deviation
    sigma <-sqrt(sigma.squared)
}
```

and the data and the initial values by

```
DATA
list( n=10, y=c(-1.76, 0.38, 1.23, -0.67, -0.47, -1.36, 1.41, -0.07,
                -1.23, 2.35) )

INITS
list( mu=1, tau=2 )
```

Initial values are needed for parameters μ and τ since they are the stochastic components of the model. Parameter σ^2, which is the one that we actually wish to estimate, is defined in WinBUGS as a deterministic node, and hence initial values for this parameter are defined indirectly via the initial values of τ.

3.4.8   Data transformations

When we wish to consider transformations of the response data, which are stochastic nodes, the simplest way is to insert directly transformed data. Nevertheless, WinBUGS allows for transformations of the response data. In fact, this is the only case where we can specify the same node twice: first, to define the transformation of the original data and then to assign the random distribution. In the example above, we can specify that the logarithm of the absolute value of y follows a normal distribution. Then the likelihood for the new transformed model can be express by the following WinBUGS syntax:

```
for (i in 1:n){
     z[i] <- log( abs( y[i]) )
     z[i] ~ dnorm( mu , tau )
}
```

These transformations are permitted only for functions of observed data. No stochastic nodes can participate in the transformations of the syntax shown above. Thus, missing values are also not allowed in the transformed data.

From the WinBUGS team, it is recommended specifying any transformations at the beginning of the model code in order to separate data transformations from the actual model definition and, in this way, simplify the model code structure.

3.5   COMPILING THE MODEL AND SIMULATING VALUES

After writing the full model code, the data and the initial values in an odc file we need to compile and run the model. This procedure is described using the following steps (also see Section 3.7 for a summary of the procedure):

1. Open model specification tool.

2. Check the model's syntax.

3. Load data.

4. Compile model.

5. Set initial values.

6. Run the MCMC algorithm.

To be more specific:

1. *Open the Model Specification Tool.* Follow the path
   Model> Specification   to open the (model) specification tool.

 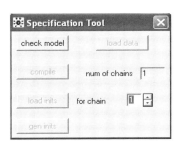

In this way, the model specification tool appears on the screen. This tool includes all the basic operations needed to initialize the MCMC algorithm (checking the model's code syntax, loading the data, compiling the model, and setting the initial values) and specify the number of chains that we wish to generate.

2. *Check the syntax of the model.* Highlight the command model and press the check model box of the model specification tool.

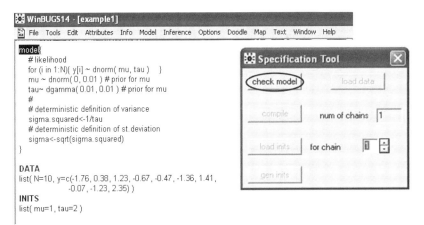

WinBUGS checks the model's syntax starting from the first character of the high-lighted command. If no command is highlighted, then the check begins at the top of the opened file or window. We recommend always highlighting the desired model command to avoid problems, especially if multiple model codes are included in the same odc file or window. If a problem in the syntax exists, then an indication is given in the lower left of the WinBUGS window while the cursor is placed at the location where the error was detected.

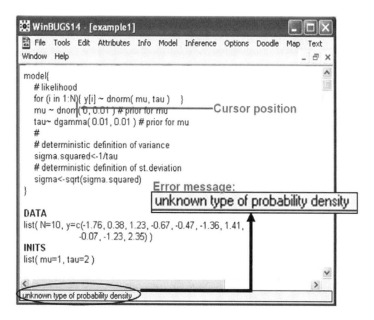

Otherwise, the message model is syntactically correct appears in the same position.

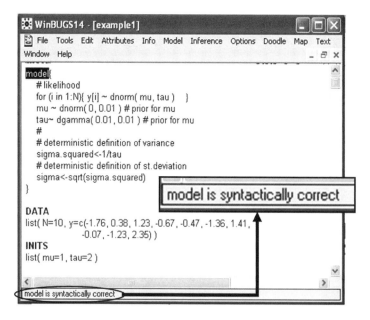

3. *Load the data.* Highlight the word list in the data list format and press the load data box of the model specification tool.

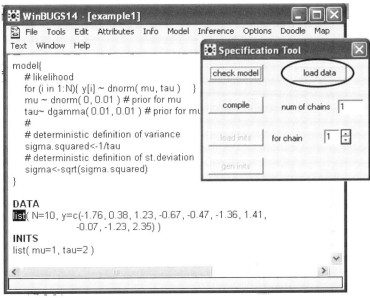

When the data are defined in a rectangular format then we highlight the first row of the data where the names are declared.

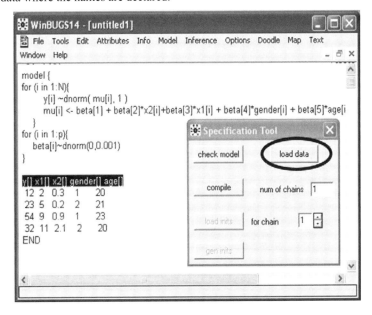

If a set of data is loaded successfully, then the message data loaded will appear at the bottom left of the WinBUGS window (status bar). Otherwise, an error message will appear in the same position.

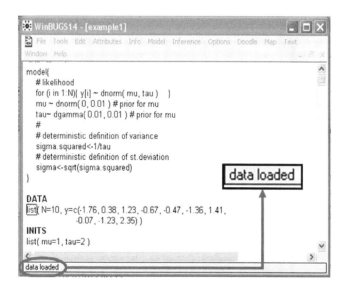

Finally, when multiple sets of data need to be declared, then we need to repeat the procedure (highlight data, load data) once for every set of data. For example, if the data are in the following format

```
list( aa=c(1,2,3,4,5,6,7,8,9,10))
list( n=4, p=5)
y[] x1[] x2[] gender[] age[]
 12  2  0.3    1     20
 23  5  0.2    2     21
 54  9  0.9    1     23
 32 11  2.1    2     20
END
```

then the procedure is as described in Figure 3.1

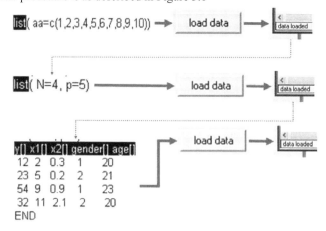

Figure 3.1 Example of multiple-dataset specification.

4. *Compile Model.* After all data are loaded, press the `compile` box in the model specification tool.

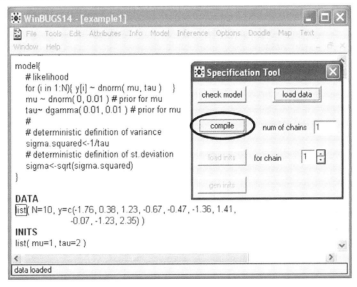

If the compilation is successful then the message `model compiled` will appear in the status bar otherwise an error message will appear in the same position.

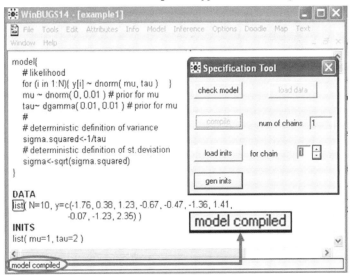

5. *Set initial values.* Initial values are set by following a procedure similar to the one used for data. We highlight the word `list` and then press the `load inits` box at the model specification tool.

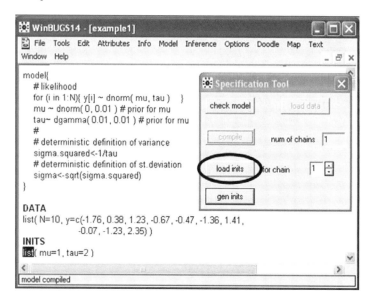

If all initial values are set successfully, the message `model is initialized` will appear in the status bar; otherwise an error message will be generated in the same position.

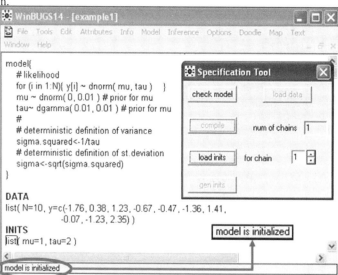

If no initial values are specified for some stochastic nodes, we get the message `the chain contains uninitialized variables` in the status bar.

In this case, we can generate random values for the remaining nodes using the `gen inits` box of the model specification tool. This box is useful when the number of stochastic nodes is large, such as in the cases of missing values, random effects, or latent data. This option should be avoided because it may generate inappropriate values that, for some nodes, will cause overflows or simply delay the convergence of the algorithm.

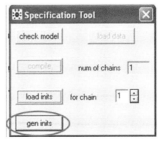

After setting (loading or generating) the initial values, the algorithm is initialized and is ready to simulate values using the `update` tool, which is described in the following step.

6. *Run the MCMC algorithm — generate random variables (burnin period).* Follow the path `Model> Update` to open the `update` tool.

The update tool has the following features:

a. `Updates`: This refers to the number of additional iterations we want to generate (and store).

b. `Refresh`: This refers to the number of iterations that WinBUGS will use to refresh the current iteration index in the `update` tool and update the online trace plots that are available via the `Sample monitor` tool. Using small refresh values is useful for monitoring the chain and for stopping or pausing the algorithm. The refresh value must be changed according to the current MCMC algorithm. For simple models (where the algorithm generates large iterations rapidly), we recommend using large refresh values, while for more complicated models (where the MCMC algorithm will be slower) we recommend selecting low refresh values.

c. `Update button`: Pressing this button will generate additional iterations equal to the number defined in the `Updates` box. If this button is pressed during the update procedure, then the algorithm will pause. During the pause of the MCMC run, specifications of the chain can be changed (e.g., the refresh rate) before restarting the algorithm by again clicking this button.

d. `Thin`: This defines the thin (lag) of iterations kept. Hence, if we set the additional updates equal to T' and the thin equal to k, then WinBUGS will generate $k \times T'$ iterations but will store only the last one in every sequence of k generated values.

e. Iteration: This is a visual counter of the current iteration of the algorithm. This is not the actual number of iterations carried out by the MCMC algorithm; rather it is the number of iterations that have already been stored. Thus, if this counter is equal to T' iterations and the thin is specified to be equal to k, then the actual number of generated values will be equal to $k \times T'$.

f. Overrelax: We may implement an overrelaxed MCMC algorithm (Neal, 1998), whenever this is feasible, by checking this box. This approach may be used to reduce high autocorrelations. This can be achieved by generating multiple sets of values and selecting the one that is negatively correlated with the values of the current iteration. Since multiple values are generated, the sampling of each iteration is considerably increased. According to Spiegelhalter et al. (2003d), "this method is not always effective and should be used with caution."

g. Adapting: This is an indicator box (like iterations) and is ticked during the tuning period when the Metropolis or the slice sampler algorithms are used. During the tuning period, parameters of the samplers are optimized in order to make the algorithms effective. The tuning period is set to 4000 and 500 iterations for Metropolis and slice sampler, respectively. The iterations of the tuning phase cannot be used for posterior inference (summaries, plots, etc.).

Note that the update tool will be available only after initializing correctly the MCMC algorithm for a given model.

7. *Set the parameters we wish to monitor.* Follow the path Inference> Samples to open the sample monitor tool.

Write the name of the parameter that we wish to monitor in the node box and then press the set button.

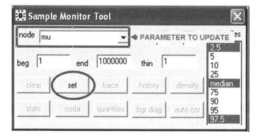

Using the procedure described above, we specify the parameters whose posterior distributions we wish to estimate via the MCMC generated values. The simulated values of these parameters will be now stored in order to produce a detailed posterior analysis. In WinBUGS terminology, this procedure is called "setting the monitored parameters".

8. *Update the MCMC algorithm — generate and store random variables.* After setting the parameter we wish to monitor, we update the MCMC sampler by repeating the procedure described in step 6. After setting the parameters of interest and generating additional random values, we can monitor the posterior distribution by extracting posterior summary statistics and plots. Analysis of the MCMC output is made via the Inference menu and mainly by the sample monitor tool. Detailed description follows in Section 3.6.

3.6   BASIC OUTPUT ANALYSIS USING THE SAMPLE MONITOR TOOL

The main tool for output analysis is the sample monitor tool. It provides a variety of options for simple output analysis of the stored random values. Most of them can be applied after setting which parameters we wish to monitor and after generating additional random values from the posterior distribution of the model (see steps 7 and 8 of the previous section).
    The following choices are available:

- node *box*: In this text box, the user inserts the name of the node to be used (set or obtain summaries). If the text does not corresponds to a name of a node, then all options will remain disabled. When a node that has not been monitored yet is typed, then only the set option is enabled.

On the other hand, when the node is included in the monitored ones, then all options are enabled. A list of the monitored nodes can be viewed by clicking on the arrow at the right of the box. Output analysis of a monitored node is obtained by selecting its name from this list.

If we wish to obtain the same summary for all monitored nodes, then the star character (*) can be used in the node text box. An internal node named `deviance` is automatically generated by WinBUGS. It is equal to minus twice times the log-likelihood evaluated in the parameter values of each iteration. This can also be set as a node name at the `sample monitor` tool. This value is frequently used to compare or evaluate the fit of the current model.

- `chains` *box*: In this box, we specify which chains will be used for calculation of summary statistics.

- `beg` *and* `end`: Iterations included in the posterior analysis are declared here.

- `thin` *text box*: The lag of the iterations used for the posterior analysis is defined in this box. It considers only one observation every k iterations. This is similar to the `thin` in the update tool. The difference is that here `thin` refers to the stored values and can be changed by the user after monitoring autocorrelations and MC errors. It refers only to the calculation of summaries and plots and not to the stored set of generated values as in the `update` tool.

- `set` *button*: This button is used to specify which nodes we wish to monitor after the current state of the algorithm. Values generated after pressing the `set` button are stored and can be used for posterior analysis.

- `trace` *button*: This button produces an online plot of the generated values against each iteration number. Points are plotted every k iteration; where k is the value declared in `refresh` button of the `update` tool. This plot is also referred as dynamic trace plot in the WinBUGS manual.

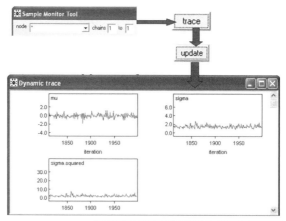

- `density` *button*: This button produces an approximate visual kernel estimate of the posterior density or probability function.

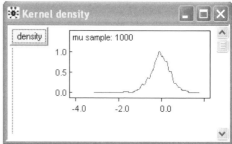

- `history` *button*: While the `trace` plot provides an online plot of the values generated, the `history` button draws a full trace plot of all stored values.

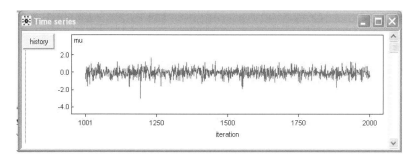

- `percentiles` *selection box*: In this text box, we specify which posterior percentiles we wish to calculate in the `stats`.

- `stats` *button*: This button estimates the posterior summary estimates by the MCMC output: mean, standard deviation, Monte Carlo (MC) error, and selected percentiles (default values are 2.5%, 50%, and 97.5%). It also provides details concerning the number of iterations discarded as burnin period and iterations finally kept for estimation. Monte Carlo error is estimated using the batch mean method; see Section 2.2.2 for details.

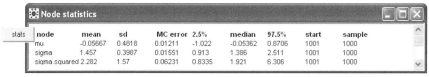

- `coda` *button*: This button produces a set windows with the sampled values in a format compatible to one used by CODA software. CODA is an add-in package for Splus and R that is used for checking convergence of the algorithm using a variety of diagnostics.

- `quantiles` *button*: A plot of the evolution for the median and the 2.5% and 97.5% percentiles for each iteration of the algorithm is obtained by this button.

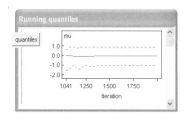

- `bgr diag` *button*: This button implements a modified version of the Gelman–Rubin convergence diagnostic (Gelman and Rubin, 1992; Brooks and Gelman, 1998). It can be applied only when multiple chains are used.

- `auto cor` *button*: This button plots the autocorrelations of the chain for each parameter of interest. The default is to plot all autocorrelation using lag from 1 to 50. Values of the autocorrelations can be monitored by double-clicking on the plot and then pressing Ctrl on the keyboard and the left mouse button.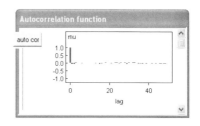

- `clear` *button*: This button clears and removes all stored sampled values of a given node. It is equivalent to restarting the algorithm with initial values equal to the ones of the current state of the algorithm.

The `density` and `stats` options are used to obtain posterior (numerical and visual) estimates of the posterior distribution, while the remaining options are used them mainly to monitor the convergence of the chain. MC error, which is calculated in `stats` option, can also be used to check the convergence of the chain. The `density`, `auto cor`, `stats`, `coda`, `quantiles`, and `bgr diag` buttons will be disabled if the MCMC algorithm is in an adapting phase of a Metropolis or slice sampler (initial 4000 and 500 iterations, respectively).

The convergence of the chain can be initially checked visually using trace plots obtained by the `History` option in the `sample monitor` tool. Values within a parallel zone without strong seasonalities will indicate convergence of the chain. Moreover, MC error can be checked. If the value is low in comparison to its posterior summaries (especially its standard error), then the posterior density is estimated with accuracy (`stats` option). For example, we can assume convergence when the MC error is lower than the 1% of the corresponding posterior standard deviation. We can further monitor convergence using autocorrelation plots. If autocorrelations are low, then convergence is obtained in a relatively low number of iterations (for details, see Section 2.2.2.4).

A formal convergence diagnostic (Brooks and Gelman, 1998) can be implemented using the option `bgr diag`, but it requires multiple chain simulation (see Section 4.3).

Finally, we can use CODA (Best et al., 1996) and BOA (Smith, 2005) packages in R to obtain additional model diagnostics. In order to extract generated values in a file format compatible to this packages, we use the `coda` option. Files required by CODA or BOA are opened in separate windows that can be saved and imported in R (or Splus).

3.7 SUMMARIZING THE PROCEDURE

We can summarize the procedure of obtaining posterior sample of model in WinBUGS by the following steps

1. Prepare the `odc` file — write the model code, specify data and initial values.

2. Compile and initialize the model.

3. Run the model/generate random values.

4. Perform output analysis using WinBUGS.

5. Apply convergence tests using BOA/CODA or other software.

6. Import WinBUGS output values to other package for additional analysis (plots, tests, etc.).

Steps 1–4 are the basic ones, while steps 5 and 6 are useful but can be omitted in simple examples where convergence is obvious and no additional analysis is needed. The procedure is also summarized in Figure 3.2

3.8 CHAPTER SUMMARY AND CONCLUDING COMMENTS

In this chapter, the user was introduced to the basic functions of WinBUGS software. Description of the code structure and commands are provided in detail, as well as guidance for running and analyzing simple models. Hence, after this chapter, the reader must be able to write simple models in WinBUGS code, compile them, generate values, and, finally, produce simple analysis of the MCMC output.

In the chapter that follows, a full example is provided, as well as more advanced or specialized functions of WinBUGS .

Problems

3.1 Consider the following data:

$$0.4 \ 0.01 \ 0.2 \ 0.1 \ 2.1 \ 0.1 \ 0.9 \ 2.4 \ 0.1 \ 0.2$$

Use the exponential distribution $Y_i \sim$ exponential(θ) to model these data and impose a prior on $\log \theta$.

a) Define the data in WinBUGS using both rectangular and list formats. Use $\theta = 1$ as initial value.

b) Write the model code.

c) Compile the model and obtain a sample of 1000 iterations after discarding initial 500 iterations as burnin.

d) Monitor the convergence graphically using trace and autocorrelations plots.

e) Calculate Monte Carlo errors using WinBUGS .

f) Obtain posterior summaries and density plots for θ, $1/\theta$, and $\log \theta$.

g) Export MCMC values using the CODA option. Import them in another statistical package and obtain the ergodic mean plot.

3.2 Consider data of Problem 3.1.

a) Use the gamma and the log-normal distributions to model the data.

b) Use the normal distribution for the logarithms of the original values to model the data.

c) In all models perform analysis similar to that in Problem 3.1.

d) Compare the results obtained under all the models above.

3.3 Use the Poisson and the negative binomial distributions to model the data of Problem 1.8.

a) Obtain a posterior sample for the model parameters.

b) Calculate the probability for each team to win a game.

3.4 Use the Poisson and the multinomial distributions to model the data of the 2×2 contingency table of Problem 1.6.

a) Define the data using both the rectangular and the list formats.

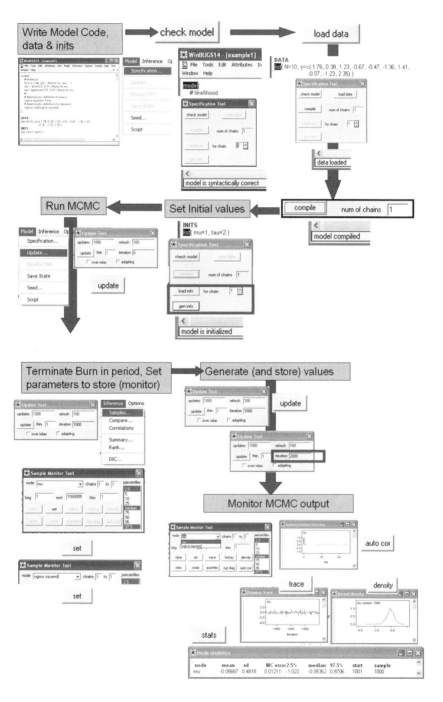

Figure 3.2 Summary chart for running a model in WinBUGS.

b) Perform output analysis and obtain posterior summaries for the cell probabilities as well as for the expected counts under both distributions.

c) Estimate the posterior distribution of the odds ratio

$$OR = \frac{E_{11}E_{22}}{E_{12}E_{21}}$$

under both formulations, where E_{ij} are the expected counts corresponding to cell i,j.

3.5   Consider the following data:

Sample 1: 1.50   0.54   0.85   2.04 -0.49 -0.65   0.98 -0.63 -1.72
          -0.06   0.60   1.25 -0.72   0.66 -0.39

Sample 2: 0.80   0.03 -0.37   0.67   0.76   2.47   1.36   2.29   2.26
           2.71 -0.28   1.70   1.88 -0.35   1.90

a) Use the normal and the Student's t distributions to model these data assuming that the parameters of each sample are different.

b) Obtain the posterior distributions for the means and variances for each sample and compare them.

c) Estimate the posterior distributions of $\mu_1 - \mu_2$ and σ_1/σ_2. Use the posterior sample to infer for the equality of the means and the variances of the two samples, where μ_k and σ_k are the mean and standard deviation, respectively, of the kth sample.

CHAPTER 4

WinBUGS SOFTWARE: ILLUSTRATION, RESULTS, AND FURTHER ANALYSIS

In this chapter we provide a full example into WinBUGS and additional, more specialized, functions, and tools in WinBUGS . The chapter is divided in five sections. In Section 4.1 we analyze the data of Example 1.4 (Kobe Bryant's's data) using the basic functions of WinBUGS introduced in the previous chapter. In Section 4.2 we proceed to describe the remaining functions of the inference menu, which offers a variety of tools for the analysis of the MCMC output. Generation of multiple chains is described in Section 4.3. Finally, control of figure properties and the remaining tools, and menus are presented in the last two sections of the chapter. All functions, tools and menus are illustrated using Example 1.4 (Kobe Bryant's's data) and the model fitted in Section 4.1.

4.1 A COMPLETE EXAMPLE OF RUNNING MCMC IN WinBUGS FOR A SIMPLE MODEL

4.1.1 The model

In this section we will consider Example 1.4. Let us examine again the Kobe Bryant success probabilities but now, instead of the original parameterization, alternatively consider the following expression of the success probabilities

$$\pi_k = \pi_{k-1} \times R_k \tag{4.1}$$

for $k = 2, \ldots, 8$ (corresponding to seasons 2000/01, $\ldots$, 2006/07), while π_1 will be estimated individually and will denote the success probability for season 1999. Quantities R_k

denote the performance of the current season (in terms of success probabilities) in comparison to the previous season. We use a beta prior for the success probability of the first season and gamma priors for the relative success measures R_k. In both cases we use low parameter values to express prior ignorance. Thus, we use

$$\pi_1 \sim \text{beta}(0.01, 0.01) \quad \text{and} \quad R_k \sim \text{gamma}(0.01, 0.01) \quad \text{for } k = 2, \dots, 8.$$

This prior setup can be simply defined in WinBUGS code using the commands

```
pi[1]  ~ dbeta( 0.01, 0.01)
R[1]   <- 1
for (k in 2:YEARS){
    R[k] ~ dgamma( 0.01, 0.01)
}
```

Note that if we use a vector to specify R_k, then we also need to define R_1. For this reason, we assign it a dummy value equal to one.

The likelihood for this model is given by

$$f(\boldsymbol{y}|\pi_1, \dots, \pi_8) = \prod_{i=1}^{8} \pi_i^{y_i} (1 - \pi)^{N_i - y_i},$$

with π_k given by (4.1). This likelihood will be written in WinBUGS code as

```
for (k in 1:YEARS){ y[k]   ~ dbinom( pi[k], N[k] ) }
```

while Eq. (4.1) will be expressed by the following syntax:

```
for (k in 2:YEARS){ pi[k]<-pi[k-1]*R[k] }
```

Hence the complete model code is given as follows:

```
model {
    # stochastic part of the likelihood
    for (k in 1:YEARS){ y[k]   ~ dbin( pi[k], N[k] ) }
    # deterministic part of the likelihood
    for (k in 2:YEARS){ pi[k]<-pi[k-1]*R[k] }
    # prior for pi[1]
    pi[1] ~ dbeta( 0.01, 0.01)
    # dummy value for R[1]
    R[1]   <- 1
    # Prior for R[k]
    for (k in 2:YEARS){R[k] ~ dgamma( 0.01, 0.01) }
}
```

In this code we can avoid multiple loops by including all commands related to the model in a single loop. Thus, the following code will have exactly the same effect as the previous code:

```
model {
    # stochastic likelihood part for the first observation
    y[1]   ~ dbin( pi[1], N[1] )
    for (k in 2:YEARS){
        # stochastic part of the likelihood
        y[k]   ~ dbin( pi[k], N[k] )
        # deterministic part of the likelihood
        pi[k]<-pi[k-1]*R[k]
        # Prior for R[k]
        R[k] ~ dgamma( 0.01, 0.01)
    }
```

```
# dummy value for R[1]
R[1]    <- 1
# prior for pi[1]
pi[1] ~ dbeta( 0.01, 0.01)
}
```

Although the final result will be the same in terms of WinBUGS, the model code cannot be followed as easily as the first one. We recommend writing model codes that are easy to follow, including detailed and descriptive comments. A clear, well-written model code accompanied by meaningful comments facilitates future reuse by its author and smooth implementation by other users.

4.1.2   Data and initial values

For illustration, we use a mixed data format: a combination of rectangular and list formatted data. In the list format, only the number of available seasons (i.e., the number of binomial experiments) is specified. The successes and the total attempts (y_i and N_i) for each season are defined as vectors in the following rectangular data format:

```
list(YEARS=8)
y[]   N[]
554   1183
701   1510
749   1597
868   1924
516   1178
573   1324
978   2173
399    845
END
```

Alternatively, a single list format can be used:

```
list(YEARS=8, y=c(554,701,749,868,516,573,978,399),
              n=c(1183,1510,1597,1924,1178,1324,2173, 845) )
```

Initial values values must be specified for π_1 and R_k ($k = 2, \ldots, 8$) using the following syntax:

```
list(pi=c(0.5, NA, NA, NA, NA, NA, NA, NA), R=c(NA,1,1,1,1,1,1,1) )
```

In this last syntax, initial values have been specified only for the first component of vector π and for all components of R except the first one. No initial values can be specified for $\pi_2, \ldots, \pi_8$ and R_1 since the they are deterministic and constant nodes, respectively. The value NA represents missing (unavailable) data.

4.1.3   Compiling and running the model

To compile and run the MCMC algorithm for 1000 burnin and 2000 iterations finally kept, we follow the steps below.

1. Highlight the word model in the model code window.

2. Open   the   model specification   tool   following   the   path Model> Specification.

3. Click on check model box.

4. If the message `model is syntactically correct` appears in the left bottom bar of the WinBUGS window, then proceed to step 5 (data load). Otherwise, correct model code according to the indication given by WinBUGS (the cursor will indicate the position of the false code).

5. Load the list format data by clicking on the `list` command and then press the `load data` button at the model specification tool. The message `data loaded` must appear if the data syntax is correct.

6. Load the rectangular formatted data by clicking on the header row (i.e., the names of the columns — vectors) and then press the `load data` button at the model specification tool. The message `data loaded` must appear (again) if the data syntax is correct.

7. Click on the `compile` button of the model specification tool. If no error is found, then the message `model compiled` will appear in the bottom left of the window. Otherwise correct the error indicated by WinBUGS and rerun everything (from the beginning).

8. Load initial values by highlighting the word `list` and then press the button `load inits` at the model specification tool.

9. If the message `model initialized` appears in the bottom left of the screen, then the algorithm is ready to run.

10. Open the `update` tool by following the path `Model>Update`.

11. Select the number of burnin iterations (updates), the refresh lag and thin, and press the `update` button (for this example we use updates=1000, refresh=100, thin=1). The index `iteration` will begin counting the number of iterations. It will stop when the number of iterations specified in the `updates` text box is reached.

12. Open the `sample monitor` tool in the path `Inference>Samples`.

13. Set the parameters that we wish to monitor (vector π and components 2–8 of vector R). In the sample monitor tool, write in the `node` text box pi (the name of π) and press the `set` button. Repeat the same for components 2–8 of vector R by inserting the name `R[2:8]` in the `node` text box. Repeat the same procedure for all additional nodes and parameters that you wish to monitor.

14. Open the online trace plots. In the sample monitor tool, type the star character (asterisk, *) in the `node` text box and press the `trace` button. A window with the online plots will appear.

15. Return to the update tool and set the number of iterations equal to 2000. Then press the `update` button. The index `iteration` will start (again), increasing until the desired number iterations is achieved. During the update the online plot will depict the generation of the monitored values.

16. We can now use the `sample monitor` tool to obtain some initial output analysis or extract data in other statistical software.

In order to get analysis for a single monitored parameter, we need to type the name of the corresponding Win-BUGS node in the Node text box of the `sample monitor` tool and then press the option or button we wish to implement. The names of the monitored parameters 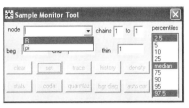 can be also selected from the list of monitored nodes that is available at the right of the node text box.

If we wish to select specific components of a matrix or array, we can extract them using the I : J syntax. For example, R[2:8] will select components 2–8 of vector R.

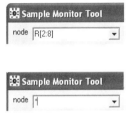

If we wish to implement the same analysis for all monitored parameters, then we type the asterisk (*) character in the node text box instead of the name of a specific node.

4.1.4 MCMC output analysis and results

4.1.4.1 Checking convergence.   The following analysis can be obtained using `sample monitor` tool in WinBUGS .

We first produce trace plots for all monitored parameters by typing the asterisk (*) in the node text box of the monitor tool. We then press the `History` button. The trace plot of π_1 follows for illustration.

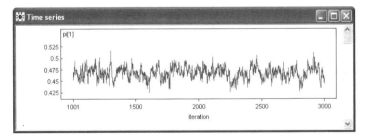

No patterns or irregularities are observed, and therefore convergence can be assumed. (see sections 2.2.2.4 and 3.6 for details).

We then monitor autocorrelations by pressing the `auto cor` button.

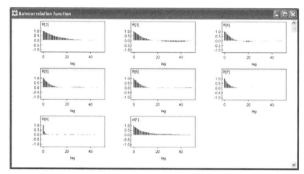

From this window, we observe that autocorrelations for all parameters become low only after considering a lag equal to 30. Thus, an independent sample can be obtained by rerunning the algorithm with `thin` set equal to 30 at the update tool.

Numerical values of the calculated autocorrelations can be obtained by selecting the autocorrelation plot (by double-clicking on its picture) and then pressing the keyboard Control key and the left mouse button. A window with all autocorrelation values of the selected parameter will appear on the screen.

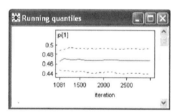

Then we monitor the evolution of selected quantiles using the `quantiles` button. This plot indicates that the requested quantiles have been stabilized, implying that the algorithm has converged in terms of π_1. Note that in this example we cannot use the option `bgr diag` since only a single chain was generated.

Using the `stats` option, Monte Carlo errors can be calculated. They measure the variation of the mean of the parameter of interest due to the simulation. If MC errors are low in comparison to the corresponding estimated posterior standard deviations, then the estimated the posterior mean was estimated with high precision. Increasing the number of iterations will decrease MC error. Here for parameter π_1 we get MC error$= 0.00116$, while the posterior standard deviation is 0.0135 (for a sample of 2000 iterations after discarding an additional 1000 iterations as burnin), which is equal to the 8.6% of the posterior standard deviation.

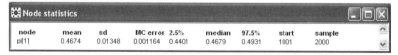

node	mean	sd	MC error	2.5%	median	97.5%	start	sample
pi[1]	0.4674	0.01348	0.001164	0.4401	0.4679	0.4931	1001	2000

Increasing the number of iterations to 10,000 will considerably decrease MC error to 0.00056 (48% decrease), while the posterior standard deviation will be relatively stable and equal to 0.0150. Note that the Monte Carlo variability of the remaining posterior summaries is minimal.

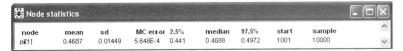

node	mean	sd	MC error	2.5%	median	97.5%	start	sample
pi[1]	0.4687	0.01449	5.648E-4	0.441	0.4688	0.4972	1001	10000

Finally, the CODA button creates two windows for each chain. The first one is the output window (or file) with all the generated values that are stored sequentially for each monitored parameter with the corresponding iteration number. Hence the values generated for the first node will be stored first followed by the second ones and so on. This window must be stored as a text file with extension out to enable the user to import it directly in CODA.

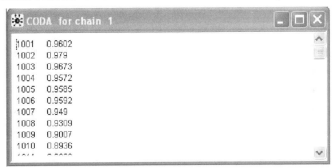

The second window refers to an index file with details related to the position of each node stored in the out file. Thus, the name of each node and the number of the starting and ending lines of the corresponding output are given in this window which must be stored as a text file with extension ind for direct use with the CODA package.

Generally, transforming CODA-type data to usual R or Splus data frames is not difficult for a user of medium experience. Nevertheless, functions and routines are available on the Web for converting CODA-type files in R/Splus data frames directly.

4.1.4.2 *Calculation of posterior summaries.*
The main tool for the calculation of the posterior summaries in WinBUGS is the stats option, which provides estimates of the posterior mean, standard deviation, and quantiles (including the median) for the given generated sample. The total number of iterations (generated sample size) and the number of iterations that the generated sample started (hence the burnin period) are also provided.

node	mean	sd	MC error	2.5%	median	97.5%	start	sample
R[2]	0.994	0.03967	0.003786	0.923	0.9918	1.074	1001	2000
R[3]	1.011	0.0359	0.002278	0.9383	1.01	1.079	1001	2000
R[4]	0.9615	0.03426	0.001879	0.8925	0.9618	1.026	1001	2000
R[5]	0.9725	0.03851	0.002836	0.9014	0.9723	1.047	1001	2000
R[6]	0.9908	0.04484	0.003078	0.9026	0.9902	1.079	1001	2000
R[7]	1.038	0.04279	0.002523	0.9617	1.036	1.126	1001	2000
R[8]	1.051	0.04557	0.001521	0.9676	1.049	1.147	1001	2000
pi[1]	0.4674	0.01348	0.001164	0.4401	0.4679	0.4931	1001	2000

From these results we can infer that Kobe Bryant's expected success field rate was equal to 46.7% for season 1999–2000. For the next two seasons (2000/01, 2001/02), his success rate was about the same (posterior means for $R_2 = 0.994$ and $R_3 = 1.011$). For seasons

2002/03 and 2003/04, his success rate was decreased by 4% and 3%, respectively (posterior expectations for R_4 and R_5 equal to 0.9615 and 0.9725). For the next season, the success rate was constant (posterior mean for $R_6 = 0.99$), and finally for the last two seasons we observe increase of the success rate by 4% and 5%, respectively (posterior means of R_7 and R_8 equal to 1.038 and 1.051).

We may also report the 95% posterior credible intervals for all R_i that quantify the relative performance of a current season in comparison to the previous one. All intervals lie around the reference value of one, which indicates a constant and stabilized performance of the player over the seasons studied.

The density option provides a graphical representation of the posterior density estimate for each node. For parameter π_1 in the Kobe Bryant's data, see Figure 4.1.

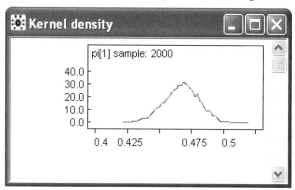

Figure 4.1   Density plot of Kobe Bryant's success rate of field goals for the first season (π_1).

Additional analysis can be performed using the remaining available options of the inference menu. These menus are described in detail in Section 4.2. Moreover, using the CODA option, the generated MCMC output can be exported in a text format and then imported in any statistical software for a more detailed output analysis.

4.2 FURTHER OUTPUT ANALYSIS USING THE INFERENCE MENU

In this section, details are provided concerning further output analysis within WinBUGS. We focus on the tools that are available in the inference menu while generating multiple chains, and the remaining operations available in WinBUGS are described in the sections that follow.

All procedures are described using the example of Kobe Bryant's field goal success rates presented in Section 4.1.

The inference menu offers a variety of tools for a more detailed output analysis. In summary, the following tools are available:

Compare: This tool is used to compare graphically the elements of a monitored vector node.

Correlations: This tool enables the user to obtain the correlation values and the scatterplots between two or more nodes.

Summary: This tool provides summary statistics for a node. The results are similar to those obtained with stats option of the sample monitor tool.

`Rank`: This tool provides useful analysis of the posterior distribution of ranks of the elements of a vector node. This tool is convenient when focus is on the evaluation of ranking of experimental units (organizations, teams, or subjects).

`DIC`: This tool offers the possibility of calculating the deviance information criterion (DIC) (Spiegelhalter et al., 2002) used to compare competitive models.

4.2.1 Comparison of nodes

In order to compare the elements of a vector node, we need to open the `comparison` tool following the path `Inference>Compare`.

We can plot the elements of a monitored vector by typing its name at the `node` text box and then pressing the `box plot` or the `caterpillar` buttons. The first will draw a boxplot for each node; the second one will draw horizontal lines or bars representing the 95% credible intervals of each node. For graphical representation of the procedure, see Figures 4.2 and 4.3.

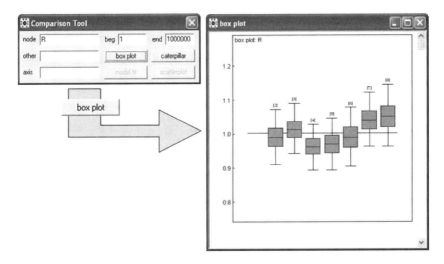

Figure 4.2   Comparison of posterior distributions of elements of vector R (relative performance in field goals) using boxplots.

Boxplots produced in WinBUGS are slightly different from the traditional one. The limits of each box represent the posterior quartiles, while the middle bar the posterior mean (by default, there is an option that allows the user to depict the posterior medians instead). The ending of the whisker lines represent the 2.5% and 97.5% posterior percentiles. The WinBUGS caterpillar plot essentially differs from the boxplot in two ways: (1) in the caterpillar plot, the box (referring to quartiles) is omitted, and (2) the bars are horizontal (parallel to the x axis). In both plots, the horizontal reference line represents the posterior mean estimated from all nodes depicted in the plot.

The `axis` box can be used to define which values will be used in the x axis. The node used in this box must be of the same length as the corresponding node that defined the `node` box. It can be either stochastic or constant (e.g., the values of an explanatory variable). In cases where the `axis` node is stochastic, the posterior means are used as values for the x axis.

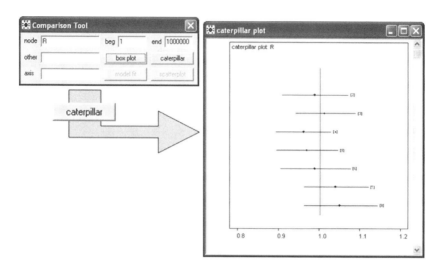

Figure 4.3 Comparison of posterior 95% credible intervals of elements of vector R (relative performance in field goals).

To illustrate the last two options, we have also defined a new deterministic node called index simply indicating the year sequence (taking values from one to eight) using the command

```
for (k in 1:YEARS){index[k] <-k }
```

After compiling and running the new model, we activate the comparison tool. We can now plot the parameter pi (Kobe Bryant's success rate) versus the time index by typing the corresponding names at the text boxes node and axis. Boxplot and caterpillar plots will not be affected by putting an additional node in the axis text box, but now the buttons model fit and scatterplot will also be activated. The first one creates a plot of the 95% credible interval of pi (node argument) for each element with the corresponding x axis values taken by index (axis argument). The 2.5% and 97.5% percentiles and mean points for successive values of index will be also connected in a plot that is equivalent to using a piecewise linear model.

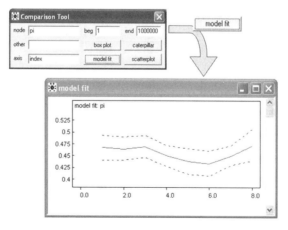

If we additionally type a node name in text box others, the values of this vector will be also printed on the plot. For example, using node R in the others text box will create the following plot

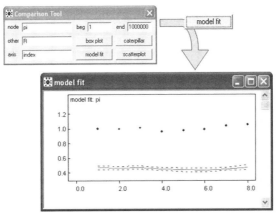

including the posterior means of node R. Note that the constant value of one for R[1] is also plotted. Generally the three arguments (node, other, axis) must be vectors of equal length. When stochastic nodes are used in other and axis then the posterior means are used. If we type R, R, and index in the text boxes node, other, and axis, respectively, we obtain the following plot:

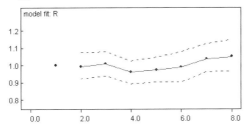

No posterior credible interval is plotted for R[1] since it refers to a constant node. If we eliminate node R[] by the others text box, then the corresponding points will not be plotted in the graph.

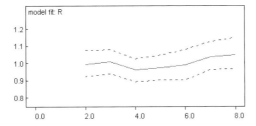

The scatterplot option provides a similar plot, with the values of node plotted against the values of axis. When stochastic nodes are used, then posterior means are used. Additionally, an exponentially weighted smoothed lined is fitted and plotted. Medians, bars, linear regression lines, and additional features can be plotted by changing the properties of the graph (click the right mouse button, then select properties and then special; for more details, see Section 4.4); for examples of such plots, see Figure 4.4 . Finally, beg and end text boxes are used to specify the the beginning and the ending iterations used for the visual comparisons.

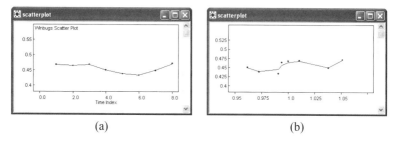

(a)   (b)

Figure 4.4   Scatterplots of (a) success probabilities versus time index (pi vs. index) and (b) success probabilities versus relative performance (pi vs. R) for Example 1.4 (Kobe Bryant's data).

4.2.2   Calculation of correlations

Correlations between the elements of a monitored vector node or between two monitored nodes can be obtained using the correlation tool (follow the path Inference> Correlations).

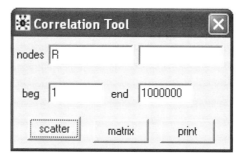

In the Correlation tool, the following fields and buttons and options are available:

- nodes *text boxes*: Here, the names of the nodes for which we wish to calculate posterior correlations are specified by the user. The second box can be left empty, provided that the node typed in the first one is a vector. In this case, correlations between all elements of the typed vector node are calculated. If two vector nodes are defined (one in each text box), then posterior correlations between the elements of the first and second nodes are calculated.

- beg *and* end *text boxes*: As usual, the range of the iterations that will be used for calculation of the posterior correlations can be defined here.

- scatter *button*: This produces a scatterplot matrix between the elements of the requested nodes.

- matrix *button*: Produces a visual representation of the correlation matrix using different shades of black instead of the actual values. This simplifies comparisons of the correlations in multidimensional problems.

- `print` *button*: This provides the numerical estimates of the posterior correlations between the elements of the requested nodes.

In Figure 4.5 we provide results of this tool for a single vector node (`R`; plots a–c) and for two vector nodes (`R` and `pi`; plots d and e).

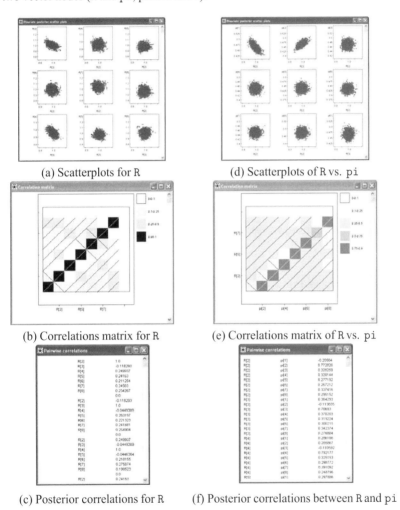

(a) Scatterplots for `R`   (d) Scatterplots of `R` vs. `pi`

(b) Correlations matrix for `R`   (e) Correlations matrix of `R` vs. `pi`

(c) Posterior correlations for `R`   (f) Posterior correlations between `R` and `pi`

Figure 4.5   Results of the `correlation` tool using data of Example 1.4 (Kobe Bryant's data).

4.2.3 Using the `summary` tool

The `summary monitor` tool can be used independently from the `sample monitor` tool. We can open the corresponding dialog box by following the path `Inference>Summary`. Using this tool, summary statistics of the posterior distribution of a node can be obtained. It is equivalent to the `stats` option in `sample monitor` tool. The main difference from the latter is that no values are stored here and hence less storage is required. This is convenient in high-dimensional problems where only some summary statistics may be needed for specific nodes.

When we open the summary monitor tool, all four buttons are inactive. In order to activate them, we must define again which nodes we wish to monitor via this tool and then generate some additional iterations. Implementation of this procedure in Example 1.4 for pi can be summarized by the following steps:

- Type pi in the node text box and then press the set button.

- Return to the update tool and generate additional values (e.g., 1000).

- Retype the node's name (pi) and press the stats button. Then the following results will appear:

node	mean	sd	2.5%	median	97.5%	sample
pi[1]	0.4681	0.01396	0.4152	0.4668	0.5069	1000
pi[2]	0.4633	0.01283	0.423	0.466	0.5091	1000
pi[3]	0.4695	0.01239	0.4217	0.469	0.5143	1000
pi[4]	0.4516	0.01151	0.4156	0.4508	0.4972	1000
pi[5]	0.4366	0.01398	0.3837	0.4366	0.4823	1000
pi[6]	0.4323	0.01329	0.3771	0.4333	0.4747	1000
pi[7]	0.4494	0.01101	0.4008	0.4489	0.4938	1000
pi[8]	0.4729	0.01727	0.4163	0.4731	0.5226	1000

As it is evident, the resulting output is the same as the corresponding one given by the stats option in the sample monitor tool.

- For pi node, press the mean button. Then only the posterior means of this vector node will appear on a separate window.

If the second step is skipped (i.e., the generation of additional iterations after set), then no action will be performed when pressing either the stats or the means button.

4.2.4   Evaluation and ranking of individuals

In many problems there is interest in the evaluation and ranking of individual or units under study. Such problems typically arise in the evaluation of healthcare units (hospitals or medical centers), educational institutes (schools, colleges, universities, postgraduate programs), students, or athletic teams

WinBUGS provides the facility to monitor some rank-related statistics on the basis of specific stochastic vector nodes that represent estimated or predicted performance. This can be done automatically using the `rank monitor` tool, which can be invoked following the path `Inference>Rank`. Using this tool, in each iteration of the MCMC algorithm the values of the vector node are rearranged in ascending order and the corresponding ranks are calculated and stored. Therefore, ranks indicate the sequence of the stochastic vector elements in ascending order; for example, rank 1 may indicate that this element had the lowest value. This is also an independent tool, where we need to specify separately which nodes will be monitored in terms of rank. The procedure (for Example 1.4) for evaluating Kobe Bryant's performance using the ranks of node `pi` can be summarized by the following steps.

- Type `pi` in the `node` text box and then press the `set` button (in `rank monitor` tool).

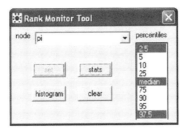

- Return to the `update` tool and generate additional values (e.g., 1000).

- Retype the node's name (`pi`) and press the `stats` button. Then the following results will appear:

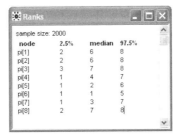

As you can see, the median of the rank for each p_i is given within the 2.5–97.5% percentiles. By this window we can infer which seasons were the best in terms of the player's performance. For example, seasons 3 and 8 have the higher median rank (equal to 7). We can further infer that season 3 was the best since the 2.5% percentile is higher than the corresponding one for season 8 (3 vs. 2). In this way we can obtain an informal ordering of the player's performance for each season.

- The button `histogram` will provide a visual representation of the rank distribution for every node. Thus, using the `histogram` button, we obtain the following plots:

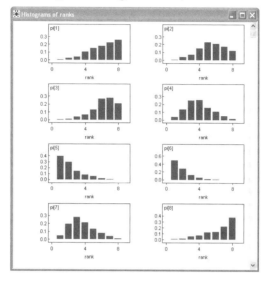

- Finally, the `clear` button deletes all simulated rank values for the selected node. If we wish to rerun some new rank values for the same node, then we have to repeat the procedure outlined above.

Note that more detailed analysis of ranks can be obtained if we include corresponding code within the model definition; for details see the surgical example in Spiegelhalter et al. (1996b, p. 18). In terms of storage, the rank monitor tool is more efficient since it reserves lower storage space than does the corresponding one, which calculates ranks analytically within the model's code.

4.2.5 Calculation of deviance information criterion

The deviance information criterion (DIC) was introduced by Spiegelhalter et al. (2002) as a measure of model comparison and adequacy. It is given by the expression
$$\text{DIC}(m) = 2\overline{D(\boldsymbol{\theta}_m, m)} - D(\overline{\boldsymbol{\theta}}_m, m) = D(\overline{\boldsymbol{\theta}}_m, m) + 2p_m,$$

where $D(\boldsymbol{\theta}_m, m)$ is the usual deviance measure, which is equal to minus twice the log-likelihood
$$D(\boldsymbol{\theta}_m, m) = -2\log f(\boldsymbol{y}|\boldsymbol{\theta}_m, m)$$

and $\overline{D(\boldsymbol{\theta}_m, m)}$ is its posterior mean, p_m can be interpreted as the number of "effective" parameters for model m given by

$$p_m = \overline{D(\boldsymbol{\theta}_m, m)} - D(\overline{\boldsymbol{\theta}}_m, m),$$

and $\overline{\boldsymbol{\theta}}_m$ is the posterior mean of the parameters involved in model m. Smaller DIC values indicate a better-fitting model. DIC must be used with caution since it assumes that the posterior mean can be used as a "good" summary of central location for description of the posterior distribution. Problems have been reported in cases where posterior distributions are not symmetric or unimodal. Details concerning DIC limitations and possible problems can be found in Spiegelhalter et al. (2003*d*, section entitled "Tricks: Advanced Use of the BUGS Language"). A detailed illustration of DIC can be found in Section 6.4.3; for additional details on the use of DIC and other methods for model comparison, see Chapter 11.

The DIC tool can be opened following the path Inference>DIC. The procedure for calculating DIC is similar to those for summary and ranks monitor tools. We first press the set button to start storing DIC values, then update the sampler and finally obtain the results by pressing the DIC button.

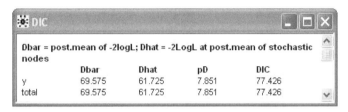

4.3 MULTIPLE CHAINS

4.3.1 Generation of multiple chains

The procedure described in Section 4.1.3 is followed when we wish to generate a single chain. If we wish to generate additional samples, then we follow steps 1–6 as in Section 4.1.3. The remaining of the steps are slightly modified according to the following:

7. Set the number of chains you wish to generate in the text box next to the compile button and then click on compile. For this example, three chains were used.

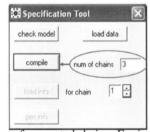

8. Load initial values by highlighting the word list and then pressing the button load inits in the model specification tool.
   Repeat the procedure as many times as the number of generated chains. For instance, for three chains we need to set three different sets of initial values. The generate button might be used for some chains but is recommended for specifying different starting values: with appropriate dispersion in order to ensure convergence. For Kobe Bryant's data, we have used the following three sets of initial values.

```
list(pi=c(0.5,NA,NA,NA,NA,NA,NA,NA), R=c(NA,1,1,1,1,1,1,1) )
list(pi=c(0.1,NA,NA,NA,NA,NA,NA,NA), R=c(NA,1,1,1,1,1,1,1) )
list(pi=c(0.9,NA,NA,NA,NA,NA,NA,NA), R=c(NA,1,1,1,1,1,1,1) )
```

9–12. If the model is initialized, then update all chains by the given number of iterations (the same as in Section 4.1.3). Hence 1000 iterations means that each chain will be updated by 1000 iterations.

13. Set the parameters to be monitored. By default, observations from all chains are monitored, but this can be modified by the chain...to... text boxes that are

present on the upper right of the `sample monitor` tool. These text boxes can be used to extract posterior summaries and output analysis only for selected chains.

14. Open the online trace plots. Now one line for each chain is outlined on the dynamic trace plot.

15. Update the sampler by 2000 iterations.

16. The `sample monitor` tool can now be used to obtain initial output analysis and export data to other statistical programs.

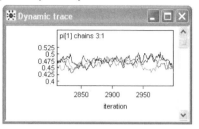

4.3.2   Output analysis

In the output analysis we observe the following differences:

- In all plots (autocorrelations, trace, history, quantiles), results for all chains are visually separated in the same graph. For example, in the `history` plot one line for each chain is printed; see Figure 4.6 for the trace plot resulting from the model of Kobe Bryant's data. Only in the `density` plot, a single estimate of the posterior distribution is calculated (and plotted) using values from all chains.

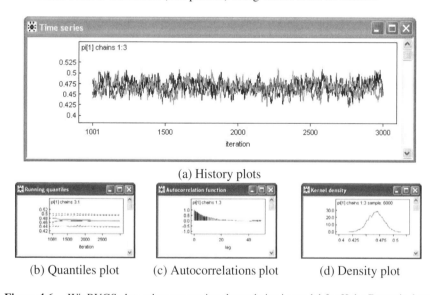

(a) History plots

(b) Quantiles plot   (c) Autocorrelations plot   (d) Density plot

Figure 4.6   WinBUGS plots when generating three chains in model for Kobe Bryant's data.

- In the `stats` option, the posterior summaries using values from all requested chains are calculated. For an update of 2000 iterations using three chains, we obtain the following results.

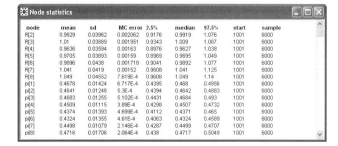

In this output, 6000 iterations in total were used to calculate the posterior summaries (2000 from each of three generated samples).

- In the `coda` option, one output window/file with the generated values of each chain is obtained. The index window/file is common for all chains; see Figure 4.7

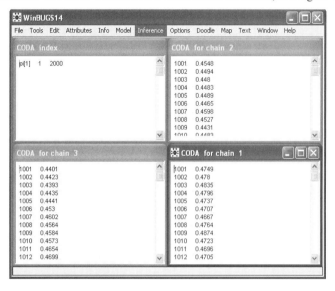

Figure 4.7   CODA output when generating three chains in model for Kobe Braynt's data.

- The Gelman–Rubin diagnostic (Gelman and Rubin, 1992) is now available via the `bgr diag` option (see Section 4.3.3 for details).

4.3.3   The Gelman–Rubin convergence diagnostic

The Gelman–Rubin diagnostic (Gelman and Rubin, 1992) of is available in WinBUGS via the `bgr diag` option, when multiple chains are generated in parallel, each one starting from different initial values. Then an ANOVA-type diagnostic test is implemented by calculating and comparing the between-sample and the within-sample variability (i.e., intersample and intrasample variability). The statistic R can be estimated by

$$\hat{R} = \frac{\hat{V}}{\text{WSS}} = \frac{T'-1}{T'} + \frac{\text{BSS}/T'}{\text{WSS}} \frac{\kappa+1}{\kappa}$$

where κ is the number of generated samples/chains, T' is the number of iterations kept in each sample/chain, BSS/T' is the variance of the posterior mean values over all generated samples/chains (between-sample variance), WSS is the mean of the variances within each sample (within-sample variability), and

$$\hat{V} = \frac{T'-1}{T'}\text{WSS} + \frac{\text{BSS}}{T'}\frac{\kappa+1}{\kappa}$$

is the pooled posterior variance estimate. When convergence is achieved and the size of the generated data is large, then $\hat{R} \to 1$. Brooks and Gelman (1998) adopted a corrected version of this statistic given by

$$\hat{R}_c = \frac{d+3}{d+1}\hat{R}$$

where d is the estimated degrees of freedom for the pooled posterior variance estimate $\hat{V}$; for more details see Brooks and Gelman (1998, sec. 1.3).

Using a line plot, the `bgr diag` option available in the `sample monitor` tool plots the evolution of the pooled posterior variance $\hat{V}$ (in green color), average within-sample variance WSS (in blue), and their ratio $\hat{R}$. The dashed line denotes the reference value of one. Note that in WinBUGS the estimates are based on the ranges of the 80% posterior credible intervals. These quantities are calculated every 50 iterations.

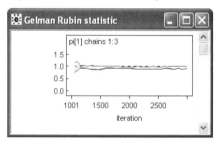

We generally expect $\hat{R}$ to be higher than one at the initial stage of the algorithm, provided that the starting points are suitably scattered over the parameter space. As the number of iterations increase, $\hat{R}$ tends to one and $\hat{V}$ and WSS will stabilize, indicating the convergence of the algorithm.

Numerical estimates can be obtained by double clicking the left mouse button on the plot (opening the figure) and then pressing the left mouse and the keyboard Control buttons to open the window with the associated information.

Values of Gelman Rubin statistic

| | | | -------- 80% interval -------- | | |
| | Unnormalized | | | Normalized as plotted | | |
End iteration of bin	of pooled chains	mean within chain	of pooled chains	mean within chain	BGR ratio
1051	0.03409	0.0298	0.8464	0.7399	1.144
1101	0.04027	0.03421	1.0	0.8494	1.177
1151	0.03786	0.03672	0.9401	0.9117	1.031
1201	0.03832	0.03722	0.9516	0.9243	1.029
1251	0.03771	0.03744	0.9365	0.9296	1.007
1301	0.03647	0.03619	0.9056	0.8987	1.008
1351	0.0364	0.03549	0.9037	0.8812	1.026
1401	0.03587	0.03539	0.8907	0.8788	1.014
1451	0.03532	0.03467	0.8769	0.8609	1.019
1501	0.03609	0.03543	0.8961	0.8797	1.019

Normalized values are used so that all lines can be plotted in the same figure. These normalized values are calculated by dividing the original $\hat{V}$ and WSS by their maximum value (i.e., $\max\{\hat{V}, \text{WSS}\}$).

4.4 CHANGING THE PROPERTIES OF A FIGURE

4.4.1 General graphical options

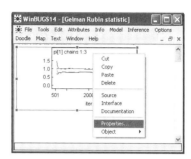

Options of each graph can be changed by following the steps below:

1. Select the graph (left mouse click).

2. Right mouse click to open the drop-down menu.

3. Select properties in the menu.

After this procedure, the `plot Properties` menu opens with five different tabs: `margins`, `axis bounds`, `titles`, `all plots`, and `fonts`.

The `margins`, `axis bounds`, and `titles` tabs are straightforward to use. The `fonts` tab can be used to change the font family and size in titles and in the axes. Finally, by pressing the `all plots` tab, the user can apply the settings of the current plot (all of them or specific ones) to all graphs in the WinBUGS working window.

Two buttons are also available in the `plot properties` menu: `apply` and `special`. The first one executes a requested change on the plot properties while the latter generates special figure properties that vary according to the type of the plot. In simple plots, this option is inactive since no special options are available. Special options of specific plots are illustrated below.

4.4.2 Special graphical options

Special graphical options are available for the density plot of the `inference` tool, and for boxplots, caterpillar, model fit, and scatterplot in the `compare` tool.

For the density plot, the only available special option is the smoothing parameter of the plot (default value = 0.2). Increasing the value of this parameter decreases the smoothing producing a higher number of spikes on the plot, while decreasing this value increases the smoothing; see example in Figure 4.8

The available special properties in boxplots allow for several modifications in the graph. From this menu, we can

- Eliminate the reference line or change its value (the default is the mean over all components).

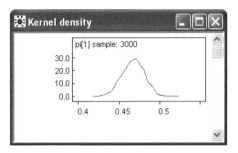
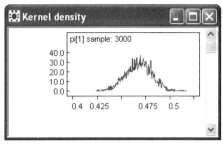

(a) Smoothing parameter = 0.1   (b) Smoothing parameter = 0.4

Figure 4.8   Density plots using different smoothing parameters.

- Eliminate the boxplot labels.

- Depict the posterior mean (default) or the posterior median using the middle line of the boxplot.

- Order the boxplots according to their posterior mean/median from the lowest (left) to the highest (right).

- Plot the boxplots parallel to either the y axis (vertical plot) or the x axis (horizontal plot).

- Use the logarithmic scale.

- Change the color of the boxes.

See Figure 4.9, for a graphical description of the menu.

Special properties in caterpillar plots are similar to the box-plot special properties.

The only available model fit special options are related to the use of logarithmic scale on x and/or y axis.

Finally the scatterplot special properties allow us to

- Plot the posterior means or medians as points on the graph.

- Change the color of the points.

- Draw or eliminate the fitted line.

- Select the color of the fitted line.

- Select the type of fitted line (linear or smoothed) and their corresponding parameter values.

- Use logarithmic scale on x and/or y axis.

- Plot credible intervals (for means of the variable plotted on y axis) as bars around the points.

See also Figure 4.10 for a description of the scatterplot special menu.

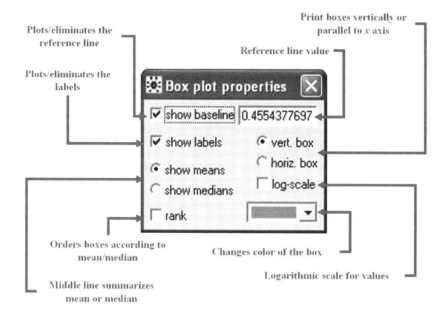

Figure 4.9   Special properties of WinBUGS boxplots.

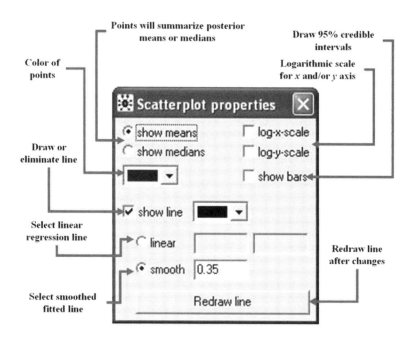

Figure 4.10   Special properties of WinBUGS scatterplots.

4.5 OTHER TOOLS AND MENUS

4.5.1 The node info tool

Information concerning the current state of a node can be extracted using the node info tool (follow the path Info>Node info).

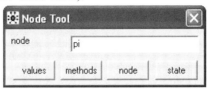

Using this tool, the current values of a node and the update method used for stochastic nodes is provided by the corresponding buttons. For example, for Kobe Bryant's model, the current state for node pi (after 3000 iterations) as follows:

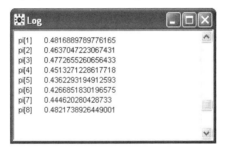

while the slice Gibbs method is used to update π_1 and $R_2, \ldots, R_8$ nodes according to the information we obtain using the method option.

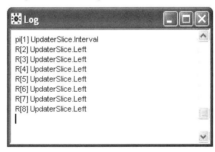

4.5.2 Monitoring the acceptance rate of the Metropolis–Hastings algorithm

The monitor Metropolis tool is activated only when the Metropolis method is used by WinBUGS. It can be opened following the path Model>Monitor Met. It calculates statistics (minimum, maximum, and average) related to the acceptance rate of the random-walk Metropolis during the adaptive phase used by WinBUGS to tune the variance of the symmetric normal proposal. The adaptive phase is set by default equal to 4000 iterations, and all acceptance rates are calculated for batches of 100 iterations. The aim of the adaptive phase is to achieve an acceptance rate between 20% and 40%; for details, see Spiegelhalter et al. (2003d, pp. 6, 25).

4.5.3   Saving the current state of the chain

Frequently, we get an initial sample from the MCMC algorithm and check whether it has converged. If there is an indication of nonconvergence, then we run the algorithm for an additional number of iterations. This is equivalent to starting the algorithm using as initial values the ones from the last iteration of the previous run. On some occasions, it is convenient to save the current state of the chain in case we wish to generate additional iterations in the future. This can be achieved in WinBUGS by the `Model>Save State` path. The output is given in a list format and hence can be used directly as initial values for the next run of the model. For the example of Koby Bryant's data, we get the following output:

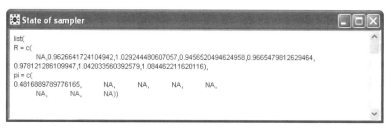

4.5.4   Setting the starting seed number

All simulation algorithms that generate a sequence of pseudorandom numbers are called *random generators*. They require a starting number that is used to generate a sequence of pseudorandom numbers. Every starting seed generates a single sequence of random numbers. This property is useful if we wish to reproduce exactly the same results multiple times. This starting seed can be set in WinBUGS following the path `Model>Seed`.

4.5.5   Running the model as a script

As we have already seen, in order to run a model, we need to repeat a number of actions (check model, define data, compile model, etc.). This procedure is laborious and time-consuming, especially when various different models are fitted. For this reason, Win-BUGS provides the facility to write a script code that generates all the actions required to simulate and monitor posterior values. This is equivalent to the `backbugs` command in the classic version of BUGS for DOS and UNIX. More details on this issue are provided in Appendix B.

4.6   SUMMARY AND CONCLUDING REMARKS

In this chapter, we have illustrated both basic and more specialized functions of Win-BUGS using a simple example based on Kobe Bryant's success rate in field goals. After this chapter, the reader must be able to construct simple models in WinBUGS, produce more specialized analysis (including boxplots of the posterior densities, analysis of ranks, correlations, and calculation of DIC) and generate multiple chains and must be able to change the technical details and options of WinBUGS plots. Finally, brief details of the remaining tools and menus are also provided.

Details about the construction of models using directed acyclic graphs and illustration of running WinBUGS models using scripts (instead of menus) are briefly discussed in this chapter. Further details can be found in Appendices A and B, respectively.

In the chapter that follows, the normal linear regression model is introduced and illustrated in detail.

Problems

4.1   Following the guidelines presented in Section 4.1, perform the same analysis in WinBUGS for the following survival times

$$51\ 3\ 17\ 13\ 5\ 4\ 17\ 1\ 5\ 3\ 8\ 22\ 0\ 1\ 13\ 8\ 15\ 3\ 1\ 13$$

assuming

a) Weibull distribution

b) Gamma distribution

c) Log-normal distribution

4.2   Compare the models of Problem 4.1 using the DIC. Which model is preferable for those data?

4.3   For each model of Problem 4.1, run five chains simultaneously and calculate the Gelman–Rubin convergence diagnostic.

4.4   From WinBUGS examples, volume 1 (Spiegelhalter et al., 2003*a*), run the example `Surgical`.

a) Produce a sample of 2000 iterations after discarding an additional 500 iterations as burnin.

b) Check for the convergence of the chain.

c) Analyze the MCMC output and make inferences concerning the model parameters.

d) Produce an a posteriori analysis of the hospital ranking (using the `rank` tool).

e) Compare the posterior distributions of the mortality rates of all hospitals, using the `compare` tool.

4.5   For the data and implemented models of Problem 3.5

a) Use the `compare` tool to produce scatterplots of the sampled values and estimate the posterior correlations between the parameters of interest.

b) Run the MCMC chain for 5000 iterations plus 1000 burnin iterations. Save the current state of the algorithm and use these values to rerun the sampler using these values as initial values (no burnin is needed in the new sample).

c) Follow the instructions in the WinBUGS manual (Spiegelhalter et al., 2003*d*) and in Appendix B and rerun your MCMC in the script/background mode.

CHAPTER 5

INTRODUCTION TO BAYESIAN MODELS: NORMAL MODELS

5.1 GENERAL MODELING PRINCIPLES

Statistical models are nowadays used to describe parsimoniously real life problems observed under uncertainty. A *statistical model* is a collection of probabilistic statements (and equations) that describe and interpret present behavior or predict future performance. It consists of three important components: the response variable (or variables) Y, the explanatory variables $X_1, X_2, \ldots, X_p$, and a linking mechanism between the two sets of variables.

The response variables Y are the main study variables, and they represent the stochastic part of the model. By the term `stochastic` we refer to random variables whose outcome is uncertain before it is observed. Concerning these variables, we are frequently interested in describing the mechanism underlying or leading to the appearance of a certain outcome of Y and predict a future outcome of Y. Since the response variable is the stochastic component of the model, we can write

$$Y|X_1, X_2, \ldots, X_p \sim \mathcal{D}(\boldsymbol{\theta})$$

where $\mathcal{D}(\boldsymbol{\theta})$ is a distribution with parameter vector $\boldsymbol{\theta}$. For example, for normal regression models, the response (stochastic component of the model) is written as

$$Y|X_1, X_2, \ldots, X_p \sim N(\mu, \sigma^2),$$

where $N(\mu, \sigma^2)$ is the normal distribution with mean μ and variance σ^2. Models with one response variable are called *univariate*, while models with more than one response variables are called *multivariate*. In this book we focus on univariate models.

Bayesian Modeling Using WinBUGS, by Ioannis Ntzoufras
Copyright © 2009 John Wiley & Sons, Inc.

As explanatory variables $X_1, \ldots, X_p$, we consider all variables that potentially influence the response variable Y. Inference concerning the significance, the type (negative or positive), and the magnitude of the effect of each X_i on Y is the main focus in such models. Usually X_i are considered as fixed, nonstochastic components, that is, deterministic nodes in WinBUGS. Hence, it is more precise to define the distribution of Y conditional on the observed explanatory variables

$$ Y|X_1, \ldots, X_p \sim \mathcal{D}\Big(\boldsymbol{\theta}(\boldsymbol{\beta}, \boldsymbol{\phi}, X_1, \ldots, X_p)\Big) . $$

Parameter vector $\boldsymbol{\theta}$ is expressed as a function of the explanatory variables and a new alternative set of parameters $(\boldsymbol{\beta}, \boldsymbol{\phi})$ that substitutes the original ones in terms of estimation and inference. Concerning the new set of parameters, vector $\boldsymbol{\beta}$ summarizes the association between the response and the explanatory variables, while $\boldsymbol{\phi}$ refers to other characteristics of the distribution such as the variance or the shape. Usually, the mean of the response model is associated with the response variables, but in more complicated models, the variance or other moment functions can also be estimated via the explanatory variables. The function used to connect the stochastic and the deterministic part of the model (variables Y and X_i) can be referred to as the "generalized linking" function. The terminology and principles described above were originally introduced to define generalized linear models (McCullagh and Nelder, 1989), but they can be adopted for a wide range of models. For example, the simpler case of the above mentioned general linking function is to express the mean of the response variable given the explanatory variables as a function of the linear combination of the explanatory variables; hence we can write

$$ E(Y|X_1, X_2, \ldots, X_p) = \mu(\boldsymbol{\beta}, X_1, \ldots, X_p) = g^{-1}\left(\beta_0 + \sum_{j=1}^{p} \beta_j X_j\right) . $$

In this equation, the linear combination of of of all explanatory variables is used to predict the expected value of the response variable Y and is often called the *linear predictor* η of the model. This setup is introduced in generalized linear models and, within this context, $g(\mu)$ is referred as the *link function*.

To complete the formulation presented above, a prior distribution must be defined for the parameters under estimation. Thus, in this formulation a prior distribution $f(\boldsymbol{\beta}, \boldsymbol{\phi})$ for the parameters $(\boldsymbol{\beta}, \boldsymbol{\phi})$ remains to be specified.

Finally, explanatory variables are usually defined as deterministic (fixed) quantities. In practice these variables are frequently random. Within the general framework of statistics it is generally easy to extend the model by either simply considering X_i terms as random variables with additional parameters under estimation; see Ryan (1997, pp. 34–35) for a discussion. This can be specified in a straightforward manner within the Bayesian framework using additional hierarchical levels in our model (see Chapter 9 for details).

5.2 MODEL SPECIFICATION IN NORMAL REGRESSION MODELS

Normal regression models are the most popular models in statistical science. They are based on the initial work of Sir Francis Galton in the late years of the 19th century (Stanton, 2001). In normal regression models, the response variable Y is considered to be a continuous random variable defined in the whole set of real numbers following the normal distribution with mean μ and variance σ^2. Therefore, the model can be summarized by the following equations

$$Y|X_1, \ldots, X_p \sim N\left(\mu(\boldsymbol{\beta}, X_1, \ldots, X_p), \sigma^2\right) \tag{5.1}$$

with

$$
\begin{aligned}
\mu(\boldsymbol{\beta}, X_1, \ldots, X_p) &= \beta_0 + \beta_1 X_1 + \cdots + \beta_p X_p \\
&= \beta_0 + \sum_{j=1}^{p} \beta_j X_j,
\end{aligned} \tag{5.2}
$$

where σ^2 and $\boldsymbol{\beta} = (\beta_0, \beta_1, \ldots, \beta_p)^T$ is the set of regression parameters under estimation. Frequently, the following alternative representation of the regression model is adopted

$$Y = \beta_0 + \beta_1 X_1 + \ldots + \beta_p X_p + \varepsilon; \quad \varepsilon \sim N(0, \sigma^2).$$

Although this formulation has a nice interpretation since the response variable is directly expressed as a function of the explanatory variables plus a random normal error with variance σ^2, the initial model expression given by (5.1) and (5.2) is more general and follows the model building principles described in Section 5.1.

In order to simplify notation and make it compatible with the corresponding Win-BUGS notation, we hereafter remove the condition on the explanatory variables. Hence we denote $Y|X_1, \ldots, X_p$ simply by Y and the corresponding expected value $E(Y|X_1, \ldots, X_2)$ by $E(Y)$ or μ.

5.2.1 Specifying the likelihood

Let us observe a sample of size n with response values $\boldsymbol{y} = (y_1, \ldots, y_n)^T$ and $x_{i1}, \ldots, x_{ip}$, the values of the explanatory variables $X_1, \ldots, X_p$ for individuals $i = 1, \ldots, n$. Then the model is expressed as

$$
\begin{aligned}
Y_i &\sim N(\mu_i, \sigma^2) \\
\mu_i &= \beta_0 + \beta_1 x_{i1} + \ldots + \beta_p x_{ip} \quad \text{for } i = 1, \ldots, n \,.
\end{aligned} \tag{5.3}
$$

Within WinBUGS the normal distribution is defined in terms of its precision $\tau = \sigma^{-2}$. Hence the above likelihood in WinBUGS will be written as

```
for (i in 1:n){
    y[i] ~ dnorm( mu[i], tau )
    mu[i] <- beta0 + beta1 * x1[i] + ... + betap*xp[i]
}
s2<-1/tau
s <-sqrt(s2)
```

The last two commands are used to deterministically specify the connection between the variance, the standard deviation, and the precision parameter $\tau = \sigma^{-2}$ used by the normal distribution in WinBUGS. Moreover, in this code all parameters β_j are defined separately as single scalar nodes, while explanatory variables are specified as vector nodes with names x1, ..., xp of length n. When monitoring these parameters, each one of them must be set separately in the sample monitor tool of WinBUGS.

5.2.2 Specifying a simple independent prior distribution

In normal regression models, the simplest approach is to assume that all parameters are a priori independent having the structure

$$f(\boldsymbol{\beta}, \tau) \quad = \quad \prod_{j=0}^{p} f(\beta_j) f(\tau),$$

$$\beta_j \quad \sim \quad N(\mu_{\beta_j}, c_j^2) \ \text{ for } \ j = 0, \dots, p \ \text{ and} \qquad (5.4)$$

$$\tau \quad \sim \quad \text{gamma}(a, b) \, .$$

The gamma prior of the precision parameter induces prior mean and variance given by

$$E(\tau) = \frac{a}{b} \ \text{ and } \ \text{Var}(\tau) = \frac{a}{b^2},$$

respectively. In this prior setup, we have substituted the variance σ^2 by the corresponding precision parameter τ in order to make it compatible to the WinBUGS notation. The gamma prior used for τ corresponds to an inverse gamma prior distribution for the original variance parameter with prior mean and variance given by

$$E(\sigma^2) = \frac{b}{a-1} \ \text{ and } \ \text{Var}(\sigma^2) = \frac{b^2}{(a-1)^2(a-2)},$$

respectively.

When no information is available, a usual choice for the prior mean is the zero value ($\mu_{\beta_j} = 0$). This prior choice centers our prior beliefs around zero, which corresponds to the assumption of no effect of X_j on Y. In this way, we express our prior doubts about the effect of X_j on Y, prompting Spiegelhalter et al. (2004, pp. 90, 158–160) to call this a "sceptical" prior. The prior variance c_j^2 of the effect β_j is set equal to a large value (e.g., 10^4) to represent high uncertainty or prior ignorance. Similarly, for τ we use equal low prior parameter values, setting in this way its prior mean equal to one and its prior variance large. For example, we may use $a = b = 0.01$ which results in $E(\tau) = 1$ and $V(\tau) = 100$. This approach is also adopted in all illustrations of the WinBUGS manual and example volumes. More details concerning the specification of more complicated prior distributions are provided in Sections 5.3.2 and 5.3.3.

Within WinBUGS, the prior setup described above can be incorporated by simply adding

```
beta0  ~ dnorm( 0.0, 1.0E-4 )
beta1  ~ dnorm( 0.0, 1.0E-4 )
.........................
betap  ~ dnorm( 0.0, 1.0E-4 )
tau    ~ dgamma( 0.01, 0.01 )
```

In this syntax, value 1.0E-4 is the scientific notation for $1.0 \times 10^{-4} = 0.001$, which is the prior precision of each β_j and corresponds to prior variance equal to 10^4. The definition above can be considerably simplified by using vectors instead of single nodes; see Section 5.3.3 for details.

5.2.3 Interpretation of the regression coefficients

Each regression coefficient pertains to the effect of explanatory variable X_j on the expectation of the response variable Y adjusted for the remaining covariates. The inference concerning the model parameter can be divided into three basic stages:

1. Is the effect of X_j important for the prediction or description of Y?

2. What is the association between Y and X_j (positive, negative, or other)?

3. What is the magnitude of the effect of X_j on Y ?

Concerning query 1, we initially focus on examining whether the posterior distribution of β_j is scattered around zero (or not). Posterior distributions far away from the zero value will indicate an important contribution of X_j on the prediction of the response variable. Although formal Bayesian hypothesis testing is not based on simply examining the posterior distribution and their credible intervals, such analysis can offer a first and reliable tool for tracing important variables.

In stage (query) 2, we identify whether the relationship is positive or negative. This can be based on the signs of the posterior summaries of central and relative location (e.g., mean, median, 2.5% and 97.5% percentiles). If all of them are positive or negative, then the corresponding association can be concluded. *Positive association* means that changes of the explanatory variable X_j cause changes of the same direction for variable Y while *negative association* means that changes of the explanatory variable X_j cause changes of the opposite direction for variable Y. Within this analysis, we can a posteriori calculate the posterior probability:

$$\pi_0 = \min \left\{ f(\beta_j < 0|\boldsymbol{y}), f(\beta_j > 0|\boldsymbol{y}) \right\}.$$

When the zero value lies at the center of the posterior distribution, then the value shown above will be close to $\frac{1}{2}$ indicating that there is no clear positive or negative effect of X_j on Y. When π_0 is low (e.g., lower than 2.5%, 1%, or 0.5%), then we may conclude positive or negative association depending on the sign of the posterior location summaries. Within WinBUGS we can calculate the posterior probability $f(\beta_j > 0|\boldsymbol{y})$ using the syntax

```
p.betaj <- step( betaj )
```

which creates a binary node p.betaj taking values equal to one when β_j is positive and zero otherwise. Obtaining the posterior mean via the sample monitor tool provides us the estimate of the posterior probability $f(\beta_j > 0|\boldsymbol{y})$.

In WinBUGS , it is also convenient to calculate the deviance information criterion (DIC) (Spiegelhalter et al., 2002) to compare models with different covariates and, in this way, evaluate their importance concerning their effect on Y. A brief description of DIC as well as illustration of its calculation in WinBUGS is provided in Section 4.2.5. In order to use DIC in regression models, we need to fit models including and excluding the variable of interest and then select the one with the lower value of DIC. In the case of a large number of covariates, this procedure can be quite tedious since a large number of models must be fitted before drawing conclusions regarding the model with the lowest value of DIC. More formal approaches concerning model checking, comparison, and selection are described in Chapters 10 and 11.

Finally, the magnitude of the effect of variable X_j on Y is given by the posterior distribution of β_j (for $j = 1, \ldots, p$) since

$$\begin{aligned} \Delta \mu_{X_j} &= \mu(\boldsymbol{\beta}, X_1, \ldots, X_{j-1}, X_j = x + 1, X_{j+1}, \ldots, X_p) \\ &\quad -\mu(\boldsymbol{\beta}, X_1, \ldots, X_{j-1}, X_j = x, X_{j+1}, \ldots, X_p) \\ &= \beta_0 + \beta_j(x+1) + \sum_{k \neq j; k=1}^{p} \beta_k X_k - \beta_0 - \beta_j x - \sum_{k \neq j; k=1}^{p} \beta_k X_k \\ &= \beta_j. \end{aligned}$$

Hence the posterior mean or median of β_j will correspond to the corresponding posterior measures of the expected change of the response variable Y. Hence, an increase of one unit of X_j, given that the remaining covariates will remain stable, induces an a posteriori average change on the expectation of Y equal to the posterior mean of β_j; see Table 5.1 for further details.

Table 5.1   Summary interpretation table for regression coefficients β_j

---

INTERPRETATION OF MODEL COEFFICIENTS β_j ($j = 1, \ldots, p$)

$$\Delta\mu_{X_j} = \beta_j$$

- $\Delta\mu_{X_j}$ denotes the expected difference of Y if X_j increases by one unit and the rest of the covariates remain the same.
- If $\beta_j = 0 \Rightarrow$ no effect on Y.
- If $\beta_j < 0 \Rightarrow$ negative effect on Y (Y is expected to decrease when X_j increases and vice versa).
- If $\beta_j > 0 \Rightarrow$ positive effect on Y (Y is expected to increase when of X_j increases and decrease when X_j decreases).
- β_j is the expected change (increase or decrease) when X_j increases by one unit and the rest of the covariates remain unchanged.

Concerning the constant parameter β_0, its interpretation corresponds to the expected value of the response variable Y when the observed values of all covariates are equal to zero. Frequently such combination lies outside the range of the observed covariate values. In such cases, the interpretation of β_0 is not reliable since we infer or predict the behavior of Y for values of X_j that have not been observed. In such cases, direct interpretation of β_0 may not lead to realistic and sensible interpretation. An alternative is to center around zero all explanatory variables X_j by subtracting their sample mean. In this case, the constant β_0^c represents the expected value of Y when all covariates are equal to its sample means, representing in this way the expected response Y for an "average" or "typical" subject according to our sample. In WinBUGS this quantity can be directly estimated using the command

```
typical.y<- beta0 + beta1 * mean(x1[]) + ... + betap * mean(xp[])
```

without changing the parametrization of the original model. The approach described above can also be used to calculate the expected values of Y for any combination of values of X_j.

Parameter precision τ (and the variance σ^2) indicates the precision of the model. If the precision τ is high (σ^2 low), then the model can accurately predict (or describe) the expected values of Y. Therefore, we can rescale this quantity using the sample variance of the response variable Y, namely, s_Y^2, using the R_B^2 statistic given by

$$R_B^2 = 1 - \frac{\tau^{-1}}{s_Y^2} = 1 - \frac{\sigma^2}{s_Y^2},$$

where s_Y^2 is the sample variance of Y. This quantity can be interpreted as the proportional reduction of uncertainty concerning the response variable Y achieved by incorporating the explanatory variables X_j in the model. Moreover, it can be regarded as the Bayesian analog of the adjusted coefficient of determination R^2_{adj} (used in the frequentistic approach of the normal regression model), given by

$$R^2_{\text{adj}} = 1 - \frac{\hat{\sigma}^2}{s^2_Y},$$

where

$$\hat{\sigma}^2 \;=\; \frac{1}{n-p} \sum_i^n (y_i - \hat{y}_i)^2 \;\text{ with }\; \hat{y}_i = \hat{\beta}_0 + \sum_{i=1}^p X_{ij}\hat{\beta}_j,$$

where $\hat{\beta}_j$ are the maximum likelihood estimates of β_j.

In order to calculate R^2_B in WinBUGS, we can use the commands

```
sy2 <- pow( sd(y[]) , 2)
R2B  <- 1 - s2/sy2
```

or directly using the precision parameter τ and the syntax

```
R2B  <- 1 - 1/(tau*sy2)
```

5.2.4   A regression example using WinBUGS

Example 5.1 . Soft drink delivery times. The following example deals with the quality of the delivery system network of a soft drink company; see example 4.1 in Montgomery and Peck (1992). In this problem, we are interested in estimation of the required time needed by each employee to refill an automatic vending machine owned and served by the company. For this reason, a small quality assurance study was set up by an industrial engineer of the company. As the response variable, the engineer considered the total service time (measured in minutes) of each machine, including its stocking with beverage products and any required maintenance or housekeeping. After examining the problem, the industrial engineer recommended two important variables that affect delivery time: the number of cases of stocked products and the distance walked by the employee (measured in feet). A dataset of 25 observations was finally collected. This dataset is reproduced in the book's Website with permission of John Wiley and Sons, Inc.

Setting up the data and the model code.   Following the approach described in Sections 5.2.1 and 5.2.2, we define the data in either a rectangular or a list format. The rectangular format of the data are provided in Table 5.2, while the full model code of the example, including the list data format and the initial values, is given in Table 5.3. All three variables used in the model (time, cases, distance) are defined as separate vectors in the list data format, while in the initial values, each parameter τ, β_0, β_1, and β_2 was initialized separately.

Table 5.2 Rectangular WinBUGS format of "soft drink delivery times" example (Example 5.1)

```
time[]   cases[] distance[]
 16.68   7 560
 11.5    3 220
 12.03   3 340
 14.88   4 80
 13.75   6 150
 18.11   7 330
  8      2 110
 17.83   7 210
 79.24  30 1460
 21.5    5 605
 40.33  16 688
 21     10 215
 13.5    4 255
 19.75   6 462
 24      9 448
 29     10 776
 15.35   6 200
 19      7 132
  9.5    3 36
 35.1   17 770
 17.9   10 140
 52.32  26 810
 18.75   9 450
 19.83   8 635
 10.75   4 150
END
```

Table 5.3   Full model code for "soft drink delivery times" example (Example 5.1)

```
model{
    # model's likelihood
    for (i in 1:n){
        time[i] ~ dnorm( mu[i], tau ) # stochastic componenent
        # link and linear predictor
        mu[i] <- beta0 + beta1 * cases[i] + beta2 * distance[i]
    }
    # prior distributions
    tau ~ dgamma( 0.01, 0.01 )
    beta0 ~ dnorm( 0.0, 1.0E-4)
    beta1 ~ dnorm( 0.0, 1.0E-4)
    beta2 ~ dnorm( 0.0, 1.0E-4)
    # definition of sigma
    s2 <-1/tau
    s <-sqrt(s2)
    # calculation of the sample variance
    for (i in 1:n){ c.time[i]<-time[i]-mean(time[]) }
    sy2 <- inprod( c.time[], c.time[] )/(n-1)
    # calculation of Bayesian version R squared
    R2B <- 1 - s2/sy2
    # Expected y for a typical delivery time
    typical.y <- beta0 + beta1 * mean(cases[]) + beta2 * mean(distance
        [])
}

INITS
list( tau=1, beta0=1, beta1=0, beta2=0 )

DATA (LIST)
list( n=25,
    time = c(16.68, 11.5, 12.03, 14.88, 13.75, 18.11,  8, 17.83,
            79.24, 21.5, 40.33, 21, 13.5, 19.75, 24, 29, 15.35,
            19, 9.5, 35.1, 17.9, 52.32, 18.75, 19.83, 10.75),
    distance = c(560, 220, 340, 80, 150, 330, 110, 210, 1460,
                605, 688, 215, 255, 462, 448, 776, 200, 132,
                36, 770, 140, 810, 450, 635, 150),
    cases = c( 7, 3, 3, 4, 6, 7, 2, 7, 30, 5, 16, 10, 4, 6, 9,
            10, 6, 7, 3, 17, 10, 26, 9, 8, 4) )
```

Results.   Posterior summaries and densities, after running the MCMC algorithm for 3000 iterations and discarding the initial 1000 ones, are provided in Table 5.4 and Figure 5.1, respectively. Descriptive analysis of the posterior distribution of R_B^2 indicates a considerable improvement of the precision (posterior mean equal to 0.95) in the prediction of delivery times when including in the model covariates cases and distance.

Table 5.4   WinBUGS posterior summaries for Example 5.1 after 2000 iterations and additional discarded 1000 burnin iterations

node	mean	sd	MC error	2.5%	median	97.5%	start	sample
R2B	0.9511	0.01743	5.118E-4	0.9064	0.9548	0.9734	1001	2000
beta0	2.356	1.188	0.03076	-0.03996	2.372	4.635	1001	2000
beta1	1.61	0.1806	0.003737	1.272	1.609	1.968	1001	2000
beta2	0.01447	0.003812	8.476E-5	0.006872	0.01446	0.02211	1001	2000
p.beta0	0.974	0.1591	0.004037	0.0	1.0	1.0	1001	2000
p.beta1	1.0	0.0	2.236E-12	1.0	1.0	1.0	1001	2000
p.beta2	1.0	0.0	2.236E-12	1.0	1.0	1.0	1001	2000
s	3.386	0.5695	0.0168	2.531	3.302	4.749	1001	2000
typical.y	22.38	0.683	0.01701	21.09	22.37	23.78	1001	2000

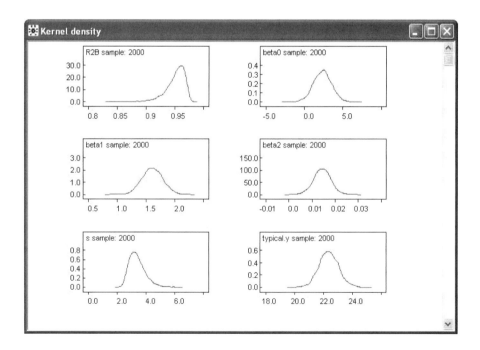

Figure 5.1   Posterior densities of model parameters for Example 5.1 (soft drink delivery times).

Concerning the posterior distribution of σ, we observe that with the current model we can predict the expected delivery time with an a-posteriori expected error of 3.4 minutes.

Considering as point estimates the posterior means, we end up with model

$$\text{Expected time} = 2.36 + 1.6 \times \text{ cases } + 0.015 \times \text{ distance}.$$

Minor changes are observed in the regression equation if posterior medians are used as point estimates instead.

Observing all parameters, we can infer that the effect of both explanatory variables (cases and distance) have an important contribution to the prediction of delivery time. All summary statistics and the posterior densities indicate that zero is far away from the posterior distribution with posterior probability of having positive association between each X_j and Y equal to one.

Furthermore, for each additional case stocked by the employee, the expected delivery time is a posteriori expected to increase by 1.6 minutes (96 seconds). The increase in expected delivery time for each additional case, lies between 1.3 and 2.0 minutes (76 and 118 seconds) with probability 95%. For every increase of the walking distance by one foot, the delivery time is a posteriori expected to increase by 0.87 seconds, while every 100 feet of additional walking distance increases by 1.5 minute the posterior expected delivery time (ranging between 0.7 and 2.2 minutes with probability 95%). In terms of meters, for every 100 m of walking distance the expected delivery time will increase by 4.7 minutes on posterior average (one foot is equal to 0.3048 m, resulting in an increase of the expected delivery time by $100/0.3048 * 0.01447 = 4.747$ minutes).

Parameter β_0 has no sensible interpretation in this example since the zero value is non-sense for both explanatory variables (the delivery employee will always have to stock some cases of products in the machine and walk at least a small distance to reach the delivery location). For this reason, no interpretation of this parameter is attempted. We only observe that the zero value lies at the left tail of the posterior distribution within the range of the 95% posterior interval. Moreover, the posterior probability of positive β_0 is equal to 97.4%.

Since the interpretation for β_0 is meaningless, we can focus on the predicted value for the a typical or representative delivery route. According to the posterior summaries of node `typical.y`, a typical delivery route will take 22.4 minutes on average and will range from 21.1 to 23.8 minutes with probability 95%.

5.3 USING VECTORS AND MULTIVARIATE PRIORS IN NORMAL REGRESSION MODELS

5.3.1 Defining the model using matrices

The regression model described above can be defined in WinBUGS using vectors and matrices instead of scalar nodes. To achieve that, within the likelihood loop, we substitute the mean specification code line with the following syntax:

```
mu[i] <- beta0 + beta[1] * x[i,1] + ... + beta[p]*x[i,p]
```

Matrix x (denoted by `x[,]` in WinBUGS) is a $n \times p$ matrix. Each column x_j of x corresponds to each explanatory variable X_j, while each row $x_{(i)}$ corresponds to the explanatory variable values of the ith subject of the sample. Concerning the data definition, each column of the matrix can be defined using the rectangular data format with header `x[,1] ... x[,p]`. Otherwise, in list format, x must be defined as an array with dimensions n and p.

Moreover, when the number of variables p is large, it is convenient to use the command `inprod( b[], x[i,])` to express the linear combination of the explanatory variables $\sum_{j=1}^{p} \beta_j x_{ij}$. Hence, the mean can be defined by the syntax

```
mu[i] <- beta0 + inprod( b[], x[i,])
```

where b[] is vector $\boldsymbol{\beta}_{\{\backslash 0\}} = (\boldsymbol{\beta}_1, \dots \boldsymbol{\beta}_p)$.

Usually we incorporate the constant term in a matrix $\boldsymbol{X} = [\mathbf{1}_n, \boldsymbol{x}]$ of dimension $n \times (p+1)$, which is called the *data matrix*; where $\mathbf{1}_n$ is a vector of length n with all elements equal to one corresponding to the "constant" term. The linear predictor in (5.3) can be now written as $\mu_i = \boldsymbol{X}_{(i)}\boldsymbol{\beta}$, where $\boldsymbol{X}_{(i)}$ is the ith row of $\boldsymbol{X}$. This expression can be coded in WinBUGS by

```
x[i,1]<-1
mu[i] <- inprod( X[i,], beta[] )
```

In this syntax, X[,] is a matrix of dimension $n \times (p+1)$ with all components of the first column x[,1] equal to one. Using this definition, beta[1] represents the constant coefficient β_0, X[,j] refers to the vector of values of X_{j-1}, and beta[j] corresponds to coefficient β_{j-1} for $j = 1, \dots, p+1$. In the rectangular format, data are defined in the usual manner with header names referring to columns from 2 to $p+1$, that is, X[,2] ... X[,p+1]. This approach enables us to monitor all regression coefficients simultaneously by simply considering the vector node beta.

5.3.2 Prior distributions for normal regression models

Conjugate analysis for the normal regression model has been presented in Section 1.5.5. The conjugate prior for the normal regression model is considered if we specify $[\boldsymbol{\beta}, \sigma^2]$ to a priori follow a normal–inverse gamma distribution. Hence we can write

$$\boldsymbol{\beta}|\sigma^2 \sim N_P(\boldsymbol{\mu}_\beta, c^2 \boldsymbol{V}\sigma^2) \text{ and } \sigma^2 \sim IG(a,b) , \qquad (5.5)$$

where $P=p+1$ and c^2 is a parameter controlling the overall magnitude of the prior variance.

A special case of the family of prior distributions described above is the popular is popular Zellner (1986) g-prior, in which

$$\boldsymbol{V} = (\boldsymbol{X}^T \boldsymbol{X})^{-1}.$$

Parameter c^2 was denoted by g in Zellner's original publication. The default choice of $c^2 = n$ is usually adopted when no information is available since it has an interpretation of adding prior information equivalent to one data point (Kass and Wasserman, 1995; Fouskakis et al., 2008). This prior has been widely used because it considerably simplifies posterior computations and reduces the number of prior variance parameters that remain to be specified down to one.

In the case where no prior information is available, we may simplify the prior by considering independent normal distributions by setting

$$\boldsymbol{V} = c^2 \boldsymbol{I}_P$$

with c^2 set large to express prior ignorance (e.g., $c = 100$). Hence we can simply rewrite the prior as

$$\beta_j|\sigma^2 \sim N(\mu_{\beta_j}, c^2\sigma^2) \text{ for } j = 0, 1, \dots, p, \qquad (5.6)$$

where μ_{β_j} are the components of the prior mean vector $\boldsymbol{\mu}_\beta$.

Generally, the conjugate prior setup described above is very convenient for implementing Bayesian variable selection (Raftery et al., 1997). Moreover, Zellner's g-priors were widely used within this context since they allow us for a sensible default choice of prior distributions;

see Fernandez et al. (2000) for comparison between different values of c^2 and Liang et al. (2008) for discussion and extensions concerning the g-priors.

Another alternative is to consider the simpler prior setup of Section 5.2.2, where all parameters are a priori independent. This choice is usually selected when no information is available. It is not conjugate, and hence MCMC methods need to be implemented in order to estimate the posterior distribution. Nevertheless, this prior setup is conditionally conjugate, resulting in conditional posterior distributions for β and τ that can be calculated analytically, allowing us to construct an efficient Gibbs sampler.

Finally, other type of prior distributions for β have been proposed in the related literature. For example, the Student t distribution or the Cauchy distribution can be used instead of the normal prior, but obvious differences are seldom observed when no prior information is available.

5.3.3 Multivariate normal priors in WinBUGS

In the case that vector nodes are used in the model specification, the independence prior (5.4) can be defined in WinBUGS by specifying the priors for β_j terms within a loop. Hence the $P = p + 1$ lines that define the prior distribution for each component of β can be substituted by

```
for (j in 1:P){ beta[j] ~ dnorm( 0.0, 1.0E-4 ) }
```

If we wish to use the prior distribution (5.6), which is a conjugate prior, assuming that the elements of β vector are a priori independent, then we use the following the syntax:

```
precision <- tau/c2
for (j in 1:P){ beta[j] ~ dnorm( beta0[j], precision ) }
```

In order to define the conjugate prior (5.5), we first need to specify in WinBUGS the precision matrix used in the multivariate normal prior distribution. The elements of the prior precision matrix T can be expressed as

$$T_{lj} = \frac{\tau}{c^2}[V^{-1}]_{lj} \qquad (5.7)$$

for $l, j \in \{1, 2, \ldots, p + 1\}$, where V^{-1} is the inverse of matrix V and $[V^{-1}]_{lj}$ is the lth row and jth column element of V^{-1}. Hence, in WinBUGS , we need to first calculate the V^{-1} and then the elements of the prior precision matrix using (5.7) within a double `for` loop. The multivariate prior can be specified by the following syntax:

```
# calculation of the inverse matrix of V
inverse.V[1:P,1:P] <- inverse(V[,])
# calculation of the elements of prior precision matrix
for(l in 1:P){ for (j in 1:P){
    prior.T[l,j] <- inverse.V[l,j] * tau /c2
}}
# multivariate prior for the beta vector
beta[1:P] ~ dmnorm( mu.beta[], prior.T[,] )
# gamma prior for the precision
tau     ~ dgamma( 0.01, 0.01 )
# deterministic calculation of variance
s2      <- 1/tau
```

In this syntax P stands for the length of beta. The prior values V, μ_β, and c^2 can be specified either within the list of the data or directly within the WinBUGS model code. For example, the syntax

```
c2 <- 100
for (j in 1:P){ mu.beta[j] <- 0.0 }
for (l in 1:P){ for (j in 1:P){
   V[l,j] <- equals(l,j)
}}
```

defines $\boldsymbol{\mu}_\beta = \mathbf{0}_P$, $c^2 = 100$, and $\boldsymbol{V} = \boldsymbol{I}_P$, resulting in a $\beta_j \sim N(0, 100\sigma^2)$ prior. Node V[l,j] will take the value one if $l = j$ (hence we have a diagonal element) and zero otherwise. Generally, it is easier to specify the components of $\boldsymbol{T}$ directly in the data than compute them inside the WinBUGS code.

Finally, Zellner's g-prior can be specified using the preceding syntax with matrix $\boldsymbol{V}^{-1}$ calculated using the syntax

```
for (l in 1:P){ for (j in 1:P){
   inverse.V[l,j] <- inprod( X[,l] , X[,j] )
}}
```

This syntax calculates matrix $\boldsymbol{A} = \boldsymbol{V}^{-1}$ with elements $A_{lj} = [\boldsymbol{V}^{-1}]_{lj} = \sum_{i=1}^{n} x_{il}x_{ij}$, which are the elements of matrix $(\boldsymbol{X}^T\boldsymbol{X})$ used in Zellner's g-prior.

5.3.4 Continuation of Example 5.1

Here we rerun the same model using Zellner's g-prior. The basic code of the model is given in Table 5.5, in which matrix $\boldsymbol{V} = (\boldsymbol{X}^T\boldsymbol{X})$ is defined within the model code. The first column X[,1] of matrix X is defined within the model code, and the remaining columns are defined within the data part. In Table 5.5, the first and last rows of the data are also given (in a rectangular data format) while the remainder have been denoted by dotted lines to save space. Additionally, the sample size n and the number of parameters P involved in the linear predictor are specified separately in a list format (hence these two types of data must be loaded separately). Alternatively, we can use the list format directly (with the help of R or Splus as described in Section 3.4.6.3). The list format that can be used alternatively in the code of Table 5.5 can be defined by the following code:

```
list( n=25, P=3,
      time = c(16.68,  11.5,  12.03,  14.88,  13.75,  18.11,    8,  17.83,
               79.24,  21.5,  40.33,  21,  13.5,  19.75,  24,  29,  15.35,
               19,  9.5,  35.1,  17.9,  52.32,  18.75,  19.83,  10.75),
      X=structure(.Data=c(NA,  7,560,NA,  3,220,NA,  3,340,NA ,4,  80,
                  NA,  6,150,NA,  7,330,NA,  2,110,NA ,7,210,NA ,30,1460,
                  NA,  5,605,NA ,16,688,NA ,10,215,NA ,4,255,NA,  6,  462,
                  NA,  9,448,NA ,10,776,NA,  6,200,NA ,7,132,NA,  3,   36,
                  NA ,17,770,NA ,10,140,NA ,26,810,NA ,9,450,NA,  8,  635,
                  NA,  4,  150), .Dim = c(25,3)) )
```

As you may observe, some values (corresponding to the first column) are equal to NA (i.e., are defined as missing values). This is because the values of the first column are defined within the model's code. Generally, if the list format is used, it is more convenient to specify the whole matrix X within the data format and remove command X[i,1] <- 1.0 (line 19 of Table 5.5) from the code.

In order to define the prior parameters $\boldsymbol{V}$, $\boldsymbol{\mu}_\beta$, and c^2 in the data structure, the model code must be slightly changed by omitting the first lines appearing in the code of Table 5.5 related to the definition of the corresponding parameters. Code for the general normal–inverse gamma prior setup in which all prior parameters are specified in a data list format is provided in Table 5.6.

Table 5.5   WinBUGS code for Example 5.1 using Zellner's g-prior and parameter vectors[a]

```
model{
    #-------------------------------------------------------------------
    # definition of prior parameters
    c2 <- 10000
    # prior means
    for (j in 1:P){ mu.beta[j] <- 0.0 }
    # calculation of xtx
    for (i in 1:P){ for (j in 1:P){
        inverse.V[l,j] <- inprod( X[,l] , X[,j] )
    }}
    # calculation of the elements of prior precision matrix
    for(l in 1:P){ for (j in 1:P){
        prior.T[l,j] <- inverse.V[l,j] * tau /c2
    }}
    #-------------------------------------------------------------------
    # model's likelihood
    # ------------------
    for (i in 1:n){
        X[i,1] <- 1.0
        # specifying the constant term in the first column
        time[i] ~ dnorm( mu[i], tau ) # stochastic componenent
        # link and linear predictor
        mu[i] <- inprod( beta[], X[i,] )
    }
    # prior distributions
    # ------------------
    # calculation of the inverse matrix of V
    # prior parameters
    # multivariate prior for the beta vector
    beta[1:P] ~ dmnorm( mu.beta[], prior.T[,] )
    # gamma prior for the precision
    tau      ~ dgamma( 0.01, 0.01 )
    # deterministic calculation of variance
    s2      <- 1/tau
    s <-sqrt(s2)
    #
}

INITS
list( tau=1, beta=c(1, 0, 0) )

DATA (RECT.)
list(n=25, P=3)
time[]   X[,2] X[,3]
 16.68   7 560
 ...     ... ...
 10.75   4 150
END
```

[a] Data are compressed to conserve space.

Table 5.6 WinBUGS code for Example 5.1 using multivariate conjugate normal–gamma prior[a]

```
model{
    #-----------------------------------------------------------------
    # definition of prior parameters
    # calculation of the inverse matrix of V
    inverse.V[1:P,1:P] <- inverse(V[,])
    # calculation of the elements of prior precision matrix
    for(l in 1:P){ for (j in 1:P){
      prior.T[l,j] <- inverse.V[l,j] * tau /c2
    }}
    #-----------------------------------------------------------------
    # model's likelihood
    # ------------------
    for (i in 1:n){
        # specifying the constant term in the first column
        time[i] ~ dnorm( mu[i], tau ) # stochastic componenent
        # link and linear predictor
        mu[i] <- inprod( beta[], X[i,] )
    }
    # prior distributions
    # ------------------
    # calculation of the inverse matrix of V
    # prior parameters
    # multivariate prior for the beta vector
    beta[1:P] ~ dmnorm( mu.beta[], prior.T[,] )
    # gamma prior for the precision
    tau    ~ dgamma( 0.01, 0.01 )
    # deterministic calculation of variance
    s2    <- 1/tau
    s <-sqrt(s2)
    #
}

INITS
list( tau=1, beta=c(1, 0, 0) )

list(n=25, P=3, c2=25,
    mu.beta=c(0,0,0),
    V=structure(.Data=c( 0.113215186112351,    -0.0044485932353407,
                        -8.3672569807385E-05, -0.0044485932353407,
                        0.00274378329085448, -4.78570865728707E-05,
                        -8.36725698073853E-05,-4.78570865728707E-05,
                        1.22874474243973E-06), .Dim = c(3,3)),
        time = c(16.68, 11.5, 12.03, 14.88, 13.75, 18.11,  8, 17.83,
                79.24, 21.5, 40.33, 21, 13.5, 19.75, 24, 29, 15.35,
                19, 9.5, 35.1, 17.9, 52.32, 18.75, 19.83, 10.75),
        X=structure(.Data=c(1, 7, 560, 1, 3, 220, 1, 3, 340, 1, 4, 80,
        1, 6, 150, 1,  7, 330, 1,  2, 110, 1, 7, 210, 1, 30, 1460,
        1, 5, 605, 1, 16, 688, 1, 10, 215, 1, 4, 255, 1,  6,  462,
        1, 9, 448, 1, 10, 776, 1,  6, 200, 1, 7, 132, 1,  3,   36,
        1,17, 770, 1, 10, 140, 1, 26, 810, 1, 9, 450, 1,  8,  635,
        1, 4, 150), .Dim = c(25,3))  )
```

[a] All prior parameters and data for matrix V are defined within the list data format.

The model was run for $c^2 = n = 25$ and for $c^2 = 10^4$, and results are presented in Tables 5.7 and 5.8, respectively. Results (especially those concerning the variance parameter σ^2) are sensitive to the choice of c^2, indicating that the unit information choice $c^2 = n = 25$ here is informative.

Table 5.7   WinBUGS posterior summaries for Example 5.1 after 2000 iterations and additional discarded 1000 burnin iterations using Zellner's g-prior ($c^2 = n = 25$)

node	mean	sd	MC error	2.5%	median	97.5%	start	sample
beta[1]	2.228	2.039	0.04583	-1.974	2.275	6.238	1001	2000
beta[2]	1.565	0.3246	0.007071	0.9216	1.563	2.205	1001	2000
beta[3]	0.01358	0.006783	1.423E-4	4.297E-5	0.01348	0.02702	1001	2000
s	6.247	0.938	0.01813	4.785	6.134	8.393	1001	2000

Table 5.8   WinBUGS posterior summaries for Example 5.1 after 2000 iterations and additional discarded 1000 burnin iterations using Zellner's g-prior ($c^2 = 10^4$)

node	mean	sd	MC error	2.5%	median	97.5%	start	sample
beta[1]	2.329	1.07	0.02317	0.1783	2.341	4.406	1001	2000
beta[2]	1.622	0.1695	0.004654	1.297	1.618	1.945	1001	2000
beta[3]	0.0143	0.003547	7.901E-5	0.007414	0.01427	0.0214	1001	2000
s	3.153	0.4811	0.01238	2.375	3.109	4.253	1001	2000

5.4   ANALYSIS OF VARIANCE MODELS

The normal models discussed in Sections 5.2 and 5.3 assess the association between continuous variables. To be more specific, normal regression models identify which and how specific continuous explanatory variables influence a continuous response variable. Analysis of variance models also assume a normal response variable, but now the explanatory variables are categorical. In this section we provide specific examples with one and two variables as well as details concerning their parametrization. A short discussion for multifactor analysis of variance closes this section, while an example of a three-way analysis of variance model is provided in Section 6.1.

5.4.1   The one-way ANOVA model

Let us assume a categorical variable A (also called *factor*) with levels $\ell = 1, 2, \ldots, L_A$ and a continuous response variable Y. When we assume that the effect of the categorical variable A is influencing the mean of the continuous variable Y, then this is equivalent to defining different means of Y for each category of A. Thus, assuming normal distributions for the response variable Y, the model can be summarized by

$$Y \sim N(\mu'_\ell, \sigma^2),$$

where $\ell = 1, 2, \ldots, L_A$ indicates the group (category) of factor A from which Y originates and μ'_ℓ indicates the mean of Y for the ℓ category. An alternative method is to write

$$\mu'_\ell = \mu_0 + \alpha_\ell \tag{5.8}$$

This expression decomposes the original mean of each category level μ'_ℓ to an overall common parameter μ_0, called *constant* and *group-specific* parameters α_ℓ, which are termed *effects* of ℓ level on the response variable Y. The interpretation of these parameters depends on the parametrization used for α_ℓ; see next subsection for details.

Let us now consider a random sample of n individuals resulting in n_ℓ subjects for each level ℓ ($\ell = 1, 2, \ldots, L_A$) of variable A. Then the model can be written as

$$Y_{\ell k} \sim N(\mu'_\ell, \sigma^2) \text{ and } \mu'_\ell = \mu_0 + \alpha_\ell \tag{5.9}$$

for $k = 1, 2, \ldots, n_\ell$ and $\ell = 1, 2, \ldots, L_A$.

In practice, we usually observe n pairs (a_i, y_i) that are realizations of the random variables (A_i, Y_i), where $a_i \in \{1, 2, \ldots, L_A\}$ is the group or level at which the ith subject belongs. In this case the model can be rewritten as

$$Y_i \sim N(\mu_i, \sigma^2) \text{ and } \mu_i = \mu'_{a_i} = \mu_0 + \alpha_{a_i} \tag{5.10}$$

for $i = 1, 2, \ldots, n$.

5.4.2   Parametrization and parameter interpretation

From (5.9) it is evident that we are interested in estimating the mean values μ'_ℓ of Y for each level of A. Thus, the original formulation $\mu_i = \mu'_{a_i}$ can be used to directly estimate the parameters of interest. Nevertheless, parametrization (5.10) is used for two reasons: (1) it separates the constant overall effect from the effect of the categorical variable A, and (2) it allows for generalization of the ANOVA formulation when additional categorical explanatory variables are involved in the model.

In the direct estimation of the mean values μ'_ℓ, we estimate L_A parameters (one for each group/level). When the alternative parametrization (5.9) is used, then we need to estimate $L_A + 1$ parameters. To make the model identifiable (i.e., the estimation feasible) and the two models equivalent, we impose one constraint on the new set of parameters. This constraint also specifies the interpretation and practical meaning of each parameter. Many parametrizations can be imposed by using different constraints, but two of them are most frequently met in statistical literature: the corner (CR) and the sum-to-zero (STZ) constraints. These two parametrizations are described here in detail.

5.4.2.1   *Corner constraints.*   In *corner constraints*, the effect of a level $r \in \{1, 2, \ldots, L_A\}$ is set equal to zero: $\alpha_r = 0$. This level r is referred to as the *baseline* or *reference category* of factor A. Usually the first or the last (in order) level is used as the reference category. In medicine, placebo or standard (old) treatment are used as baseline levels. In the following discussion, we use the first level as the reference category: $\alpha_1 = 0$. Under this parametrization the mean of Y will be summarized by

$$\begin{aligned} \mu'_1 = E(Y|A = 1) &= \mu_0 \\ \mu'_\ell = E(Y|A = \ell) &= \mu_0 + \alpha_\ell \text{ for } \ell \geq 2, \end{aligned}$$

where it is obvious that the constant parameter has a straightforward interpretation. It is simply the mean of Y for the reference category. Moreover, if we consider any difference $\mu'_\ell - \mu'_1 = \alpha_\ell$, then we obtain the effect α_ℓ of the ℓth category of factor A. Hence, parameter α_ℓ is the expected difference of Y for an individual belonging in ℓ group (or level) of variable A in comparison to an individual from the reference group or category of A.

5.4.2.2 Sum-to-zero constraints. According to the name of this parametrization, the following constraint is imposed:

$$\sum_{\ell=1}^{L_A} \alpha_\ell = 0. \tag{5.11}$$

In practice, within the likelihood we substitute one parameter (usually the first or the last one) with the function resulting from the sum-to-zero constraint (5.11). When we substitute the first level, then

$$\alpha_1 = -\sum_{\ell=2}^{L_A} \alpha_\ell . \tag{5.12}$$

The interpretation of this parametrization is different from the corresponding interpretation of the corner constrained parameters. In STZ, the constant term encapsulates an overall mean effect since

$$\sum_{\ell=1}^{L_A} \mu_\ell' \;=\; L_A\mu_0 + \sum_{\ell=1}^{L_A} \alpha_\ell = L_A\mu_0 \Leftrightarrow \mu_0 = \frac{1}{L_A}\sum_{\ell=1}^{L_A} \mu_\ell',$$

while parameter α_ℓ describes deviations of each level from this overall mean effect. Positive values induce an increased effect in comparison to the overall mean, while negative values induce effects lower than the overall mean level.

5.4.3 One-way ANOVA model in WinBUGS

In this section, we assume that the data are given by pairs (a_i, y_i) referring to the characteristics of the ith individual. The stochastic part of the likelihood is the same as the one used for normal regression models above. The deterministic part of the likelihood is slightly changed since the mean must be specified as a function of each level of A [see Eq. (5.10)]. Hence, the likelihood is defined in WinBUGS

```
for (i in 1:n){
    y[i] ~ dnorm( mu[i], tau )
    mu[i] <- mu0 + alpha[ a[i] ]
}
```

The imposed constraint must be set outside the likelihood loop. Thus, for CR parametrization, we set

```
a[1] <- 0.0
```

while, for STZ parametrization, we set

```
a[1] <- -sum( a[2:LA] )
```

Alternatively, the group means μ_ℓ' (denoted by m in the WinBUGS code) can be estimated directly. The desired effects can be simply calculated as contrasts of the group means μ_ℓ'. Thus we can write

$$\mu_0 = \mu_1', \quad \alpha_1 = 0 \text{ and } \alpha_\ell = \mu_\ell' - \mu_1' \text{ for } \ell = 2, \ldots, L_A$$

for CR parametrization and

$$\mu_0 = \overline{\mu}' = \frac{1}{L_A}\sum_{\ell=1}^{L_A} \mu_\ell' \quad \text{and } \alpha_\ell = \mu_\ell' - \overline{\mu}' \text{ for } \ell = 1, \ldots, L_A$$

for STZ parametrization. The model can be specified in WinBUGS using the following syntax

```
for (i in 1:n){
    y[i] ~ dnorm( mu[i], tau )
    mu[i] <- m[ a[i] ]
}
#
# for corner constraints
mu.cr       <- m[1]
alpha.cr[1]<-0
for (l in 2:LA){ alpha.cr[l]<- m[l]-m[1] }
#
# for STZ constraints
mu.stz      <- mean(m[])
for (l in 1:LA){ alpha.stz[l]<- m[l]-mu.stz }
```

Finally, we may adopt one parametrization and calculate the parameters of the other one as

$$\mu_0^{STZ} = \frac{1}{L_A}\sum_{\ell=1}^{L_A}\mu'_\ell = \frac{1}{L_A}\sum_{\ell=1}^{L_A}(\mu_0^{CR}+\alpha_\ell^{CR}) = \mu_0^{CR} + \frac{1}{L_A}\sum_{\ell=1}^{L_A}(\alpha_\ell^{CR})$$
$$= \mu_0^{CR} + \overline{\alpha}^{CR}$$

and

$$\alpha_\ell^{STZ} = \mu'_\ell - \overline{\mu}' = \mu_0^{CR} + \alpha_\ell^{CR} - \mu_0^{CR} - \overline{\alpha}^{CR} = \alpha_\ell^{CR} - \overline{\alpha}^{CR}$$

for $\ell = 1, \ldots, L_A$. Thus, in WinBUGS , we can calculate the STZ constraints by the model with corner constraints using the code

```
# for STZ constraints from CR
mu.stz      <- m.cr + mean(alpha.cr[])
for (l in 1:LA){ alpha.stz[l]<- alpha.cr[l]-mean(alpha.aplha[]) }
```

As a prior for μ and α_ℓ (for $\ell = 2, \ldots, L_A$), we consider a simple normal distribution with mean zero and low precision to express prior ignorance. Hence, in the WinBUGS code

```
mu0     ~ dnorm( 0.0 , 1.0E-4)
for (l in 2:LA){ alpha[i] ~ dnorm( 0.0, 1.0E-04 )}
```

the prior for the precision τ is defined as in normal regression models. In the formulation above, no prior is imposed on the constrained parameter α_1 since it is set equal to zero and therefore it does not appear in the likelihood equation. In this case, parameter α_1 is a constant node, while, in STZ parametrization, it is a logical/deterministic node since it is defined as a function of the remaining parameters.

When using the prior setup described above, we must be very careful since, under different parametrizations, we impose different prior distributions on the group means μ'_ℓ. In the case that the prior precision is small, differences due to this incompatibility of prior specification will be minor, but when prior information is used, these parameters must be specified carefully in order to lead to compatible prior beliefs.

5.4.4 A one-way ANOVA example using WinBUGS

Example 5.2. Evaluation of candidate school tutors. The director of a private school wishes to employ a new mathematics tutor. For this reason, the ability of four candidates is examined using the following small study. A group of 25 students was randomly divided into four classes. In all classes, the same mathematical topic was taught for 2 hours per day for 1 week. After completing the short course, all students had to take the same test. Their grades were recorded and compared (see Table 5.9). The administrator wishes to employ the tutor whose students attained higher performance at the given test.

Table 5.9   Data for Example 5.2 (school tutors' evaluation data)

Candidate	Students' grades
1	84 58 100 51 28 89
2	97 50 76 83 45 42 83
3	64 47 83 81 83 34 61
4	77 69 94 80 55 79

Setting up the data and the model code.   Data are coded in WinBUGS using two columns/variables: the first one with the student's grades and the second one corresponding to which candidate tutor was teaching in each student's class. Within the data we have also defined the number of cases ($n = 25$) and the number of tutors (TUTORS= 4). Data can be specified using either a list format or a rectangular format (see Table 5.10 for the list format of the data).

Table 5.10   WinBUGS list format data for Example 5.2

```
list( n=25, TUTORS=4,
     grade=c(84,  58,  100,  51,  28,  89,  97,  50,   76,  83,  45,  42,  83,
             64,  47,   83,  81,  83,  34,  61,  77,  69,   94,  80,  55,  79),
     class=c(1,  1,  1,  1,  1,  1,  2,  2,  2,  2,  2,  2,  2,
             3,  3,  3,  3,  3,  3,  3,  4,  4,  4,  4,  4,  4) )
```

The model can be defined according to the guidelines in the previous subsection. Each group (tutor) mean will be equal to a constant term plus the effect with index specified by the variable classes. Initial values will be specified for the constant mu and the effect parameter alpha except from the first (baseline) level. The code and the initial values are provided in Table 5.11. STZ parametrization is used, but commands for CR parametrization are also provided as comments.

Table 5.11   WinBUGS code and initial values for Example 5.2[a]

```
model{
    # model's likelihood
    for (i in 1:n){
        mu[i] <- m + alpha[ class[i] ]
        grade[i] ~ dnorm( mu[i], tau )
    }
    #### stz constraints
    alpha[1] <-  -sum(alpha[2:TUTORS])
    #### CR Constraints
    # alpha[1] <- 0.0

    # priors
    m~dnorm( 0.0, 1.0E-04)
    for (i in 2:TUTORS){ alpha[i]~dnorm(0.0, 1.0E-04)}
    tau ~dgamma( 0.01, 0.01)
    s <- sqrt(1/tau) # precision
}
INITS
list( m=1.0, alpha=c(NA, 0,0,0), tau=1.0 )
```

[a]Corner parametrization can be fitted by removing the comment sign # in line 10 and adding it to line 8 of the code.

Results.   After generating 3000 iterations in total and discarding the initial 1000 iterations, the posterior summaries given in Table 5.12 have been calculated. In this example, we are interested in evaluating the overall performance of each tutor that is encapsulated by each parameter α_j (alpha[j]) for $j = 1, 2, 3, 4$. Comparing the posterior means and medians of these parameters, we can see that the fourth tutor has a higher performance (close to 6 while the remaining tutors have negative effects, indicating that their performance is below the overall mean). Nevertheless, from the graphical representations of Figures 5.2, we may conclude that the posterior distributions of α_j are not clearly discriminated, indicating that the between-tutor differences are minor. Since the school needs to hire only one tutor, we recommend hiring the last one but keeping in mind that differences in the tutors' performance in this small study did not indicate clear differences between tutors' actual abilities.

Table 5.12   Posterior summaries for ANOVA parameters of Example 5.2

node	mean	sd	MC error	2.5%	median	97.5%	start	sample
alpha[1]	-0.5661	8.016	0.1488	-16.27	-0.661	15.43	1001	2000
alpha[2]	-1.218	7.437	0.1442	-15.77	-1.317	13.6	1001	2000
alpha[3]	-4.17	7.323	0.1497	-18.04	-4.158	11.04	1001	2000
alpha[4]	5.955	8.345	0.1626	-10.34	6.066	22.56	1001	2000
m	68.96	4.561	0.1109	60.0	68.98	78.11	1001	2000
s	22.21	3.638	0.08999	16.54	21.74	30.55	1001	2000

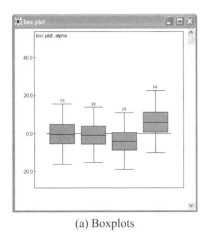
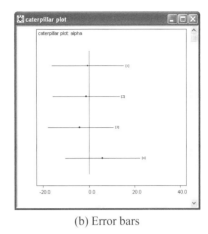

(a) Boxplots   (b) Error bars

Figure 5.2   Posterior boxplots and error bars for tutors' effects in Example 5.2 (school tutors' evaluation).

5.4.5   Two-way ANOVA models

5.4.5.1   The main effects model.   We can extend the analysis of variance model to accommodate additional categorical explanatory variables. In the following, we illustrate models with two categorical explanatory variables (two-way analysis of variance models) and then, in the Section 5.4.6 we briefly describe implementation of the general multifactor ANOVA models.

Let as consider two categorical factors A and B with L_A and L_B levels, respectively. Then a natural extension of model (5.8) is to include the additive effect of the additional variable in the expression of the mean μ'_{ab} for a and b levels of A and B categorical variables, respectively. Hence μ'_{ab} can be written as

$$\mu'_{ab} = \mu_0 + \alpha_a + \beta_b$$

for $a = 1, 2, \ldots, L_A$ and $b = 1, 2, \ldots, L_B$.

When data are given in tabular format of size $L_A \times L_B$ with n_{ab} observation per cell, then the model is given by

$$y_{abk} \sim N(\mu'_{ab}, \sigma^2) \text{ and } \mu'_{ab} = \mu_0 + \alpha_a + \beta_b$$

for $k = 1, 2, \ldots, n_{ab}$ and a, b taking values as above.

This tabular setup is restrictive and difficult to define within WinBUGS model code unless a sample with equal numbers of observations per cell is considered; see Section 5.4.5.4 for more details. A more general setup, equivalent to the formulation for one-way models [see Eq. (5.10)], is given by the following equation

$$y_i \sim N(\mu_i, \sigma^2) \text{ and } \mu_i = \mu_0 + \alpha_{a_i} + \beta_{b_i}$$

for $i = 1, 2, \ldots, n$. In this model formulation, data are given in the form of (a_i, b_i, y_i) for each subject i with a_i and b_i representing the groups/levels of factors A and B, respectively, in which i subject belongs.

5.4.5.2   *Parametrization and parameter interpretation.*   Parametrization of the model described above can be completed by using STZ or CR constraints for both variables involved in the model. Mixed parametrization (e.g., CR for one variable and STZ for the other) can be used without any difficulty, but the user must be careful with the parameter interpretation. For this reason, such practice is not recommended for users with limited experience in modeling. In the following, we briefly present the parameter interpretation for each parametrization using the same logic as in Section 5.4.2.

For the corner constraints, the constant parameter indicates the mean of Y for the reference categories of both variables A and B since $\mu'_{11} = \mu_0$ (assuming here the first categories as the baseline levels). Effect α_a indicates the expected group differences between subjects of level a and subjects belonging in the reference category of A when both being members of any group b of factor B since

$$
\begin{aligned}
\mu'_{ab} - \mu'_{1b} &= \mu_0 + \alpha_a + \beta_b - \mu_0 - \alpha_1 - \beta_b \\
&= \alpha_a - \alpha_1 = \alpha_a \, .
\end{aligned}
$$

Similarly, β_b is equal to the expected difference between individuals of level b and individuals of the reference category of factor B belonging in the same group a of variable A since

$$
\begin{aligned}
\mu'_{ab} - \mu'_{a1} &= \mu_0 + \alpha_a + \beta_b - \mu_0 - \alpha_a - \beta_1 \\
&= \beta_b - \beta_1 = \beta_b \, .
\end{aligned}
$$

For the sum-to-zero constraints, the constant parameter is equal to an overall mean (grand mean) estimate since

$$
\begin{aligned}
\sum_{a=1}^{L_A} \sum_{b=1}^{L_B} \mu_{ab} &= \sum_{a=1}^{L_A} \sum_{b=1}^{L_B} (\mu_0 + \alpha_a + \beta_b) \\
&= L_A L_B \mu_0 + L_B \sum_{a=1}^{L_A} \alpha_a + L_A \sum_{b=1}^{L_B} \beta_b = L_A L_B \mu_0 \\
\Leftrightarrow \mu_0 &= \frac{1}{L_A L_B} \sum_{a=1}^{L_A} \sum_{b=1}^{L_B} \mu_{ab} \, .
\end{aligned}
$$

Moreover, since the mean of all effects α_a is equal to zero, the effect α_a indicates group differences between level a and the overall mean effect of A within each group b of variable B. Similarly, β_b represents differences between level b and the overall mean effect of B within each group a of variable A.

In the "main effects" model defined above, the effect of the level of each factor is assumed constant across the levels of the second variable. This property is the main characteristic of the main effects model and, for this reason, such effects are called *additive*.

5.4.5.3   *The two-way interaction model.*   Frequently in practice the effect of an explanatory categorical variable A on a response variable Y is influenced by another factor B. In such cases, the two factors A and B interact in terms of their effect on Y and, therefore the additive model is no longer valid. A more complicated structure must be defined to encapsulate this "interaction" effect on the mean, which can be expressed by

$$
\mu'_{ab} = \mu_0 + \alpha_a + \beta_b + \alpha\beta_{ab}
$$

for $a = 1, 2, \ldots, L_A$ and $b = 1, 2, \ldots, L_B$. Parameters $\alpha\beta_{ab}$ now determine the way that the two categorical factors A and B interact and change their effects on Y.

Constraints on interaction term $\boldsymbol{\alpha\beta} = (\alpha\beta_{ab})$ must be set in a manner similar to that for the main effects. In order to match the $L_A \times L_B$ group means, we need to impose $L_A + L_B - 1$ constraints on the interaction parameters $\boldsymbol{\alpha\beta}$. In the CR parametrization we impose the following constraints

$$\alpha\beta_{r_a b} = \alpha\beta_{a r_b} = 0$$

for all $a = 1, 2, \ldots, L_A$ and $b = 1, 2, \ldots, L_B$, where $r_a \in \{1, 2, \ldots, L_A\}$ and $r_b \in \{1, 2, \ldots, L_B\}$ are the reference categories for categorical factors A and B, respectively. In our case, we consider $r_a = r_b = 1$ and hence $\alpha\beta_{1b} = \alpha\beta_{a1} = 0$.

Similarly, for the STZ parametrization, we impose constraints

$$\sum_{a=1}^{L_A} \alpha\beta_{ab} = \sum_{b=1}^{L_B} \alpha\beta_{ab} = 0 . \tag{5.13}$$

In order to estimate the model, we simply substitute $L_A + L_B - 1$ parameters by the corresponding equations resulting from constraints (5.13). For example, by considering the substitution of the first levels of each factor, we end up setting

$$\alpha\beta_{a1} = -\sum_{b=2}^{L_B} \alpha\beta_{ab} \text{ for all } a = 1, 2, \ldots, L_A, \tag{5.14}$$

$$\alpha\beta_{1b} = -\sum_{a=2}^{L_A} \alpha\beta_{ab} \text{ for all } b = 2, 3, \ldots, L_B . \tag{5.15}$$

Interpretation is similar to the one described in the previous section, but we now have to also consider the interaction term. In corner parametrization μ_0 is still the mean of Y for experimental units in the reference categories of both categorical factors. Effects α_a now denote the mean difference between the reference and a level of variable A when B is set to its reference category r_b since

$$\begin{aligned}
\mu'_{a r_b} - \mu'_{r_a r_b} &= \mu_0 + \alpha_a + \beta_{r_b} + \alpha\beta_{a r_b} - \mu_0 - \alpha_{r_a} - \beta_{r_b} - \alpha\beta_{r_a r_b} \\
&= \mu_0 + \alpha_a - \mu_0 = \alpha_a .
\end{aligned}$$

Similarly, β_b denotes the mean difference between the reference and b level of variable B when A is set to its reference category r_a. The interaction term $\alpha\beta_{ab}$ denotes the additional effect due to the interaction between the two levels since, for $a, b > 1$, we obtain

$$\begin{aligned}
\mu'_{ab} - \mu'_{r_a b} &= \mu_0 + \alpha_a + \beta_b + \alpha\beta_{ab} - \mu_0 - \alpha_{r_a} - \beta_b - \alpha\beta_{r_a b} \\
&= \mu_0 + \alpha_a + \beta_b + \alpha\beta_{ab} - \mu_0 - \beta_b = \alpha_a + \alpha\beta_{ab} .
\end{aligned}$$

The difference $\mu'_{ab} - \mu'_{r_a b}$ is now not the same as the corresponding difference in the reference category (i.e., for $b = r_b$). This difference is now affected by the levels of the factor B, which is not the case in the main effects model.

Interpretation of the STZ parameters is equivalent to the corresponding interpretation in CR parametrization, but now all comparisons are made with respect to the "grand mean". It is slightly more difficult and, for this reason, corner constraints are frequently used to simplify interpretation.

5.4.5.4 Data in tabular format (equal observations per cell). Frequently, data for analysis of variance models are provided in a tabular form. In this section we briefly demonstrate the WinBUGS model code when data are tabulated.

Let us consider two categorical factors A and B with levels L_A and L_B. Then the data will be presented in a $L_A \times L_B$ tabular format with K observations in each level combination (cell). Here we assume equal observations per cell. An equivalent approach can be followed in cases of unbalanced data using missing values to fill in empty cells of the defined matrix. When unbalanced data are given in raw, individual-type format, then the approach described in Section 5.4.5.5 is recommended.

Response data y_{abk} refer to the kth observation of Y for the a and b levels of the factors A and B, respectively. The data structure is depicted in Table 5.13. We use a $L_A \times L_B \times K$ array to store the data in WinBUGS. In cases where $K = 1$ (i.e., one observation per cell), we can use a $L_A \times L_B$ matrix.

Table 5.13 Tabular format for two-way ANOVA data with K observations per cell

A	Factor B					
	1	2	$\cdots$	b	$\cdots$	L_B
1	$y_{111}, \ldots, y_{11K}$	$y_{121}, \ldots, y_{12K}$	$\cdots$	$y_{1b1}, \ldots, y_{1bK}$	$\cdots$	$y_{1L_B 1}, \ldots, y_{1L_B K}$
2	$y_{211}, \ldots, y_{21K}$	$y_{221}, \ldots, y_{22K}$	$\cdots$	$y_{2b1}, \ldots, y_{2bK}$	$\cdots$	$y_{2L_B 1}, \ldots, y_{2L_B K}$
$\vdots$	$\vdots$	$\vdots$	$\vdots$	$\vdots$	$\vdots$	$\vdots$
a	$y_{a11}, \ldots, y_{a1K}$	$y_{a21}, \ldots, y_{a2K}$	$\cdots$	$y_{ab1}, \ldots, y_{abK}$	$\cdots$	$y_{aL_B 1}, \ldots, y_{aL_B K}$
$\vdots$	$\vdots$	$\vdots$	$\vdots$	$\vdots$	$\vdots$	$\vdots$
L_A	$y_{L_A 11}, \ldots, y_{L_A 1K}$	$y_{L_A 21}, \ldots, y_{L_A 2K}$	$\cdots$	$y_{L_A b1}, \ldots, y_{L_A bK}$	$\cdots$	$y_{L_A L_B 1}, \ldots, y_{L_A L_B K}$

Setting up the data. The simplest approach is to use R to set up an array of appropriate dimension, then export it to WinBUGS and finally edit the file to meet the requirements of WinBUGS (see Section 3.4.6.3 for general instructions).

To specify the data in a list format, we can use the syntax

```
list( LA=#LA#, LB=#LB#, K=#K#,
      y = structure(
             .Data=c( y111,   y112,   ..., y11K,
                      y121,   y122,   ..., y12K,   ..., ...,
                      y1LB1,  y1LB2,  ..., y1LBK,
                      y211,   y212,   ..., y21K,
                      y221,   y222,   ..., y22K,   ..., ...,
                      y2LB1,  y2LB2,  ..., y2LBK,
                      ..., ...,
                      yLA11,  yLA12,  ..., yLA1K,
                      yLA21,  yLA22,  ..., yLA2K,   ..., ...,
                      yLALB1, yLALB2, ..., yLALBK),
             .Dim = c( #LA#,#LB#,#K# )
             )
      )
```

where `#LA#`, `#LB#`, and `#K#` are fixed numbers defining the number of levels for each variable and the number of observations per cell. In the syntax above we have specified first all observations for cell $(1, 1)$ followed by observations for cell $(1, 2), \ldots, (1, L_B)$ (levels of factor B changing first). After data for all categories of B are defined, then we insert all data for the next level of A using the same structure, and so on.

A simpler way is to specify each column y[a,b,] (or y[,b,k]) separately using the rectangular data format. This is easier to follow, although it requires specification of $L_A \times L_B$ columns.

The model code. Having specified the data, we can specify the model likelihood using a triple loop in order to define all elements of the response array y. Thus the main effects model is defined by

$$y_{abk} \sim N(\mu'_{abk}, \tau^{-1}) \text{ and } \mu'_{abk} = \mu_0 + \alpha_a + \beta_b$$

for $a = 1, 2, \ldots, L_A$, $b = 1, 2, \ldots, L_B$, and $k = 1, 2, \ldots, K$, while the corresponding WinBUGS code is given by

```
for (a in 1:LA){
    for (b in 1:LB){
        for (k in 1:K){
            y[a,b,k] ~ dnorm( mu[a,b,k], tau )
            mu[a,b,k] <- mu0 + alpha[a] + beta[b]
        }}}
```

Priors and constraints are defined as in the one-way ANOVA model.

For the interaction model, we only need to change the specification of the mean since $\mu'_{ab} = \mu_0 + \alpha_a + \beta_b + \alpha\beta_{ab}$. Thus, the fifth line of the code above is substituted by the following command:

```
        mu[a,b,k] <- mu0 + alpha[a] + beta[b] + ab[a,b]
```

Note that the interaction term ab is defined as a matrix of dimension $L_A \times L_B$. Priors must be defined for all components of $\alpha\beta$ except for the ones referring to the baseline levels r_a and r_b. Hence we specify the prior

$$\alpha\beta_{ab} \sim N(0, 10^4) \text{ for } a = 2, \ldots, L_A \text{ and } b = 2, \ldots, L_B$$

when $r_a = r_b = 1$. Similarly, for the other parameters, we use the common low-information priors

$$\mu_0 \sim N(0, 10^4), \quad \alpha_a \sim N(0, 10^4) \text{ and } \beta_b \sim N(0, 10^4)$$

for all $a = 2, \ldots, L_A$ and $b = 2, \ldots, L_B$.

For the corner constraints, with the first level as baseline for both variables, we use the following syntax to specify the priors and the constraints:

```
# corner constraints for baseline levels (1st levels)
ab[1,1] <- 0
for (a in 2:LA){ ab[a,1]<-0 }
for (b in 2:LB){ ab[1,b]<-0 }
# prior distributions for the rest interaction parameters
for (a in 2:LA){
    for (b in 2:LB){
            ab[a,b]~dnorm(0.0, 1.0E-04)
}}
```

Sum-to-zero constraints will be defined by substituting the first four lines of the code above with the following syntax, which represents (5.15) and (5.14).

```
# corner constraints for baseline levels (1st levels)
for (a in 1:LA){ ab[a,1]<- -sum( ab[a,2:LB] ) }
for (b in 2:LB){ ab[1,b]<- -sum( ab[2:LA,b] ) }
```

5.4.5.5   A two-way ANOVA example.

Example 5.3. Schizotypal personality data. Let us consider the data presented in
Table 5.14, inspired by a student survey examining the association between schizotypal traits and impulsive and compulsive buying behavior of university students
(Iliopoulou, 2004). Cell values of the table represent scores of a psychometric
scale called *schizotypal personality questionnaire* [SPQ, Raine (1991)] of individuals that are close to the median values of each group. In the original study by
Iliopoulou (2004), no female students living in villages reported high economic
status (indicated by NA). In Chapter 10 we illustrate how we can deal with missing
response values using the Bayesian approach. Nevertheless, here, we have substituted the missing values with two imaginary values (indicated within parentheses)
for illustration purposes. In this example we are interested in the effect of the student's family economic status (categorical with three levels: low, medium, high),
area of residence (binary: City, Village), and gender (male, female) on Raine's
(1991) SPQ total score.
Raine's SPQ scale is a 74-item self-administered questionnaire used to measure the
concepts related to schizotypal personality. The questionnaire consists of binary
zero–one (yes–no) items. It provides subscales for nine schizotypal features as
well as an overall scale for schizotypy. These nine specific characteristics of a
"schizotypal personality" are defined in the DSM-III-R diagnostic and statistical
manual of mental disorders edited by the American Psychiatric Association (1987),
and are measured by an SPQ subscale calculated as the sum of the questionnaire
items that refer to each schizotypal characteristic.
A "schizotypal personality" suffers from minor episodes of "pseudoneurotic" problems. In general, the prevalence rate of schizotypy is about 10% in the general
population. The importance of schizotypal personality in psychiatric research is
prominent for two reasons: (1) schizotypal subjects are at increased risk of developing schizophrenia during their lifetimes, and (2) since they are healthy persons,
they can participate in psychiatric/psychological research studies (by completing
questionnaires — psychometric instruments) which tryly schizophrenic subjects
are unable to do. In this example we focus on the total SPQ score.

Table 5.14   Data for Example 5.3

	City		Village	
	Gender		Gender	
Economic status	Male	Female	Male	Female
Low (1)	9	18	14	29
Medium (2)	25	22	26	25
High (3)	23	12	NA (24)	NA (13)

Two-way interaction model. In this subsection, we ignore the "city residence factor",
assuming that we have two observations for each combination of economic status (factor A)
and gender (factor B). Hence we need to set up the data in $3 \times 2 \times 2$ array ($L_A = 3$, $L_B = 2$
and $K = 2$).

Following the directions above, we first rearrange the data in the following way:

	Gender			
	Male (1)		Female (2)	
Economic status	$K = 1$	$K = 2$	$K = 1$	$K = 2$
Low (1)	9	14	18	29
Medium (2)	25	26	22	25
High (3)	23	24	12	13

We can now define the data in WinBUGS using the following syntax:

```
list( LA=3, LB=2, K=2,
      y = structure(
            .Data=c( 9, 14, 18, 29,
                    25, 26, 22, 25,
                    23, 24, 12, 13),
            .Dim = c( 3,2,2 )
            )
      )
```

To ensure that the data have been imported correctly in WinBUGS, we compile the model and go to Info>Node Info, insert y to view the response data, and press the values box. The values of y will be printed in the log file for cross checking. The correct format in this example is as follows:

```
y[1,1,1]        9.0
y[1,1,2]       14.0
y[1,2,1]       18.0
y[1,2,2]       29.0
y[2,1,1]       25.0
y[2,1,2]       26.0
y[2,2,1]       22.0
y[2,2,2]       25.0
y[3,1,1]       23.0
y[3,1,2]       24.0
y[3,2,1]       12.0
y[3,2,2]       13.0
```

Alternatively, we can use four columns of length equal to 3 to set up the data using the following rectangular data format:

```
list( LA=3, LB=2, K=2)
 y[,1,1] y[,1,2] y[,2,1] y[,2,2]
     9      14      18      29
    25      26      22      25
    23      24      12      13
END
```

In the list data format of this syntax, we define the dimensions of the table needed for the specification of the likelihood in the model code.

The full code for the interaction model is given in Table 5.15. Note that instead of alpha, beta, and ab we have used the names econ, gender, and econ.gender to reflect the actual names of the factors.

Table 5.15   WinBUGS code and initial values for Example 5.3

```
model{
    # model's likelihood
    for (a in 1:LA){
    for (b in 1:LB){
        for (k in 1:K){
            y[a,b,k] ~ dnorm( mu[a,b,k], tau )
            mu[a,b,k] <- mu0 + econ[a] + gender[b] + econ.gender[a,b]
            #mu[a,b,k]<- mu0 + econ[a] + gender[b]
    }}}
    #### CR Constraints
    econ[1] <- 0.0
    gender[1] <- 0.0
    econ.gender[1,1] <-0.0
    for (a in 2:LA){econ.gender[a,1]<-0.0}
    for (b in 2:LB){econ.gender[1,b]<-0.0}

    # priors
    mu0~dnorm( 0.0, 1.0E-04)
    for (a in 2:LA){econ[a]~dnorm( 0.0, 1.0E-04)}
    for (b in 2:LB){gender[b]~dnorm( 0.0, 1.0E-04)}
    for (a in 2:LA){
        for (b in 2:LB){
            econ.gender[a,b]~dnorm( 0.0, 1.0E-04)
    }}
    tau ~dgamma( 0.01, 0.01)
    s <- sqrt(1/tau) # precision

    for (a in 1:LA){
    for (b in 1:LB){
            mean.spq[a,b]<- mu0 +econ[a] +gender[b] +econ.gender[a,b]
    }}


}

INITS
list( mu0=1.0, econ=c(NA, 0,0), gender=c(NA, 0),
    econ.gender=structure(.Data=c(NA, NA, NA, 0, NA, 0), .Dim=c(3,2))
        , tau=1.0 )
```

Results.   Results obtained after 3000 iterations (discarding the initial 1000 iterations as burnin) are provided in Table 5.16 and Figure 5.3 . From the boxplots we can observe that

- Both interaction terms are far away from zero, indicating that this term must be included in the model.

- The 95% posterior intervals of the two interaction parameters (econ.gender$_{22}$ and econ.gender$_{32}$) have common values, indicating that they can be considered as a posteriori equal.

- Both gender and economic status effects are far away from zero, indicating that they influence the mean SPQ score.

- Small differences are observed between the posterior distributions of the medium and high economic status effects.

Table 5.16   Posterior summaries for parameters of interaction in two-way ANOVA model of Example 5.3

node	mean	sd	MC error	2.5%	median	97.5%	start	sample
econ[2]	14.09	4.251	0.2462	5.74	14.0	22.43	1001	2000
econ[3]	12.01	4.287	0.2297	3.497	12.0	20.59	1001	2000
econ.gender[2,2]	-14.19	5.966	03869	-26.55	-14.07	-2.552	1001	2000
econ.gender[3,2]	-23.08	6.095	0.375	-36.51	-22.83	-11.74	1001	2000
gender[2]	12.08	4.285	0.2705	4.02	11.94	21.27	1001	2000
mu0	11.46	3.012	0.1542	5.57	11.43	17.62	1001	2000
s	4.081	1.4	0.06154	2.3	3.793	7.522	1001	2000

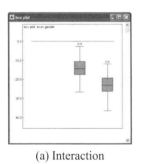 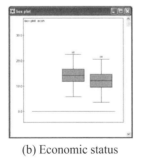 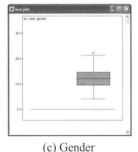

(a) Interaction   (b) Economic status   (c) Gender

Figure 5.3   Posterior boxplots for two-way ANOVA parameters in Example 5.3 (Schizotypal personality data).

To be more specific, the mean SPQ score for a male student with low economic family status is a posteriori expected to be equal to 11.5. Female students with the same economic status are a posteriori expected to score about 12 units higher (i.e., 23.5). Moreover, we observe a positive effect of economic status on male students since students with medium family economic status are a posteriori expected to score 14 points in addition to the ones scored by male students with low economic status. The corresponding increase is lower for the high economic group (about 12 points), indicating no separation in the effects of these two groups.

The effect of economic status is opposite for female students since the interaction term indicates that increase in their economic status from low to medium does not change the a posteriori expected mean SPQ score (since econ[2]+econ.gender[2,2] = 14.09 − 14.19 = −0.1). Moreover, a female student with high economic family status is a posteriori expected to get an SPQ score of 11 units lower than a female from low economic status (since econ[3]+econ.gender[3,2] = 12.01 − 23.08 = −11.07).

The preceding comments can be depicted in an interaction plot based on the posterior means of expected SPQ scores (i.e., μ values) for each combination of the categories of the two variables under consideration. In Figure 5.4 we have used one line for each gender. Note that in the main effects model, these two lines will not cross and will be parallel, indicating equal effects of the categorical variable across all levels of the other other one. From this plot it is clear that the economic status is positively associated with the mean SPQ score for males and negatively associated for females. Moreover, the difference between males and females is minor for the medium economic group. Finally, the mean SPQ score is a posteriori expected to be equal for males with low economic status and females with high economic status.

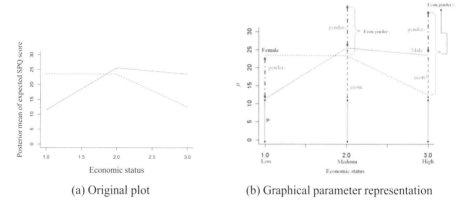

(a) Original plot   (b) Graphical parameter representation

Figure 5.4   Interaction plot based on posterior means of expected SPQ score in Example 5.3.

In Figure 5.4a the original interaction plot is provided, while in Figure 5.4b the same plot is annotated with comments on each parameter. The curly brackets refer to the inter-action terms, which indicate deviations from the main effects model. Hence the expected SPQ difference between females and male students for medium and high economic status is given from the differences econ.gender$_{22}$−gender$_2$ and econ.gender$_{32}$−gender$_2$ respectively. The plots in Figure 5.4 can be enriched by using more informative posterior boxplots or error bars for each group; see, for example, Figure 5.5.

Individual type data.   Individual type data are usually available in practice. In this format data from each variable are provided in a tabular format with n rows and $p+1$ columns. Each row corresponds to one individual and each column, to one variable (as usually in statistical packages). Using this approach, we can easily fit an ANOVA model for unbalanced data. The main effects model can be easily programed using the syntax

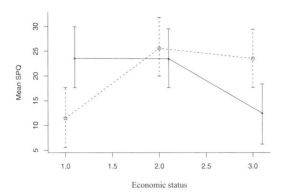

Figure 5.5   Interaction plot based on 95% posterior intervals of expected SPQ score in Example 5.3.

```
for (i in 1:n){
    y[i] ~ dnorm( mu[i], tau )
    mu[i] <- mu0 + alpha[ a[i] ] + beta[ b[i] ]
}
```

while for the interaction term we only substitute the expression for μ by

```
    mu[i] <- mu0 + alpha[ a[i] ] + beta[ b[i] ] + ab[ a[i], b[i] ]
```

in order to also accommodate the interaction term ab.

For Example 5.3 the data can be inserted using the following syntax

```
list( n=12, LA=3, LB=2,
          g=c(1,1,1,2,2,2,1,1,1,2,2,2),
          e=c(1,2,3,1,2,3,1,2,3,1,2,3),
          y=c(9, 25, 23, 18, 22, 12, 14, 26, 24 ,29 ,25, 13) )
```

in list format, or by

```
list(n=12, LA=3, LB=2)
  g[]       e[]       y[]
   1         1         9
   1         2        25
   1         3        23
   2         1        18
   2         2        22
   2         3        12
   1         1        14
   1         2        26
   1         3        24
   2         1        29
   2         2        25
   2         3        13
END
```

when using the rectangular data format. In the data format displayed above, g[] and e[] denote the gender and economic status of each student of the data. The expression for the mean μ_i will be given by

```
mu[i] <- mu0 + gender[ g[i] ] + econ[ e[i] ]
```

for the main effects model and by

```
mu[i] <- mu0 + gender[ g[i] ] + econ[ e[i] ]
              + econ.gender[ e[i], g[i] ]
```

for the interaction model.

Note that, since we can easily handle unbalanced data in this way, for Example 5.3 we can eliminate the two observations that correspond to missing values at the original dataset and rerun the MCMC algorithm. Table 5.17 presents results after removing these two values. Results are slightly different, but the main conclusions remain the same as in the full data analysis.

Table 5.17   Posterior summaries for parameters of interaction two-way ANOVA model of Example 5.3 using unbalanced data

node	mean	sd	MC error	2.5%	median	97.5%	start	sample
econ[2]	14.23	6.16	0.4092	3.204	14.07	26.48	1001	2000
econ[3]	11.32	7.72	0.4144	-3.831	11.52	26.4	1001	2000
econ.gender[2,2]	-14.63	9.385	0.7089	-34.23	-14.11	1.689	1001	2000
econ.gender[3,2]	-23.09	11.08	0.674	-45.34	-22.87	-2.465	1001	2000
gender[2]	12.32	6.563	0.4713	1.033	11.89	24.5	1001	2000
mu0	11.44	4.507	0.2476	3.277	11.41	19.8	1001	2000
s	5.473	3.123	0.222	2.583	4.741	12.43	1001	2000

The individual-type data format is recommended since it is more general and can be implemented in most models by slightly modifying the model code shown above. Moreover, unbalanced data can be easily fitted using this approach.

5.4.6 Multifactor analysis of variance

ANOVA models can be extended to accommodate more than two categorical variables. The main effects model includes no interaction, and its interpretation is similar to the one described for the two-way model. Including higher order interaction complicates the model, and the user must be careful when using them.

Although we can extend the approach described in this section for multifactor ANOVA models, it is not recommended since the constraints we need to impose considerably complicate the model structure, resulting in a long and obscure model code in WinBUGS . Such models can be fitted in a more straightforward manner using dummy variables, which are described in detail in Section 6.1.

Problems

5.1   **Hubble's constant data story.** The following data refer to the distance in megaparsecs (Mpc) of 24 galaxies from earth and the velocity [in kilometers per second (km/s)] with which they appear to be moving away from us; note that 1 parsec = 3.26 light years. They were collected by Edwin Hubble in 1929 (Hubble, 1929) during his attempt to explain the past and predict the future behavior of the universe. Hubble formulated the following law

$$\text{Recession velocity} = H_0 \times \text{distance}$$

where H_0 is Hubble's constant estimated to be equal to 75 km s^{-1} Mpc^{-1}. Both the data and the problem are available at the *Data and Story Library* (DASL) at http://lib.stat.cmu.edu/DASL/Datafiles/Hubble.html.

a) Fit a simple regression model to identify whether the constant can be set equal to zero according to Hubble's data. Use a low information prior for this model.

b) Fit a simple regression model excluding the constant term and compare the posterior distribution of Hubble's constant with Hubble's estimated value. Use a low-information prior for this model.

c) Interpret the regression coefficient estimated by both models.

d) According to Hubble, the value $1/H_0$ estimates that all galaxies originated in the same place (i.e., since the big bang) and hence provides an estimate of the age of the universe. Produce its posterior distribution.

e) Produce sensitivity analysis in the model without intercept using a prior distribution with $N(0, \sigma_\beta^2)$. Compare the results for $\sigma_\beta^2 \in \{100, 1000, 10^4\}$. Is the zero prior mean a sensible choice?

f) Produce sensitivity analysis in the model without intercept using a prior distribution with $N(\mu_\beta, 10^4)$. Compare the results for $\mu_\beta \in \{0, 75, 425\}$.

g) Use different regression models for positive and negative velocities. Compare the posterior distributions of the estimated H_0 values. Are they similar?

5.2 Consider the following data

```
x: -5    -4 -3    -2    -1  0   1   2   3    4    5
y: -26.8  9 27.07 22.87 8.43 4.38 3.64 4.41 37.80 95.96 179.10
```

a) Use simple linear regression to fit data of the problem.

b) Estimate the posterior distributions of the expected values of Y_i for all i according to this model.

c) Produce an error bar comparing the expected values of Y_i for all i to the true observed values.

d) Use polynomial regression to fit the above data. How many polynomial terms are needed to adequately fit the data?

e) In the polynomial regression model, use Zellner's g-prior and compare the results with the corresponding ones when an independent prior is used (in both cases assume that no prior information exists).

5.3 Consider the simulated dataset of Dellaportas et al. (2002) (available at book's Webpage). This dataset consists of $p = 15$ covariates and $n = 50$ observations. Covariates X_j (for $j = 1, \ldots, 15$) were generated from a standardized normal distribution, while the response variable was generated from

$$Y_i \sim N(X_{i4} + X_{i5}, (2.5)^2) \text{ for } i = 1, 2, \ldots, 50.$$

a) Estimate the full regression model using a simple independent low-information prior.

    b) Compare the estimates using boxplots and error bars. Which coefficients must be removed from the model, at least using these graphs?

    c) Are the true coefficient values placed in the center of the posterior distribution?

5.4 Using the data of Problem 5.3

    a) Estimate the full regression model using the Zellner's g-prior with $c^2 = n$. Use vector and matrices whenever you can.

    b) Compare the results with the corresponding ones using an independent prior.

5.5 Use a simple ANOVA model to compare the means of the following samples

```
Group 1: 10.0  9.6  8.8  9.1 11.8  9.6  9.0 10.7 10.0 11.6
Group 2:  9.7  9.2  9.0  9.7  9.0  9.8  9.6  8.3  8.1 10.0  8.3  8.2
Group 3: 10.9 13.9  5.6 11.5 12.4  8.6 15.2  5.5
Group 4: 10.3 10.6  9.5  9.4 10.8
```

    a) Compare the posterior distributions of the expected values for the four groups. Can they be considered equal?

    b) Compare the posterior distributions of the model effects to those of the first group. Can they be considered equal to zero?

    c) Use unequal variances for each group. Compare their posterior distributions and infer their equality or inequality.

5.6 Assume the following information to be available:

Group	Sample mean	Sample variance	Group size
1	40.8	2.7	30
2	40.1	3.0	100
3	33.5	2.2	56
4	28.9	2.3	16

Use the available information tabulated above to fit in WinBUGS a simple one-way ANOVA model based on the distribution of the sample means. Produce posterior analysis of the model parameters and infer the equality or the inequality of the mean values.

5.7 Download and read carefully the paper by Kahn (2005) and the associated dataset. [1]

    a) Fit a two-way ANOVA model (without interactions) with response to the FEV measurement and explanatory variables for the categorical variables gender and smoking status.

    b) Interpret the posterior results of the model parameters.

    c) Examine whether an interaction term is needed for these data. If so, interpret the parameters of the interaction model.

5.8 **Albuquerque home prices data story.** [2] Download the home-priced data available at http://lib.stat.cmu.edu/DASL/Datafiles/homedat.html.

---

[1] Available at http://www.amstat.org/publications/jse/v13n2/datasets.kahn.html.
[2] Both the data and the problem are available from *The Data and Story Library* (DASL) (http://lib.stat.cmu.edu/DASL/) at Carnegie Mellon University.

a) Compare the mean prices for the different number of features, the northeast location, and the corner location using simple one-way ANOVA models.

b) Using a two-way ANOVA model, determine whether the effect of different numbers of features (considered as categorical here) and the northeast location impact the home prices interactively.

c) Using a two-way ANOVA model, determine whether the northeast or the corner location influence home prices.

d) In all models, use error bars and boxplots to compare the posterior distributions of interest.

e) For each model, plot error bars of the expected home prices for each observed value.

CHAPTER 6

INCORPORATING CATEGORICAL VARIABLES IN NORMAL MODELS AND FURTHER MODELING ISSUES

In this chapter we focus on the incorporation of categorical covariates in the usual regression models presented in the previous chapter. The use of dummy variables for the two parametrizations introduced in section 5.4 is first illustrated. Then we focus on ANOVA models using dummy variables and on the analysis of covariance (ANCOVA) models in which both qualitative and quantitative variables are used as covariates.

Categorical variables can be incorporated in regression models via the use of dummy variables. These dummy variables identify which parameters must be added (and how) to the linear predictor.

Let us consider the simple one-way ANOVA with a categorical factor A and L_A categories. Under the CR constraints, the mean of an individual observation i with data (y_i, a_i) is given by

$$\mu_i = \mu_0 \qquad \text{when} \quad a_i = 1$$
$$\mu_i = \mu_0 + \alpha_{a_i} \quad \text{when} \quad a_i > 1.$$

The aim is to express the mean μ_i as a linear combination of the covariates equivalent to (5.2). Thus, we can write

$$\mu_i = \mu_0 + \alpha_2 D_{i2}^A + \alpha_3 D_{i3}^A + \cdots + +\alpha_{L_A} D_{iL_A}^A,$$

where $D_{i\ell}^A$; $\ell = 1, 2, \ldots, L_A$ are dummy variables defined as

$$D_{i\ell}^A = 1 \text{ if } a_i = \ell \text{ and } D_{i\ell}^A = 0 \text{ otherwise.} \tag{6.1}$$

As we can observe, parameters $(\mu_0, \alpha_2, \ldots, \alpha_{L_A})$ play the same role as do parameters $\boldsymbol{\beta} = (\beta_0, \beta_1, \ldots, \beta_p)$ in usual regression models. In all cases the number of dummies that we need to use in order to define a model will be equal to $L - 1$, where L are the number of levels of the variable under consideration.

Similarly, for the STZ parametrization the mean of an individual observation i with data (y_i, a_i) is given by

$$\begin{aligned} \mu_i &= \mu_0 - \alpha_2 - \alpha_3 \ldots - \alpha_{L_A} \quad \text{when} \quad a_i = 1 \\ \mu_i &= \mu_0 + \alpha_{a_i} \qquad\qquad\qquad\quad \text{when} \quad a_i > 1. \end{aligned}$$

The dummy variables here are slightly more complicated since

$$\begin{aligned} D_{i\ell}^{A,\text{stz}} &= 1 \quad \text{if} \quad a_i = \ell \\ D_{i\ell}^{A,\text{stz}} &= -1 \quad \text{if} \quad a_i = 1 \\ D_{i\ell}^{A,\text{stz}} &= 0 \quad \text{otherwise.} \end{aligned} \tag{6.2}$$

Here we have expressed the first category as a function of the remaing of the parameters. If we wish to omit from the parameters a different category, then the dummy variables must be modified accordingly. Note that the STZ dummy variables can be easily expressed as the difference of the CR dummy variables. Hence, assuming that the first category will serve as baseline, we can write

$$D_{i\ell}^{A,\text{stz}} = D_{i\ell}^{A} - D_{i1}^{A} . \tag{6.3}$$

The use of dummy variables simplifies the model specification since its structure is the same regardless of the type of parametrization or data we use. Moreover, interaction terms are directly defined by the products of the corresponding dummy variables. For example, the product $D_{ia}^{A} D_{ib}^{B}$ will provide the parameter $\alpha\beta_{ab}$ of the interaction between factors A and B for the corner constraints. Hence we can write

$$D_{iab}^{AB} = D_{ia}^{A} D_{ib}^{B}, \tag{6.4}$$

where D_{iab}^{AB} is the dummy variable for the interaction between a and b levels of factors A and B for i subject. This multiplicative property is also true for higher interaction terms and is convenient in terms of WinBUGS programming since the definition of the constraints (as in Section 5.4) is avoided. Only actually used parameters are defined in the model, while omitted parameters (as in the STZ parametrization) can be directly monitored using a simple deterministic/logical node in WinBUGS model code.

Dummy variables can also be specified within WinBUGS using the following commands. For a categorical variable A with L_A levels in which i subject belongs in a_i category, we can specify the dummy variables for CR parametrization using the following syntax

```
for (i in 1:n){
    D.A2[i] <- equals(a[i],2)
    ....
    D.ALA[i] <- equals(a[i],LA)
}
```

where LA is the number of levels of factor A, L_A. The syntax displayed above can be additionally simplified if we use a matrix of dimension $n \times L_A$ to specify the dummy variables for factor A. Then we can write

```
for (i in 1:n){
    for (l in 1:LA){
        D.A[i,l] <- equals(a[i],l)
}}
```

Using this syntax, we set $D.A_{i\ell} = 1$ if $a_i = \ell$ and zero otherwise following the definition of the dummy variables given in (6.1). Moreover, we have specified L_A dummy variables instead of the required $L_A - 1$. In the linear predictor (equation for μ_i), we must use only $L_A - 1$ of the columns of matrix D.A. The column excluded by the linear predictor corresponds to the baseline (or reference) category.

Similarly, following (6.2), the syntax

```
for (i in 1:n){
    for (l in 1:LA){
        DSTZ.A[i,k] <- equals(a[i],l) - equals( a[i], 1 )
    }}
```

produces the corresponding dummies for the STZ parametrization. The difference between STZ and CR dummy variables is in the first (baseline) category. In STZ parametrization, all dummies are set equal to minus one (-1) for the reference category (DSTZ.$A_{i\ell} = -1$ for all $\ell = 2, \ldots, L_A$) instead of zero in the CR parametrization. Here, the first column must be omitted from the specification of μ_i. Changing the reference category also affects values for all dummy variables. Alternatively, the STZ dummy variable can be defined by (6.3) and, thus, specify them in WinBUGS by

```
            DSTZ.A[i,1]  <- D.A[i,1] - D.A[i,1]
```

following the syntax used for the CR constraints. Alternatively, dummy variables can be defined in another statistical package and then imported in WinBUGS model code as ready-to-use data.

Matrix X of dimension $n \times p$ with columns the constant term, the dummy variables of the factors, and their corresponding interactions involved in the model is called a *design matrix*. When mixed (dummy and continuous) types of variables are included in the linear predictor, it is called a *data matrix* as in the regression model. We can directly use matrix X to specify our model as described in Section 5.3.4 for the regression model. Such a strategy will considerably simplify the model specification. The design matrix can be easily constructed either within WinBUGS model code (following similar approach as for the construction of dummy variables) or outside WinBUGS and then by importing the design matrix in the data of the WinBUGS code; for a detailed example, see Section 6.1.

Since matrix X can be defined regardless of the the type of variables we use, all prior distributions described in Section 5.3.2 can be used without difficulty.

Moreover, the algebra related to dummy variables is intriguing since, in cases of balanced data, all design matrices can be calculated using Krönecker products. More specifically, the STZ parametrization has interesting properties resulting in independent posterior distributions of parameters related to different model terms (main effects and interaction parameters); see Ntzoufras (1999b) for a brief discussion.

6.1   ANALYSIS OF VARIANCE MODELS USING DUMMY VARIABLES

Example 6.1. A three-way ANOVA model for schizotypal personality data.
In the following illustration we specify an ANOVA model for all three variables (economic status, gender, city) involved in the unbalanced dataset presented in Table 5.14 of Example 5.3. We fit our three-way model in WinBUGS using dummy variables.

Dummy variables for both parametrizations are given in Table 6.1. Code for the full three-way model is provided in Table 6.2.

Table 6.1   Dummy variables for Example 6.1

Gender	Economic Status	City	D_2^{gender}	D_2^{econ}	D_3^{econ}	D_2^{city}	D_2^{gender}	D_2^{econ}	D_3^{econ}	D_2^{city}
			Corner constraints				STZ constraints			
1	1	1	0	0	0	0	-1	-1	-1	-1
1	2	1	0	1	0	0	-1	1	0	-1
1	3	1	0	0	1	0	-1	0	1	-1
2	1	1	1	0	0	0	1	-1	-1	-1
2	2	1	1	1	0	0	1	1	0	-1
2	3	1	1	0	1	0	1	0	1	-1
1	1	2	0	0	0	1	-1	-1	-1	1
1	2	2	0	1	0	1	-1	1	0	1
2	1	2	1	0	0	1	1	-1	-1	1
2	2	2	1	1	0	1	1	1	0	1

To avoid definition of each parameter in separate scalar nodes, the design matrix X as well as a vector node for all parameters (beta) may be used instead. This approach requires a shorter code, which is presented in Table 6.3. The user must identify the location in which each parameter is stored in the vector node beta. Specific parameters can be additionally stored in a separate deterministic node vector in order to monitor only parameters of interest (e.g., the three-way interactions).

Note that in this model, we have 10 observations and 13 parameters. This model is clearly overparametrized given the available set of data. Although such a model cannot be fitted in the classical modeling approach, within the Bayesian context the posterior distribution can be estimated for all parameters. The difference is that for the nonidentifiable parameters, their posterior coincides to their corresponding prior distribution. This implies that no additional information can be extracted from the data for estimation of these parameters, and therefore they can be assumed as redundant.

In the following, we start by fitting the full three-way interaction model and then remove the terms that are not important since their posterior distributions are scattered around zero. Starting from the three-way model, we observe that the two parameters of the three-way interaction term lie around zero with large variance, indicating that parameters of this term can be eliminated from the model.

Table 6.4 provides posterior summaries for all parameters using the matrix–vector code. In the last column we provide the corresponding parameter name if we use the first (simpler) programming approach. In this table we observe two major points: (1) all interaction terms are centered around zero, and hence we may remove them; and (2) parameters beta[12] and beta[10] (terms $\text{gender.econ.city}_{232}$ and econ.city_{32}, respectively) have posterior summaries that match with the corresponding summaries of the $N(0, 10^4)$ prior distribution. This is a clear indication that the data do not contribute any information to the posterior distributions of these two parameters. This is sensible since no observations are available for males and females living in a village and having high economic status; see Table 5.14.

In the second step, we remove the three-way interaction term (by eliminating X[i,11] and X[i,12] and beta[11:12] from the model code) and rerun the algorithm. Results are similar as those above, and therefore we proceed by further removing the interaction term econ.city. We proceed by successively fitting models: gender*econ+gender*city,

Table 6.2 WinBUGS code for the full three-way model for Example 6.1 using unbalanced data of Table 5.14

```
model{
    for (i in 1:n){
        # CR dummy variables
        D.gender[i]<-equals(g[i],2)
        D.econ2[i]<-equals(e[i],2)
        D.econ3[i]<-equals(e[i],3)
        D.city[i]<-equals(ci[i],2)
        # STZ dummy variables
        #Dstz.gender[i]<-equals(g[i],2)-equals(g[i],1)
        #Dstz.econ2[i]<-equals(e[i],2)-equals(e[i],1)
        #Dstz.econ3[i]<-equals(e[i],3)-equals(e[i],1)
        #Dstz.city[i]<-equals(ci[i],2)-equals(ci[i],1)
    }
    # model's likelihood
    for (i in 1:n){
            y[i] ~ dnorm( mu[i], tau )
            mu[i] <- mu0 + gender * D.gender[i] + econ[2]*D.econ2[i]
                         + econ[3]*D.econ3[i] + city*D.city[i]
                         + gender.econ[2]*D.gender[i]*D.econ2[i]
                         + gender.econ[3]*D.gender[i]*D.econ3[i]
                         + gender.city*D.gender[i]*D.city[i]
                         + econ.city[2]*D.econ2[i]*D.city[i]
                         + econ.city[3]*D.econ3[i]*D.city[i]
                    +gender.econ.city[2]*D.gender[i]*D.econ2[i]*D.city[i]
                    +gender.econ.city[3]*D.gender[i]*D.econ3[i]*D.city[i]
    }
    #### set zero all nuisance parameters
    econ[1] <- 0.0
    econ.city[1]<-0.0
    gender.econ[1]<-0.0
    gender.econ.city[1]<-0.0
    # priors
    mu0~dnorm( 0.0, 1.0E-04)
    gender  ~ dnorm( 0.0, 1.0E-04)
    city    ~ dnorm( 0.0, 1.0E-04)
    gender.city ~ dnorm( 0.0, 1.0E-04)
    for (k in 2:3){
        econ[k] ~ dnorm( 0.0, 1.0E-04)
        gender.econ[k] ~ dnorm( 0.0, 1.0E-04)
        econ.city[k] ~ dnorm( 0.0, 1.0E-04)
        gender.econ.city[k] ~ dnorm( 0.0, 1.0E-04)
    }
    tau ~dgamma( 0.01, 0.01)
    s <- sqrt(1/tau) # precision
}
INITS
list( mu0=1.0, econ=c(NA, 0,0), gender=0, city=0, gender.city=0,
      gender.econ=c(NA,0,0), econ.city=c(NA,0,0),
      gender.econ.city=c(NA,0,0), tau=1.0)

DATA (LIST)
list(n=10)
 g[]    e[]    ci[]    y[]
  1      1      1       9
  1      2      1      25
  1      3      1      23
  2      1      1      18
  2      2      1      22
  2      3      1      12
  1      1      2      14
  1      2      2      26
  2      1      2      29
  2      2      2      25
END
```

Table 6.3   WinBUGS code for the full three-way model for Example 6.1 using design matrix X and parameter vector beta

```
model{
    #
    # Creating the design matrix X
    for (i in 1:n){
        X[i,1]<- 1.0 # constant term
        # CR dummy variables
        X[i,2]<-equals(g[i],2) # gender
        X[i,3]<-equals(e[i],2) # econ2
        X[i,4]<-equals(e[i],3)  # econ3
        X[i,5]<-equals(ci[i],2) # city
        # STZ dummy variables
        # X[i,2]<-equals(g[i],2)-equals(g[i],1)
        # X[i,3]<-equals(e[i],2)-equals(e[i],1)
        # X[i,4]<-equals(e[i],3)-equals(e[i],1)
        # X[i,5]<-equals(ci[i],2)-equals(ci[i],1)
        #
        # specification of interaction terms
        X[i,6]<-X[i,2]*X[i,3]    # gender*econ2
        X[i,7]<-X[i,2]*X[i,4]    # gender*econ3
        X[i,8]<-X[i,2]*X[i,5]    # gender*city
        X[i,9]<-X[i,3]*X[i,5]    # econ.city2
        X[i,10]<-X[i,4]*X[i,3]   # econ.city3
        X[i,11]<-X[i,2]*X[i,3]*X[i,5] # gender.econ.city2
        X[i,12]<-X[i,2]*X[i,4]*X[i,5] # gender.econ.city3
    }
    #
    # model's likelihood
    for (i in 1:n){
            y[i] ~ dnorm( mu[i], tau )
            mu[i] <- inprod( X[i,], beta[] )
    }
    # priors
    for (j in 1:12){ beta[j]~dnorm( 0.0, 1.0E-04) }
    tau ~dgamma( 0.01, 0.01)
    s <- sqrt(1/tau) # precision
}

INITS
list( beta=c(0,0,0,0,0,0,0,0,0,0,0,0), tau=1.0)
```

Table 6.4 Posterior summaries for parameters of full three-way interaction model of Example 6.1

```
node       mean     sd    MC error 2.5%   median 97.5%  term
beta[1]    9.204   7.776  0.08657 -6.435  9.003  26.28  mu0
beta[2]    8.735   10.29  0.1017  -14.26  9.001  29.78  gender(female)
beta[3]    15.59   10.79  0.1359  -8.699  16.0   36.85  econ[2] - med
beta[4]    13.48   11.35  0.1528  -12.59  13.98  35.85  econ[3] - high
beta[5]    4.688   10.74  0.1097  -18.52  4.98   25.98  city
beta[6]   -11.47   13.98  0.1669  -39.8  -11.99  21.0   gender.econ[2]
beta[7]   -19.36   15.4   0.229   -48.3  -20.0   15.45  gender.econ[3]
beta[8]    6.32    14.25  0.145   -21.74  6.014  37.41  gender.city
beta[9]   -3.453   14.85  0.1623  -33.49 -3.991  30.36  econ.city[2]
beta[10]   0.1041  99.89  1.039   -196.9  0.445  193.3  econ.city[3]
beta[11]  -4.548   19.01  0.2206  -48.27 -4.024  34.26  gender.econ.city[2]
beta[12]  -0.3412  99.84  1.041   -196.9 -1.349  194.2  gender.econ.city[2]
s          4.995   8.582  0.5114   0.102  1.607  30.57  st.dev.
```

gender*econ+city, and gender*econ. The asterisk indicates that all lower interaction terms (and effects) are also included in the model. For example, gender*econ means that the main effects gender and econ are also included in the model. Boxplots of parameters of the fitted models are provided in Figure 6.1. From this boxplot we observe that both models gender*econ+city and gender*econ seem to be satisfactory in the sense that the zero value lies at the tail areas for all parameters. Here we adopt model gender*econ+city since only for one parameter the zero value lies within the 95% posterior interval, and this is the lower boundary of this interval. Note that model gender*econ is the same as in the two-way analysis, but here we have considered the unbalanced data (10 observations instead of 12). Formal model comparisons follow in Chapters 10 and 11.

In this strategy we start from the full model, including all higher order interaction terms, and proceed by removing higher-order interaction terms first. We do not remove lower-order interaction terms if these are nested within higher-order ones (hierarchically structured models). This approach leads to models that are easier to interpret and is provided only for convenience. The user may remove main effects or lower order interaction terms but must be very careful with interpretation. This approach it is not recommended unless the problem setup forces us to follow such strategy.

Posterior summaries of the suggested model are provided in Table 6.5. Interpretation of the parameters is similar to that in the two-way analysis. Additionally, in this model, city resident students are a posteriori expected to score five SPQ points ($\beta_5 = 5$) higher than students of the same gender and economic status living in rural areas.

6.2 ANALYSIS OF COVARIANCE MODELS

Using both qualitative and quantitative variables as explanatory variables results in *analysis of covariance* (ANCOVA) models. ANCOVA models initially were used as an extension of ANOVA models in order to test for differences in the means of the response variable after adjusting for the effect of one or more numerical variables. In this way, variability predicted by numerical covariates can be removed before comparing between-groups differences. In practice such models do not demonstrate any difficulties in terms of estimation

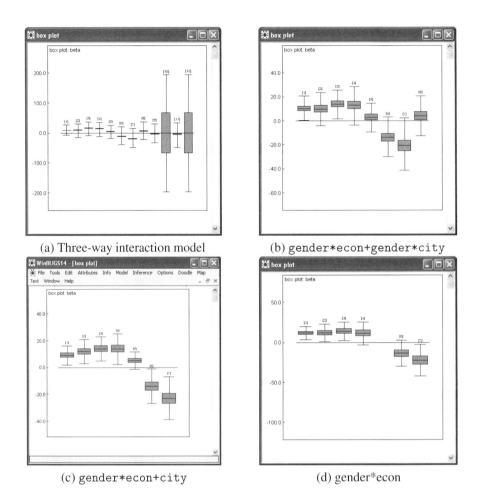

(a) Three-way interaction model

(b) gender*econ+gender*city

(c) gender*econ+city

(d) gender*econ

Figure 6.1 Boxplots of parameters for interaction models for Example 6.1; asterisk indicates that all lower interaction terms are also included in the model.

Table 6.5 Posterior summaries for parameters of model gender*econ+city of Example 6.1 with 2000 MCMC Iterations

node	mean	sd	MC error	2.5%	median	97.5%	start	term
beta[1]	9.014	3.524	0.07379	2.146	9.075	16.27	1001	(mu0)
beta[2]	11.91	4.604	0.09902	2.496	11.96	20.76	1001	(gender)
beta[3]	13.97	4.375	0.08849	4.647	13.99	22.74	1001	(econ2)
beta[4]	14.06	5.673	0.127	2.833	14.1	25.13	1001	(econ3)
beta[5]	5.079	3.297	0.06297	-1.631	5.001	11.91	1001	(city)
beta[6]	-13.92	6.065	0.101	-26.04	-14.0	-0.4573	1001	(gender.econ2)
beta[7]	-23.0	7.878	0.1796	-37.84	-23.25	-8.817	1001	(gender.econ3)
s	3.979	2.298	0.1045	1.741	3.377	10.03	1001	(st.dev.)

or model building. The interesting part is the interpretation of models and their structures when we incorporate interaction terms between categorical and numerical variables. Such interactions control the slopes of the regression lines for different groups. The two most important models are the *parallel lines* and *different slopes* models.

We will briefly explain the results obtained with ANCOVA models by describing in more detail the simpler case with one quantitative variable and one qualitative variable. Additional explanatory variables can be incorporated in the model's linear predictor similarly.

6.2.1   Models using one quantitative variable and one qualitative variable

Let us assume that we have available data (X_i, A_i, Y_i) with X_i, where A_i represents the numerical and categorical explanatory variables, respectively, and Y_i is the continuous response variable for subject i. Then the following models can be fitted:

1. Constant model: $\mu_i = \beta_0$

2. Common line for all groups: $\mu_i = \beta_0 + \beta_1 X_i$

3. Constant mean within each group: $\mu_i = \beta_0 + \alpha_{a_i}$

4. Parallel line (or common slope) model: $\mu_i = \beta_0 + \beta_1 X_i + \alpha_{a_i}$

5. Common intercept model: $\mu_i = \beta_0 + \beta_1 X_i + \delta_{a_i} X_i$

6. Separate regression lines for each group: $\mu_i = \beta_0 + \beta_1 X_i + \alpha_{a_i} + \delta_{a_i} X_i$

Models 1–3 are already known from simple regression and ANOVA models. More specifically, model 1 is the simple constant model that "estimates" the grand mean of the sample. Model 2 is a simple regression model since it ignores the grouping variable A and fits a single regression line to the data. Finally, model 3 is a usual ANOVA model that estimates the means for each category of A by ignoring the numerical variable X. Here we focus on models 4–6, which are the actual ANCOVA models. Model 4 assumes that the effect of each explanatory variable is additive. Moreover, each effect is not influenced by the effect of the other variable. On the contrary, model 6 assumes that the effect of the numerical variable is different for each group of the categorical variable. When considering ANCOVA models, we may also construct a common intercept model by making specific assumptions about parameters α_ℓ resulting from model 5. This aspect is discussed in further detail in the paragraphs that follow. Graphical representations of these models are provided in Figure 6.2.

6.2.2   The parallel lines model

The parallel lines model can be summarized using the following equations

$$
\begin{aligned}
Y_i &\sim N(\mu_i, \sigma^2) \\
\mu_i &= \beta_0 + \beta_1 X_i + \alpha_{a_i} \quad \text{for } i = 1, \dots, n \, .
\end{aligned}
\tag{6.5}
$$

The mean of the response variable in this model can be effectively summarized by

$$
\mu_i = \beta_{0,a_i}^* + \beta_1 X_i,
\tag{6.6}
$$

where $\beta_{0,\ell}^* = \beta_0 + \alpha_\ell$. It is clear that the regression lines for each group of factor A will have different intercepts and share a common slope (i.e., will be parallel). Such models are also termed *common slope models*.

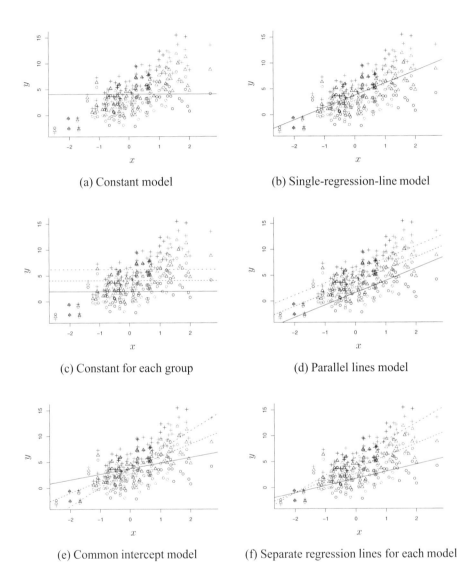

(a) Constant model

(b) Single-regression-line model

(c) Constant for each group

(d) Parallel lines model

(e) Common intercept model

(f) Separate regression lines for each model

Figure 6.2   Illustration of the possible models with one quantitative explanatory variable and one qualitative explanatory variable.

As in ANOVA models, we need to impose one constraint on the parameters α_ℓ ($\ell = 1, 2, \ldots, L_A$) of the categorical variable. We can use either CR or STZ constraints depending on the parameter interpretation we prefer. Hence the model parameters involved in the linear predictor in this model will be

$$\boldsymbol{\beta} = (\beta_0, \beta_1, \alpha_2, \ldots, \alpha_{L_A}),$$

omitting the α_1 as the parameter corresponding to the baseline or the reference category.

Using the CR parametrization with the first level as the reference category, we simplify the regression line for the reference category to

$$\mu_i = \beta_0 + \beta_1 X_i,$$

providing a straightforward interpretation for β_0 that corresponds to the expected Y for the reference group when $X = 0$. Using similar arguments, we can interpret α_ℓ ($\ell \geq 2$) as the expected difference of Y between a subject of group ℓ and a subject in the reference category having the same X since

$$E(Y|X, A = \ell) - E(Y|X, A = 1) = \beta_0 + \beta_1 X + \alpha_\ell - \beta_0 - \beta_1 X = \alpha_\ell .$$

Similarly, β_1 is the expected difference in Y when comparing two subjects differing by one unit in X and belonging in the same group since

$$E(Y|X = x+1, A = \ell) - E(Y|X = x, A = \ell) = \beta_0 + \beta_1(x+1) + \alpha_\ell - \beta_0 - \beta_1 x - \alpha_\ell = \beta_1 .$$

Interpretation for STZ parametrization is similar to that for CR, but our comparisons can be based on the property

$$\frac{1}{L_A} \sum_{\ell=1}^{L_A} E(Y|X, A = \ell) = \beta_0 + \beta_1 X .$$

This equation indicates that in STZ parametrization we assume as a reference category the one that represents the average behavior across all factor levels. Following the relationships presented above, β_0 under STZ parametrization provides an estimate of the expected value of Y when $X = 0$ averaged across all levels of A, while α_ℓ provides the expected deviation of Y for a subject belonging in ℓ group with $X = x$ from the corresponding expected value of Y averaged across all levels of A and also having $X = x$. The STZ parametrization is equivalent to using covariates centered around zero (i.e., using $Z_i = X_i - \overline{X}$) where the constant parameter β_0^c represents the expected Y for a "typical" individual of the observed data; see Section 5.2.3 for more details.

The main property of this model is that the effect of X is the same in each group (equal to β_1) and the effect of the grouping variable is the same when subjects of the same X are compared. This is not the case in the *different slopes model*, in which the effect of each covariate on Y depends on the values of the other covariate. Graphical representation and interpretation of the parallel lines model is provided in Figure 6.3.

It is relatively straightforward to express the likelihood of the above model in WinBUGS. We can use either the direct approach, with the data in three vectors for $\boldsymbol{y}$, $\boldsymbol{x}$, and $\boldsymbol{a}$, or the design matrix of the model as described in the previous sections. In the first approach the likelihood in WinBUGS can be written as

```
for (i in 1:n){
    y[i] ~ dnorm( mu[i], tau )
    mu[i] <- beta0 + beta1 * x[i] + alpha[ a[i] ]
}
```

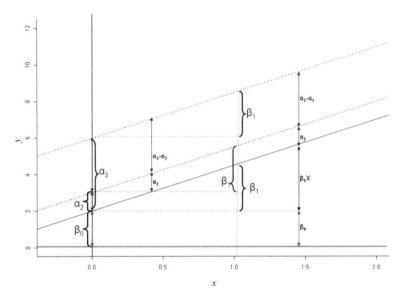

Figure 6.3 Graphical representation of parallel lines model using corner constrained parameters; different lines represent the fitted regression models for different levels of the categorical variable under consideration.

The constraints can be easily incorporated in the model using the same commands as in ANOVA models, namely

```
alpha[1] <- 0.0
```

for CR constraints and

```
alpha[1] <- -sum( alpha[2:LA] )
```

for STZ constraints. We can avoid imposing constraints by expressing the linear predictor as in (6.6) and hence write directly

```
mu[i] <-  beta.star[ a[i] ] + beta1 * x[i]
```

Finally, when using dummy variables, the linear predictor is written in the following form

$$
\begin{aligned}
\mu_i &= \beta_0 + \beta_1 X_i + \alpha_2 D_{i2} + \ldots + \alpha_{L_A} D_{iL_A} \\
&= \beta_0 + \beta_1 X_i + \sum_{\ell=2}^{L_A} \alpha_\ell D_{i\ell},
\end{aligned}
$$

which can be easily coded in WinBUGS using the following expression to define the linear predictor

```
mu[i]<- beta0 +beta1*x[i] +alpha[2]*D[i,2]+...+alpha[LA]*D[i,LA]
```

or, alternatively, using the `inprod` command

```
mu[i]<- beta0 + beta1*x[i] + inprod( alpha[2:LA], D[i,2:LA] )
```

In this syntax, we additionally need to specify `alpha[1]` by imposing the restriction corresponding to the desired parametrization; for example, we set $\alpha_1 = 0$ for the corner parametrization. The `inprod` syntax can be also generalized for all variables included in the model as in the usual regression model. In such a case all data must be stored in a data matrix X of dimension $n \times (L_A + 1)$.

6.2.3   The separate lines model

When assuming different regression lines for each level of the categorical variable, the mean of Y can be defined using the expression

$$\mu_i = \beta_0 + \beta_1 X_i + \alpha_{a_i} + \delta_{a_i} X_i \, . \tag{6.7}$$

Alternatively, the model can be rewritten as

$$\mu_i = \beta_{0,a_i}^* + \beta_{1,a_i}^* X_i, \tag{6.8}$$

where $\beta_{0,\ell}^* = \beta_0 + \alpha_\ell$ and $\beta_{1,\ell}^* = \beta_1 + \delta_\ell$ for $\ell = 1, 2, \ldots, L_A$. Thus, this model is equivalent to fitting one regression model for each group.

Adopting a single ANCOVA model for the whole dataset instead of fitting different regression lines for each group allows us to monitor and/or check for deviations from specific assumptions such as, for example, the common slopes assumption. Parameters δ_ℓ denote the interaction parameters between variable X and factor A. Their magnitude reflects deviations from the parallel lines model and encapsulates changes in the effect of the quantitative covariate X on Y across the groups of factor A. STZ or CR parametrization may also be imposed to obtain an identifiable model.

The parameters involved in the linear predictor are given by

$$\boldsymbol{\beta} = (\beta_0, \beta_1, \alpha_2, \ldots, \alpha_{L_A}, \delta_2, \ldots, \delta_{L_A}) \, .$$

Under the CR parametrization, constraints $\alpha_1 = \delta_1 = 0$ are imposed that result in

$$\mu_i = \beta_0 + \beta_1 X_i$$

for the first reference category. Hence parameters β_0 and β_1 will have the usual regression interpretation referring to subjects of the reference category.

The main effects α_ℓ denote the difference in the expected values of Y between subjects belonging in the ℓth and the reference category when $X = 0$ since

$$E(Y|X = 0, A = \ell) - E(Y|X = 0, A = 1) = \beta_0 + \alpha_\ell - \beta_0 = \alpha_\ell \, .$$

Finally, $\beta_1 + \delta_\ell$ is the expected difference of Y when comparing two subjects differing by one unit in X and belonging in the ℓth group since

$$
\begin{aligned}
E(Y|X = x, A = \ell) - E(Y|X = 0, A = \ell) \;=\; & \beta_0 + \beta_1(x+1) + \alpha_\ell + \delta_\ell(x+1) \\
& - [\beta_0 + \beta_1 x + \alpha_\ell + \delta_\ell x] \\
=\; & \beta_0 + \delta_\ell \, .
\end{aligned}
$$

Therefore δ_ℓ provides the difference between the effects of one unit increase of X on Y in the ℓth group and the reference category.

Similarly, we can interpret model parameters in the STZ parametrization by comparing all parameters to the "mean" regression line (averaged over all groups) given by

$$\frac{1}{L_A} \sum_{\ell=1}^{L_A} E(Y|X, A = \ell) = \beta_0 + \beta_1 X \ .$$

Parameters α_ℓ and δ_ℓ will now represent differences for the intercept (i.e., for $X = 0$) and for the slope of each group from this "mean" regression line.

Graphical representation and interpretation of the separate lines model is depicted in Figure 6.4.

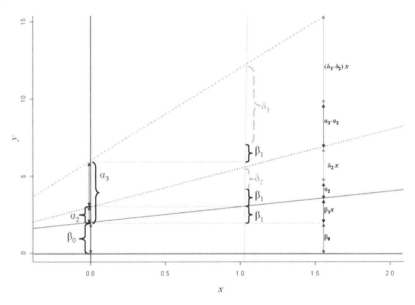

Figure 6.4   Graphical representation of separate regression lines model using corner constrained parameters; different lines represent the fitted regression models for different levels of the categorical variable under consideration.

The general ANCOVA model can be fitted in WinBUGS using the following syntax

```
mu[i] <- beta0 + beta1*x[i] + alpha[ a[i] ] + delta[ a[i] ]*x[i]
```

for the linear predictor expression (6.7). The intercept and the slope of each regression line can be calculated using the syntax

```
for (l in 1:LA){
    beta0.star[l] <- beta0 + alpha[l]
    beta1.star[l] <- beta0 + delta[l]
    }
```

Constraints for α_1 and δ_1 must be imposed as usual. Alternatively, we may directly use equation (6.8) by writing

```
mu[i] <- beta0.star[ a[i] ] + beta1.star[ a[i] ] * x[i]
```

The latter has the advantage that no constraints are needed since the model is directly identifiable. The original parameters of (6.7) can be extracted by adding the following commands in our model code

```
beta0 <- beta0.star[1]
beta1 <- beta1.star[1]
for (l in 2:LA){
    alpha[l] <- beta0.star[l] - beta0
    delta[l] <- beta1.star[l] - beta1
    }
```

for CR parametrization. For STZ parametrization, we need to define β_0 and β_1 as the mean of all intercepts and slopes, respectively

```
beta0 <- mean(beta0.star[1:LA])
beta1 <- mean(beta1.star[1:LA])
```

followed by the loop used for the specification of the CR parametrization.

Using dummies is also straightforward since the multiplicative property (6.4) also holds for the interaction term between a quantitative variable and a qualitative variable. Hence each interaction parameter δ_ℓ will be the coefficient of a variable defined as

$$D_{i\ell}^{AX} = D_{i\ell}^{A}X_i \ .$$

The model can now be written

$$\mu_i = \beta_0 + \beta_1 X_i + \sum_{\ell=2}^{L_A} \alpha_\ell D_{i\ell}^A + \sum_{\ell=2}^{L_A} \delta_\ell D_{i\ell}^A X_i,$$

which can be coded in WinBUGS using the syntax

```
mu[i]<-beta0 +beta1*x[i] +alpha[2]*DA[i,2] +...+ alpha[LA]*DA[i,LA]
            +delta[2]*DA[i,2]*x[i] +...+ delta[LA]*DA[i,LA]*x[i]
```

The `inprod` command can be used to simplify the linear predictor

```
mu[i]<- beta0 + beta1*x[i] + inprod(alpha[2:LA]*DA[i,2:LA])
                           + inprod(delta[2:LA]*DAX[i,2:LA])
```

where `DAX` is a matrix with each column representing the interaction term between factor A and variable X and can be calculated using the following loop

```
for (l in 1:LA){ DAX[i,l] <- DA[i,l] * x[i] }
```

nested within the likelihood loop (i.e., embedded within a loop with i taking values from 1 to n). Use of the design matrix, in combination with the `inprod` command, further simplifies the model code as in usual regression models.

Finally, removing the main effects α_ℓ will result in the common intercept model, which assumes equal expected values of Y for all groups when $X = 0$. If this model is combined with centering X round zero, then it assumes that people with average X are expected to have equal Y in all groups. Common intercept models are rarely used in practice — they are usually adopted only if a scientific scenario supports them; see Section 6.3.2 for an example.

6.3 A BIOASSAY EXAMPLE

Bioassays are experiments performed in the preclinical stage of a drug experiment. They ensure drug safety and determine appropriate drug dosage. The main objective is to estimate the unknown concentration of a substance of interest in a new drug in comparison to the standard preparation (drug) of known concentration. This unknown concentration is called the *drug potency* and is usually estimated by the ratio of of the mean concentration of a standard treatment over the mean of the (new) test preparation:

$$\text{Relative potency} = \frac{\text{mean concentration of standard treatment}}{\text{mean concentration of test treatment}}.$$

The mean corresponds to a response variable that may measure volume, weight, or dosage of each treatment required to achieve the same results. If the relative potency > 1, then a lower dose of the new drug produces the same results as does a higher dosage of the standard drug. Therefore, the new drug is more potent than the standard treatment.

The response is usually quantitative or binary. In the following example we deal with an indirect bioassay in which we replicate an experiment in prespecified levels of dosage for the two treatments. In such assays, one or more drugs at different dosage levels are administered to experimental units such as cell cultures, tissues, organs, or living animals. Here we consider the case where the response is quantitative, hence ANCOVA is used to analyze the data and estimate the relative potency. The aim is, using statistical models, to estimate the relative potency of the new preparation. Two different approaches are popular: the parallel line analysis and the slope ratio analysis. Details concerning the parallel and slope ratio analysis can be found in Chen (2007) and references cited therein. Bayesian approaches to the parallel line and slope ratio analysis have been investigated by Darby (1980) and Mendoza (1990), respectively.

> **Example 6.2. Factor 8 example.** Factor 8 (F8) is a blood clotting agent. Deficiency of F8 in the human body leads to hemophilia. The (fictitious) data of an experiment are given in Table 6.6, in which we wish to compare a new/test preparation with the standard preparation of F8 having potency equal to 1.2 international units (IU). As a response Y we use the time until clotting occurs within seconds. Three dilutions (doses) were used (1:40, 1:20, 1:10) with four measurements per dose (quantitative variable X) and two test drugs (standard and test) are compared (using the categorical explanatory variable in our ANCOVA model).

Table 6.6   Data for bioassay example (Example 6.2)

Dose		Blood clotting time (in seconds)	
		Standard	Test (new)
1:40	0.025	68.8, 67.6, 68.1, 67.6	69.0, 67.9, 68.6, 68.3
1:20	0.050	61.4, 59.8, 62.3, 60.6	60.9, 60.3, 61.6, 61.8
1:10	0.100	53.5, 51.9, 53.6, 52.2	53.8, 54.9, 54.1, 54.2

6.3.1 Parallel lines analysis

Model formulation.   For a fixed dose d_s of a standard drug, we assume that the same effect on Y is achieved with dose $d_t = \rho d_s$ of the test preparation, where ρ is the relative potency of the two drugs. We adopt the model

$$E(Y) = \mu = \begin{cases} \beta_0 + \beta_1 \log(\text{dose}) & \text{for the standard treatment} \\ \beta_0 + \beta_1 \log(\rho \times \text{dose}) & \text{for the test treatment} \end{cases}$$

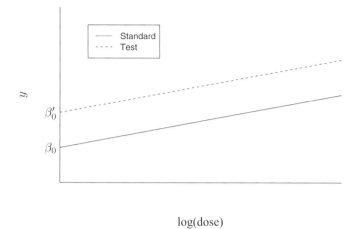

Figure 6.5 Graphical representation of parallel lines analysis model.

which is graphically represented in Figure 6.5.

The model for the test treatment can be rewritten as

$$
\begin{aligned}
\mu_T &= && \beta_0 + \beta_1 \log(\rho \text{ dose}) \\
&= && \beta_0 + \beta_1 \log(\rho) + \beta_1 \log(\text{dose}) \\
&= && \beta_0' + \beta_1 \log(\text{dose}),
\end{aligned}
$$

where $\beta_0' = \beta_0 + \beta_1 \log(\rho)$. The two models have the same slope but different intercepts. The common slope is given by β_1, while the intercept of the standard treatment is denoted by β_0 and that for the test treatment, by β_0'. Thus, the relative potency can be calculated as follows:

$$
\beta_0' = \beta_0 + \beta_1 \log(\rho) \Leftrightarrow \rho = \exp\left(\frac{\beta_0' - \beta_0}{\beta_1}\right). \tag{6.9}
$$

Using MCMC, it is straightforward to calculate the posterior distribution of ρ either directly by including it in the model or by simply calculating it from the values of a parallel lines ANCOVA model.

In Table 6.7 we provide the WinBUGS code for fitting the parallel lines model. In this model code, different intercepts have been fitted directly to avoid contraints. Alternatively, the code using the design matrix approach for the CR parametrization is given in Table 6.8. Commands for fitting the model using STZ parametrization are also provided as code comments in the same table. A normal prior distribution with mean zero and variance equal to 1000 was used for the parameters of the linear predictor and gamma with mean 1 and variance 1000 for the model's precision.

Table 6.7   WinBUGS code for parallel lines model in Example 6.2

```
model{
    # model's likelihood
    for (i in 1:n){
        y[i] ~ dnorm( mu[i], tau )
        mu[i] <- beta0[drug[i]] + beta1*log(dose[i])
    }
    #
    # relative potency
    rho <- exp( (beta0[2]-beta0[1])/beta1 )
    # potency estimate
    potency <- rho * 1.2
    #
    # prior distributions
    beta0[1] ~ dnorm( 0.0, 0.001)  # constant for standard treatment
    beta0[2] ~ dnorm( 0.0, 0.001)  # constant for test treatment
    beta1    ~ dnorm( 0.0, 0.001)  # slope
    tau      ~ dgamma( 0.001, 0.001) # precision of regression model
    s <- 1/sqrt(tau) # standard error of regression
    #
    # test rho>1 (test more potent)
    more.potent21 <- step(rho-1)
    #
    intercept.difference <- beta0[2]-beta0[1]
}

INITS
list( beta0=c(0,0), beta1=0, tau=1 )

DATA (LIST)
list( n=24,
y=c(68.8, 67.6, 68.1, 67.6,  69.0, 67.9, 68.6, 68.3,
    61.4, 59.8, 62.3, 60.6,  60.9, 60.3, 61.6, 61.8,
    53.5, 51.9, 53.6, 52.2,  53.8, 54.9, 54.1, 54.2),
dose=c(0.025, 0.025, 0.025, 0.025, 0.025, 0.025, 0.025, 0.025,
       0.050, 0.050, 0.050, 0.050, 0.050, 0.050, 0.050, 0.050,
       0.100, 0.100, 0.100, 0.100, 0.100, 0.100, 0.100, 0.100),
drug=c(1,1,1,1,2,2,2,2, 1,1,1,1,2,2,2,2, 1,1,1,1,2,2,2,2)
)
```

Table 6.8   WinBUGS code for parallel lines model in Example 6.2 using design matrix approach[a]

```
model{
    for (i in 1:n){
        #
        # creating the design matrix
        X[i,1]<-1.0              # beta1=constant term
        X[i,2]<-log(dose[i]) # beta2=log dose
        X[i,3]<- equals( drug[i], 2 ) # beta3=CR dummy for test
            treatment
        # beta3=STZ dummy
        #X[i,3]<- equals( drug[i], 2 )-equals( drug[i], 1 )
        # model likelihood
        y[i] ~ dnorm( mu[i], tau )
        mu[i] <- inprod( beta[], X[i,])
    }
    #
    rho <- exp( beta[3]/beta[2] ) # relative potency in CR
    #rho <- exp( 2*beta[3]/beta[2] ) # relative potency in stz
    # potency estimate
    potency <- rho * 1.2
    #
    # prior distributions
    for (j in 1:3){ beta[j]~dnorm( 0.0, 0.001) }
    tau       ~ dgamma( 0.001, 0.001) # precision of regression model
    s <- 1/sqrt(tau) # standard error of regression
}

INITS
list( beta=c(0,0,0), tau=1 )
```

[a] Data are as specified in Table 6.7; substitute lines 7 and 14 by lines 8 and 15, respectively, to switch from CR to STZ parametrization.

Results.   Posterior summaries of the parallel lines model have been calculated (see Table 6.9) after running the MCMC algorithm for 2000 iterations and discarding additional 1000 iterations as a burnin period. The following point estimate of the model can be provided using the posterior means:

$$y_i \sim N(\mu_i, 0.77^2), \quad \mu_i = \begin{cases} 28.79 - 10.62 \log(\text{dose}) & \text{for the standard treatment} \\ 29.46 - 10.62 \log(\text{dose}) & \text{for the test treatment} \end{cases}.$$

Table 6.9   Posterior summaries for parameters of parallel lines model in Example 6.2

node	mean	sd	MC error	2.5%	median	97.5%	start	sample
beta0[1]	28.79	0.865	0.01804	27.09	28.8	30.45	1001	2000
beta0[2]	29.46	0.8701	0.01875	27.75	29.46	31.2	1001	2000
beta1	-10.62	0.2808	0.006079	-11.18	-10.62	-10.05	1001	2000
potency	1.127	0.03341	7.865E-4	1.062	1.127	1.195	1001	2000
rho	0.939	0.02785	6.554E-4	0.8846	0.9391	0.9961	1001	2000
s	0.7731	0.1318	0.003941	0.5753	0.7532	1.085	1001	2000
tau	1.81	0.5676	0.01645	0.8504	1.763	3.027	1001	2000

Looking at the estimates of the model, we conclude the following:

- According to the intercepts β_0 and β_0', the clotting time when the dose is equal to 1 (log-dose=0) is a posteriori expected to be equal to 28.8 and 29.5 seconds, respectively. This interpretation does not have any practical meaning since this value is far away from the range of dose values used in this experiment. To get reasonable estimates, we may transform the log-dose such that the zero value corresponds to a realistic dosage level (e.g., the lower level of dosage); see Table 6.10 for new rescaled results.

Table 6.10   Posterior summaries for parameters of parallel lines model after rescaling the dose $[\log_2(40\text{dose})]$ in Example 6.2

node	mean	sd	MC error	2.5%	median	97.5%	start	sample
beta0[1]	67.96	0.3024	0.006875	67.37	67.96	68.55	1001	2000
beta0[2]	68.63	0.2973	0.006413	68.05	68.63	69.23	1001	2000
beta1	-7.353	0.1971	0.004279	-7.736	-7.354	-6.973	1001	2000
potency	1.127	0.03467	8.547E-4	1.058	1.126	1.198	1001	2000
rho	0.939	0.02889	7.123E-4	0.8816	0.9386	0.9981	1001	2000
s	0.7731	0.1318	0.003943	0.5754	0.7532	1.086	1001	2000
inter.dif	0.654	0.307	0.007264	0.0543	0.6648	1.239	1001	2000

- We observe a negative association between the drug dosage and the clotting time. Interpretation is not exactly the same as in usual regression models in terms of dosage increase, since the log-dose is used as an explanatory variable. One unit increase of the log-dose corresponds to multiplying the original dosage by e (≈ 2.7) since

$$E(Y|X = \log(\text{dose}) + 1, A = k) - E(Y|X = \log(\text{dose}), A = k) = \beta_1$$

and

$$\begin{aligned} \log(\text{new dose}) &= \log(\text{dose}) + 1 = \log(\text{dose}) + \log e = \log(e \times \text{dose}) \Rightarrow \\ \text{new dose} &= e \times \text{dose}. \end{aligned}$$

Hence we can now interpret β_1 as the expected change in clotting time when the dosage increases the original one by 1.7 times (170%). In our data, when the dosage increases by 170%, the clotting time is a posteriori expected to decrease by 10.6 seconds.

To obtain more interpretable parameters, we propose using as explanatory variable the $\log_2(40 \times \text{dose})$ instead of the $\log(\text{dose})$. In this way the rescaled explanatory variable will assume values equal to 0,1,2. Now the intercept is directly linked with the expected clotting time of the lower dosage used in the experiment, while the new slope coefficient is associated with the expected decrease of clotting time when the dosage doubles. Results of this rescaled model are provided in Table 6.10.

In order to calculate the proposed transformation in WinBUGS we need to express the new variable in terms of the natural logarithm $\log$ (or $\ln$). We write

$$x = \log_2(40 \times \text{dose}) = \frac{\log(40 \times \text{dose})}{\log(2)}$$

using the command

```
x[i] <- log( 40 *dose[i] )/log(2)
```

and then use x[i] in the linear predictor instead of log(dose[i]). A slight change is also needed for the calculation of the relative potency since the new model is now given by

$$
\begin{aligned}
\mu_i &= \beta^*_{0a_i} + \beta^*_1 \frac{\log(40 \, \text{dose}_i)}{\log(2)} \\
&= \beta^*_{0a_i} + \beta^*_1 \frac{\log(40)}{\log(2)} + \frac{\beta^*_1}{\log(2)} \log(\text{dose}_i).
\end{aligned}
$$

The original parameters β_{0k} and β_1 are associated with the parameters of the new rescaled model by the equations

$$\beta_{0k} = \beta^*_{0k} + \beta^*_1 \frac{\log(40)}{\log(2)} \quad \text{and} \quad \beta_1 = \frac{\beta^*_1}{\log(2)},$$

resulting to relative potency

$$\rho = \exp\left(\frac{\beta_{02} - \beta_{01}}{\beta_1}\right) = \exp\left(\log(2)\frac{\beta^*_{02} - \beta^*_{01}}{\beta^*_1}\right) = 2^{(\beta^*_{02} - \beta^*_{01})/\beta^*_1}.$$

Interpretation of the model parameters is more straight forward in this model. The clotting time for the lower dosage is a posteriori expected to be equal to 68 and 68.6 seconds for the standard and the test treatment, respectively. The difference between the two drugs is about 0.65 second with about 98% of the posterior values to be positive.

When we double the dosage, we a posteriori expect a decrease of the clotting time by 7.4 seconds.

From the posterior densities for the two intercepts (see Figure 6.6), we observe that they lie in the same range of values, and thus we deduce equal potency between the two drugs. This comparison might be misleading since the generated values for the two parameters are highly correlated (see Figure 6.7). To get a more reliable picture, we consider the posterior density of their difference ($\beta'_0 - \beta_0$). Indeed, from the posterior density of the difference (see Figure 6.8), we observe the zero value (equal intercepts) lying at the right tail area of the

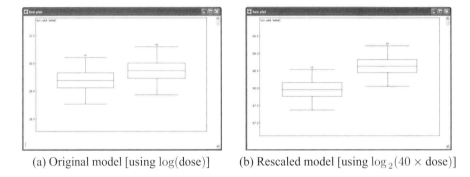

(a) Original model [using $\log(\text{dose})$]   (b) Rescaled model [using $\log_2(40 \times \text{dose})$]

Figure 6.6   Posterior boxplots for intercepts under original and rescaled parallel lines models in Example 6.2.

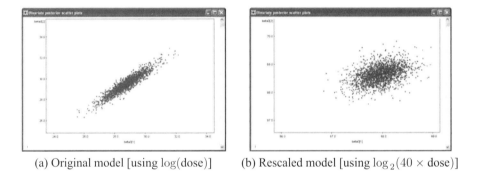

(a) Original model [using $\log(\text{dose})$]   (b) Rescaled model [using $\log_2(40 \times \text{dose})$]

Figure 6.7   Posterior scatterplots for intercepts under original and rescaled parallel lines models in Example 6.2.

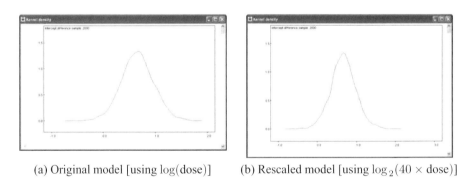

(a) Original model [using $\log(\text{dose})$]   (b) Rescaled model [using $\log_2(40 \times \text{dose})$]

Figure 6.8   Posterior densities for intercept difference under original and rescaled parallel lines models in Example 6.2.

distribution. This value corresponds to the 1.85% percentile of the posterior distribution. This indicates that the two treatments differ in terms of potency.

We can reach similar conclusions if we monitor directly the relative potency ρ since it is a simple function of the above mentioned difference. We observe that the test treatment is a posteriori expected to be about 6% less potent than the standard treatment, ranging from 0.88 to 0.998 with probability 95%. The actual potency of the test treatment was a posteriori expected to be equal to 1.13 IU, ranging from 1.058 to 1.198 IU with probability equal to 95%; see Figure 6.9.

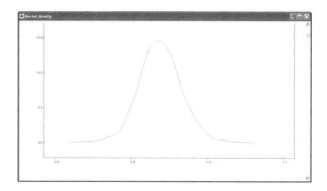

Figure 6.9   Posterior density of relative potency of new compared to standard treatments using the parallel lines model in Example 6.2.

Checking the parallel lines assumption.   We may check the parallelism assumption by simply fitting the separate lines model and comparing the posterior distribution of the differences between slopes. In order to fit this model in WinBUGS , we simply change the linear predictor to

```
mu[i] <- beta0[drug[i]] + beta1[drug[i]]*log(dose[i])
```

and define normal prior distributions for beta0[1], beta0[2], beta1[1], and beta1[2]. We also define the slope difference using the command

```
slope.difference<-beta1[2]-beta1[1]
```

in order to monitor its posterior distribution and

```
p2 <- 1-step(slope.difference)
```

which calculates the posterior probability that the slope difference is lower than zero value (parallel lines assumption). Very high or low values provide an indication that these slopes are different. Results using 2000 iterations (using the rescaled log-dose) are given in Table 6.11; see also Figure 6.10 for the posterior density plot. From these results we observe that the a posteriori expected slope difference is equal to 0.51, ranging from -0.22 to 1.28

Table 6.11   Posterior summaries for parameters of separate lines model in Example 6.2 using $\log(40 \times \text{dose})/\log(2)$ as explanatory variable

node	mean	sd	MC error	2.5%	median	97.5%	start	sample
p2	0.084	0.2774	0.005963	0.0	0.0	1.0	1001	2000
slope.difference	0.5096	0.3783	0.008266	-0.2171	0.4961	1.277	1001	2000

with probability 95%. We also estimate $P(\text{slope difference} < 0) = 0.084$, indicating that, although the zero value lies at the left tail area of the posterior distribution, it does not provide strong evidence against the parallel lines assumption.

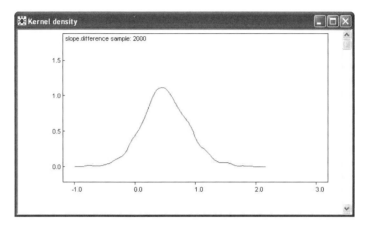

Figure 6.10   Posterior density of slope difference in separate lines model in Example 6.2.

6.3.2   Slope ratio analysis: Models with common intercept and different slope

Model formulation.   In this approach we assume again that for any fixed dose d_s of the standard preparation we have the same effect with the dose $d_t = \rho d_s$ of the test preparation, but now we adopt the model

$$E(Y) = \mu = \begin{cases} \beta_0 + \beta_1 \text{dose} & \text{for the standard treatment} \\ \beta_0 + \beta_1 \rho\, \text{dose} & \text{for the test treatment} \end{cases}$$

which is graphically represented in Figure 6.11.

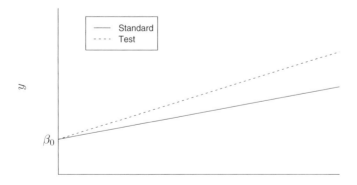

Dose

Figure 6.11   Graphical representation of slope ratio analysis model.

The model for the test treatment can be rewritten as

$$\mu_T = \beta_0 + \beta_1' \text{dose}$$

where $\beta_1' = \beta_1 \rho$. The two models now have the same intercept but different slopes. The common intercept is given by β_0 while the slope of the standard treatment is given by β and that of the test treatment, by β_1'. Therefore we can calculate the relative potency by

$$\rho = \frac{\beta_1'}{\beta_1} . \tag{6.10}$$

It is straightforward to calculate the relative potency when the common intercept model is fitted using interaction terms under CR and STZ parametrization. The model can be generally expressed using the linear predictor

$$\mu_i = \beta_0 + \beta_1 \text{dose}_i + \delta_{a_i} \text{dose}_i$$

with $\delta_1 = 0$ and $\delta_1 = -\delta_2$ for CR and STZ parametrizations, respectively. The slopes are equal to β_1 and $\beta_1 + \delta_2$ in CR, resulting in

$$\rho = \frac{\beta_1 + \delta_2}{\beta_1}$$

while for STZ the slopes are equal to $\beta_1 - \delta_2$ and $\beta_1 + \delta_2$ with relative potency given by

$$\rho = \frac{\beta_1 + \delta_2}{\beta_1 - \delta_2} .$$

Before we proceed, we rescale the dose in order to attain a model with parameters of simple interpretation. For this reason, in this analysis, we use the transformation

$$x = \frac{\text{dose}}{\min(\text{dose})} . \tag{6.11}$$

In this way, the two regression slopes provide the expected decrease of the clotting time when the dosage is increased by a quantity equal to the minimum dose used in the experiment. Note that in this analysis we must not change the zero point of the new transformed variable since such a transformation has a direct effect on the estimated model. For example, the transformation

$$x = \frac{\text{dose} - \min(\text{dose})}{\min(\text{dose})} .$$

results in $x = 0$ for dosage equal to the minimum dose of the experiment. With this transformation, the common intercept β_0 of the model corresponds to the expected clotting time for the lower dose. Therefore this model assumes equal effects of the lower dose for both treatments instead of the more realistic assumption of equal effect for the zero dosage of both treatments imposed by the original slope ratio analysis model.

Under transformation (6.11), the relative potency is still given by the ratio of the two slopes. The minimum value of a vector is calculated in WinBUGS using the command `ranked(dose[],1)`. Hence the command

```
x[i] <-  dose[i]/ranked(dose[],1)
```

calculates the new rescaled dose used in the linear predictor of our model.

WinBUGS code for the common intercept model is provided in Table 6.12 using the rescaled dosage (6.11). Different slopes have been used to avoid constraints. Alternatively, our model can be constructed using the design matrix approach as in the parallel lines model. The corresponding code for the CR parametrization is provided in Table 6.13. Commands for fitting the model using STZ parametrization are also provided in the same table as comments within the model code. Finally, syntax for specifying improper flat priors $f(\beta_j) \propto 1$ are also provided as model code comments. These improper priors were used to check the sensitivity of the posterior distribution. Minor differences were observed between the posteriors resulted by the improper and the $N(0, 10^3)$ prior distributions.

Table 6.12   WinBUGS code for common intercept/different slopes model in Example 6.2

```
model{
    for (i in 1:n){
        # calculate the rescaled dose
        x[i] <- dose[i]/ranked(dose[],1)
        # model's likelihood
        y[i] ~ dnorm( mu[i], tau )
        mu[i] <- beta0 + beta1[drug[i]]*x[i]
    }
    #
    # relative potency
    rho <- beta1[2]/beta1[1]
    # potency estimate
    potency <- rho * 1.2
    #
    # prior distributions
    # normal priors
    # beta0     ~ dnorm( 0.0, 0.001)  # constant for standard treatment
    # beta1[1] ~ dnorm( 0.0, 0.001)  # constant for test treatment
    # beta1[2] ~ dnorm( 0.0, 0.001)  # slope
    ## flat improper priors
    beta0     ~ dflat()  # constant for standard treatment
    beta1[1] ~ dflat()  # constant for test treatment
    beta1[2] ~ dflat()  # slope
    tau       ~ dgamma( 0.001, 0.001) # precision of regression model
    s <- 1/sqrt(tau) # standard error of regression
    #
    # test rho>1 (test more potent)
    more.potent21 <- step(rho-1)
}
INITS
list( beta1=c(0,0), beta0=0, tau=1 )
```

Results.   A normal prior distribution with mean zero and variance equal to 1000 was used for the parameters of the linear predictor and gamma with mean 1 and variance 1000 for the model's precision. Results using 2000 iterations and discarding the initial 1000 iterations are provided in Table 6.14. When using the original dose, estimated posterior means and variances are found large for specific parameters. The normal prior distributions were proved to be informative in this case since results were sensitive to different values of the prior variance. Therefore the flat improper distribution [using the command dflat()] was finally used.

Table 6.13   WinBUGS code for common intercept/different slopes model in Example 6.2 using design matrix approach and rescaled dose[a]

```
model{
    for (i in 1:n){
        #
        # creating the design matrix
        X[i,1]<-1.0                # beta1=constant term
        # beta2=rescaled dose for standard treatment
        X[i,2]<-dose[i]/ranked(dose[],1)
        # beta3=CR dummy for interaction (slope difference)
        X[i,3]<- equals( drug[i], 2 )*X[i,2]
        ## beta3=STZ dummy for interaction (slope difference)
        #X[i,3]<- (equals( drug[i], 2 )-equals( drug[i], 1 ))*X[i,2]
        # model likelihood
        y[i] ~ dnorm( mu[i], tau )
        mu[i] <- inprod( beta[], X[i,])
    }
    #
    rho <- (beta[3]+beta[2])/beta[2] # relative potency in CR
    #rho <-(beta[2]+beta[3])/(beta[2]-beta[3]) # relative potency in
        stz
    # potency estimate
    potency <- rho * 1.2
    #
    # prior distributions
    # for (j in 1:3){ beta[j]~dnorm( 0.0, 0.0001) } # normal priors
    for (j in 1:3){ beta[j]~dflat() }                # flat improper
        priors
    tau         ~ dgamma( 0.001, 0.001) # precision of regression model
    s <- 1/sqrt(tau) # standard error of regression
}
INITS
list( beta=c(0,0,0), tau=1 )
```

[a] Data are specified as in Table 6.7; substitute lines 8–9 and 17 by lines 10–11 and 18, respectively, to switch from CR to STZ parametrization.

Table 6.14   Posterior summaries for parameters of common slope model in Example 6.2[a]

node	mean	sd	MC error	2.5%	median	97.5%	start	sample
beta0	71.98	0.6029	0.01443	70.75	71.99	73.2	1001	2000
beta1[1]	-4.885	0.2482	0.005378	-5.37	-4.886	-4.39	1001	2000
beta1[2]	-4.58	0.2508	0.005515	-5.078	-4.58	-4.065	1001	2000
more.potent21	0.0805	0.2721	0.005348	0.0	0.0	1.0	1001	2000
potency	1.126	0.05168	0.001109	1.027	1.125	1.232	1001	2000
rho	0.9382	0.04307	9.243E-4	0.8555	0.9378	1.027	1001	2000
s	1.362	0.2325	0.006969	1.014	1.33	1.915	1001	2000
tau	0.5829	0.1828	0.005289	0.2727	0.5657	0.9737	1001	2000

[a] Results using code of Table 6.12 (flat priors).

A point estimate of the model based on the posterior means is given by

$$Y_i \sim N(\mu_i, 0.77^2), \quad \mu_i = \begin{cases} 71.98 - 4.88x_i & \text{for the standard treatment} \\ 71.98 - 4.58x_i & \text{for the test treatment} \end{cases},$$

where x_i is defined by (6.11).

Thus we can infer that the clotting time is a posteriori expected to be equal to 72 seconds when no preparation is used but to be reduced by 4.9 seconds when we increase the dose of the standard treatment by the minimum quantity (1:40) of the experiment. Moreover, the clotting time is a posteriori expected to be reduced by 4.6 seconds when increasing the dose of the test treatment by a quantity equal to the minimum quantity (1:40) of the experiment.

The posterior distribution of the relative potency under the common intercept model is given in Figure 6.12. The new (test) treatment is a posteriori expected to be 6.2% less potent than the standard treatment with values ranging from 0.86 to 1.03 with 95% probability. Although the posterior mean of the relative potency is similar to the corresponding one in the parallel analysis, the 95% posterior interval is wider in the current analysis, including the value of 1 which corresponds to the assumption of equal treatment potency. The posterior probability that the second treatment is more potent than the standard one was found equal to 0.08, which is considerably higher than the corresponding probability in the parallel lines analysis. Finally, The new (test) treatment is a posteriori expected to have potency equal to 1.13 IU, with values ranging from 1.03 to 1.23 IU with 95% posterior probability.

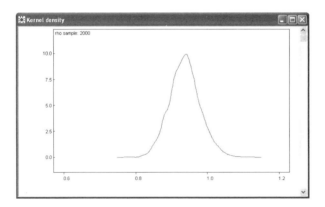

Figure 6.12 Posterior density of relative potency using common intercept model in Example 6.2.

Checking the assumption of common intercepts. In the preceding model we need to check for the equal intercepts assumption. The separate lines model can be used to check this assumption. From the posterior summaries of Table 6.15 and the posterior density plot of the intercept difference in Figure 6.13 we observe that the zero value (which corresponds to the parallel lines assumption) is close to the center of the posterior distribution since the posterior mean of the intercept difference is equal to -0.21, and the 95% percentile ranges from -2.56 to 2.18 . Moreover, the zero value corresponds to the 42% percentile of the posterior distribution, indicating that this assumption is a posteriori sensible.

Table 6.15 Posterior summaries for intercept difference of separate lines model in Example 6.2[a]

node	mean	sd	MC error	2.5%	median	97.5%
intercept.difference	-0.2147	1.199	0.02616	-2.564	-0.2198	2.184
p2	0.424	0.4942	0.01217	0.0	0.0	1.0

[a] Rescaled dose (6.11) is used as explanatory variable.

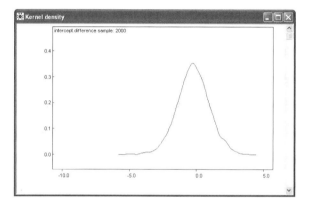

Figure 6.13 Posterior density of intercept difference in separate lines model for Example 6.2.

6.3.3   Comparison of the two approaches

Two different approaches based on ANCOVA models have been implemented in order to estimate the relative potency of a new drug using a simple bioassay example. Both adopted models fitted above indicate that the new drug is about 6% less potent than the standard. The first analysis indicates clear differences between the two drugs since the posterior density of the estimated relative potency is far away from the value of 1, which corresponds to equally potent drugs. From the second model, the difference is not so clear since the posterior distribution of the estimated potency is more dispersed, indicating that the equal potency assumption might be plausible. In order to reach our final conclusion, we need to determine which model describes better the random behavior that we are studying. For this reason we may use a "naive" approach based on the Bayesian versions of R^2 quantities described earlier in this chapter or more sophisticated techniques based on the predictive distributions to check the assumptions of each model and its goodness of fit. Also, Bayesian model comparison techniques can be used to identify which model is more appropriate in this case. All these issues are described in Chapters 10 and 11.

Before closing, we provide the Bayesian measures of R_B^2 in Table 6.16 in order to monitor the goodness of fit of the models. This measure indicates that the models of the parallel lines analysis fit the data slightly better than the models of the slope ratio analysis. Moreover, the non-common-slope model (in parallel analysis) only slightly increases the posterior mean of this measure (posterior densities are identical). This is similar to the result in the non-common-intercept model in the slope ratio analysis since the posterior mean of R_B^2 is slighlty lower that the corresponding one in the common slope model, while all posterior summaries are very close.

Table 6.16 Posterior summaries for R_B^2 in Example 6.2

Analysis	model	mean	sd	MC error	2.5%	median	97.5%
Parallel lines	1 Parallel lines	0.9839	0.005795	1.715E-4	0.9693	0.9852	0.9914
	2 Non common slope	0.9847	0.005355	1.341E-4	0.9714	0.9858	0.9919
Slope Ratio	3 Common intercept	0.9501	0.01801	5.352E-4	0.9042	0.9538	0.9732
	4 Non common intercept	0.9479	0.01827	4.56E-4	0.9016	0.9515	0.9726

6.4 FURTHER MODELING ISSUES

6.4.1 Extending the simple ANCOVA model

In practice, the number of quantitative and qualitative variables involved in a regression model is high, and the applied statistical analyst must select between them. Extending the simple ANCOVA models by incorporating interaction terms between more than two variables or factors results in highly complicated models. Nevertheless, their interpretation can be based on the three simple ANCOVA models (parallel lines, common intercept, and separate regression lines) presented in Section 6.2.

When dealing with multiple regression/ANCOVA models, the user must bear in mind that

- The most frequently used model is the main effects model, in which we do not include any interaction. This model corresponds to the simple model of parallel lines described in Section 6.2. It is simple to interpret and easily understood by specialists in other scientific fields.

- Generally interactions must be avoided unless the data or a scientific scenario of the problem at hand supports such an action. By avoiding interactions, we simplify the model and make it easily understandable to scientists not familiar with statistics.

- When we decide to use interactions in multiple regression/ANCOVA models, we usually restrict this use to two-way interactions. This is due mainly to their interpretation, which is simpler than in models with higher-order interactions. Moreover, the size of the data seldom allow us to extend the model by including higher-order interactions.

- An interaction between categorical variables imposes different intercepts for the level combinations of the factors involved in the corresponding interaction term.

- Higher-order interaction terms that involve one quantitative variable impose different slopes (effects) of the quantitative variable for each level combination of the factors involved in this interaction term.

Usually, model selection methods are implemented to identify well-fitted models. Stepwise-like methods may be used within the Bayesian context, but such procedures will be intensive because of the computational effort needed to fit Bayesian models. This can be simplified in normal models since any posterior distribution of interest can be calculated analytically when using conjugate prior distributions. Alternatively, "automatic" model comparison and selection methods based on MCMC schemes can be applied. These methods are briefly presented and discussed in Chapter 11.

6.4.2   Using binary indicators to specify models in multiple regression

When we wish to fit various models by trying to exclude or include different variables in the linear predictor, we need to write in WinBUGS different model codes for each model. A much simpler approach can be based on incorporating a binary vector $\gamma = (\gamma_1, \ldots, \gamma_p)$ to the linear predictor η_i, which can be now written as

$$\mu_i = \beta_0 + \sum_{j=1}^{p} \gamma_j \beta_j X_{ij}$$

and defined in WinBUGS using the syntax

```
mu[i] = beta0 + gamma[1]*beta[1]*x[i,j]+...+ gamma[p]*beta[p]*x[i,p]
```

where X_{ij} (and x[i,j]) is the j covariate or dummy variable. The binary indicator γ_j is set equal to one if we wish to include the X_j variable in our model and zero otherwise. This can be specified in the WinBUGS data section without changing the model code itself. Although we may extend the expression above using a binary indicator also for the constant term, β_0 is usually included in the model. Moreover, in multiple regression models with large number of covariates it is convenient to use the data matrix X in combination with the inprod command to define the model. In this case the linear predictor can be expressed as

$$\mu_i = \sum_{j=1}^{P} \gamma_j \beta_j X_{ij} = \sum_{j=1}^{P} \beta_{\gamma,j} X_{ij},$$

where $\beta_{\gamma,j} = \gamma_j \beta_j$ and $P = p + 1$ the number of parameters involved in the linear predictor. The first column of X corresponds to the constant term, while the remaining columns represent the quantitative or dummy variables considered as covariates.

The vector $(\beta_{\gamma,j})$ can be specified in WinBUGS using the syntax

```
for (j in 1:P) { beta.g[j] <- gamma[j]*beta[j] }
```

while the linear predictor will now be given using the inprod command by

```
mu[i] = inprod( beta.g[], X[i,] )
```

The binary vector γ is a model indicator and is used in Bayesian variable selection methods [see, e.g., George and McCulloch (1993)]. It is relatively easy to extend our model to incorporate uncertainty concerning the inclusion of each covariate. We only have to use a Bernoulli prior for each γ_j. For example, $\gamma_j \sim$ Bernoulli($\frac{1}{2}$) can be used to express prior indifference concerning the inclusion or exclusion of the X_j variable from the model. Other technical issues such as the selection of the prior distribution and the Bartlett–Lindley paradox (Lindley, 1957; Bartlett, 1957) further complicate implementation of the Bayesian variable selection. This issue is discussed in more detail in Chapter 11.

6.4.3   Selection of variables using the deviance information criterion (DIC)

As we have already mentioned in Section 4.2.5, the deviance information criterion (DIC) is defined as
$$\mathrm{DIC}(m) = 2\overline{D(\boldsymbol{\theta}_m, m)} - D(\overline{\boldsymbol{\theta}}_m, m) = D(\overline{\boldsymbol{\theta}}_m, m) + 2p_m \ . \tag{6.12}$$

It is used as a measure of model comparison and adequacy (Spiegelhalter et al., 2002). Low-value DICs indicate better-fitted models. Quantity p_m is the number of "effective" parameters for model m. In normal models, presented in this chapter, p_m is approximately equal to the true number of parameters d_m. For this reason, the DIC here is approximately equal to Akaike's information criterion (AIC) (Akaike, 1973, 1974). Generally DIC is considered as a generalization of AIC since in more complicated models, such as the hierarchical presented in Chapter 9, p_m is different from the number of parameters used in the model.

We must further note that DIC must be used carefully since it assumes that the posterior mean is a good measure of central location. For this reason, DIC must not be used when the posterior distributions are highly skewed or bimodal (Spiegelhalter et al., 2003d, section entitled "Tricks: Advanced use of the BUGS language"). In normal models, the posterior distribution of parameters of the linear predictor is usually symmetric. Some problems may appear when the precision parameter τ is low.

According to Section 4.2.5, the minimum DIC model identifies the model which offers the best short-term predictions. Nevertheless, Spiegelhalter et al. (2002) suggest the following rule of thumb: that models with DIC difference within the minimum value lower than two (2) deserve to be considered as equally well, while models with values ranging within 2--7 have considerably less support. For this reason, we may report all models with DIC difference from the best one lower than a threshold value (e.g., 2, 5, or 7) depending on the number of models we wish to report.

In normal linear models, the simplest strategy is to fit all 2^p models and identify the one with the minimum DIC value. This procedure is painful and time consuming within WinBUGS, even for a medium sized dataset. For example, when 5, 10, or 20 variables are used as covariates, then the number of models under consideration is equal to 32, 1024, and $1,048,576$. Here we present a simple stepwise strategy that is suboptimal but can identify "well-fitted" models more rapidly.

6.4.3.1   A stepwise method for DIC based variable selection in WinBUGS.
First, we specify the starting model. Suitable starting models are the constant or null model (without any covariate in the model) and the full model (with all p covariates in the model). The latter can be used only when the number of model parameters is lower than the sample size ($p + 2 < n$). A good starting point can be also obtained after fitting the full model and then adopt as the initial model the one including covariates for which the posterior distribution of the corresponding coefficients is away from zero.

The procedure can be summarized by the following steps:

1. Select initial model with binary variable indicator $\gamma^{(0)}$ (see Section 6.4.2). Set $\gamma = \gamma^{(0)}$.

2. Consider the currently selected model with binary indicator γ such that

$$Y_i \sim N(\mu_i, \tau^{-1}) \text{ and } \mu_i = \beta_0 + \sum_{j=1}^{p} \gamma_j \beta_j X_{ij}. \qquad (6.13)$$

Further consider p candidate models characterized by their binary indicators $\gamma_k^c = (\gamma_{1,k}^c, \gamma_{2,k}^c, \ldots, \gamma_{p,k}^c)^T$ for $k = 1, \ldots, p$. Each of these p candidate models is given by

$$Y_i \sim N\left(\mu_{i,k}^c, \frac{1}{\tau_k^c}\right) \text{ and } \mu_{i,k}^c = \beta_{0,k}^c + \sum_{j=1}^{p} \gamma_{j,k}^c \beta_{j,k}^c X_{ij} \qquad (6.14)$$

with

$$\gamma_{j,k}^c = \gamma_{j,k}^c I(j \neq k) + (1 - \gamma_{j,k}^c) I(j = k). \qquad (6.15)$$

Equation (6.15) defines each of the p different models such that it differs from γ only in the status of the kth variable. For example, if X_j is included in model γ ($\gamma_j = 1$), then it is removed from jth candidate model ($\gamma_{j,j}^c = 1 - \gamma_j = 0$) but is included in all other candidate models ($\gamma_{j,k}^c = \gamma_j = 1$ for all $k \neq j$).

The models under consideration in this step will be denoted by

$$\mathcal{M}^{\text{step}} = \left\{ \gamma, \gamma_1^c, \dots, \gamma_p^c \right\}.$$

3. Run the MCMC code and estimate DIC values for all models of step 2 (γ and γ_k^c for $k = 1, \dots, p$).

4. From the fitted models of step 2, select model γ^{opt} with the lowest DIC value defined as

$$\gamma^{\text{opt}} = \left\{ \gamma' \in \mathcal{M}^{\text{step}} : \text{DIC}(\gamma') = \min \left\{ \text{DIC}(\gamma'') \text{ for } \gamma'' \in \mathcal{M}^{\text{step}} \right\} \right\}.$$

5. If $\gamma^{\text{opt}} = \gamma$ (i.e., the selected model is the same as the one indicated by the previous cycle of the procedure), then terminate the process, otherwise set $\gamma = \gamma^{\text{opt}}$ and return to step 2.

This procedure can be set up within WinBUGS in a relatively straightforward manner. We only need to define the response variable of each model in a different vector in order to obtain separate DIC values for all fitted models in each cycle of the procedure described above. The use of the binary vectors γ simplifies the implementation of this procedure in WinBUGS. We only need to set up the current model γ and then run the associated WinBUGS code to identify the best model (γ^{opt}) in each cycle of the proposed procedure.

Example 6.3. Oxygen uptake experiment. The data of this example come from an experiment that was conducted to examine the oxygen uptake (variable name: O2UP) in milligrams of oxygen per minute given the following measurements:

- X_1 =BOD: biological oxygen demand

- X_2 =TKN: total Kjeldahl nitrogen (TKN)

- X_3 =TS : total solids (TS)

- X_4 =TVS: total volatile solids (TVS)

- X_5 =COD: chemical oxygen demand (COD)

Each variable is measured in milligrams per liter. Data of this example were taken from Weisberg (2005, pp. 230–231). They are available in the book's Website and are used and reproduced with permission of John Wiley and Sons, Inc.
The aim here is to develop an efficient equation relating the logarithm of O2UP with the other available measurements for use in future predictions.

In this example we facilitate DIC to identify which equation is needed for the prediction of $\log(\text{O2UP})$. In each step we fit p models running in parallel within the same WinBUGS model code and then select the model with the lowest DIC value.

WinBUGS model code for the stepwise procedure.   We initially define the data of the model using the following syntax:

```
list(n=20, p=5, gamma=c(1,1,1,1,1),
BOD = c(1125, 920, 835, 1000, 1150, 990, 840, 650, 640, 583,
        570, 570, 510, 555, 460, 275, 510, 165, 244, 79),
TKN = c(232, 268, 271, 237, 192, 202, 184, 200, 180, 165,
        151, 171, 243, 147, 286, 198, 196, 210, 327, 334),
TS = c(7160, 8804, 8108, 6370, 6441, 5154, 5896, 5336, 5041, 5012,
       4825, 4391, 4320, 3709, 3969, 3558, 4361, 3301, 2964, 2777),
TVS = c(85.9, 86.5, 85.2, 83.8, 82.1, 79.2, 81.2, 80.6, 78.4, 79.3,
        78.7, 78, 72.3, 74.9, 74.4, 72.5, 57.7, 71.8, 72.5, 71.9),
COD = c(8905, 7388, 5348, 8056, 6960, 5690, 6932, 5400, 3177, 4461,
        3901, 5002, 4665, 4642, 4840, 4479, 4200, 3410, 3360, 2599),
O2UP = c(36, 7.9, 5.6, 5.2, 2, 2.3, 1.3, 1.3, 0.6, 0.7, 1, 1,
         0.8, 0.6, 0.4, 0.7, 0.6, 0.4, 0.3, 0.9))
```

Here, we define each variable that will be used in the model (BOD, TKN, TS, TVS, COD, O2UP), the samples size (n), the number of covariates (p), and the structure of the current model (gamma). Within the model code we specify the transformed variable that will be used as response, and we also simplify the names of the covariates by setting them equal to x_{ij} using the syntax

```
# initial definition of variables for the current model
for (i in 1:n){
    y[i] <- log(O2UP[i])
    x[i,1]<-BOD[i]
    x[i,2]<-TKN[i]
    x[i,3]<-TS[i]
    x[i,4]<-TVS[i]
    x[i,5]<-COD[i]
    }
```

Following the description of the stepwise procedure described above, we need to define model (6.13) using the following WinBUGS code

```
# model specificication of the current model
for (i in 1:n){
    y[i] ~ dnorm( mu[i], tau )
    mu[i] <- beta0 + beta[1] * gamma[1]*x[i,1]
                   + beta[2] * gamma[2]*x[i,2]
                   + beta[3] * gamma[3]*x[i,3]
                   + beta[4] * gamma[4]*x[i,4]
                   + beta[5] * gamma[5]*x[i,5]
    }
```

and the corresponding prior distribution by

```
# prior distributions
beta0~dnorm(0, 0.001)
for (j in 1:p){ beta[j]~dnorm(0, 0.001) }
tau~dgamma( 0.001, 0.001)
```

For the candidate models we first define their model structure as expressed by (6.15) using the syntax

```
# definition of gammas for candidate models
for (k in 1:p){
    for (j in 1:p){
        gamma.can[j,k] <- gamma[j]*(1-equals( k,j ))
                          + (1-gamma[j])*equals( k,j )
    }
}
```

Furthermore, one vector for each fitted model in each cycle must be defined in order to calculate each DIC separately (note that if a matrix is used instead, then a single DIC value will be calculated for all models). Hence we specify the response variables in WinBUGS by

```
# response data for candidate models
for (i in 1:n){
      y1[i] <- y[i]
      y2[i] <- y[i]
      y3[i] <- y[i]
      y4[i] <- y[i]
      y5[i] <- y[i]
      y1[i] ~ dnorm( mu.can[i,1], tau.can[1] )
      y2[i] ~ dnorm( mu.can[i,2], tau.can[2] )
      y3[i] ~ dnorm( mu.can[i,3], tau.can[3] )
      y4[i] ~ dnorm( mu.can[i,4], tau.can[4] )
      y5[i] ~ dnorm( mu.can[i,5], tau.can[5] )
  }
```

Finally, the linear predictor and the prior are defined using the following syntax:

```
# linear predictors for each model
for (k in 1:p){ for (i in 1:n){
      mu.can[i,k]<-beta0.can[k]+beta.can[1,k]*gamma.can[1,k]*x[i,1]
                              +beta.can[2,k]*gamma.can[2,k]*x[i,2]
                              +beta.can[3,k]*gamma.can[3,k]*x[i,3]
                              +beta.can[4,k]*gamma.can[4,k]*x[i,4]
                              +beta.can[5,k]*gamma.can[5,k]*x[i,5]
      }
}
#
# prior distributions for parameters of candidate models
for (k in 1:p){
      beta0.can[k]~dnorm(0, 0.001)
      for (j in 1:p){ beta.can[j,k]~dnorm(0, 0.001) }
      tau.can[k]~dgamma( 0.001, 0.001)
}
```

This syntax is similar to the one used for the current model with the difference that a double loop is used to define all $p = 5$ models fitted in each cycle. Hence, parameters are now stored in a matrix form instead of the separate vectors that we have used thus far.

Results. Initially, the proposed stepwise procedure was implemented starting from the full model [$\gamma^{(0)} = (1, 1, 1, 1, 1)$]. The procedure was terminated in four cycles, fitting $6 + 3*5 = 21$ models in total. The model TS+COD was indicated as the best one with estimated DIC value equal to 40.47 and as covariates, the total solids (TS) and the chemical oxygen demand (COD). Raw WinBUGS results are provided in Table 6.17, and the procedure is summarized in Table 6.18.

The same model was also selected when starting from the constant model. The procedure was terminated in three cycles fitting 16 models. First, the third variable (TS) was added followed by the fifth one (COD) in the second step. Results are summarized in Table 6.19

When we fit the full model, then the zero value is included in the 95% posterior interval of the parameter effects of all covariates. Hence the ad hoc rule of starting from the model with covariates whose posterior intervals do not include zero indicates the null model as a good starting point.

All starting points used here indicate model TS+COD ($X_3 + X_5$) as the best one according to DIC. The posterior distributions can be obtained after monitoring parameters of the current model in the final step of the stepwise procedures implemented here; see Table 6.20 for

Table 6.17 Stepwise WinBUGS DIC results for Example 6.3 starting from full model

DIC STEPWISE STARTING FROM FULL MODEL

Step 1: $\gamma = (11111)$

```
Dbar Dhat pD DIC
y 38.164 30.659 7.505 45.669
y1 36.655 30.273 6.382 43.037
y2 37.986 31.707 6.279 44.265
y3 40.191 33.874 6.318 46.509
y4 37.030 30.687 6.343 43.374
y5 41.274 34.941 6.333 47.607
total 231.301 192.141 39.160 270.460
```

Step 2: REMOVE $X_1 = BOD$: $\gamma = (01111)$

```
Dbar Dhat pD DIC
y 36.660 30.269 6.391 43.051
y1 38.030 30.632 7.398 45.428
y2 37.076 31.887 5.189 42.265
y3 39.900 34.622 5.278 45.178
y4 35.663 30.401 5.262 40.925
y5 42.506 37.260 5.246 47.752
total 229.835 195.070 34.765 264.599
```

Step 3: REMOVE $X_4 = TVS$: $\gamma = (01101)$

```
y 35.737 30.425 5.312 41.049
y1 37.109 30.722 6.387 43.496
y2 36.246 32.086 4.160 40.406
y3 41.814 37.722 4.092 45.907
y4 36.688 30.260 6.427 43.115
y5 42.357 38.173 4.185 46.542
total 229.952 199.389 30.563 260.515
```

Step 4: REMOVE $X_2 = TKN$: $\gamma = (00101)$

```
Dbar Dhat pD DIC
y 36.283 32.094 4.189 40.471
y1 37.189 31.918 5.271 42.460
y2 35.643 30.408 5.235 40.878
y3 42.237 39.158 3.079 45.317
y4 37.219 31.907 5.312 42.531
y5 42.059 38.913 3.146 45.205
total 230.630 204.398 26.232 256.862
```

FINAL MODEL: X3+X5, $\gamma^{opt} = (00101)$

Table 6.18 Tabulated summary of stepwise procedure for Example 6.3, starting from full model

Step	(1)	(2)	(3)	(4)
Previous model: γ	—	11111	01111	01101
Action	Initialize	$-X_1$	$-X_4$	$-X_2$
Current model				
DIC	45.7	43.1	41.0	**40.5**
γ	11111	01111	01101	00101
Candidate model variables				
X_1: BOD	$(-)$ **43.0**	$(+)$ 45.4	$(+)$ 43.5	$(+)$ 42.5
X_2: TKN	$(-)$ 44.3	$(-)$ 42.3	$(-)$ **40.4**	$(+)$ 40.9
X_3: TS	$(-)$ 46.5	$(-)$ 45.2	$(-)$ 45.9	$(-)$ 45.3
X_4: TVS	$(-)$ 43.4	$(-)$ **40.9**	$(+)$ 43.1	$(+)$ 42.5
X_5: COD	$(-)$ 47.6	$(-)$ 47.8	$(-)$ 46.5	$(-)$ 45.2

Key: $(+)$ = add a variable to the current model; $(-)$ = remove a variable from current model.

Table 6.19 Tabulated summary of stepwise procedure for Example 6.3, starting from null/constant model

Step	(1)	(2)	(3)
Previous model: γ	—	00000	00100
Action	Initialize	$+X_3$	$+X_5$
Current Model			
DIC	66.8	45.2	**40.5**
γ	00000	00100	00101
Candidate Models Variable			
$X_1 : BOD$	$(+)$ 50.8	$(+)$ 46.2	$(+)$ 42.5
$X_2 : TKN$	$(+)$ 68.8	$(+)$ 46.4	$(+)$ 40.8
$X_3 : TS$	$(+)$ **45.0**	$(-)$ 66.8	$(-)$ 45.3
$X_4 : TVS$	$(+)$ 54.9	$(+)$ 46.2	$(+)$ 42.6
$X_5 : COD$	$(+)$ 45.3	$(+)$ **40.6**	$(-)$ 45.2

Key: $(+)$ = add a variable to the current model; $(-)$ = remove a variable from current model.

details. Since β_1, β_2 and β_4 are not involved in the corresponding likelihood of the finally selected model, their posterior coincides with the corresponding prior distribution with zero prior mean and variance equal to 1000 (or standard deviation equal to 31.62).

Table 6.20 Posterior summaries for parameters of model $X_3 + X_5$: TS+COD for Example 6.3[a]

node	mean	sd	MC error	2.5%	median	97.5%	start	sample
beta[1]	-0.06103	32.14	0.5393	-63.22	0.2142	63.51	1001	4000
beta[2]	-0.408	31.83	0.4559	-63.2	-0.4789	61.82	1001	4000
beta[3]	3.448E-4	1.308E-4	2.152E-6	8.6E-5	3.465E-4	5.991E-4	1001	4000
beta[4]	-0.09077	31.83	0.5583	-62.09	-0.792	63.23	1001	4000
beta[5]	3.259E-4	1.283E-4	2.022E-6	6.738E-5	3.254E-4	5.76E-4	1001	4000
beta0	-3.162	0.493	0.007998	-4.14	-3.165	-2.179	1001	4000
tau	2.947	1.031	0.02212	1.302	2.843	5.29	1001	4000

[a] Parameters β_1, β_2, and β_4 are not included in the model; hence their posterior summaries are equal to their prior distribution, $N(0, 1000)$.

6.5 CLOSING REMARKS

This chapter focuses on the extension of the simple normal regression model incorporating categorical variables, with emphasis on the simplest case of one qualitative variable and one quantitative variable. Further issues, such as model comparison using the deviance information criterion (DIC), are also discussed.

After becoming familiar with the construction and analysis of the normal model, the reader can easily extend its formulation using different distributions for the response variable (e.g., Student's t distribution), incorporating nonlinear components (such as polynomial functions of the original functions), higher order interactions of specific interest, or building models based on transformations of the original variables.

In the following chapters we proceed with a description of the generalized linear models used for response variables of different types (quantitative continuous or discrete, qualitative, binary) and then extend our discussion to the Bayesian hierarchical models that account for data with more complicated structures.

Problems

6.1 Construct the models described in Problem 5.5 using dummy variables under corner and sum-to-zero constraints.

6.2 Use dummy variables for corner and sum-to-zero parametrizations to fit the models of Problem 5.6.

6.3 Consider the paper of Kahn (2005) and the associated dataset, described in Problem 5.7.

    a) Use WinBUGS to obtain posterior estimates of the parameters of the five models presented by Kahn (2005).

    b) Compare results of the Bayesian analysis with the corresponding ones presented in Kahn's paper.

    c) Interpret the estimated model coefficient for all models under consideration.

6.4   For Albuquerque home prices data presented in Problem 5.8, fit a three-way ANOVA model for the home prices using as factors the features of the house, the northeast location and the corner location, Use dummy variables for either the corner or the sum-to-zero parametrization.

    a) Fit the model with no interaction terms and compare it with the model with all interaction terms using DIC.

    b) Start from the full model and remove higher order interaction terms using DIC.

6.5   For Albuquerque home prices data presented in Problem 5.8, include all available variables to model the price of a house:

    a) Fit in WinBUGS a model with no interaction terms.

    b) Interpret the model parameters.

    c) Remove variables that do not contribute to the model using DIC.

    d) Consider the model with all two-way interaction terms. Remove all terms that are not important using DIC. Interpret the parameters of the finally selected model.

6.6   Consider the following data for the bioassay example (Example 6.2)

		Blood clotting time (in seconds)		
Dose		Standard	Test (new)
1:40	0.025	67.5 70.5 67.5	68.8 67.1 67.0 —
1:20	0.050	— 62.1 62.4	— 60.1 59.4 62.4
1:10	0.100	53.7 51.6 54.5	53.9 51.7 53.4 52.7

    a) Perform the same analysis as in Section 6.3.

    b) Use γ vector binary indicators to combine models in one WinBUGS code as described in Section 6.4.2.

    c) Compare models using DIC. Be careful: All DICs must refer on the same response Y with the same scale.

6.7   For Problem 6.3, incorporate all fitted models in a single WinBUGS model code file using the γ binary vector described in Section 6.4.2. Compare all fitted models using DIC.

6.8   For Problems 5.2 and 5.3, use DIC to select the appropriate variables that must be included in the model.

6.9   Download the mshop1 dataset from the book's Website (www.stat-athens.aueb.gr/~jbn/winbugs_book). The data were compiled from a customer satisfaction survey contacted in Chios, Greece (Sarantinidis, 2003). The given dataset is only part of the original dataset. Explanation of the variables is provided in file mshop1.txt.

    a) Build an ANCOVA model to identify which variables influence the total money spent.

    b) Use DIC to finally select which variables are appropriate for the model.

    c) Interpret the parameters of the model on the basis of their posterior distributions.

CHAPTER 7

INTRODUCTION TO GENERALIZED LINEAR MODELS: BINOMIAL AND POISSON DATA

7.1 INTRODUCTION

Generalized linear models (GLMs) constitute a wide class of models encompassing stochastic representations used for the analysis of both quantitative (continuous or discrete) and qualitative response variables. They can be regarded as the natural extension of normal linear regression models and are based on the exponential family of distributions, which includes the most common distributions such as the normal, binomial and, Poisson. Generalized linear models have become very popular because of their generality and wide range of application. They can be considered as one of the most prominent and important components of modern statistical theory. They have provided not only a family of models that are widely used in practice but also a unified, general way of thinking concerning the formulation of statistical models.

As we have already mentioned, three are the components of a GLM: the random/stochastic component, the systematic component (or linear predictor), and the link function; see also Section 5.1.

The stochastic component contains the response variable Y_i and its assumed distribution $\mathcal{D}(\boldsymbol{\theta})$, which, within the GLM framework, is a member of the exponential dispersion family as defined in (1.3) of Section 1.5.6. The systematic component is a function of the explanatory variables (or covariates) similarly as in normal regression models. Usually a linear combination of these variables is used, and for this reason this component is also called a *linear predictor*. Finally, the link function $g(\boldsymbol{\theta})$ is the mathematical expression which connects the parameters of the response Y with the linear predictor and the covariates. In

Bayesian Modeling Using WinBUGS, by Ioannis Ntzoufras
Copyright ©2009 John Wiley & Sons, Inc.

GLM a location parameter (e.g., the mean) is usually linked with the linear predictor. Hence a GLM can be summarized by the following expressions

$$
\begin{aligned}
Y_i &\sim \mathrm{expf}\Big(\vartheta_i, \phi, a(), b(), c()\Big) & \text{(stochastic component)} \\
\eta_i &= \boldsymbol{X}_{(i)}\boldsymbol{\beta} = \beta_0 + \sum_{j=1}^{p} x_{ij}\beta_j & \text{(systematic component)} \\
\vartheta_i &= R(\theta_i) & \text{(canonical — distribution parameter function)} \\
g(\theta_i) &= g(R^{-1}(\vartheta_i)) = g_\vartheta(\vartheta_i) = \eta_i & \text{(link function)} \\
\boldsymbol{\theta}_m &= (\boldsymbol{\beta}^T, \phi)^T & \text{(model parameters)},
\end{aligned}
\tag{7.1}
$$

where $\mathrm{expf}\Big(\vartheta_i, \phi, a(), b(), c()\Big)$ denotes the exponential family with location and dispersion parameters ϑ_i and ϕ, respectively, and $a()$, $b()$, $c()$ are functions needed to specify the structure of the specific distribution with density or probability function given by

$$
f(y|\vartheta, \phi) = \exp\left(\frac{y\vartheta - b(\vartheta)}{a(\phi)} + c(y, \phi)\right).
\tag{7.2}
$$

In this setup, we denote by θ the original location parameter of the distribution, by ϑ the canonical parameter of the exponential family, and by $R(\theta)$ the function that connects the two parameters. Moreover, we denote by $g(\theta)$ and $g_\vartheta(\vartheta)$ the link functions that associate the location parameter θ and the canonical parameter ϑ, respectively, with the linear predictor η. Note that the latter is given by $g_\vartheta(\vartheta) = g(R^{-1}(\vartheta))$.

In this section we introduce the user to the basic notions of the generalized linear models. In the following two sections we provide details concerning the exponential family and some common distributions that belong in this family. The section continues with a presentation of the most common link functions used and discussion of some of the more complicated ones proposed for binomial data. Finally, we close this section with a short description of the most common models for every type of data.

Additional details concerning generalized linear models can be found in a variety of well-written books related to the topic such as McCullagh and Nelder (1989), Lindsey (1997), and Fahrmeir and Tutz (2001). A detailed illustration of Bayesian inference and analysis focusing on GLMs can be found in Dey et al. (2000).

7.1.1   The exponential family

A member of the exponential family has a probability or density function that can be expressed by the following form

$$
f(y|\vartheta) = \exp\Big\{ R(\vartheta)T(y) + B(\vartheta) + C(y)\Big\},
\tag{7.3}
$$

where $\vartheta = R(\theta)$ is the canonical parameter and both ϑ and θ are location parameters; for this expression, see Bernardo and Smith (1994, p. 266), Lindsey (1997, p. 10), Gamerman and Lopes (2006, pp. 51–54), and Agresti (2002, p. 116) for equivalent expressions.

The more general expression (7.2) of McCullagh and Nelder (1989) also incorporates in the formulation a scale parameter ϕ in addition to the location parameter ϑ. These parameters are collectively called the *exponential dispersion family*. When the dispersion parameter ϕ is known, then (7.2) becomes equal to (7.3) if we set $R(\vartheta) = \vartheta/a(\phi)$, $T(y) = y$, $B(\vartheta) = -b(\vartheta)/a(\phi)$, and $C(y) = c(y, \phi)$. This expression is more convenient when one parameter distributions that are members of the exponential family are adopted.

The mean and the variance of Y with distribution in the exponential family with parameters ϑ and ϕ are equal to

$$E(Y) = \frac{db(\vartheta)}{d\vartheta} = b'(\vartheta) \;\; \text{and} \;\; V(Y) = \frac{d^2 b(\vartheta)}{d\vartheta^2} a(\phi) = b''(\vartheta)a(\phi) \, . \qquad (7.4)$$

7.1.2 Common distributions as members of the exponential family

Members of the exponential family are popular distributions such as the normal, the binomial, the Poisson, the gamma, and the inverse Gaussian distributions. We may further include distributions that are special cases of these distributions (e.g., the exponential and the Pareto distributions) or distributions that result as transformations of the above mentioned random variables (such as the log-normal or the inverse gamma distributions), which can be treated as GLMs without any methodological complications. In the following paragraphs we provide details concerning the most popular distributions of the exponential family; see Table 7.1 for a tabulated summary.

Normal distribution.   The density function of the normal distribution can be written as

$$f(y|\mu, \sigma^2) = \exp\left(\frac{y\mu - \mu^2}{\sigma^2} + \left\{ -\frac{1}{2}\log(2\pi) - \frac{1}{2}\log\sigma^2 - \frac{y^2}{2\sigma^2} \right\} \right) \, .$$

Hence the canonical parameter ϑ is equal to the mean of the normal distribution μ, the scale parameter ϕ is equal to the variance σ^2, $b(\vartheta) = b(\mu) = \mu^2/2$, $a(\phi) = a(\sigma^2) = \sigma^2$, and $c(y, \phi) = c(y, \sigma^2) = -\frac{1}{2}\log(2\pi) - \frac{1}{2}\log\sigma^2 - \frac{1}{2}y^2/\sigma^2$.

Binomial distribution.   For the binomial distribution binomial(π, N), we can express the probability function as

$$f(y|\pi, N) = \exp\left(y\log\frac{\pi}{1-\pi} - N\log\frac{1}{1-\pi} + \log\binom{N}{y} \right) \, .$$

The canonical parameter is now given by $\vartheta = \log[\pi/(1-\pi)]$ while $a(\phi) = \phi = 1$ and

$$c(y, \phi) = \log\binom{N}{y}$$

which does not depend on either y or ϑ. Solving the expression of the canonical parameter in terms of π we have that $\pi = e^{\vartheta}/(1+e^{\vartheta})$ resulting in $b(\vartheta) = N\log(1+e^{\vartheta})$. Using (7.4), we obtain $E(Y) = Ne^{\vartheta}/(1+e^{\vartheta}) = N\pi$ and $V(Y) = Ne^{\vartheta}/(1+e^{\vartheta})^2 = N\pi(1-\pi)$ as expected in the binomial case.

Negative binomial distribution.   Similar to this is the logic for the negative binomial distribution $\mathrm{NB}(\pi, k)$ (for fixed k), where we can write $\vartheta = \log(1-\pi)$ and $b(\vartheta) = -k\log(\pi) = -k\log(1-e^{\vartheta})$, $a(\phi) = 1$ and

$$c(y, \phi) = \log\binom{y+k-1}{y} \, .$$

Poisson distribution. The probability function of the Poisson(λ) distribution can be written as

$$f(y|\pi, N) = \exp\left(y\log(\lambda) - \lambda - \log(y!)\right).$$

The canonical parameter is now given by $\vartheta = \log \lambda$ while $a(\phi) = \phi = 1$, $c(y, \phi) = \log(y!)$ and finally $b(\vartheta) = \lambda = e^{\vartheta}$. Using (7.4), we obtain both the mean and the variance equal to λ as expected in the Poisson distribution.

Gamma distribution. The probability function of the gamma$(a, a/\mu)$ distribution with mean μ and variance μ^2/a can be written as

$$
\begin{aligned}
f(y|a, \mu) &= f_G(y; a, b = a/\mu) = \frac{(a/\mu)^a}{\Gamma(a)} y^{a-1} e^{-ay/\mu} \\
&= \exp\left((a-1)\log y - \frac{a}{\mu}y + a\log a + a\log\frac{1}{\mu} - \log\Gamma(a)\right) \\
&= \exp\left(\frac{(-\mu^{-1})y - (-\log(-[-\mu^{-1}]))}{a^{-1}} + \left\{a\log(ay) - \log y - \log\Gamma(a)\right\}\right).
\end{aligned}
$$

The canonical parameter is now given by $\vartheta = -\mu^{-1}$ (or $-b/a$ in the original parametrization) while $\phi = a^{-1}$, $a(\phi) = \phi = a^{-1}$ and $c(y, \phi) = \phi^{-1}\log(y\phi^{-1}) - \log y - \log\Gamma(1/\phi)$. Finally, $b(\vartheta) = -\log(-[-1/\mu]) = -\log(-\vartheta)$. Using (7.4), we obtain $E(Y) = \mu$ and $V(Y) = \mu^2/a$ as defined above.

Inverse Gaussian distribution. The density of the inverse Gaussian distribution $Y \sim$ IGaussian(μ, λ) is given by

$$
\begin{aligned}
f(y|\mu, \lambda) &= \left(\frac{\lambda}{2\pi y^3}\right)^{1/2} \exp\left(-\frac{\lambda(y - \mu)^2}{2\mu^2 y}\right) \\
&= \exp\left(-\frac{\lambda y^2}{2\mu^2 y} - \frac{\lambda\mu^2}{2\mu^2 y} + \frac{2\lambda y\mu}{2\mu^2 y} + \frac{1}{2}\left\{\log\lambda - \log(2\pi) - 3\log y\right\}\right) \\
&= \exp\left(-\frac{\lambda y}{2\mu^2} - \frac{\lambda}{2y} + \frac{\lambda}{\mu} + \frac{1}{2}\left\{\log\lambda - \log(2\pi) - 3\log y\right\}\right) \\
&= \exp\left(\frac{y(-1/\mu^2) - (-1/\mu)}{\lambda^{-1}} - \frac{1}{2}\left\{\frac{\lambda}{y} - \log\lambda + \log(2\pi) + 3\log y\right\}\right),
\end{aligned}
$$

where the canonical parameter is given by $\vartheta = -\mu^{-2}$ and the scale parameter by $\phi = \lambda^{-1}$. Moreover we have $b(\vartheta) = -(\mu)^{-1} = -(-2\vartheta)^{1/2}$ since $\mu = (-2\vartheta)^{-1/2}$, $a(\phi) = \phi = \lambda^{-1}$, and $c(y, \phi) = -\frac{1}{2}\left\{(\phi y)^{-1} + \log(2\pi\phi y^3)\right\}$. The mean and the variance of the preceding parametrized inverse Gaussian distribution is given by $E(Y) = \mu$ and $V(Y) = \mu^3/\lambda$. In GLMs the alternative parametrization $IGaussian(\mu, \lambda = 1/\sigma^2)$ is frequently used.

Other well-known distributions. Other well known distributions that are reported as members of the exponential family are the beta, Weibull, multinomial, and he Dirichlet distributions. The last two can be regarded as members of the multivariate extension of the exponential family, while the Weibull distribution can be rewritten as in (7.3) for a given dispersion parameter.

Table 7.1 Details of most common members of exponential dispersion family

Distribution	Notation	Values of Y	Mean	Variance	ϑ	$b(\vartheta)$	$a(\phi)^a$	$\mu(\vartheta)$	$c(y,\phi)$
Normal	$N(\mu,\sigma^2)$	$\mathbb{R}$	μ	σ^2	μ	$\vartheta^2/2$	σ^2	ϑ	$-\frac{1}{2}\log(2\pi\sigma^2)$ $-\frac{1}{2}y^2/\sigma^2$
Binomial	binomial(π,N)	$\{0,1,\dots,N\}$	$N\pi$	$N\pi(1-\pi)$	$\log[\pi/(1-\pi)]$	$N\log(1+e^\vartheta)$	1	$N/(1+e^{-\vartheta})$	$\log(N!/y!)$ $-\log(N-y)!$
Negative Binomial	NB(π,k)	$\mathbf{N}=\{0,1,2,\dots\}$	$k(1-p)/p$	$k(1-p)/p^2$	$\log(1-\pi)$	$-k\log(1-e^\vartheta)$	1	$ke^\vartheta/(1-e^\vartheta)$	$\log(y+k-1)!$ $-\log[y!(k-1)!]$
Poisson	Poisson(λ)	$\mathbf{N}=\{0,1,2,\dots\}$	λ	λ	$\log\lambda$	e^ϑ	1	e^ϑ	$\log(y!)$
Gamma	gamma$(a,b)^b$	$(0,\infty)$	$\mu=a/b$	$\mu^2/a=a/b^2$	$-\mu^{-1}=-b/a$	$-\log(-\vartheta)$	a^{-1}	$-\vartheta^{-1}$	$\phi^{-1}\log(y\phi^{-1})$ $-\log\left[y\Gamma(1/\phi)\right]$
Inverse Gaussian	IGaussian(μ,λ)	$(0,\infty)$	μ	μ^3/λ	$-\mu^{-2}$	$-(-2\vartheta)^{1/2}$	λ^{-1}	$(-2\vartheta)^{-1/2}$	$-\frac{1}{2}(\phi y)^{-1}$ $-\frac{1}{2}\log(2\pi\phi y^3)$

[a] $a(\phi)=\phi$.

[b] $b=a/\mu$.

7.1.3 Link functions

7.1.3.1 *Common link functions.* The link function is a monotonic and differentiable function used to match the parameters of the response variable with the systematic component, namely, the linear predictor and the associated covariates. Usually no restriction lies on the definition of such variables, but often we focus on the mean of the distribution because the measures of central location are usually of main interest. GLM-based extensions in which dispersion or shape parameters are linked with covariates also exist in statistical literature [e.g., see Rigby and Stasinopoulos (2005)]. A desirable property of the link function is to map the range of values in which the parameter of interest lies with the set of real numbers $\mathbb{R}$ in which the linear predictor takes values. For example, in the binomial case we wish to identify link functions that map the success probability π from $[0, 1]$ to $\mathbb{R}$.

The simplest link function is the one that sets the linear predictor equal to the mean μ. This is indeed the usual link function for the normal models. This function is not appropriate for other distributions such as the Poisson distribution since their mean is positive while $\eta \in \mathbb{R}$. In cases where the parameter of interest is positive, we usually adopt the log-link, which implies a multiplicative relation between covariates and the parameter of interest.

Usually the default choice of link function is provided by the *canonical link*, in which we set the canonical parameter (as an expression of the mean or other parameters of the distribution) equal to the linear predictor. The canonical link function for common distributions are summarized in Table 7.2.

Table 7.2 Canonical link functions of most common members of exponential dispersion family

Distribution	Link name	Link function $g(\mu)$	$g(\boldsymbol{\theta})$
Normal	Identity	μ	
Binomial	Logit	$\log\left[(\mu/N)/(1-\mu/N)\right]$	$\log\left[\pi/(1-\pi)\right]$
Negative binomial	Complementary log	$\log\left[\mu/(k+\mu)\right]$	$\log(1-\pi)$
Poisson	Logarithmic	$\log\lambda$	
Gamma	Reciprocal	$1/\mu$	
Inverse Gaussian	Squared reciprocal	$1/\mu^2$	

For binomial (and negative binomial) models, a wide variety of link functions exist. The canonical link is the so-called logit link defined as

$$g(\pi) = \log\left(\frac{\pi}{1-\pi}\right).$$

Other popular alternatives are the probit link (frequently used in econometrics) defined as

$$g(\pi) = \Phi^{-1}(\pi)$$

and the complementary log–log link function given by

$$g(\pi) = \log\left\{-\log(1-\pi)\right\},$$

where $\Phi(x)$ is the cumulative probability function of the standardized normal distribution and $\Phi^{-1}(\pi)$ is its corresponding inverse function.

General links can be adopted by considering a family of functions indexed by one or more (unknown) continuous-valued parameters. For a binomial response, such functions can be easily obtained by considering the inverse cumulative probability function $F^{-1}(\pi; \theta)$ of a random variable $Z \sim D(\theta)$. Parameter θ gives rise to a range of possible link functions that can be used. Even the link functions mentioned above (logit and the complementary log–log) can be obtained by the inverse cumulative probability functions of the logistic and the extreme value distributions.

Other link functions that may be used are the log–log link for binomial data, the square root $g(\mu) = \sqrt{\mu}$ (for distributions with positive mean), and the exponent $g(\mu) = (\mu + c_1)^{c_2}$ when $\mu > -c_1$ (Lindsey, 1997, p. 21).

7.1.3.2 *More complicated link functions for binomial data.*

A wide variety of link distributions have been proposed for binomial models. Here we briefly present some of the possible alternative link functions described in the related literature. For example, a straigtforward extension of the probit link is provided by considering the inverse t-link family proposed by Albert and Chib (1993) and the log-gamma link family proposed by Genter and Farewell (1985); see Ntzoufras et al. (2003) for the Bayesian implementation. The inverse t-link family is given by $g(\pi) = F_T(\pi; \nu)$, where $F_T(y; \nu)$ is the distribution function of a t distribution with $\nu > 1$ degrees of freedom. This link family includes as a special case the probit link function for large values of ν ($\nu \to \infty$). Furthermore, Albert and Chib (1993) argue that the t-link with $\nu = 8$ is a reasonable approximation to the logit link. The log-gamma link family is defined as

$$g(\pi) = \frac{1}{|\theta|} \log \left\{ \theta^2 F_G^{-1} \left(I(\theta > 0) \pi + I(\theta < 0)(1 - \pi); \ a = \theta^{-2}, b = 1 \right) \right\}.$$

This family includes as special cases the probit link ($\theta \to 0$), the log–log link ($\theta = -1$) and the complementary log–log link ($\theta = 1$). The link parameter θ controls the symmetry of the link function in contrast to the t-link which is symmetric for all values of ν.

Further link functions are provided by using mixtures of beta distributions as prosed by Mallick and Gelfand (1994) and normal scale mixtures as proposed by Basu and Mukhopadhyay (2000). Lang (1999) considers the link function

$$\pi = g^{-1}(\eta) = \sum_{k=1}^{3} w_k(\rho) F_k(\eta),$$

where $w_k(\rho)$ are mixing proportions, ρ is a mixing parameter to be estimated, and $F_k(y)$ are the cumulative distribution functions of the extreme minimum value, the logistic and extreme maximum value distribution for $k = 1, 2, 3$, respectively. Aranda-Ordaz (1981) and Albert and Chib (1997) used a family of symmetric links defined as

$$g(\lambda) = \frac{2}{\lambda} \times \frac{\pi^\lambda - (1 - \pi)^\lambda}{\pi^\lambda + (1 - \pi)^\lambda},$$

where $\lambda = 0.0, 0.4, 1.0$ correspond to the logit, (approximately) probit, and linear link functions, respectively. Aranda-Ordaz (1981) also proposed an asymmetric family given by

$$g(\pi) = \frac{(1 - \pi)^{-\lambda} - 1}{\lambda}.$$

Other approaches include the link family based on the inverse distribution function of the logarithm of an F-distributed random variable (Prentice, 1976), the two-parameter link family of Pregibon (1980) defined as

$$g(\pi) = \frac{\pi^{\lambda_1 - \lambda_2} - 1}{\lambda_1 - \lambda_2} - \frac{(1 - \pi)^{\lambda_1 + \lambda_2} - 1}{\lambda_1 + \lambda_2},$$

the Box–Cox transformation based link family (Guerrero and Johnson, 1982)

$$g(\pi) = \frac{1}{\lambda} \left\{ \left(\frac{\pi}{1 - \pi} \right)^{\lambda} - 1 \right\},$$

the generalization of the logit link suggested by Stukel (1988), and a family of robust link functions proposed by Haro-López et al. (2000). More details concerning available link functions and their Bayesian analysis and comparison for binomial models are provided by Ntzoufras et al. (2003) and Czado and Raftery (2006).

7.1.4   Common generalized linear models

Different models of the exponential family are appropriate for different types of response variables. In this section, we summarize which are the most common models for each type of response variable.

Response variables defined in $\mathbb{R}$.   In the case of continuous response with values defined over the whole range of real numbers, the normal regression model described Chapters 5 and 6 is the most popular choice. When the normality of error assumption is not appropriate, the normal model can be extended using errors that follow the Student's t distribution. Although this model cannot be considered as a member of the exponential family, it can be easily fitted using WinBUGS .

Positive continuous response variables.   When positive defined continuous response variables are considered, then one initial approach is to transform them using the Box–Cox transformation or the logarithm. Using a common (normal) regression model for the transformed response variable may lead to a plausible model. Such models can be treated in the same way as usual regression models since, after transforming the response variable, inference is exactly the same. It is common practice to first consider the logarithm of the original response variable as a possible transformation since in several cases it may eliminate problems related to the assumptions of the model such as the normality or errors, linearity of the mean, or homoscedasticity. This very simple strategy (i.e., using the logarithm of a variable as the response in a normal model) is equivalent to assuming the log-normal distribution for the original response variable. Other common distributional choices for positive continuous responses are the gamma, the exponential, the inverse Gaussian, and the Weibull distributions.

   A positive response variable is the survival time, or more generally the time until an event of interest occurs, which is of central interest in medical studies, especially in clinical trials. When modeling survival times, censoring is an additional characteristic that must be considered in the model. A survival time is considered as censored when part of its information is not available. For example, we may know that a patient was alive for the 50 first days of the study, but we might be ignorant of the exact time of failure. The Weibull distribution is commonly used for modeling such response data.

Binary (success/failure) responses. Binary (zero–one) data (i.e., $y \in \{0, 1\}$) can be modeled using the Bernoulli distribution. Moreover, when the response variable measures the number of successes after the repetition of n such experiments, then the binomial distribution with success probability π and n replications must be considered. In such cases, the response variable takes values from zero to n, $y \in \{0, 1, \ldots, n\}$. Note that in the case of binary data the Bernoulli distribution is equal to a binomial distribution with $n = 1$.

The canonical link is the logit function $\log\left(\pi/(1 - \pi)\right)$, which models the log-odds of success as linear combination of the covariates (Berkson, 1944, 1951). Logit models are the most popular stochastic formulations for such data and are cited as *logistic regression models*. Another popular link is the probit link (Bliss, 1935), which gives results similar to those for the logit link; see also Albert and Chib (1993) for the Bayesian implementation. Finally, complementary log–log link (Fisher, 1922) is a less popular link, but it models more efficiently the tails of the distribution, especially when asymmetry between low and high probability values is observed.

Counts and responses defined in **N** . Response variables defined in **N** $= \{0, 1, 2, \ldots, \}$ frequently represent number of events occurred within a prespecified time interval, namely, counts or frequencies. The Poisson distribution is naturally adopted in such cases resulting in Poisson regression models that are also referred to as *Poisson log-linear* (or simply *log-linear*) *models*, due to the canonical log-link adopted in most cases. Poisson log-linear models are also used for the analysis of high-dimensional contingency data, which result from the cross-classification of several categorical variables as introduced by Birch (1963).

The Poisson distribution implies the restrictive assumption of equal mean and variance, which has generated much discussion within the statistical community, leading to less restrictive models that allow for overdispersion (larger variance than mean) or underdispersion [e.g., see Lindsey (1997, sec. 2.3) and Agresti (2002, pp. 130–131) for a related discussion]. A popular distribution that allows for overdispersion is the negative binomial distribution.

Other response variables. For response variables that do not belong in the general cases described above, we propose as initial step transforming them in such way that they can be fitted using the preceding models. Otherwise, special models must be constructed following the logic of the generalized models.

For example, for variables that are defined in a range $y \in (a, b)$, we may rescale them in the zero–one interval by setting $y^* = (y - a)/(b - a)$ and use the beta distribution for the stochastic component. Another alternative is to use a logit-like transformation by setting $y^{**} = \log\left\{(y - a)/(b - y)\right\}$ and use normal regression models.

For categorical responses with $k > 2$ levels, the multinomial distribution may be used as a natural extension of the binomial models. The same distribution can be used for grouped categorical variables where the frequencies of k different outcomes will be recorded as responses. Finally, such responses can be modeled indirectly using the Poisson log-linear models for contingency tables when all covariates are categorical [see, e.g., Fienberg (1981, chaps. 6, 7)].

For response variables defined in the set of integer numbers $\left(y \in \mathbf{Z} = \{\ldots, -3, -2, -1, 0, 1, 2, 3, \ldots\}\right)$ a model based on the differences of Poisson latent variables has been developed by Karlis and Ntzoufras (2006, 2008). This model cannot be considered as a member of the exponential family, but the conditional likelihood when the latent (under estimation) Poisson variables are known is a simple Poisson likelihood. To fit the model, we can use the EM algorithm to estimate the posterior mode (or the MLE estimate) or simple MCMC algorithms as described by Karlis and Ntzoufras (2006).

Finally, ordinal variables can be modeled using a variety of alternative approaches that have been introduced in the literature. A natural extension of the usual log-linear model used for contingency models can be adopted by Goodman's (1979) association models, which were originally used for two-way contingency models. Such models cannot be considered as GLMs because of the multiplicative expression between the model parameters and Poisson's expected values.

7.1.5   Interpretation of GLM coefficients

Interpretation of GLM's coefficients is equivalent to the corresponding interpretation of the parameters in usual normal regression models. Thus, interest lies in (1) whether the effect of X_j is important for the prediction or description of Y, (2) the type of association between Y and X_j (positive, negative, linear, or other), and (3) the magnitude of the effect of X_j on Y.

Concerning the importance of the effect, we can simply monitor the posterior distribution and report whether the zero value is away from its center. Moreover, DIC may also be used to identify which model is more appropriate for the description of the available data, as implemented in the previous chapter; see Chapters 10 and 11 for other, more formal, approaches of model checking and comparison.

The type of the association (negative or positive) is simply indicated by the corresponding sign of the posterior summaries for each coefficient as in common regression models.

Concerning the interpretation of model parameters, interest lies in quantifying the effect of each covariate X_j on the corresponding parameter of interest of the response variable Y (usually on the mean of Y). Although this is straightforward in normal regression models, since the canonical link is used and, therefore, the effect of each X_j is linear to the mean of Y, it is slightly more complicated in GLMs and depends on the form of the link function. For this reason, we may interpret the effect of each covariate using one (or more) of the following approaches:

- Use the first-order differences or relative differences of the means.

- Use the marginal effect of X_j on the mean.

- Calculate and present the means of Y for specific plausible combinations of X_j values.

The mean of Y can be substituted by any other parameter of interest. Further details on these approaches can be found in Futing Liao (1994, pp. 6–9) and in Aldrich and Nelson (1984, pp. 75–80).

Using first-order differences is the simplest approach. With this approach, we express model parameters β_j as the effect of one unit increase on the mean (or other parameter of interest). For example, in the simple case of the normal regression model with one covariate we have

$$
\begin{aligned}
E(Y|X = x + 1) &= \beta_0 + \beta_1(x + 1) = \beta_0 + \beta_1 x + \beta_1 = E(Y|X = x) + \beta_1 \Leftrightarrow \\
\Delta\mu &= E(Y|X = x + 1) - E(Y|X = x) = \beta_1,
\end{aligned}
$$

where $E(Y|X = x)$ is the mean of Y when $X = x$. When the link function is a logarithmic expression of the parameter of interest, then the effect on this parameter is multiplicative and is usually expressed as a relative difference or as a percentage change. For example, in a Poisson log-linear model with one covariate, $E(Y|X = x + 1) = E(Y|X = x)e^{\beta_1}$.

In specific cases, it is difficult to express the difference or the relative difference as a direct function of the model coefficients β_j. In such cases, it is useful to consider the *marginal effect* of X_j on a parameter of interest of Y, which is given by the first derivative of the parameter of interest over X_j

$$d\theta_{X_j} = \frac{\partial \theta}{\partial X_j} = \frac{\partial \theta}{\partial \eta}\frac{\partial \eta}{\partial X_j} = \frac{\partial g^{-1}(\eta)}{\partial \eta}\beta_j \, ,$$

where $g(\theta)$ is the link function of the parameter of interest θ. This expresses the rate of increase or decrease of θ when the covariate is equal to x. It can be considered as a rough approximation of the increase of θ when X_j increases by one unit and the other covariates remain the same; for details, see Futing Liao (1994, pp. 6–9) and Aldrich and Nelson (1984, pp. 77).

In the case that we consider the canonical parameter ϑ and the corresponding canonical link, the marginal effect on the canonical parameter is equal to β_j for all members of the exponential family since $\theta = \vartheta$ and $g(\vartheta) = \vartheta$.

Finally, it is convenient to provide the predicted means (or other parameters) under the estimated model for several scenarios or profiles of individual observations. Plausible scenarios can be the average, the median, and the extreme low or high profiles (worst and best-case scenarios). The first two profiles refer to the calculation of θ values for observations with all covariates equal to the sample mean or medians. Alternatively, population means and medians can also be used if available. This produces an estimate of θ for an average or typical person in our sample (or population). The worst-case scenario (low profile) involves the computation of θ by setting all covariates to the minimum or maximum values for covariates with positive or negative effect on Y, respectively. Similarly, the best-case scenario (high profile) involves the computation of θ by setting all covariates to the maximum or minimum values for covariates with positive or negative effect on Y, respectively. Similar estimates can be estimated by calculating all θ_i ($i = 1, \ldots, n$) and reporting the covariate values for $\min_{i=1}^{n} \theta_i$ and $\max_{i=1}^{n} \theta_i$, respectively.

Additional intuition concerning the effect of each X_j on Y can be also obtained if we consider increasing or decreasing one X_j at a time from a basic profile (e.g., the mean one). Although this does not directly provide the effect of the model parameters, it does provide an overall view of how Y is affected by each X_j. An advantage of this approach is that it can be easily comprehended by people with no quantitative background, in contrast to the other two approaches, which may appear as more technical.

7.2   PRIOR DISTRIBUTIONS

In this section we focus on prior distributions of parameters β involved in the linear predictor of (7.1) given the dispersion parameter ϕ.

Usually, in normal prior distributions of the type

$$\beta_j | \phi \sim N(\mu_{\beta_j}, \sigma^2_{\beta_j}\phi) \, .$$

The variance depends on the dispersion parameter ϕ in order to achieve an appropriate scaling of the prior distribution. In the normal case, if $\phi = \sigma^2 \sim \mathrm{IG}(a, b)$ then we endup with the corresponding conjugate prior distribution discussed previously. When no prior information is available, the prior mean is set equal to zero, while the corresponding variance is set large to express prior ignorance. Alternatively, a prior independent to the dispersion parameter can be considered. When the variance is set large to express prior ignorance, then no differences in the resulting posterior distribution will be observed.

Independent priors are plausible when the design or data matrix is orthogonal since, in such cases, model parameters have similar interpretation over all models. We can easily incorporate such priors in ANOVA-type models with sum-to-zero constraints. When we are interested in prediction rather than description of variable relations, we may orthogonalize the design matrix and proceed with model selection in the new orthogonal model space (Clyde et al., 1996). In nonorthogonal cases, especially when high dependences among covariates exist, the use of such a prior setup may result in undesirable influence on the posterior distribution and hence must be avoided. Independent prior distributions also result if we consider the prior of Knuiman and Speed (1988) for Poisson log-linear models in high dimensional contingency tables using the sum-to-zero parametrization.

This prior assumes prior independence between model parameters. This may be problematic when prior information is available or when collinear covariates exist, resulting in an undesirable effect to the posterior distribution. For this reason, a multivariate normal prior can be considered instead with

$$\boldsymbol{\beta}|\phi \sim N(\boldsymbol{\mu}_\beta, \boldsymbol{\Sigma}_\beta) \,.$$

An extension of the Zellner's g-prior considered in normal models can also be adopted here if we set the prior variance covariance matrix equal to

$$\boldsymbol{\Sigma}_\beta = c^2 \Big(-H(\widehat{\boldsymbol{\beta}}) \Big)^{-1}, \tag{7.5}$$

where $\widehat{\boldsymbol{\beta}}$ is the maximum-likelihood estimate and $H(\boldsymbol{\beta})$ is the second derivative matrix of $\log f(\boldsymbol{y}|\boldsymbol{\beta}, \phi)$, which in GLMs is given by

$$-H(\boldsymbol{\beta}) = \boldsymbol{X}^T \boldsymbol{H} \boldsymbol{X},$$

where $\boldsymbol{H}$ is a $n \times n$ diagonal matrix with elements

$$h_i = \left(\frac{\partial \mu_i}{\partial \eta_i} \right)^2 \frac{1}{a_i(\phi) b''(\vartheta)} \,. \tag{7.6}$$

Details concerning h_i for some popular distributions are provided in Table 7.3.

Table 7.3 Generalized linear model weights h_i

Model	Link	GLM weights h_i
Normal	Identity	σ^{-2}
Poisson	Log	λ_i
Binomial	Logit	$N_i \pi_i (1 - \pi_i)$
	Probit[a]	$N_i \big[\pi_i (1 - \pi_i) \{\varphi(\pi_i)\}^2 \big]^{-1}$
	clog–log	$-N_i (1 - \pi_i) \{\log(1 - \pi_i)\}^2 \pi_i^{-1}$

[a] $\varphi(z)$ is the density function of standardized normal distribution evaluated at z.

A special case of the preceding multivariate normal prior distribution when the variance–covariance matrix (7.5) is the *unit information prior* for $c^2 = n$. This prior has precision approximately equal to the precision provided by one data point. More detailed discussion of this prior can be found in Spiegelhalter and Smith (1988) and Kass and Wasserman (1995).

For the normal distribution, since h_i are all equal to the precision σ^{-2} of the regression model, the resulting prior will be equivalent to Zellner's g-prior. For the remaining models, h_i depends on estimated parameter values for each case [e.g., h_i will be equal to $\widehat{\lambda}_i = \exp\left(\boldsymbol{X}_{(i)}\widehat{\boldsymbol{\beta}}\right)$ in the Poisson case and a function of the success rate $\widehat{\pi}_i = [1 + \exp(-\boldsymbol{X}_{(i)}\widehat{\boldsymbol{\beta}})]^{-1}$ in the binomial case], resulting in a data-dependent prior. When the variance inflation parameter c^2 is set equal to large, the effect of this dependence will be minimal since the prior will be essentially noninformative. To avoid this data dependence, Ntzoufras et al. (2003) proposed using the prior mean to obtain rough prior estimates of h_i. For example, in binomial logistic regression models the prior model weight is set equal $h_i = N_i \exp\left(\boldsymbol{X}_{(i)}\boldsymbol{\mu}_\beta\right)\left[1 + \exp\left(\boldsymbol{X}_{(i)}\boldsymbol{\mu}_\beta\right)\right]^{-2}$, while for the common case of a zero mean $h_i = N_i/4$. The latter is even more simplified to a prior variance covariance matrix equal to $\boldsymbol{\Sigma}_\beta = 4N^{-1}c^2\left(\boldsymbol{X}^T\boldsymbol{X}\right)^{-1}$ when $N_i = N$ for all $i = 1, 2, \ldots, n$.

7.3 POSTERIOR INFERENCE

7.3.1 The posterior distribution of a generalized linear model

For generalized linear models under the general setup (7.1) and responses Y_i with probability or density function (7.2), the likelihood is given by

$$f(\boldsymbol{y}|\boldsymbol{\beta}, \phi) = \exp\left(\sum_{i=1}^{n} \frac{y_i g_\vartheta^{-1}\left(\boldsymbol{X}_{(i)}\boldsymbol{\beta}\right) - b\left(g_\vartheta^{-1}\left(\boldsymbol{X}_{(i)}\boldsymbol{\beta}\right)\right)}{a(\phi)} + \sum_{i=1}^{n} c(y_i, \phi)\right).$$

where we have assumed a common dispersion parameter for all observations. When different dispersion parameters are assumed, then we simply substitute ϕ by ϕ_i in the likelihood.

Using the multivariate normal prior described in Section 7.2, we end up with the posterior

$$f(\boldsymbol{\beta}, \phi|\boldsymbol{y}) \quad \propto \quad \exp\left(\sum_{i=1}^{n} \frac{y_i g_\vartheta^{-1}\left(\boldsymbol{X}_{(i)}\boldsymbol{\beta}\right) - b\left(g_\vartheta^{-1}\left(\boldsymbol{X}_{(i)}\boldsymbol{\beta}\right)\right)}{a(\phi)} + \sum_{i=1}^{n} c(y_i, \phi)\right.$$
$$\left. -\frac{1}{2}\log|\boldsymbol{\Sigma}_\beta| - \frac{1}{2}(\boldsymbol{\beta} - \boldsymbol{\mu}_\beta)^T\boldsymbol{\Sigma}_\beta^{-1}(\boldsymbol{\beta} - \boldsymbol{\mu}_\beta)\right) f(\phi),$$

which simplifies to

$$f(\boldsymbol{\beta}|\phi, \boldsymbol{y}) \propto \exp\left(\sum_{i=1}^{n} \frac{y_i g_\vartheta^{-1}\left(\boldsymbol{X}_{(i)}\boldsymbol{\beta}\right) - b\left(g_\vartheta^{-1}\left(\boldsymbol{X}_{(i)}\boldsymbol{\beta}\right)\right)}{a(\phi)} - \frac{1}{2}(\boldsymbol{\beta} - \boldsymbol{\mu}_\beta)^T\boldsymbol{\Sigma}_\beta^{-1}(\boldsymbol{\beta} - \boldsymbol{\mu}_\beta)\right)$$

when the dispersion parameter is fixed, as, for example, in binomial models, where $f(\phi)$ in the full posterior is the prior of the dispersion parameter ϕ.

This posterior and its corresponding summaries cannot be evaluated analytically, except for the normal model when using the conjugate prior described in Section 1.5.5. Although approximation methods can be used, MCMC methods are now available and widely used for the computation of the posterior distribution. Specifically, the Gibbs sampler can be easily applied because of the result obtained by Dellaportas and Smith (1993), which allowed for implementation of the adaptive rejection method of Gilks and Wild (1992) since the posterior distributions of the parameters in specific GLMs is log-concave. Alternatively, Metropolis–Hastings algorithms or the slice sampler can be used; see Chapter 2 for an example in logistic regression models. This method is also used in WinBUGS for the generation of random values from the posterior distribution of GLMs.

7.3.2   GLM specification in WinBUGS

WinBUGS code for the specification of a GLM is similar to the corresponding one for the specification of a normal regression model. Differences lie in the specification of the stochastic component and the link function.

The stochastic component will be now defined using the appropriate distribution for the response variable. Details concerning the distributions of the most popular GLMs are summarized in Table 7.4. Note that the inverse Gaussian distribution is not included in the standard distributions of WinBUGS ; however, it can be modeled using an alternative approach presented in Section 8.1.

Four link functions are available in WinBUGS: `log`, `logit`, `probit`, and the `cloglog`. Commands for the specification of link functions in WinBUGS can be used only in the left part of the definition of the linear predictor . The remaining link functions can be defined by setting the parameter of interest θ_i equal to $g^{-1}(\eta_i)$.

7.4   POISSON REGRESSION MODELS

In this section we focus on Poisson regression models for response variables defined in $\mathbf{N}$. Such variables usually express the number of successes (visits, telephone calls, number of scored goals in football) within a fixed time interval. They are frequently called *Poisson log-linear models* because of the canonical log-link, which is widely used.

The Poisson log-linear model is summarized by the following expression:

$$Y_i \sim \text{Poisson}(\lambda_i) \text{ with } \log \lambda_i = \beta_0 + \sum_{j=1}^{p} \beta_j x_{ij} = \mathbf{X}_{(i)}\boldsymbol{\beta} \ .$$

7.4.1   Interpretation of Poisson log-linear parameters

In Poisson log-linear models, the effect of each X_j is linear to the log-mean of Y, resulting in an exponential effect of X_j on the mean of Y.

Let us first consider the simplest case where only one covariate involved. Then the mean of Y can be expressed as

$$
\begin{aligned}
\log \lambda_i &= \beta_0 + \beta_1 X_i \Leftrightarrow \\
\lambda_i &= e^{\beta_0} e^{\beta_1 x_i} \\
&= B_0 B_1^{x_i} \text{ where } B_j = e^{\beta_j} \text{ for } j = 0, 1,
\end{aligned}
$$

Table 7.4 WinBUGS commands for distributions within the exponential family[a]

Distribution name	WinBUGS syntax	Probability or density function $f(x)$	Mean	Variance
1. Normal	y ~ dnorm(mu,tau)	$\sqrt{\tau/(2\pi)}\exp\left[-\frac{1}{2}\tau(y-\mu)^2\right]$	μ	$1/\tau$
(Log-normal)[b]	y ~ dlnorm(mu,tau)	$\sqrt{\tau/(2\pi)}y^{-1}\exp\left[-\frac{1}{2}\tau(\log y - \mu)^2\right]$	$e^{\mu+1/(2\tau)}$	$(e^{1/\tau}-1)e^{2\mu+1/\tau}$
2. Binomial	y ~ dbin(p,N)	$N!p^y(1-p)^{N-y}/[y!(N-y)!]$	Np	$Np(1-p)$
(Bernoulli)[c]	y ~ dbern(p)	$p^y(1-p)^{1-y}$	p	$p(1-p)$
3. Negative binomial	y ~ dnegbin(p,r)	$(y+r-1)!p^r(1-p)^y/[y!(r-1)!]$	$r(1-p)p^{-1}$	$r(1-p)p^{-2}$
4. Poisson	y ~ dpois(lambda)	$e^{-\lambda}\lambda^y/y!$	λ	λ
5. Gamma	y ~ dgamma(a,b)	$b^a y^{a-1}e^{-by}/\Gamma(a)$	a/b	a/b^2
(Chi-squared)[d]	y ~ dchisqr(k)	see gamma$(k/2,\frac{1}{2})$	k	$2k$
(Exponential)[e]	y ~ dexp(lambda)	$\lambda e^{-\lambda y}$	$1/\lambda$	$1/\lambda^2$

[a] Terms in parentheses can be considered as special cases of the distributions shown above.
[b] $\log(y)$ follows the normal distribution.
[c] Binomial with $N=1$.
[d] Gamma with $a=k/2$ and $b=\frac{1}{2}$.
[e] Gamma with $a=1$ and $b=\lambda$.

where B_0 denotes the expected counts (or Y) when the covariate is equal to zero ($X = 0$). Interpretation of β_1 is slightly different from the corresponding one in normal models since relative mean differences are considered in the Poisson case. Let us denote by $\lambda(x) = E(Y|X = x)$ the expected counts (Y) for covariate with $X = x$. Then

$$\log\big(\lambda(x + 1)\big) - \log\big(\lambda(x)\big) = \beta_1,$$

resulting in

$$\lambda(x + 1) = B_1 \lambda(x) = e^{\beta_1}\lambda(x).$$

Hence, when the covariate X is increased by one unit, then the expected Y becomes equal to B_1 times the corresponding value of Y for $X = x$. An even more comprehensive interpretation can be based on the percentage change of the expected Y given by $(B_1 - 1) \times 100$ when X increases by one unit. The type of association between X and Y is highlighted by the sign of β_1 (and $B_1 - 1$) as in the normal regression models.

When X is categorical with K levels, then the linear predictor is expressed as a linear function of $K - 1$ dummy variables denoted by D_j for $j = 2, \ldots, K$. Let us consider the simpler case of corner parametrization with the first one ($j = 1$) as the reference/baseline category and hence setting $\beta_1 = 0$. Then we express the model by

$$\begin{aligned}
\log \lambda_i &= \beta_0 + \sum_{j=2}^{K} \beta_j D_{ij} \Leftrightarrow \\
\lambda_i &= e^{\beta_0} \exp\left(\sum_{j=2}^{K} \beta_j D_{ij}\right) \\
&= B_0 \prod_{j=2}^{K} B_j^{D_{ij}} \text{ where } B_j = e^{\beta_j} \text{ for } j \in \{0, 2, 3, \ldots, K\}.
\end{aligned}$$

When individual i belongs in the first category of X (i.e., $X_i = 1$), then $\lambda_i = B_0$, while when individual i belongs in the kth category ($k > 1$) of X (i.e., $X_i = k$), then

$$\lambda(X = k) = B_0 B_k = B_k \lambda(X = 1).$$

Therefore, quantity $B_k = e^{\beta_k}$ can be now interpreted as the relative change of the Poisson expectation λ when an individual belongs in k category of X compared to the baseline/reference category. The interpretation for the STZ parametrization is similar, but all coefficients express the relative change of the current level compared with an overall "average" level instead of the baseline category used in corner parametrization.

This is similar to the interpretation of the parameters in the multiple Poisson regression case. The difference here is that in every change of a single explanatory variable (say, X_j), other covariates need to remain the constant since

$$\lambda_i = B_0 \prod_{j=1}^{p} B_j^{x_{ij}} \text{ with } B_j = e^{\beta_j} \text{ for } j = 0, 1, 2, \ldots, p.$$

In Bayesian inference, a usual point estimate for the model parameters is provided by the posterior means (or medians). Concerning the estimation of relative difference, the posterior distribution of B_j and its posterior summaries may be considered directly by using a simple deterministic/logical node in WinBUGS . Alternatively, the exponent of the posterior mean or median of β_j can be used as sensible estimates since they correspond to the posterior harmonic mean and the posterior median of B_j, respectively. Moreover, exponentiating either the posterior mode or any posterior quantile provides the corresponding posterior summaries for B_j. Finally, calculation of the posterior standard deviation itself must be calculated directly by using deterministic (logical) nodes in WinBUGS given by $B_j = e^{\beta_j}$.

7.4.2   A simple Poisson regression example

Example 7.1. Aircraft damage dataset. Here we consider the aircraft damage dataset of Montgomery et al. (2006). The dataset refers to the number of aircraft damages in 30 strike missions during the Vietnam war. Hence it consists of 30 observations and the following four variables:

- `damage`: the number of damaged locations of the aircraft

- `type`: binary variable which indicates the type of plane (0 for A4; 1 for A6)

- `bombload`: the aircraft bomb load in tons

- `airexp`: the total months of aircrew experience

In this example we can use the Poisson distribution to monitor the number of damages after each mission.

Data of this example are available in the book's Website and are reproduced with permission of John Wiley and Sons, Inc.

7.4.2.1 *Model specification in WinBUGS*.   The initial model will have the following structure

$$
\begin{aligned}
\text{damage}_i &\sim \text{Poisson}(\lambda_i) \\
\log \lambda_i &= \beta_1 + \beta_2 \, \text{type}_i + \beta_3 \, \text{bombload}_i + \beta_4 \, \text{airexp}_i \\
&\quad \text{for} \quad i = 1, 2, \ldots, 30 \,.
\end{aligned}
$$

Here, the index of β_j takes values from 1 to 4 (instead from 0 to 3 as in the previous section) to be in concordance with the WinBUGS the code that follows.

We follow the same structure as in the linear regression model with the difference that the likelihood is now defined using the following syntax:

```
for (i in 1:30){
    damage[i] ~ dpois( lambda[i] )
    log(lambda[i]) <- beta[1] + beta[2] * type[i]
                    + beta[3] * bombload[i] + beta[4] * airexp[i]
}
```

Moreover, the exponentiated parameters B_j can be easily defined using the syntax

```
for (j in 1:4){ B[j] <- exp( beta[j] ) }
```

The usual independent normal prior with large variance ($\tau_{\beta_j} = \sigma_{\beta_j}^{-2} = 10^{-4}$) is considered as prior distribution for β_j. The full code is available in this book's Webpage.

7.4.2.2 Results. Posterior summaries of model parameters are given in Table 7.5, while 95% posterior intervals are depicted in Figure 7.1.

Table 7.5 Posterior summaries of Poisson model parameters for Example 7.1[a]

node	mean	sd	MC error	2.5%	median	97.5%	start	sample	harmonic
beta[1]	-0.766	1.089	0.1762	-3.168	-0.835	1.619	1001	1000	
beta[2]	0.580	0.466	0.0513	-0.302	0.584	1.537	1001	1000	
beta[3]	0.177	0.068	0.0099	0.040	0.177	0.308	1001	1000	
beta[4]	-0.011	0.010	0.0015	-0.033	-0.010	0.007	1001	1000	
B[1]	0.862	1.221	0.1829	0.042	0.434	5.050	1001	1000	0.465
B[2]	1.993	0.996	0.1050	0.739	1.793	4.652	1001	1000	1.786
B[3]	1.197	0.081	0.0118	1.041	1.193	1.360	1001	1000	1.194
B[4]	0.989	0.010	0.0015	0.968	0.990	1.007	1001	1000	0.989

[a]The harmonic means of B_j are calculated outside WinBUGS using the posterior means of β_j.

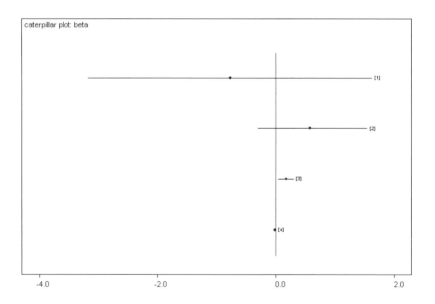

Figure 7.1 95% posterior intervals of Poisson model parameters for Example 7.1.

A point estimate of the model can be based on the posterior means. Hence the a posteriori estimated model can be summarized by

$$\log \lambda_i = -0.77 + 0.58\,\text{type}_i + 0.18\,\text{bombload}_i - 0.011\,\text{airexp}_i \ .$$

From the 95% posterior intervals of β_j, we observe that only the posterior distribution of the bombload coefficient is away from zero, indicating a significant effect of this variable on the amount of aircraft damage.

7.4.2.3 Interpretation of the model parameters. Interpretation of the model parameters can be directly based on B_j values (B[] in WinBUGS). From Table 7.5, we may conclude the following

- The expected amount of damage for A6 (type=1) aircraft is twice as much as the corresponding damage for A4 (type=0) aircraft when the two aircraft return from missions with aircrew of the same experience and both carry the same bombload. If we base our inference on the medians or the harmonic means, then the A6 is a posteriori expected to have 79% more extensive damage than an A4 aircraft with the same bombload and aircrew exprerience.

- Every tone of bombload increases the expected number of damaged aircraft locations by 20%.

- Every additional month of aircrew experience reduces the number of damaged aircraft locations by 1%.

7.4.2.4 *Estimating specific profiles.*

The expected amount of damage for the two types of aircraft for the minimum, maximum, mean and median profiles have also been calculated. In the minimum and maximum profiles, the maximum and the minimum values of crew experience was considered, respectively, since these variables are negatively associated with the number of damaged locations.

Calculation of the expected value for a profile can be easily accommodated in WinBUGS. For example, for a profile of an A6 aircraft, the expected amount of damage is calculated by

```
a6.profile <- exp( beta[1] + beta[2] + beta[3] * bombload.profile
                                     + beta[4] * airexp.profile
```

where `bombload.profile` and `airexp.profile` are the values of the two explanatory variables for the profile that we wish to consider. Substitution of these nodes by appropriate values provides the desired profiles; see Table 7.6 for the corresponding code. The profiles for A4 are obtained similarly by removing parameter β_2. Note that the minimum and maximum values of a vector v can be obtained using the commands `ranked(v[],1)` and `ranked(v[],n)`, respectively. Similarly, the median profile can be calculated using the command

Table 7.6   WinBUGS syntax for calculation of expected number of damaged locations for each profile for Example 7.1

```
# profiles
# values for bombload
profiles[1,1]<- ranked( bombload[], 1 )   # minimum of bombload
profiles[2,1]<- mean(bombload[])          # mean of bombload
profiles[3,1]<- 0.5*(ranked( bombload[],15)+ranked(bombload[],16))
        #median
profiles[4,1]<- ranked( bombload[], 30 ) # max
# values for airexp
profiles[1,2]<- ranked( airexp[], 1 )   # max experience
profiles[2,2]<- mean(airexp[])          # mean
profiles[3,2]<- 0.5*(ranked( airexp[], 15)+ranked(airexp[], 16))
      # median
profiles[4,2]<- ranked( airexp[], 30 ) # min experience

for (k in 1:4){
    a4.profile[k]<-exp(beta[1] + beta[3]*profiles[k,1] + beta[4]*
        profiles[k,2])
    a6.profile[k]<-a4.profile[k]*exp( beta[2] )
}
```

```
ranked(v[],(n+1)/2)
```

if n is odd and by

```
0.5*(ranked(v[],n/2)+ranked(v[],n/2+1))
```

if n is even.

Posterior means and the corresponding standard deviations of these profiles are provided in Table 7.7. For a typical mission with A4 aircraft we expect 0.8 damaged locations, while for A6 the corresponding number of damaged locations is about 1.3. Note that the worst-case scenario (maximum profile) where missions with 14 tons of bombload and crew with the minimum flying experience (50 months) corresponds to an expected number of 3.7 and 5.9 damaged locations for A4 and A6 aircrafts, respectively.

Table 7.7 Posterior means (standard deviations) of expected number of damaged locations for minimum, mean, median, and maximum profiles for Example 7.1

| | | | Expected damage | |
| | | | A4 | A6 |
Profile	Bombload	Experience	mean (SD)	mean (SD)
Minimum	4.0	120.00	0.27 (0.13)	0.50 (0.27)
Median	7.5	80.25	0.75 (0.24)	1.22 (0.94)
Mean	8.1	80.77	0.83 (0.27)	1.33 (0.40)
Maximum	14.0	50.00	3.68 (2.11)	5.90 (1.75)

7.4.2.5 Selection of variables using DIC.

Here only three covariates are considered, resulting in eight possible models. All models can be simultaneously fitted in Win-BUGS and then be identified as the best one (i.e. with the lowest DIC value). Code for fitting all models in a single run is provided in this book's Website. Results are summarized in Table 7.8 after 10,000 burnin and 10,000 additional iterations. Be careful to consider a sufficiently long burnin period because DIC is sensitive to initial values.

Table 7.8 DIC values for all eight models under consideration for Example 7.1[a]

	Dbar	Dhat	pD	DIC	Model
y1	108.6	107.6	1.01	109.6	Constant
y2	94.0	92.0	1.99	96.0	Type
y3	84.8	82.9	1.91	86.7	Bombload
y4	85.3	82.4	2.95	88.3	Type + Bombload
y5	106.2	104.3	1.97	108.2	Airexp
y6	88.9	85.9	3.01	92.0	Type + Airexp
y7	83.9	81.0	2.92	86.9	Bombload + Airexp
y8	83.7	79.7	3.98	87.7	Type + Bombload + Airexp
total	735.6	715.9	19.76	755.4	

[a]Burnin=10,000; iterations kept=10,000.

According to the lowest DIC (86.7) value, only the bombload must be retained in the linear predictor. Moreover, the DIC value (86.9) of the model with covariates for both the bombload and the crew experience is very close to the lowest DIC value. This is an indication that the two models have similar predictive abilities, and therefore crew experience may also be an important determinant of the number of damaged locations.

7.4.3   A Poisson regression model for modeling football data

Example 7.2. Modeling the English premiership football data. Modeling of football scores is becoming increasingly popular nowadays. In the present example we use the English premiership data for the season 2006–2007 to fit a simplified Poisson log-linear model for the prediction of model outcomes. Data were downloaded from the Webpage `http://soccernet-akamai.espn.go.com`.

7.4.3.1   *Background information and the model.*   The model was by Maher (1982) and was used by other authors, including Lee (1997) and Karlis and Ntzoufras (2000). Let us denote by y_{i1} and y_{i2} respectively the goals scored by home and away teams (HT and AT) in the ith game. Then the model can be expressed by

$$
\begin{aligned}
Y_{ij} &\sim \text{Poisson}(\lambda_{ik}) && \text{for } j = 1, 2 \\
\log(\lambda_{i1}) &= \mu + \text{home} + a_{\text{HT}_i} + d_{\text{AT}_i} \\
\log(\lambda_{i2}) &= \mu \qquad\quad + a_{\text{AT}_i} + d_{\text{HT}_i} && \text{for } i = 1, 2, \ldots, n,
\end{aligned}
$$

where n is the number of games, μ is a constant parameter; home is the home effect; HT_i and AT_i are the home and away teams, respectively, competing in the ith game; a_k and d_k are the attacking and defensive effects–abilities of k team for $k = 1, 2, \ldots, K$; and K is the number of teams in the dataset under consideration (here $K = 20$).

For attacking and defensive parameters (a_k and d_k), we use the sum-to-zero constraints in order to make the model identifiable and compare the ability of each team with an overall level of attacking and defensive abilities. Hence we set

$$
\sum_{k=1}^{K} a_k = 0 \text{ and } \sum_{k=1}^{K} d_k = 0. \tag{7.7}
$$

According to this parametrization, all parameters have a straightforward interpretation. Parameter μ denotes an overall level of log–expected goals scored in away games, while parameter $home$ encapsulates the home effect denoted by the difference between the log–expected goals scored when two teams of equal strength compete with each other. Attacking and defensive parameters a_k and d_k can be interpreted as deviations of the attacking and defensive abilities from the average level in the league. Hence, a positive attacking parameter indicates that the team under consideration has an offensive performance that is better than the average level of the teams competing in the league. Similarly, negative defensive parameters indicate teams with defensive performance better than the average level of the teams competing in the league.

The Poisson regression model adopted for the goals scored by each team conceals two important assumptions that are questionable in the sports modeling literature: the independence between home and away goals and the equality between the mean and the variance. Empirical evidence and exploratory analysis has shown a (relatively low) correlation between the goals in a football game. This correlation can be incorporated in the analysis by modeling the full score using the bivariate Poisson distribution (Karlis and Ntzoufras,

2003a) or by modeling the goal differences using Skellam's distribution (Karlis and Nt-zoufras, 2008). These models are natural extensions of the simplified model presented here. Concerning the Poisson assumption, slight overdispersion has been reported in literature; see Karlis and Ntzoufras (2000) for a discussion. This can be incorporated to the model using a negative binomial distribution; see, for example, Reep and Benjamin (1968), Reep et al. (1971), and Baxter and Stevenson (1988).

Another problem appearing in football data is the excess of specific scores, especially the 0–0 and 1–1 draws. Dixon and Coles (1997) provided an extension based on the Poisson model allowing for extra probabilities in these scores. In a similar fashion, Karlis and Ntzoufras (2003a) have proposed using a diagonal inflated bivariate Poisson model and more recently a zero inflated model for the goal differences (Karlis and Ntzoufras, 2008).

Other models include the dynamic model for paired data by Fahrmeir and Tutz (1994) and the state space model of Rue and Salvesen (2000). Related models and optimal prediction schemes have been applied by Kuonen (1996, 1997) for European national football club tournaments and by Kuonen and Roehrl (2000) for the France'98 World cup data.

7.4.3.2 *Model specification in WinBUGS*.

In this example we use four variables: the home and away goals and the home and away teams (using codes from 1 to 20), termed goals1, goals2, ht, and at in WinBUGS code which follows. The model can be specified by

```
for (i in 1:n){
    # stochastic component
    goals1[i] ~ dpois( lambda1[i] )
    goals2[i] ~ dpois( lambda2[i] )
    # linear predictor
    log(lambda1[i]) <- mu + home + a[ ht[i] ] + d[ at[i] ]
    log(lambda2[i]) <- mu        + a[ at[i] ] + d[ ht[i] ]
}
```

Note that the STZ constraints (7.7) can be imposed in WinBUGS by setting one set of parameters effects (e.g., here we use $k = 1$) equal to

$$a_1 = -\sum_{k=2}^{K} a_k \text{ and } d_1 = -\sum_{k=2}^{K} d_k$$

using the WinBUGS syntax

```
a[1] <- -sum( a[2:K] )
d[1] <- -sum( d[2:K] )
```

Prior distributions must be defined for the remaining parameters: μ, *home* and a_k, d_k for $k = 2, \ldots, K$. The usual normal low-information prior with zero mean and large prior variance are used here (with prior precision equal to 10^{-4}). A large amount of historical data are available in sports. Such information can be used to form a plausible prior distribution, which will be extremely useful in the first weeks of a competition when a limited amount of data are available. Elicitation of such historical data and their potential usefulness, especially in the beginning of each season (where limited data are available), must be carefully examined.

7.4.3.3 *Results*.

Posterior summaries of the Poisson log-linear model parameters are provided in Table 7.9. Posterior credible intervals of attacking and defensive parameters for each team are depicted in Figures 7.2 and 7.3. As we can see from the estimated model parameters, Manchester United had the highest attacking parameter while Chelsea had the lowest (i.e., best) defensive parameter. In a game where two average teams are competing each other, then the expected numbers of goals are equal to 1.32 for the home team and 0.90 for the away team, resulting in an increase of 46% of each team scoring mean when playing in its home field.

Table 7.9 Posterior summaries of expected number of goals for Example 7.2

	Team[a]	Node	Posterior Mean	SD	percentiles 2.5%	97.5%	Node	Posterior Mean	SD	percentiles 2.5%	97.5%
1.	Arsenal	a[1]	0.33	0.12	0.08	0.57	d[1]	-0.23	0.16	-0.57	0.08
2.	Aston Villa	a[2]	-0.05	0.15	-0.34	0.23	d[2]	-0.10	0.16	-0.42	0.19
3.	Blackburn	a[3]	0.16	0.14	-0.13	0.42	d[3]	0.20	0.14	-0.08	0.45
4.	Bolton	a[4]	0.06	0.14	-0.23	0.32	d[4]	0.15	0.14	-0.13	0.41
5.	Charlton	a[5]	-0.26	0.16	-0.61	0.04	d[5]	0.28	0.13	0.03	0.52
6.	Chelsea	a[6]	0.33	0.12	0.08	0.56	d[6]	-0.62	0.20	-1.04	-0.25
7.	Everton	a[7]	0.14	0.14	-0.13	0.41	d[7]	-0.22	0.17	-0.56	0.09
8.	Fulham	a[8]	-0.14	0.16	-0.47	0.16	d[8]	0.29	0.13	0.02	0.53
9.	Liverpool	a[9]	0.22	0.13	-0.04	0.47	d[9]	-0.51	0.19	-0.90	-0.16
10.	Man City	a[10]	-0.44	0.18	-0.82	-0.09	d[10]	-0.04	0.15	-0.34	0.24
11.	Man Utd	a[11]	0.60	0.11	0.38	0.81	d[11]	-0.47	0.19	-0.85	-0.12
12.	Middlesbrough	a[12]	-0.02	0.15	-0.33	0.26	d[12]	0.09	0.14	-0.20	0.35
13.	Newcastle	a[13]	-0.16	0.16	-0.48	0.13	d[13]	0.03	0.14	-0.26	0.31
14.	Portsmouth	a[14]	0.00	0.15	-0.29	0.29	d[14]	-0.07	0.15	-0.37	0.22
15.	Reading	a[15]	0.15	0.14	-0.13	0.40	d[15]	0.05	0.15	-0.24	0.33
16.	Sheff Utd	a[16]	-0.33	0.17	-0.68	-0.01	d[16]	0.19	0.13	-0.08	0.44
17.	Tottenham	a[17]	0.26	0.13	-0.00	0.51	d[17]	0.20	0.13	-0.06	0.45
18.	Watford	a[18]	-0.42	0.18	-0.78	-0.07	d[18]	0.25	0.13	-0.01	0.50
19.	West Ham	a[19]	-0.24	0.16	-0.56	0.07	d[19]	0.27	0.13	0.01	0.51
20.	Wigan	a[20]	-0.19	0.16	-0.51	0.11	d[20]	0.27	0.13	0.01	0.51
	home		0.38	0.07	0.25	0.51	μ	-0.10	0.05	-0.20	0.003

Abbreviations: Man = Manchester; Utd = United; Sheff = Sheffield; Ham = Hampshire.

7.4.3.4 *Prediction of future games.*
Models in sports are used mainly for prediction. Here we briefly illustrate how we can obtain predictions for two future games. The approach can be easily generalized to additional games using the same approach.

To illustrate the implementation in WinBUGS, we have substituted the scored goals in the last two games (Tottenham – Manchester City and Watford – Newcastle) by NA. WinBUGS automatically will generate values for the missing goals from the predictive distribution (see Chapter 10 for more details) and will provide estimates for each score by monitoring the nodes goals1 and goals2.

Posterior summaries of the predicted scores are given in Table 7.10. As we can see in both of these games, posterior means indicate that the observed actual score was expected under the fitted model.

Table 7.10 Posterior summaries of the expected scores for last two games of Example 7.2

	Home team	Away team	Actual score	Posterior Median	Mean	Posterior summaries of goal difference Mean	SD	95% CI
379.	Tottenham	Manchester City	2–1	1–1	1.65 – 0.72	0.93	1.59	(-2, 4)
380.	Watford	Newcastle	1–1	1–1	0.93 – 1.02	-0.09	1.43	(-3, 3)

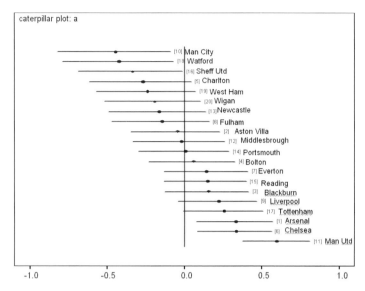

Abbreviations: Man = Manchester; Utd = United; Sheff = Sheffield; Ham = Hampshire.

Figure 7.2   95% posterior intervals for team attacking parameters for Example 7.2.

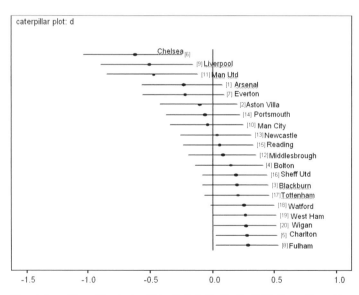

Abbreviations: Man = Manchester; Utd = United; Sheff = Sheffield; Ham = Hampshire.

Figure 7.3   95% posterior intervals for team defensive parameters for Example 7.2.

Interest also lies in calculating in the probability of each outcome (win/draw/loss), which can be easily accommodated in WinBUGS using the following syntax:

```
# calculation of the predicted differences
pred.diff[1] <- goals1[379]-goals2[379]
pred.diff[2] <- goals1[380]-goals2[380]
#
# probability of each game outcome (win/draw/loss)
for (i in 1:2){
    outcome[i,1] <- 1 - step( -pred.diff[i] )    #home wins (diff>0)
    outcome[i,2] <- equals( pred.diff[i] , 0.0 )#draw (diff=0)
    outcome[i,3] <- 1-step( pred.diff[i] )        #home loses (diff<0)
}
```

In this syntax, the elements of outcome are binary indicators denoting the win, draw, and loss of the home team in each column, respectively. Using similar syntax we can also estimate the probabilities of the expected differences. The syntax now is given by

```
# calculation of the probability of each difference
for (i in 1:2){
    pred.diff.counts[i,1]<- 1-step(pred.diff[i]+5) # less than -5
    # equal to k-7 (-5 to 5)
    for (k in 2:12){
        pred.diff.counts[i,k]<-equals(pred.diff[i],k-7)}
    pred.diff.counts[i,13]<-step(pred.diff[i]-6) # greater than 5
}
```

In this syntax pred.diff.counts is again a matrix with binary elements indicating which difference appears in each MCMC iteration. Elements 2–12 denote differences from -5 to 5, while the first and last elements denote differences lower than -5 and higher than 5, respectively.

Posterior probabilities of each predicted outcome and each value of the goal difference are summarized in Tables 7.11 and 7.12. Outcome probabilities indicate that Tottenham's probability of winning the game against Manchester City was about 60%, with a posterior mode of one goal difference. Concerning the second game (Watford vs. Newcastle), the posterior model probabilities confirm that the two teams have about equal probabilities of winning the game.

Table 7.11 Posterior probabilities of each game outcome for last two games of Example 7.2

| | | | | Posterior Probability | | |
| | | | Actual | Home | | Away |
	Home team	Away team	score	wins	Draw	wins
379.	Tottenham	Manchester City	2–1	0.59	0.24	0.17
380.	Watford	Newcastle	1–1	0.33	0.30	0.37

7.4.3.5 Regeneration of the full league. Interest also lies in reconstructing the league using the predictive distribution. Such practice is useful to evaluate whether the final observed ranking was plausible under the fitted model. It can be interpreted as the uncertainty involved in the final ranking if the league is repeated and the model is true; see Karlis and Ntzoufras (2008).

In order to reconstruct the full table in WinBUGS, we need to replicate the full scores in a tabular $K \times K$ format and then calculate the number of points for each team. The replicated league (in tabular $K \times K$ format) is calculated using the following syntax:

Table 7.12 Posterior probabilities of each game goal difference for last two games of Example 7.2

Home Team	Away Team	Actual Score	Posterior Probability of goal difference[a]								
			≤ -3	-2	-1	0	1	2	3	4	≥ 5
379. Tottenham	Man City	2–1	0.012	0.036	0.119	0.242	**0.257**	0.185	0.090	0.037	0.023
380. Watford	Newcastle	1–1	0.042	0.108	0.218	**0.303**	0.210	0.086	0.024	0.006	0.002

[a] Boldface indicates the maximum probability and the corresponding posterior mode of the difference.
Abbreviations: Man = Manchester.

```
for (i in 1:K){ for (j in 1:K){
    # replicated league
    goals1.rep[i,j]~dpois(lambda1.rep[i,j])
    goals2.rep[i,j]~dpois(lambda2.rep[i,j])
    # link and linear predictor
    log(lambda1.rep[i,j])<-  mu + home + a[ i ] + d[ j ]
    log(lambda2.rep[i,j])<-  mu            + a[ j ] + d[ i ]
    # replicated difference
    goal.diff.rep[i,j] <- goals1.rep[i,j]-goals2.rep[i,j]
}
```

The total number of points is calculated using the following syntax:

```
for (i in 1:K){ for (j in 1:K){
    # points earned by each home team (i)
    points1[i,j]<-3 * (1-step(-goal.diff.rep[i,j]))
                  + equals(goal.diff.rep[i,j],0)
    # points earned by each away team (j)
    points2[i,j]<-3 * (1-step( goal.diff.rep[i,j]))
                  + equals(goal.diff.rep[i,j],0)
}}
# calculation of the total points for each team
for (i in 1:K){
    total.points[i] <- sum( points1[i,1:20] ) -  points1[i,i]
                     + sum( points2[1:20,i] ) -  points2[i,i] }
```

In this syntax, `points1` and `points2` calculate the number of points in each game for the home and the away teams, respectively (arranged in $i = 1, 2, \ldots, K$ rows and $j = 1, 2, \ldots, K$ columns). In the total points we calculate the total number of points for each team i earned in home games (sum of i row of node `points1`) and in away games (sum of j column of node `points2`). Diagonal elements `points1[i,i]` and `points2[i,i]` are removed since they refer to each team playing against itself.

Posterior summaries of the total predicted earned points are obtained as usual (`sample monitor tool`), while posterior summaries of the ranks are obtained using the `rank monitor tool`. Results are summarized in Table 7.13 after 5000 iterations kept (and an additional 1000 iterations removed). *Ranks* refer to the total number of points in ascending order. Hence 20 refers to the team with the highest number of collected points (i.e., the champion), while one (1) refers to the team with the lowest number of collected points (i.e., the worst team in the league). The league was reproduced successfully. No differences are observed in the first four teams, while minor changes are observed for the remaining positions.

Further, probabilities for each ranking can be obtained by calculating the ranks using the commands

```
for (i in 1:K){ ranks[i] <- 21-rank(total.points[], i) }
```

and then calculating

Table 7.13 Observed and predicted (by model) points and rankings for all teams

Predicted (actual) ranking[a]	Team	Actual points	Posterior summaries for total points					Posterior percentiles for points ranks[b]		
			Mean	SD	2.5%	Median	97.5%	2.5%	Median	97.5%
1 (1)	Man Utd	89	84.7	8.1	68	85	99	16	20	20
2 (2)	Chelsea	83	78.3	8.8	60	79	94	14	19	20
3 (3)	Liverpool	68	72.6	9.2	54	73	90	12	18	20
4 (3)	Arsenal	68	69.9	9.4	51	70	88	10	17	20
5 (6)	Everton	58	62.8	9.4	44	63	81	7	15	19
6 (8)	Reading	55	55.8	9.6	37	56	75	4	13	18
7 (5)	Tottenham	60	56.0	9.6	36	55	73	4	12	18
8 (9)	Portsmouth	54	54.3	9.6	36	54	73	3	12	18
9 (11)	Aston Villa	50	53.4	9.7	34	53	73	3	12	18
10 (10)	Blackburn	52	51.9	9.6	33	52	70	3	11	17
11 (7)	Bolton	56	49.6	9.3	32	49	69	2	10	17
12 (12)	Middlesbrough	46	49.2	9.4	32	49	68	2	10	17
13 (13)	Newcastle	43	46.0	9.4	28	46	65	1	8	16
14 (14)	Man City	42	40.8	9.0	24	40	59	1	6	14
15 (16)	Fulham	39	39.3	8.9	22	39	57	1	5	13
16 (17)	Wigan	38	38.8	9.1	22	39	57	1	5	13
17 (15)	West Ham	41	37.5	8.7	21	37	55	1	4	13
18 (18)	Sheff Utd	38	37.2	8.8	21	37	55	1	4	12
19 (19)	Charlton	34	36.4	8.9	19	36	55	1	4	12
20 (20)	Watford	28	33.2	8.6	17	33	51	1	3	11

[a] Predicted ranks are calculated using the median rank and then the mean points.
[b] Ranks here refers to the number of points in ascending order (e.g., 20 denotes the best team and 1, the worst team in terms of collected points).
Abbreviations: Man = Manchester; Utd = United; Sheff = Sheffield; Ham = Hampshire.

```
for (i in 1:K){ for (j in 1:K){ rank.probs[i,j]<-equals(ranks[i],j)
    }}
```

which can be used to calculate the probability of each team's ranking. From the results provided in Table 7.14, we observe that Manchester United was clearly better than the other teams since its probability of winning the league was about 60% versus 25% for Chelsea, which ended up second in the league. More detailed analysis on the topic, using the Skellam's Poisson difference distribution, can be found in Karlis and Ntzoufras (2008).

7.5 BINOMIAL RESPONSE MODELS

Binomial data are frequently encountered in modern science, especially in the field of medical research, where the response is usually binary, indicating whether a person has a specific disease. The most popular model in this case is the logistic regression model, in which the usual logit link is adopted. The logit link is not only the obvious choice since it is the canonical link but also has a smooth and nice interpretation based on the ratio $\pi/(1-\pi)$, which is denoted the odds of $Y = 1$ versus $Y = 0$, where π is the probability of success

Table 7.14   Posterior mean, standard deviation of final league ranks, and posterior probabilities of each position (in %)

Node	Posterior mean	SD	1	2	3	4	5	6	7	8	9	10	11	12	13	14	15	16	17	18	19	20
Man Utd	1.7	1.1	**60**	23	10	4	1	1														
Chelsea	2.6	1.6	25	**33**	21	11	6	2	1	1												
Liverpool	3.7	2.0	10	21	**25**	18	11	7	3	3	1	1										
Arsenal	4.3	2.3	6	15	**21**	**21**	13	9	6	4	2	1	1	1								
Everton	6.2	2.9	1	5	10	16	**17**	13	10	8	6	5	3	2	1	1	1					
Reading	8.5	3.6		2	4	6	10	**12**	11	11	9	9	7	6	4	3	3	2	1	1	1	
Tottenham	8.8	3.6		1	3	6	10	11	**12**	10	10	9	7	6	4	4	3	2	1	1	1	
Portsmouth	9.0	3.7		1	2	6	8	11	10	**11**	9	9	8	6	5	4	3	2	2	1	1	
Aston Villa	9.5	3.8		1	2	5	7	9	10	**11**	10	9	8	7	6	5	4	3	2	2	1	
Blackburn	9.9	3.8		1	2	4	7	8	9	9	**10**	**10**	9	8	7	5	5	3	2	1	1	1
Bolton	10.8	3.8			1	2	4	6	7	**9**	**9**	**9**	**9**	**9**	8	7	6	4	3	2	2	1
Middlesbrough	11.0	3.9			1	2	4	6	7	8	**9**	**9**	**9**	**9**	**9**	7	5	5	4	3	2	1
Newcastle	12.3	3.9				1	2	3	5	6	8	8	9	9	**10**	8	9	6	6	4	4	2
Man City	14.4	3.7					1	1	2	3	4	6	6	7	8	9	10	**11**	10	9	8	6
Fulham	14.9	3.6						1	2	2	3	4	5	7	8	9	9	10	**11**	**11**	10	8
Wigan	15.1	3.6					1	1	2	2	4	4	5	5	8	10	9	10	**11**	10	10	10
West Ham	15.6	3.4						1	1	1	2	3	4	6	7	8	9	11	11	**12**	**12**	11
Sheff Utd	15.8	3.3							1	2	3	3	4	5	7	8	9	10	12	**13**	12	12
Charlton	16.0	3.3							1	2	2	3	3	5	6	8	8	10	12	12	14	**15**
Watford	17.1	2.9								1	1	2	2	3	4	5	8	8	10	14	18	**24**

[a] Boldface indicates highest probability for each team. Posterior percentages were rounded to the closest integer, while percentages $< 0.5\%$ were omitted. Sums of probabilities for each column are slightly higher than 100%, due to ties.

Abbreviations: Man = Manchester; Utd = United; Sheff = Sheffield; Ham = Hampshire.

for Y. The logistic regression model model can be summarized by

$$Y_i \sim \text{binomial}(\pi_i, N_i), \quad \log \frac{\pi_i}{1 - \pi_i} = \beta_0 + \sum_{j=1}^{p} \beta_j x_{ij} = \mathbf{X}_{(i)} \boldsymbol{\beta}$$

for $i = 1, 2, \ldots, n$. For $N_i = 1$, we have the case where Y_i is Bernoulli. Other frequently used link functions are the probit and clog–log links.

7.5.1   Interpretation of model parameters in binomial response models

7.5.1.1   *Odds and odds ratios.*   Interpretation of the parameters in logistic regression models is based in the notion of odds and odds ratios. We define as *odds* the relative probability of success ($Y = 1$) compared to the probability of failure ($Y = 0$) when we refer to binomial data. Hence

$$\text{odds} = \frac{\pi}{1 - \pi}$$

while the logistic model can be rewritten as

$$Y_i \sim \text{binomial}\left(\frac{\text{odds}_i}{1 + \text{odds}_i}, N_i \right), \quad \log(\text{odds}_i) = \beta_0 + \sum_{j=1}^{p} \beta_j x_{ij} = \mathbf{X}_{(i)} \boldsymbol{\beta}$$

using the odds representation.

The interpretation of odds is relatively simple and straightforward. It provides the number we need to multiply the probability of failure in order to calculate the probability of success. For example, odds $= 2$ implies that the success probability is twice as high as the failure probability while odds $= 0.6$ implies that the success probability is equal to 60% of the failure probability. The value of 1 is of central interest since it implies that the probabilities of both outcomes are equal to 0.5. Values of odds higher than one (> 1) indicate an increased probability of success in contrast to the failure probability ($\pi > 0.5$), while values lower than one (< 1) indicate a probability of success lower than the probability of failure ($\pi < 0.5$). Quantity $(\text{odds} - 1) \times 100$ provides the percentage increase or decrease (depending on the sign) of the success probability in comparison to the failure probability. For example, a value of odds $= 1.6$ indicates that the success probability is 60% higher than the corresponding failure probability. Similarly, a value of odds $= 0.6$ indicates that the success probability is 40% lower than the corresponding failure probability; see also Table 7.15, which summarizes the interpretation of odds.

The ratio of two odds of two different outcomes are called *odds ratios* (OR) and provide the relative change of the odds under two different conditions (denoted by $X = 1, 2$ and subscripts 1 and 2):

$$\text{OR}_{12} = \frac{\text{odds}(X = 1)}{\text{odds}(X = 2)},$$

where $\text{odds}(X = x)$ denotes the conditional success odds given that $X = x$:

$$\text{odds}(X = x) = \frac{P(Y = 1 | X = x)}{P(Y = 0 | X = x)}.$$

When $\text{OR}_{12} = 1$, then the conditional odds under comparison are equal, indicating no difference in the relative probability of Y under $X = 1$ and $X = 2$. Using similar approach, $(\text{OR}_{12} - 1) \times 100$ provides the percentage change of the odds for $X = 1$ compared with the corresponding odds when $X = 2$; see Table 7.16 for additional details.

Table 7.15   Summary interpretation table for odds

INTERPRETATION OF ODDS

$$\text{odds} = \frac{\pi}{1-\pi} = a \Leftrightarrow \pi = \frac{\text{odds}}{1+\text{odds}} = \frac{a}{1+a}.$$

- If $a = 1 \Rightarrow \pi = 1 - \pi = 0.5$.
- If $a < 1 \Rightarrow \pi < 0.5 < 1 - \pi$.
- If $a > 1 \Rightarrow \pi > 0.5 > 1 - \pi$.
- The probability of success ($Y = 1$) is a times as high as the corresponding probability of failure ($Y = 0$).
- If $a > 1$, then the success probability ($Y = 1$) is $(a - 1) \times 100\%$ times higher than the corresponding probability of failure ($Y = 0$).
- If $a < 1$, then the success probability ($Y = 1$) is $(1 - a) \times 100\%$ times lower than the corresponding probability of failure ($Y = 0$).

Table 7.16   Summary interpretation table for odds ratios

INTERPRETATION OF ODDS RATIOS

$$\text{OR}_{12} = \frac{\text{odds}(X = 1)}{\text{odds}(X = 2)} = a .$$

- If $a = 1 \Rightarrow \text{odds}(X = 1) = \text{odds}(X = 2)$.
- If $a < 1 \Rightarrow \text{odds}(X = 1) < \text{odds}(X = 2)$.
- If $a > 1 \Rightarrow \text{odds}(X = 1) > \text{odds}(X = 2)$.
- The success odds when $X = 1$ is a times as high as the corresponding odds for $X = 2$
- If $a > 1$, then the success odds when $X = 1$ are $(a - 1) \times 100\%$ times higher than the corresponding odds for $X = 2$.
- If $a < 1$, then the success odds when $X = 1$ are $(1 - a) \times 100\%$ times lower than the corresponding odds for $X = 2$.

Odds ratios (and logistic regression models) became very popular, especially in medical research, because they are good approximations of the relative risk when the prevalence of the disease is low [e.g., $< 10\%$ (Rosner, 2005)]. Hence we can summarize relative risk as follows:

$$\text{RR}_{12} = \frac{\pi_1}{\pi_2} \approx \text{OR}_{12} \quad \text{when} \quad \pi_1, \pi_2 \ \text{small}.$$

Moreover, the OR can be estimated in both prospective and retrospective studies while RR is limited to prospective studies; for more details on this aspect, see Schlesselman (1982, chap. 8) and Hosmer and Lemeshow (2000). These two facts founded odds ratios and logistic regression models as the main tools in medical research and biostatistics.

7.5.1.2 *Logistic regression parameters and odds ratios.*

Interpretation can be based on the exponents B_j of the original model parameters β_j in the same way as in the Poisson log-linear models. Now B_j are directly associated with odds ratios since

$$
\begin{aligned}
\log\big(\text{odds}(x)\big) &= \beta_0 + \beta_1 x \Rightarrow \\
\text{odds}(x) &= B_0 B_1^x \Rightarrow \\
\text{OR}_{x+1,x} &= \frac{\text{odds}(x+1)}{\text{odds}(x)} = \frac{B_0 B_1^{x+1}}{B_0 B_1^x} = B_1 = e^{\beta_1}
\end{aligned}
$$

in the simple logistic regression case with one arithmetic covariate. Therefore $B_1 = e^{\beta_1}$ denotes the relative odds magnitude when X increases by one unit. Note that for $X_i = -\beta_0/\beta_1$, we can calculate the value of X for which both probabilities are equal to 0.5. This value may be used as a threshold for prediction or, for example, for diagnosing future patients using the X variable directly.

Similarly, when X is a categorical variable with K levels and the corner parametrization is adopted with the first level as baseline/reference category, then

$$
\begin{aligned}
\log\big(\text{odds}(x)\big) &= \beta_0 + \sum_{j=2}^{K} \beta_j I(x=j) = \beta_0 + \sum_{j=2}^{K} \beta_j D_j \Rightarrow \\
\text{odds}(x) &= B_0 \prod_{j=2}^{K} B_j^{D_j} \Rightarrow \\
\text{OR}_{j1} &= \frac{\text{odds}(j)}{\text{odds}(1)} = \frac{e^{\beta_0+\beta_j}}{e^{\beta_0}} = \frac{B_0 B_j}{B_0} = B_j = e^{\beta_j} .
\end{aligned}
$$

Therefore, in logistic regression models, parameter B_j is the success odds ratio for the jth category of X versus the reference category of the same variable.

Extension of the interpretation above to multiple logistic regression models is straightforward. We only need to interpret each B_j as the change of Y when a single covariate X_j increases by one unit while the other covariates remain constant. Odds ratios estimated via multiple logistic regression models are often reported as "odds ratios adjusted for" the other covariates or "odds ratios after controlling for the effect" of the other covariates. Adjusted odds ratios estimate the joint effect of all covariates X_j ($j = 1, 2, \ldots, p$) on Y, and in this way we essentially calculate the effect of each covariate after the elimination of the effect of the other covariates.

7.5.1.3 *Parameter interpretation in probit models.*

Parameter interpretation in probit models is not as straightforward as the corresponding interpretation in logit models.

Interpretation in probit models can be based on the latent variable representation of the model [see, e.g., Powers and Xie (1999, pp. 59–61)]. With this approach, we assume a standardized normal latent unobserved variable

$$Z_i = \eta_i + \varepsilon_i = \beta_0 + \sum_{j=1}^{p} \beta_{ij} X_{ij} + \varepsilon_i = \boldsymbol{X}_{(i)}\boldsymbol{\beta} + \varepsilon_i, \quad \text{with } \varepsilon_i \sim N(0,1), \qquad (7.8)$$

(for $i = 1, 2, \ldots, n$) which specifies the observables Y_i by setting $Y_i = I(Z_i > 0)$ (i.e., $Y_i = 1$ for $Z_i > 0$ and $Y_i = 0$ when $Z_i < 0$). From this, we obtain

$$
\begin{aligned}
P(Y_i = 1|\boldsymbol{X}_{(i)}) &= P(Z_i > 0) = 1 - P(Z_i \leq 0) = 1 - P\big(Z_i - \boldsymbol{X}_{(i)}\boldsymbol{\beta} \leq -\boldsymbol{X}_{(i)}\boldsymbol{\beta}\big) \\
&= 1 - \Phi\big(-\boldsymbol{X}_{(i)}\boldsymbol{\beta}\big) = \Phi\big(\boldsymbol{X}_{(i)}\boldsymbol{\beta}\big) = \Phi(\eta_i) \,.
\end{aligned}
$$

Note that in the simple case with one covariate, for $\pi = 0$ we have the critical value of $x_c = -\beta_0/\beta_1$ as in the logit function.

The latent variable approach can also be used for the logit link (using the logistic distribution) and for the clog–log link (using the extreme value distribution), which follows.

Under the normal latent variable approach, the coefficients of the probit model indicate the expected change of the latent variable Z with one unit increase of X. Generally we can write

$$\pi(x+1) = \Phi\Big(\beta_0 + \beta_1(x+1)\Big) = \Phi\Big(\beta_0 + \beta_1 x + \beta_1\Big) = \Phi\Big(\Phi^{-1}\big(\pi(x)\big) + \beta_1\Big),$$

where $\pi(x) = P(Y = 1|X = x)$. A plausible interpretation of β_1 can be obtained if we consider the value of $x = x_c = -\beta_0/\beta_1$, for which we obtain

$$\pi\big(X = x_c + 1\big) = \pi\left(X = -\frac{\beta_0}{\beta_1} + 1\right) = \Phi\left(\Phi^{-1}\left(\frac{1}{2}\right) + \beta_1\right) = \Phi(\beta_1)\,.$$

For example, if $\beta_1 = 0.5, 1, 2, 3$ the probability of Y becomes equal to 0.69, 0.84, 0.977, and 0.99865, respectively, corresponding to an increase in the probability by 38%, 68%, 95%, and 100% when X increases from $-\beta_0/\beta_1$ to $-\beta_0/\beta_1 + 1$. Therefore β_1 is directly associated with the probability of Y when $X = x_c + 1$.

This logic can be easily generalized for calculating the deviations for changes of the linear predictor by one unit of X_j in the multiple probit model. In such a case the linear predictor from η^* will become equal to $\eta^* + \beta_j$, resulting in

$$\pi\big(X_j = x + 1, \boldsymbol{X}_{\backslash j}\big) = \Phi(\eta^* + \beta_j),$$

where $\boldsymbol{X}_{\backslash j}$ denotes all variables except X_j and $\pi(\boldsymbol{x}) = P(Y = 1|\boldsymbol{X} = \boldsymbol{x})$. For $\eta^* = 0$ (i.e., the corresponding success is probability is equal to $\pi = \frac{1}{2}$), then the success probability becomes equal to $\pi(\eta = \beta_j) = \Phi(\beta_j)$. Hence β_j is associated with the success probability when X_j increases by one unit compared to an individual with success probability $\frac{1}{2}$ (and linear predictor $\eta = 0$).

Alternatively, we may base our inference on the marginal effect of X_j on π given by

$$\frac{\partial \pi}{\partial X_j} = \frac{\partial \Phi(\eta)}{\partial \eta}\beta_j = \phi(\eta)\beta_j \,.$$

This quantity approximates the effect of one unit increase of X_j on the success probability π. Since this approximation is valid only for small deviations of X_j from the current value of x, we may consider

$$\Delta\pi \approx \phi(\eta)\beta_j \Delta X_j \quad \text{for small } \Delta X_j,$$

for example, $\Delta X_j = 0.1$. For the case where $\eta = 0$, then the equation above becomes equal to

$$\Delta\pi\left(\pi = \frac{1}{2}\right) \approx \phi(0)\beta_j\Delta X_j = \frac{1}{\sqrt{2\pi}}\beta_j\Delta X_j = 0.3989\,\beta_j\Delta X_j$$

where $\pi = 3.14159$. Hence, β_j is now directly related to the change of probability when $\pi = 0.5$, and X_j increases by a small change denoted by ΔX_j. For the values of $0.5, 1, 2$, and 3 used above, the increase in probability is approximately equal to $0.0199, 0.0399$, 0.080, and 0.120 for an increases of X_j by 0.1 while the corresponding actual increases is equal to $0.0199, 0.0398, 0.079$, and 0.118.

In WinBUGS, the posterior distributions of both quantities discussed above can be calculated by specifying the corresponding logical/deterministic nodes.

7.5.1.4 *Relationship between logit and probit parameters.* Both link functions have the same symmetry properties. Differences in the estimated probability values π_i of the two models are very close.

Following Aldrich and Nelson (1984, p. 41), we can obtain sufficient approximation of the logit parameters multiplying the probit parameters by a factor equal to $\pi/\sqrt{3} = 1.81$. This approximation results from considering the variance of the the latent variables for the logit and probit models, which are equal to $\pi^2/3$ and one, respectively. Dividing the latent variable of the logit model by the factor $\pi/\sqrt{3}$ provides the same variance as in the probit model. Then, assuming that $Z^{\mathrm{probit}} \approx Z^{\mathrm{logit}}/(\pi/\sqrt{3})$ results in

$$\beta_j^{\mathrm{logit}} = \frac{\pi}{\sqrt{3}}\beta_j^{\mathrm{probit}} = 1.81 \times \beta_j^{\mathrm{probit}}.$$

A more accurate approximation can be based on the Taylor expansion, resulting in

$$\beta_j^{L_1} = \frac{g'_{L_1}(\pi_0)}{g'_{L_2}(\pi_0)}\beta_j^{L_2} \quad \text{for } j = 1, \ldots, p, \tag{7.9}$$

where L_1 and L_2 are the two link functions we wish to compare, $g_L(\pi)$ and $g'_L(\pi)$ are the link function and the corresponding first derivative for L link, and π_0 is a constant used in the approximation; see eq. (6) in Ntzoufras et al. (2003). For the the logit and the probit link functions, this results in

$$\beta_j^{\mathrm{logit}} = \frac{\phi(\pi_0)}{\pi_0(1 - \pi_0)}\beta_j^{\mathrm{probit}} \quad \text{for } j = 1, \ldots, p,$$

which for the default choice of $\pi_0 = \frac{1}{2}$

$$\beta_j^{\mathrm{logit}} = \frac{4}{\sqrt{2\pi}}\beta_j^{\mathrm{probit}} = 1.596 \times \beta_j^{\mathrm{probit}} \quad \text{for } j = 1, \ldots, p,$$

which is the value also supported by Amemiya (1981, eq. 2.7). This approximation works reasonably well when no extreme probabilities exist. An even better approximation results using as π_0 the proportion estimated by the available sample. Using either of these approximations, we can directly obtain an odds ratio interpretation for the parameters of the probit link (or any other link).

Note that the association of the constant terms is slightly different [see eq. 5 in Ntzoufras et al. (2003)] but it is not of central interest, and hence we do not pursue this issue further in this section.

7.5.1.5 *Parameter interpretation in log–log and clog–log models.*   The *complementary log–log (clog–log) model* is given by the expression

$$\log\left(-\log(1-\pi_i)\right) = \eta_i = \beta_0 + \sum_{j=1}^{p}\beta_j X_{ij}.$$

If we consider this model for the failure probability (i.e., set $Y_i^* = 1 - Y_1$), then the model becomes equal to

$$\log\left(-\log(\pi_i)\right) = \eta_i = \beta_0 + \sum_{j=1}^{p}\beta_j X_{ij},$$

which is referred as the *log–log model.* Using the clog–log model for the success probability is equivalent to using the log–log link for the failure probability and vice versa.

Complementary log–log link model assumes a latent variable Z_i of the form (7.8) with errors ε_i following a standard extreme value distribution with density and distribution function

$$f(\varepsilon_i) = e^{\varepsilon_i}\exp\left(-e^{\varepsilon_i}\right) \ \text{ and } F(\varepsilon_i) = 1 - \exp\left(-e^{\varepsilon_i}\right),$$

respectively. To be more specific, $-\varepsilon_i$ follows the Gumbel (or extreme value) distribution with parameters $a = 0$ and $b = 1$ and mean and variance equal to 0.577 and $\pi^2/6 = 1.645$, respectively (Agresti, 2002, pp. 248–250).

For both the logit and the probit links, the corresponding errors of the latent variables are symmetric. This assumption is restrictive, and in cases where it is not realistic, the clog–log link may be more appropriate (Agresti, 2002, p. 248).

Interpretation can be obtained by the expression

$$\begin{aligned}
\left[1-\pi\left(x+1\right)\right] &= \exp\left(-e^{\beta_0+\beta_1(x+1)}\right) = \exp\left(-e^{\beta_0+\beta_1 x+\beta_1}\right)\\
&= \exp\left(-e^{\beta_0+\beta_1 x}e^{\beta_1}\right) = \exp\left(-e^{\beta_0+\beta_1 x}\right)^{e^{\beta_1}}\\
&= \left[1-\pi\left(x\right)\right]^{e^{\beta_1}}
\end{aligned}$$

in the case with one covariate and by

$$\left[1-\pi\left(X_{\backslash j}, X_j = x+1\right)\right] = \left[1-\pi\left(X_{\backslash j}, X_j = x\right)\right]^{e^{\beta_j}}$$

in case of p covariates. Hence for each probability $\pi(X)$ with covariate values X, an increase of variable X_j by one unit changes the failure probability by

$$\left(\left[1-\pi\left(X_{\backslash j}, X_j = x\right)\right]^{e^{\beta_j}-1} - 1\right) \times 100\% .$$

For $\pi = \frac{1}{2}$ and $\beta_1 = 0.5, 1, 2$ and 3, the failure probability becomes equal to 0.32, 0.15, 0.006 and 0.9×10^{-6}, decreasing by 36%, 69.6%, 98.8%, and 99.9%, respectively.

Association between the logit and clog–log link can be obtained using (7.9) with $\pi_0 = \frac{1}{2}$, resulting in

$$\beta_j^{\text{clog}-\text{log}} = 2\log 2\beta_j^{\text{logit}} = 1.386 \times \beta_j^{\text{logit}}.$$

This approximation is accurate only for samples with no extreme values. Finally, the marginal effect is given by

$$\frac{\partial \pi}{\partial \eta}\beta_j = \frac{\partial\Big(1 - \exp(-e^{\eta})\Big)}{\partial \eta}\beta_j = \exp(-e^{\eta} + \eta)\beta_j = (1 - \pi)\log(1 - \pi)\beta_j,$$

which for $\pi = \frac{1}{2}$ and a change equal to ΔX_j becomes equal to

$$\Delta\pi\left(\frac{1}{2}\right) \approx -\frac{1}{2}\log 2\beta_j\Delta X_j = -\frac{1}{2}\log 2\beta_j\Delta X_j = -0.34657 \times \beta_j\Delta X_j .$$

A summary of the interpretation details for binomial response model coefficients presented in Sections 7.5.1.2–7.5.1.5 is provided in Table 7.17.

7.5.2   A simple example

Example 7.3. Analysis of senility symptoms data using WinBUGS (data were originally analyzed in Example 2.3). In Section 2.3 we have illustrated the Metropolis–Hastings algorithm using a simple example in which 54 elderly people completed a subtest of the Wechsler Adult Intelligence Scale (WAIS) resulting to a discrete score with ranging from 0 to 20. The aim of this study was to identify people with senility symptoms (binary variable) using the WAIS score. Moreover, we were interested in calculating the threshold value of X for which $\pi > 0.5$ to enable us to identify possible patients directly using X.

7.5.2.1   *Model specification in WinBUGS.*   Here the response (senility symptoms) is binary, and hence the Bernoulli or the binomial with $n = 1$ distributions can be used to model the response variable. The explanatory variable x is the WAIS score, which is a discrete quantitative variable.

Specification of the likelihood for the logit model proceeds as usual by the following syntax

```
for (i in 1:n){
    senility[i] ~ dbin( pi[i], 1 )
    logit( pi[i] ) <- beta0 + beta1 * wais[i]
}
```

where $n = 54$. Alternatively, the Bernoulli distribution (command dbern(pi[i])) can be used instead of the binomial with $n = 1$ without any problem.

Moreover, we may also define the e^{β_j} in order to interpret directly the posterior distributions of the odds and odds ratios. This can be easily achieved in WinBUGS by defining the corresponding logical (or deterministic) nodes using the commands

```
odds0 <- exp( beta0 )
or    <- exp( beta1 )
```

The value of $X = x(\pi = 0.5)$, which corresponds to disease probability equal to 0.5, can be defined using the same manner since solving the equation $0 = \beta_0 + \beta_1 X = x(\pi = 0.5)$ results in $X = x(\pi = 0.5) = -\beta_0/\beta_1$. WinBUGS syntax is now given by

```
wais.half.prob <- - beta0/beta1
```

Table 7.17   Summary interpretation table for binomial response model coefficients.

INTERPRETATION OF MODEL PARAMETERS β_j

For all models: latent variable interpretation

- β_j ($j \geq 1$): Expected increase of the latent variable when X increases by one unit.

Logit models

- β_0: odds of $Y = 1$ vs. $Y = 0$ when all covariates are equal to zero.
- β_j ($j \geq 1$): odds ratio of $Y = 1$ vs. $Y = 0$ when X increases by one unit.

Probit models

- β_0: mean of the normal latent variable when all covariates are equal to zero.
- $\Phi(\beta_0)$: success probability when all covariates are equal to zero.
- β_j ($j \geq 1$):

  1. **Success probability when $\pi = \frac{1}{2}$ and X_j is increased by one unit** = $\Phi(\beta_j)$ is the probability when increasing X_j by one unit from a profile with $\eta = 0$ ($\pi = \frac{1}{2}$).

  2. **Marginal effect from $\pi = \frac{1}{2}$** $= 0.4 \times \beta_j \Delta X_j$ is the change in probability when increasing X_j by a small value equal to ΔX_j from a profile with $\eta = 0$ ($\pi = \frac{1}{2}$).

  3. **Approximate OR interpretation**: $e^{\xi \times \beta_j}$ is approximately equal to the odds ratio of $Y = 1$ vs. $Y = 0$ when X increases by one unit; possible values for $\xi = 1.8$ (Aldrich and Nelson, 1984, p. 41), $\xi = 1.6$ [when no extreme values exist (Ameniya, 1981, eq. 2.7)], and $\xi = \phi(\hat{\pi})/\hat{\pi}(1 - \hat{\pi})$ (Ntzoufras et al., 2003, eq. 6).

clog–log models

- $\beta_0 - 0.577$: mean of latent variable when all covariates are equal to zero.
- $1 - \exp\left(-e^{\beta_0}\right)$: success probability when all covariates are equal to zero.
- β_j ($j \geq 1$):

  1. **Failure probability when $\pi = \frac{1}{2}$ and X_j is increased by one unit** = $2^{-e^{\beta_j}}$ is the probability when increasing X_j by one unit from a profile with $\pi = \frac{1}{2}$.

  2. **Marginal effect from $\pi = \frac{1}{2}$** is equal to $-0.35 \times \beta_j \Delta X_j$. This is the change in probability when increasing X_j by a small value equal to ΔX_j from a profile with $\pi = \frac{1}{2}$.

  3. **Approximate OR interpretation**: $e^{1.39 \times \beta_j}$ is approximately equal to the odds ratio of $Y = 1$ vs. $Y = 0$ when X increases by one unit.

Using the same approach, we may define X values for other probabilities (e.g., 0.25 or 0.01).

To define the probit and clog–log models, we only need to substitute the WinBUGS function `logit(pi[i])` by the corresponding link commands `probit(pi[i])` and `cloglog(pi[i])`, respectively. Other, more complicated, link functions can be defined by expressing π_i as a function of the linear predictor η_i. Note that, for this example, arithmetic overflows occurred when using the probit link of WinBUGS. In order to avoid them, we have truncated the tails at $(-\xi, \xi), \xi > 0$, of the probit link; see the computational note that follows.

COMPUTATIONAL NOTE (probit function in WinBUGS)

To avoid overflow problems appearing in the probit function of WinBUGS, we propose truncating the tails at $(-\xi, \xi)$ of the probit link using the following syntax[1]

```
probit(pi[i])  <-  eta[i] *(1-step( abs(eta[i])-xi ))
                - xi*step( -xi - eta[i] )+ xi *step( eta[i]-xi)
```
where `eta[i]` is the linear predictor defined by `eta[i] <- beta0 + beta1 x[i]` and $\xi \geq 5$ is the truncation value. The greater is ξ, the better is the approximation to the actual probit link. The syntax displayed above implements the following function

$$
\begin{aligned}
\Phi^{-1}(\pi) &= \eta\left(1 - I(|\eta| - \xi \geq 0)\right) - \xi I(-\xi - \eta \geq 0) + \xi I(\eta - \xi \geq 0) \\
&= \eta\left(1 - I(\eta \geq \xi \text{ or } \eta \leq -\xi)\right) - \xi I(\eta \leq -\xi) + \xi I(\eta \geq \xi) \\
&= \eta I\left(-\xi < \eta < \xi\right) - \xi I(\eta \leq -\xi) + \xi I(\eta \geq \xi) \\
&= \begin{cases} -\xi & \text{if } \eta \leq \xi \\ \eta & \text{if } -\xi < \eta < \xi \ , \\ \xi & \text{if } \eta \geq \xi \end{cases}
\end{aligned}
$$

resulting in

$$
\pi = \begin{cases} 0.0000003 & \text{if } \eta \leq -5 \\ \Phi(\eta) & \text{if } -5 < \eta < 5 \\ 0.9999997 & \text{if } \eta \geq 5 \end{cases}
$$

for $\xi = 5$.

[1] The following simpler syntax was suggested by Mike Meredith (Malaysia):
```
probit(pi[i]) <- max( min( eta[i], xi), -xi).
```

To facilitate parameter interpretation in probit models, we calculate the following quantities:

- Disease probability when $\pi = \frac{1}{2}$ and X_j is increased by one unit = $\Phi(\beta_j)$ using the syntax

```
incr.prob.pi.half <- phi(beta1)
```

- Marginal effect from $\pi = \frac{1}{2}$ is defined to be equal to $0.4 \times \beta_j$ for $\Delta X_j = 1$ using the syntax

```
marg.effect.pi.half <- 0.4 * beta1
act.marg.effect.pi.half <- phi(beta1)-0.5
```

The latter expression is the actual increase of the probability when $\pi = 0.5$ and x increases by one unit.

- Approximate OR interpretation: $\xi_2 \times \beta_j$ is given by the syntax

```
xi2 <- 1.6
approx.or <- xi2 * beta1
```

The threshold value $x_c = -\beta_0/\beta_1$ for the probit link is the same as in the logit one and, therefore, can be obtained using exactly the same commands.

Concerning the clog–log model, overflow problems appeared similar to those occurring for the probit link. Hence, we also propose to truncate the clog–log function outside the interval $(-\xi_1, \xi_2)$, $\xi_1 >> \xi_2 > 0$. This interval is not symmetric around zero since the complementary log–log function approaches zero much more slowly than it approaches one. The syntax needed for this truncation as well as details are presented in the following computational note.

COMPUTATIONAL NOTE (clog–log function in WinBUGS)

To avoid overflow problems appearing in the clog–log function of WinBUGS we propose truncating the tails at $(-\xi_1, \xi_2)$, $\xi_1 >> \xi_2 > 0$ of the clog–log link using the following syntax[1]

```
cloglog(pi[i]) <- eta[i]*(1-step(-xi1-eta[i]))*(1-step(eta[i]-xi2))
                  - xi1*step(-xi1-eta[i])+ xi2*step(eta[i]-xi2)
```

where eta[i] is the linear predictor and ξ_1 (xi1) and ξ_2 (xi2) are the truncation values. The greater are ξ_1 and ξ_2, the better is the approximation to the actual clog–log link. The syntax displayed above implements the following function

$$
\begin{aligned}
\text{clog--log}(\pi) &= \eta\left(1 - I(-\xi_1 - \eta \geq 0)\right)\left(1 - I(\eta - \xi_2 \geq 0)\right) \\
&\quad -\xi_1 I(-\xi_1 - \eta \geq 0) + \xi_2 I(\eta - \xi_2 \geq 0) \\
&= \eta I(-\xi_1 - \eta < 0) I(\eta - \xi_2 < 0) - \xi_1 I(\eta \leq -\xi_1) + \xi_2 I(\eta \geq \xi_2) \\
&= \begin{cases} -\xi_1 & \text{if } \eta \leq \xi_1 \\ \eta & \text{if } -\xi_1 < \eta < \xi_2 \\ \xi_2 & \text{if } \eta \geq \xi_2 \end{cases},
\end{aligned}
$$

resulting in

$$
\pi = \begin{cases} 0.0000454 & \text{if } \eta \leq -10 \\ 1 - \exp(-e^{\eta}) & \text{if } -10 < \eta < 3 \\ 1 - 1.89 \times 10^{-9} & \text{if } \eta \geq 3 \end{cases}
$$

for $\xi_1 = 10$ and $\xi_2 = 3$.

[1] The following simpler syntax was suggested by Mike Meredith (Malaysia):

```
cloglog(pi[i]) <- max( min( eta[i], xi2), xi1).
```

Finally, to facilitate interpretation of the clog–log parameters, we use the following WinBUGS syntax

- `incr.prob.pi.half <- exp( -exp(beta1) )`

- `marg.effect.pi.half <- -0.35 * beta1`
  `act.marg.effect.pi.half <- exp( -exp(beta1) ) - 0.5`

- `approx.or <- 1.39 * beta1`

Using this syntax, we can calculate the disease probability when $\pi = \frac{1}{2}$ and WAIS is increased by one unit, the marginal effect of WAIS when $\pi = \frac{1}{2}$ and the approximate odds ratio.

Finally, the threshold value x_c is now given by $x_c = \big(\log(\log 2) - \beta_0\big)/\beta_1$ and is specified in WinBUGS using the syntax

```
wais.half.prob <- ( log(log(2))-beta0 )/beta1
```

Finally, for comparison reasons, we may calculate the odds ratios and probability difference for various values or of X. Here the range of values is limited to integer values from zero to 20. Hence we can try all values to obtain a full overview of how the success probability changes for each value of x. The syntax for the probit link is

```
for (k in 1:21) {
      probit( pi.model[k] ) <- beta0 + beta1 * (k-1)
      odds[k]            <- pi.model[k]/(1-pi.model[k])
}
for (k in 1:20) {
      Dpi[k] <- pi.model[k+1] - pi.model[k]
      or[k]  <- odds[k+1]/odds[k]
}
```

while for the clog–log link we only need to change commands in the second line of this code by substituting `probit` with `cloglog`. Estimated probabilities for all possible values of x and hence all possible individual "profiles" of this example are also calculated using the syntax displayed above (given by node `pi.model`). The usual low information priors $\beta_j \sim N(0, 1000)$ are used in this example.

7.5.2.2 Results and parameter interpretation.
Posterior summaries of the parameters for each link are provided in Table 7.18. The a posteriori expected success probabilities along with the 2.5% and 97.5% posterior values for all three link functions are depicted in Figure 7.4.

Table 7.18 Posterior summaries for model parameters for each link function for Example 2.3

Node	Logit		Probit		clog–log	
	Mean	SD	Mean	SD	Mean	SD
β_0	2.507	1.229	1.402	0.661	1.447	0.721
β_1	-0.339	0.119	-0.191	0.061	-0.260	0.076
OR[a]	0.718	0.083	0.748	0.074	0.700	0.073
WAIS($\pi = 0.5$)	6.975	2.104	6.677	3.195	6.752	1.575
DIC	55.105		54.997		54.998	

[a] Exact odds ratio in logit ($= e^{\beta_1}$); approximate for probit and clog–log.

All models indicate significant negative association between the WAIS score and the presence of senility symptoms. From the logit model, the odds of senility symptoms for individual scoring zero in WAIS are a posteriori expected to be equal to 12.27. Moreover, for each additional point of the WAIS score, a decrease in disease probability by 38% is a posteriori expected. Similarly, the corresponding approximate posterior odds for the other two models indicate decrease of 25% and 30% for the probit and clog–log links, respectively. These approximations are satisfactory summaries of the overall picture since the posterior means of the actual odds ratios for all WAIS values range from 0.61 to 0.74 for the probit link. Similar is the picture for most of the values in clog–log link with the odds ratios to range from 0.61 to 0.77 for $x > 4$. Generally, this approximation is more successful for the central π values as expected.

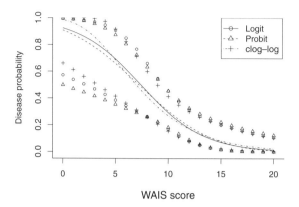

Figure 7.4 Estimated binomial models for Example 7.3. Lines represent model based on posterior means; points represent 2.5% and 97.5% posterior percentiles for disease probabilities.

Concerning the threshold disease value, all three models indicate that cases are implied for values $X \leq 6$; see Table 7.19 for a summary of posterior statistics. If we consider the median values, then the two first models indicate that $X \leq 7$ (corresponding medians equal to 7.35 and 7.29) while for the clog–log link the corresponding median is equal to 6.998 marginally indicating $X \leq 6$. Finally, we may use the 95% posterior intervals to construct a more complicated decision rule. For example, for the logit link we may say that someone whose score ≤ 2 (lower than or equal to 2) is a case, or is healthy if $X \geq 10$. Values within 3–9 form a nondecision zone since any value in this interval can be reasonably considered as a threshold value. Similarly, for the probit link we can form the following decision rule: $X = 0$ for case, $X \geq 10$ for healthy, $1 \leq X \leq 9$ for neither case or healthy (i.e., we cannot decide), whereas for the clog–log link, the corresponding decision rule is $X \leq 3$ for case, $X \geq 9$ for healthy, and $4 \leq X \leq 8$ for neither (i.e., cannot decide). Note that for the last link, the neutral zone is narrower.

Table 7.19 Posterior summaries for threshold value WAIS($\pi = 0.5$) and corresponding discrimination rules for each link function for Example 2.3

| | WAIS($\pi = 0.5$) | | | Decision rule | | | | | |
| | | | | Logit | | Probit | | clog–log | |
Node	Logit	Probit	clog–log	Case	Healthy	Case	Healthy	Case	Healthy
Mean	6.975	6.677	6.752	$X \leq 6$	$X \geq 7$	$X \leq 6$	$X \geq 7$	$X \leq 6$	$X \geq 7$
Median	7.353	7.291	6.998	$X \leq 7$	$X \geq 8$	$X \leq 7$	$X \geq 8$	$X \leq 6$	$X \geq 7$
95% posterior interval	2.149	0.028	3.262	$X \leq 2$	$X \geq 10$	$X = 0$	$X \geq 10$	$X \leq 3$	$X \geq 9$
	9.457	9.469	8.968						

Finally, the DIC value for probit is the lowest with minor differences from clog–log, while logit seems to be the worst of the three models. Nevertheless, all DIC values are quite close, indicating minor differences in the fit of the three models.

7.6 MODELS FOR CONTINGENCY TABLES

Both binomial logit and Poisson log-linear models are also used for modeling the frequencies (counts) of high-dimensional contingency tables. The two models are associated, and using the Poisson log-linear models, we can derive the corresponding logit model.

To be more specific, let us consider the Poisson log-linear model assuming corner constraints in a 2×2 contingency table resulting from the cross-classification of two binary variables X and Y. Then the Poisson log-linear model is given by

$$N_{ij} \sim \text{Poisson}(\lambda_{ij}), \quad \log(\lambda_{ij}) = \lambda^\emptyset + \lambda_i^X + \lambda_j^Y + \lambda_{ij}^{XY} \tag{7.10}$$

for $i, j = 1, 2$, where N_{ij} are the frequencies for any combination of $X = i$ and $Y = j$; λ_{ij} is the corresponding expected number of counts; $\lambda^\emptyset$, λ_i^X, λ_j^Y, and λ_{ij}^{XY} are the corresponding model parameters for the constant term, the effects X, the effect of Y and the interaction effect XY, respectively.

The Poisson likelihood can be written as

$$
\begin{aligned}
f(\boldsymbol{N}|\boldsymbol{\lambda}) &= \prod_{i=1}^{2} f(N_{i1}, N_{i2}|N_{i+}) f(N_{i+}) \\
&= \prod_{i=1}^{2} \frac{\prod_{j=1}^{2} \left\{ e^{-\lambda_{ij}} \lambda_{ij}^{N_{ij}} / N_{ij}! \right\}}{e^{-\lambda_{i+}} \lambda_{i+}^{N_{i+}} / N_{i+}!} e^{-\lambda_{i+}} \frac{\lambda_{i+}^{N_{i+}}}{N_{i+}!} \\
&= \prod_{i=1}^{2} \left\{ \frac{N_{i+}!}{N_{ij}!} \prod_{j=1}^{2} \left(\frac{\lambda_{ij}}{\lambda_{i1} + \lambda_{i2}} \right)^{N_{ij}!} \right\} \prod_{i=1}^{2} e^{-\lambda_{i+}} \frac{\lambda_{i+}^{N_{i+}}}{N_{i+}!} \\
&= \prod_{i=1}^{2} f_B(N_{i2}; \pi_i, N_{i+}) f_P(N_{i+}; \lambda_{i+}),
\end{aligned}
$$

where $N_{i+} = \sum_{j=1}^{2} N_{ij} \sim \text{Poisson}(\lambda_{i+})$, $\lambda_{i+} = \sum_{j=1}^{2} \lambda_{ij}$, and $\pi_i = \lambda_{i2}/(\lambda_{i1} + \lambda_{i2})$ and is equal to the conditional probability of $P(Y = 2|X = i)$. Hence the likelihood of a Poisson log-linear can be written as a product of two independent binomial models for N_{i2} times a Poisson model for the marginal distribution of X with expected counts $\lambda_{i+} = \lambda_{i1} + \lambda_{i2}$. Hence, model (7.10) is also referred to as the *product binomial model*. The parameters involved in the two parts of the likelihood are different, resulting in two independent likelihoods. Moreover, the logit expression resulting from the Poisson model above is given by

$$
\begin{aligned}
\text{logit}(\pi_i) &= \log \frac{\pi_i}{1 - \pi_i} = \log \frac{\lambda_{i2}}{\lambda_{i1}} = \lambda^\emptyset + \lambda_i^X + \lambda_2^Y + \lambda_{i2}^{XY} - \lambda^\emptyset - \lambda_i^X \\
&= \lambda_2^Y + \lambda_{i2}^{XY} = \lambda_2^Y + \lambda_{22}^{XY} I(i = 2).
\end{aligned}
$$

Hence $\lambda_2^Y = \beta_0$ and $\lambda_{22}^{XY} = \beta_1$ in the corresponding logistic regression model for the same data given by

$$N_{i2} \sim \text{binomial}(\pi_i, N_{i+}), \quad \log\left(\frac{\pi_i}{1 - \pi_i} \right) = \beta_0 + \beta_1 X_i, \quad i = 1, 2.$$

This logic can be generalized for more than two levels and for higher-dimensional contingency tables that involve additional categorical variables. If Y has more than two levels, then the model is a product multinomial following the same reasoning. In fact, it is common practice to use log-linear models in order to fit the corresponding multinomial model for the response Y. The parameters of the corresponding logit model are given by all λ^S, where S is a set of variables with $Y \in S$. Moreover, parameters of the multinomial model for any term T are given by the corresponding parameter of the log-linear term for term $T \cup Y$, namely, the interaction term between T and Y. For example, if $T = X_1 X_2$, then the corresponding multinomial parameter of the interaction between X_1 and X_2 is given by the log-linear parameters $\lambda^{Y X_1 X_2}$. In simpler words, we consider log-linear parameters for all terms, including Y, and we then match them to the logit parameters by eliminating Y from each term.

Note that the Poisson log-linear model is a undirected model that does not assume any of the available variables as the response one. Associations between variables are indicated via the interaction terms included in the model. On the contrary, binomial (or multinomial) logit models are directed models assuming one variable as response. Additional details concerning this issue can be found in Agresti (2002).

Examples of Poisson log-linear models for contingency tables using WinBUGS can be found in Spiegelhalter et al. (2003a, dogs data example), Congdon (2003; 2005a; 2006b), Dellaportas et al. (2000), and Ntzoufras (2002).

Problems

7.1   Using the following simulated data

X_1	X_2	X_3	X_4	X_5	Y
1	7	4.7	3	1.8	0
2	5	2.3	4	1.6	3
3	3	6.4	2	2.8	64,235
4	7	7.6	2	2.5	5
5	6	3.0	1	1.1	0
6	4	6.6	3	2.6	3,269
7	1	4.9	4	1.6	7,045
8	6	5.5	3	2.1	5
9	2	1.6	1	2.6	162,124
10	2	4.1	0	1.6	979
11	4	5.7	2	3.1	38,319
12	2	7.1	1	3.7	36,901,820
13	2	2.0	0	1.7	1,373
14	3	1.9	3	1.9	641
15	5	7.3	5	1.7	4
16	2	5.2	3	1.0	50
17	4	4.4	4	1.2	2
18	2	5.1	3	1.7	1,800
19	6	10.0	4	3.7	14,041
20	5	5.8	1	2.6	377

a) Model y in WinBUGS using a Poisson log-linear model.

b) Use DIC to select which model is more appropriate for the case.

c) Compare your results with the true expression

$$Y_i \sim \text{Poisson}(\lambda_i) \text{ with } \log \lambda_i = 3 - 2X_{i2} + 5X_{i5}$$

used to simulate Y_i.

7.2 Download the `mshop1` dataset from the book's Website (`www.stat-athens.aueb.gr/~jbn/winbugs_book`). The data were compiled from a customer satisfaction survey conducted in Chios, Greece (Sarantinidis, 2003). The given dataset is only part of the original dataset. Explanation of the variables is given in file `mshop1.txt`.
  a) Build a Poisson log-linear model to identify which variables influence the number of purchased items.
  b) Use DIC to select which variables are finally appropriate for the model.
  c) Interpret the parameters of the model on the basis of their posterior distributions.

7.3 Use the Poisson log-linear model presented in Section 7.4.3 to model the water polo data available in the book's Webpage (`www.stat-athens.aueb.gr/~jbn/winbugs_book`); see Karlis and Ntzoufras (2003a; 2005) for more details.
  a) Comment the performance of each team according to the model parameters.
  b) Interpret the model parameters for each team.
  c) Compare the model parameters, using the `compare` tool of WinBUGS.
  d) Estimate the final ranking, assuming that the tournament was based on the fact that all teams were playing against all teams once.

7.4 Use the Poisson log-linear model presented in Section 7.4.3 to model the soccer data from the Italian football/soccer championship for season 2000–2001. The data are available in the book's Webpage (`www.stat-athens.aueb.gr/~jbn/winbugs_book`).
  a) Comment the performance of each team according to the model parameters.
  b) Interpret the model parameters for each team.
  c) Compare the model parameters using the `compare` tool of WinBUGS.
  d) Regenerate the full results. Use the `ranks` tool to monitor the ranking of the teams. Were the final observed rankings in agreement with the results obtained by the assumed model?

7.5 Use the Poisson distribution to model the WinBUGS data example `salm`; see Spiegelhalter et al. (2003a).
  a) Fit the model described in Spiegelhalter et al. (2003a), ignoring the random effect λ_{ij}.
  b) Interpret the model coefficients of the fitted model.
  c) Use DIC to decide whether X_i of $\log X_i$ must be removed from the model.

7.6 Use the binomial distribution to model the WinBUGS data example `seeds`; see Spiegelhalter et al. (2003a).
  a) Use dummy variables under both corner and sum-to-zero parametrizations.
  b) Fit the model with and without the interaction term between the seed and the type of root extract. Compare the two models using DIC. In all models, ignore the random effect b_i used in the corresponding description of the problem.
  c) Interpret the model coefficients under both models. What is the difference between the two models?

d) Fit a probit model and a clog–log model. Interpret and compare the model parameters with the ones obtained from the logit model.

e) Compare the posterior distributions of the success probabilities obtained under the different link functions.

7.7 Using the following simulated data

X_1	X_2	X_3	X_4	X_5	Y	N
6	12	1	0.0	0.8	6	36
12	17	1	0.3	2.0	10	32
9	13	1	1.5	-0.5	3	26
8	19	1	0.8	1.1	4	41
4	18	1	1.1	-0.5	3	36
14	12	0	0.2	1.0	13	37
10	21	0	0.9	0.7	4	33
2	16	0	0.6	0.7	1	26
10	13	1	0.9	0.4	1	29
12	12	0	0.0	3.3	18	31
10	19	1	0.6	0.6	3	36
8	20	1	1.1	1.4	3	27
9	19	1	1.1	0.6	4	25
9	10	0	0.2	0.5	4	30
14	15	1	1.3	0.6	3	30
11	15	1	0.0	0.2	14	40
10	18	1	0.0	-2.2	3	23
2	18	1	3.4	1.0	0	34
6	15	1	0.2	1.7	2	21
5	10	1	0.6	1.7	0	26

a) Model y in WinBUGS using a logistic regression model.

b) Use DIC to select which model is more appropriate for the case.

c) Compare your results with the true expression

$$Y_i \sim binomial(\pi_i, N_i) \text{ with } logit(\pi_i) = -3 + 0.2 * X_{i1} - 0.2 * X_{i3} - X_{i4}$$

used to simulate Y_i.

d) Use the probit and clog–log links to model the data presented above. Compare the estimated probabilities of the three models.

e) Use DIC to compare the logit, probit, and clog–log models.

7.8 Consider the covariates $X_1, \ldots, X_5$ of Problem 7.7, but for the response Y, assume the following binary data:

1 1 0 0 0 1 0 0 0 1 0 0 0 0 1 0 0 0 0 0

a) Model this response in WinBUGS using a logistic regression model.

b) Use DIC to select which model is more appropriate for the case.

c) Compare your results with the true expression

$$Y_i \sim binomial(\pi_i, N_i) \text{ with } logit(\pi_i) = -3 + 0.2 * X_{i1} - 0.2 * X_{i3} - X_{i4}$$

used to simulate Y_i and with the posterior distributions obtained in Problem 7.7.

d) Use the probit and clog–log links to model these data. Compare the estimated probabilities of the three models.

e) Use DIC to compare the logit, probit, and clog–log models.

7.9 Use the binomial distribution to model the data of WinBUGS example `beetles`; see Spiegelhalter et al. (2003b).

a) Use the logit, probit, and clog–log links.

b) Interpret the model coefficients for all models.

c) Compare the posterior distributions of the model parameters and the success probabilities obtained using the three models.

7.10 Use the water polo data of Problem 7.3 to model the probability for a team of winning a game.

a) Comment the performance of each team according to the model parameters.

b) Interpret the model parameters for each team.

c) Compare the model parameters using the `compare` tool of WinBUGS

d) Compare the results with the ones obtained from the Poisson log-linear analysis of Problem 7.3.

7.11 Consider the 2×2 contingency table data of Problem 1.6.

a) Use the following Poisson log-linear model

$$n_{ij} \sim \text{Poisson}(\lambda_{ij}) \quad \text{and} \quad \log \lambda_{ij} = \mu + a_i + b_j + ab_{ij},$$

where n_{ij} are the frequencies for the ith row and jth column of the Problem 1.6 contingency table.

b) Use data in a tabular format and a "nested double `for`" syntax to code in Win-BUGS the model of (a) using both corner and sum-to-zero parametrizations.

c) Use dummy variables to code in WinBUGS the model of (a) using both corner and sum-to-zero parametrizations.

d) Calculate the posterior distributions of the odds ratio under the assumed model given by

$$\text{OR} = \log \frac{\lambda_{11}\lambda_{22}}{\lambda_{12}\lambda_{21}}$$

under the two parametrizations. Can the OR be considered equal to one?

e) Interpret the parameters of the model.

f) Fit the independence model by removing the interaction term. Compare it with the full model using DIC. What is your conclusion?

7.12 Consider the 2×2 contingency table data of Problem 1.6.

a) Use the binomial distribution to model the probability of having the disease under examination.

b) Use dummy variables to code in WinBUGS the model and both corner and sum-to-zero parametrizations.

c) Calculate the posterior distributions of the odds ratio according to the assumed model given by

$$\text{OR} = \log \frac{\pi(X=1)/\{1 - \pi(X=1)\}}{\pi(X=0)/\{1 - \pi(X=0)\}}$$

with the two parametrizations, where $\pi(X = 1)$ and $\pi(X = 0)$ are the probabilities of having the disease when exposed or not exposed to the risk factor. Can the OR considered to be equal to zero?

d) Interpret the parameters of the model.

e) Compare the posterior distributions of the parameters with the corresponding ones obtained by the log-linear model. Are they similar? If so, Why?

## CHAPTER 8

MODELS FOR POSITIVE CONTINUOUS DATA, COUNT DATA, AND OTHER GLM-BASED EXTENSIONS

In this chapter we focus on models used for positive continuous data and on additional models used for count data.

In the first case we include the log-normal, the gamma, and the inverse Gaussian models. Survival analysis data can be considered as a special case of such data. The Weibull distribution is frequently used to model such data, and, for this reason, it is also described here.

We describe two models used for overdispersed count data: the negative binomial and the generalized Poisson models. The zero inflated versions of the Poisson and these two models are also described in detail. Bivariate Poisson models and the Skellam distribution for integer-valued variables are also discussed in this chapter.

The chapter closes with a brief discussion of the extensions based on GLM, which can be easily fitted using WinBUGS .

8.1 MODELS WITH NONSTANDARD DISTRIBUTIONS

WinBUGS allows for modeling of nonstandard distributions (i.e., for a distribution that is not listed in WinBUGS ' prespecified distributions) using the zero–ones trick. Here we provide general guidelines and then illustrate this approach using an example by fitting the inverse Gaussian model.

8.1.1 Specification of arbitrary likelihood using the zeros–ones trick

We can use either the Bernoulli or the Poisson distribution to indirectly specify any arbitrary model likelihood. Let us assume a model with log-likelihood $l_i = \log f(y_i|\boldsymbol{\theta})$. Then the model likelihood can be written as

$$f(\boldsymbol{y}|\boldsymbol{\theta}) = \prod_{i=1}^{n} e^{l_i} = \prod_{i=1}^{n} \frac{e^{-(-l_i)}(-l_i)^0}{0!} = \prod_{i=1}^{n} f_P(0; -l_i)$$

Hence, the model likelihood can be written as the product of the densities of new pseudo-random variables Ξ_i $(i = 1, \dots, n)$, which follow the Poisson distribution with mean equal to minus the log-likelihood, and all observed values are set equal to zero. To ensure the positivity of the mean of each Ξ_i, we add a positive constant term C to the mean. This is equivalent to multiplying each likelihood term by e^{-C}. This action does not affect the likelihood since it is equivalent to multiplying the resulting (unnormalized) posterior distribution by a constant term equal to e^{-nC}. With this approach, the likelihood becomes equal to

$$f(\boldsymbol{y}|\boldsymbol{\theta}) = \prod_{i=1}^{n} \frac{e^{-(-l_i+C)}(-l_i+C)^0}{0!} = \prod_{i=1}^{n} f_P(0; -l_i + C) \;;$$

C must be selected in such way that $-l_i + C > 0$ for all $i = 1, 2, \dots, n$.

This approach can by fitted in WinBUGS using the following code:

```
C <- 10000
for (i in 1:n) {
    zeros[i] <- 0
    zeros[i] ~ dpois( zeros.mean[i])
    zeros.mean[i] <- -l[i] + C
    # write here the expression of the log-likelihood for i
        observation
    l[i] <- ...
}
```

In this syntax, `l[i]` must be specified accordingly for each model. For example, the normal model can be obtained if we set

$$l_i = -0.5\log(2\pi) - 0.5\log(\sigma^2) - \frac{(y_i - \mu_i)^2}{2\sigma^2}$$

using the syntax

```
l[i] <- -0.5*log(2*3.14) -0.5*log(s2) -0.5*pow( y[i]-mu[i], 2 )/s2
```

Instead of the Poisson–zeros strategy, the Bernoulli distribution can be also used for the same purpose. With this approach

$$f(\boldsymbol{y}|\boldsymbol{\theta}) = \prod_{i=1}^{n} e^{l_i} = \prod_{i=1}^{n} \left(e^{l_i}\right)^1 \left(1 - e^{l_i}\right)^0 = \prod_{i=1}^{n} f_B(1; e^{l_i}, 1),$$

where $f_B(1; e^{l_i}, 1)$ is the binomial probability function with success probability e^{l_i} and $N = 1$. Hence, the model likelihood can be expressed as the product of the densities of new pseudorandom variables Ξ_i, which now follow the Bernoulli distribution with success probability equal to e^{l_i} with all observed values set equal to one. To ensure that this probability is lower than one, we multiply each likelihood term by e^{-C}, where C is a positive large number. Now the likelihood is given by

$$f(\boldsymbol{y}|\boldsymbol{\theta}) = \prod_{i=1}^{n} \left(e^{l_i-C}\right)^1 \left(1 - e^{l_i-C}\right)^0 = \prod_{i=1}^{n} f_B(1; e^{l_i-C}, 1).$$

Now, the corresponding WinBUGS code is given by

```
C <- 100
for (i in 1:n) {
    ones[i] <- 1
    ones[i] ~ dbern( ones.p[i])
    ones.p[i] <- exp( l[i] - C )
    # log-likelihood for i observation
    l[i] <- ...
}
```

In this syntax, `l[i]` must be specified accordingly for each model as in the Poisson–zeros approach described above.

Although both approaches have the same effect, we recommend using the Poisson–zeros approach, which avoids overflow problems due to the simpler likelihood expression.

In this section, we have demonstrated how to define a new sampling distribution (i.e., likelihood) function for distribution not included in the prespecified ones within WinBUGS. The same approach can be followed to specify a prior distribution of nonstandard form (Spiegelhalter et al., 2003*d*).

8.1.2 The inverse Gaussian model

The density of the inverse Gaussian distribution $Y \sim \text{IGaussian}(\mu, \lambda)$ is given by

$$f(y|\mu, \lambda) = \left(\frac{\lambda}{2\pi y^3}\right)^{1/2} \exp\left(-\frac{\lambda(y - \mu)^2}{2\mu^2 y}\right) \quad \text{for } y > 0.$$

The mean and the variance of this parametrized inverse Gaussian distribution is given by $E(Y) = \mu$ and $V(Y) = \mu^3/\lambda$. In GLMs, the parametrization $\mu, \sigma^2 = \lambda^{-1}$ is frequently encountered.

When λ tends to infinity (or σ^2 to zero), the inverse Gaussian distribution becomes similar to a normal (Gaussian) distribution. As it is obvious from the fact that Y is positive, the inverse does not result in the inverse of a normal (Gaussian) distribution (Seshadri, 1993).

The model can be given by

$$Y_i \sim \text{IGaussian}(\mu_i, \lambda),$$

where μ_i linked with the linear predictor using the canonical (squared reciprocal) link which is given by

$$\mu_i^{-2} = \eta_i \Leftrightarrow \mu_i = \sqrt{1/|\eta_i|}$$

or the log-link which is given by

$$\log \mu_i = \eta_i \Leftrightarrow \mu_i = e^{\eta_i}.$$

The log-link is preferred because of its the easier interpretation, which is similar to the one in Poisson log-linear models.

In WinBUGS we may use either of the two parametrizations proposed above. The log-likelihood for the original parametrization is given by

```
l[i] <-    0.5*( log( lambda ) - log(2*3.14) - 3*log(y[i]) )
           - 0.5* lambda * pow( (y[i]-mu[i])/mu[i], 2 )/y[i]
```

while the mean μ_i is specified by

```
mu[i] <- sqrt( 1/abs(eta[i]) )
```

when the inverse squared link function is used, or by

```
log(mu[i]) <- eta[i]
```

when the log-link function is used. Parameter σ^2 is defined as a logical (deterministic) function using the following command

```
s2 <- 1/lambda
```

while the priors can be defined as typically normal for the model coefficients β_j; for λ we may use a gamma prior similar to the one used for the precision of the normal regression model.

> **Example 8.1. An inverse Gaussian simulated dataset.** To illustrate the approach described above, we have generated 100 random values from IGaussian($\log \mu = 3 + 2X_1 - 1X_2, \lambda = 10$). Four standardized normal variables X_j, $j = 1,2,3,4$ were generated as possible covariates. Data are available at the book's Website.

Both the Poisson–zeros and the Bernoulli–ones approaches worked satisfactorily with similar results. Results for the first model using 4000 burnin and an additional 5000 iterations finally kept are presented in Table 8.1. The constant C was set equal to $10,000$ and 100 for the Poisson and the Bernoulli approaches, respectively. The log-link was used in both cases.

As we can see, the estimated model based on the posterior means is given by

$$Y_i \sim \text{IGaussian}(\mu = e^{3.22+2.15X_1-0.95X_2+0.08X_3+0.01X_4}, \lambda = 12.14),$$

which is very closed to the actual model

$$Y_i \sim \text{IGaussian}(\mu = e^{3+2X_1-X_2}, \lambda = 10)$$

used to generate the data.

Table 8.1 Posterior summaries of inverse Gaussian model parameters for simulated data of Example 8.1

node	mean	sd	MC error	2.5%	median	97.5%	start	sample
beta[1]	3.220	0.156	0.014	2.936	3.215	3.539	4001	5000
beta[2]	2.151	0.083	0.007	1.998	2.151	2.324	4001	5000
beta[3]	-0.950	0.111	0.006	-1.183	-0.945	-0.745	4001	5000
beta[4]	-0.081	0.083	0.004	-0.242	-0.083	0.096	4001	5000
beta[5]	0.013	0.122	0.004	-0.242	0.020	0.247	4001	5000
lambda	12.140	1.779	0.033	8.913	12.070	15.82	4001	5000
s2	0.084	0.013	0.0002	0.063	0.083	0.112	4001	5000

WinBUGS code of the model using the zeros trick is given in Table 8.2. Additionally, the corresponding code using the ones trick is available at the Website of this book.

Table 8.2   WinBUGS code for inverse Gaussian model used for simulated data of Example 8.1

```
model{
    C <- 10000
    for (i in 1:n) {
        zeros[i] <- 0
        zeros[i] ~ dpois( zeros.mean[i])
        zeros.mean[i] <- -l[i] + C
        l[i] <-   0.5*( log( lambda ) - log(2*3.14) - 3*log(y[i]) )
                    - 0.5* lambda * pow( (y[i]-mu[i])/mu[i], 2 )/y[i]
        log(mu[i]) <- beta[1] + beta[2]*x1[i] + beta[3]*x2[i]
                            + beta[4]*x3[i] + beta[5]*x4[i]
    }
    # priors
    for (j in 1:5){ beta[j] ~ dnorm( 0.0 ,    0.001 ) }
    lambda ~ dgamma( 0.01, 0.01)
    s2 <- 1/lambda
}
```

8.2   MODELS FOR POSITIVE CONTINUOUS RESPONSE VARIABLES

In this section we focus on modeling positive continuous response variables. The normal distribution may be used without any major problems when the sampling distribution $f(y|\theta)$ is relatively symmetric and far away from zero. The simplest approach is to use the logarithm of the original variable assuming that $\log Y_i \sim N(\mu_i, \sigma^2)$, which is equivalent to using the log-normal distribution for the original response variable Y. Alternatively, the gamma, the exponential (which is special case of the gamma), and the inverse Gaussian (described in the previous section) distributions can be adopted. Other distributions that may be used are the inverse gamma (which can be fitted via the gamma distribution) and the Weibull distribution. The latter is frequently used in the analysis of survival data where the response measures the time until an event of interest (usually death) occurs. Survival analysis models additionally have to account for incomplete data, which are called *censored*. Such data arise when specific timepoints (where the event of interest has occurred) are known, but the exact occurrence time is not available.

8.2.1   The gamma model

The gamma model is given by

$$Y_i \sim \text{gamma}(\mu_i\tau, \tau)$$
$$\log(\mu_i) = \eta_i = \beta_0 + \beta_1 X_{i1} + ... + \beta_p X_{ip} .$$

In this expression, the mean and the variance are given by $E(Y_i) = \mu_i$ and $\text{Var}(Y_i) = \tau^{-1}$ and the log-link is used. Covariate structure can be further imposed on the precision (or the variance) parameter. Alternatively, the canonical link $\mu_i^{-1} = \eta_i$, but problems may appear in the MCMC algorithms because μ_i must be positive. Moreover, interpretation is not as straightforward as in the log-link case.

In WinBUGS, the gamma model can be implemented by

```
for (i in 1:n){
    y[i] ~ dgamma( a[i], tau )
    a[i] <- mu[i]*tau
```

```
      log(mu[i]) <- eta[i]
   }
   s2 <- 1/tau
```

If the canonical-link is used instead, then the fourth line of this code is substituted by the command `mu[i] <- 1/(eta[i])`.

8.2.2 Other models

The log-normal model is essentially the same as the normal model. It can be easily implemented by transforming the response variable in WinBUGS using the syntax

```
      ly[i] <- log( y[i] )
      ly[i] ~ dnorm( mu[i], tau )
```

or by directly imposing the log-normal distribution on Y using the command

```
      y[i] <- dlnorm( mu[i], tau )
```

Interpretation of such models is similar to the usual regression models, but now we focus on the mean of $\log Y$ instead or the original variable Y.

Concerning the exponential distribution, exponential(λ), we can either use directly the `dexp` command or define it as a special case of the gamma distribution since exponential(λ) = $\Gamma(1, \lambda)$. Covariates can be imposed on the mean λ^{-1} of the exponential distribution using either the inverse or the log-link, resulting in $\lambda_i = \eta_i$ and $\log \lambda_i = -\eta_i$, respectively. The case for the χ^2 distribution is similar. We can either use the `dchisqr` WinBUGS command or define it as a gamma($\mu/2, \frac{1}{2}$) with mean μ and variance 2μ.

The inverse gamma distribution is simply defined as $1/Y$ with $Y \sim$ gamma($\mu\tau, \tau$). Hence in WinBUGS we can easily define the inverse gamma distribution by transforming the original response variable. The corresponding WinBUGS syntax is given by

```
      inv.y[i] <- 1/ y[i]
      inv.y[i] ~ dgamma( a[i], tau )
```

In this model we can still model the mean of the inverse random variable. Note that the mean of Y^{-1} is equal to $b/(a - 1)$ and is specified only for $a > 1$.

Finally, the Weibull distribution is used with probability density function

$$f(y|\lambda, v) = v\lambda y^{v-1} e^{-\lambda y^v},$$

mean $E(Y) = \lambda^{-1/v}\Gamma(1 + v^{-1})$, and variance $V(Y) = \lambda^{-2/v}\left(\Gamma(1 + 2/v) - \Gamma(1 + 1/v)^2\right)$. It is implemented in WinBUGS using the command `y[i] ~ dweib(v, lambda)`. The exponential distribution is also a special case of the Weibull distribution for $v = 1$. Parameter λ is usually log-linked with the linear predictor. Alternatively, the mean can be directly log-linked with the linear predictor. This can be achieved by the following WinBUGS syntax

```
      lambda[i] <-  exp( v*loggam(1+1/v) - v*eta[i] )
      eta[i] <- beta0 + beta1 * cases[i] + beta2 * distance[i]
```

since $\lambda_i = \left[\Gamma\left(1 + 1/v\right)/E(Y_i)\right]^v$, resulting in $\log(\lambda_i) = v \log \Gamma\left(1 + 1/v\right) - v\eta_i$ with η_i the usual linear predictor log-linked with $E(Y_i)$, specifically, $\log E(Y_i) = \eta_i$.

The two models are equivalent to $\beta_0^* = v \log \Gamma(1 + 1/v) - \beta_0$ and $\beta_j^* = -v\beta_j$, where β_j^* and β_j denote the regression coefficients imposed on $\log(\lambda)$ and $\log(EY)$, respectively. Nevertheless, the prior distributions must be specified with caution in order to obtain the same posterior distributions. Exactly the same posterior distributions will be obtained only when we specify equivalent prior distributions under both parametrizations.

8.2.3 An example

Example 8.2. Soft drink delivery data (Example 5.1 revisited). In the following we implement all the models described in this section and use the DIC to identify which model is the best on the soft drink delivery data presented in Chapter 5 (Example 5.1).

Note that DIC as calculated in WinBUGS can be used only if the same response Y is used. If transformation of the original variable is used, then DIC values are not comparable since the response used is essentially different.

Table 8.3 presents the DIC values for various models fitted to the soft drink delivery data of Example 5.1. DIC values indicate the gamma and the Weibull models as the best ones. For the Weibull distribution, two models were fitted using covariates on the $log(\lambda)$ (parameter of the Weibull) and using covariates directly on the log-mean of the Weibull distribution. DIC values are slightly different, possibly because different priors are imposed on common parameters. Moreover, DIC values can be used for comparison only if we use exactly the same response variable. Hence, when using transformed data, DIC values reported by WinBUGS refer to the transformed and not to the original response variable. For this reason, a constant term based on the first derivative of the transformation function must be added. This term is given by $C = -2\sum_{i=1}^{n} \log |J_i|$, where J_i is the first derivative of the transformation function. For the inverse gamma and the log-normal (when log-transformed data were used) distributions, WinBUGS DIC values were found equal to -12.51 and -159.150, respectively, in which the terms $C = 2\sum_{i=1}^{n} \log y_i = 147.71$ and $C = 4\sum_{i=1}^{n} \log y_i = 295.42$ must be added to obtain the correct values. Finally, when the $\log -\mu$ representation was used, a precision limit was set to avoid arithmetic overflows of λ.

Table 8.3 Deviance values of various models fitted at soft drink delivery data of Example 8.2

	Model	$\overline{D}^a$	$\widehat{D}^b$	p_m	DIC
1.	Normal	131.2	127.0	4.2	135.4
2.	Log-normal[c,e]	131.0	126.8	4.2	135.2
3.	Exponential	201.4	198.5	2.9	204.3
4.	χ^2	145.5	142.4	3.0	148.5
5.	Gamma	129.4	125.4	4.1	133.5
6.	Inverse gamma[d,e]	136.3	132.3	4.0	140.3
7a.	Weibull ($\log \lambda$)	129.6	125.9	3.7	133.4
7b.	Weibull ($\log \mu$)	130.2	126.1	4.0	134.2

[a] $\overline{D} = \overline{D(\boldsymbol{\theta}_m, m)}$ = posterior mean of deviance.
[b] $\widehat{D} = D(\overline{\boldsymbol{\theta}}_m, m)$ = deviance evaluated at posterior means of parameters.
[c] When we use $Z_i = \log(\text{time}_i)$, then WinBUGS DIC estimate $=-12.51$. The term $C = 2\sum_{i=1}^{n} \log y_i = 147.71$ must be added in each deviance measure.
[d] WinBUGS DIC estimate $=-159.150$. The term $C = -2\sum_{i=1}^{n} \log |J_i| = 4\sum_{i=1}^{n} \log y_i = 295.42$ must be added in each deviance measure.
[e] When the response variable is transformed, the term $C = -2\sum_{i=1}^{n} \log |J_i|$ must be added in each deviance in order to obtain a comparable DIC, where J_i is the first derivative of the transformation function.

Table 8.4 presents the posterior summaries for the gamma and the Weibull model. The Weibull model is based on modeling $\log(\mu)$ to simplify interpretation. Results from both models are quite close, indicating no major differences.

	Gamma model			Weibull model		
	Posterior	Posterior percentiles		Posterior	Posterior percentiles	
Parameter	mean	2.5%	97.5%	mean	2.5%	97.5%
β_0	2.382	2.276	2.487	2.299	2.151	2.44
β_1	0.047	0.031	0.060	0.051	0.035	0.070
$\beta_2 \times 100$	0.046	0.016	0.078	0.053	0.024	0.085
τ or υ	1.912	0.940	3.192	6.878	4.724	9.335
σ	0.749	0.560	1.032	—	—	—
e^{β_0}	10.850	9.739	12.020	9.990	8.592	11.470
e^{β_1}	1.048	1.032	1.062	1.053	1.036	1.072
$e^{100\beta_2}$	1.047	1.016	1.081	1.055	1.024	1.088
typical.y[a]	19.67	18.39	21.07	19.43	18.01	20.76

[a]typical.y$= \exp\left(\beta_0 + \beta_1 \overline{X}_1 + \beta_2 \overline{X}_2\right)$.

Table 8.4   Deviance values of various models fitted at soft drink delivery data of Example 8.2.

Interpretation can be based on the exponents of β_j coefficients of the gamma model. Posterior summaries of expected times for typical deliveries (with covariates set equal to their sample means) are also provided. For these values we can claim that

- When no cases of stocked products need to be replaced in the machine and no distance is walked, then the worker needs about 10 minutes to complete the delivery ($9.7 - 12$ with the gamma model and $8.6 - 11.5$ with Weibull model). Note that this estimate is quite different from the corresponding one for the normal linear model.

- Every additional case that needs to be stocked in the machine increases the delivery time by almost 5% (4.7 and 5.1 using the gamma and the Weibull models, respectively).

- Every additional 100 feet walking distance increases the delivery time by approximately 5% (4.6 and 5.3 using the gamma and the Weibull models, respectively).

- A typical delivery is completed in about 19 minutes.

8.3   ADDITIONAL MODELS FOR COUNT DATA

As we have already mentioned, the Poisson distribution is frequently used for modeling discrete count data. This distribution imposes an important and restrictive assumption: the variance is restricted to be equal to the mean. This assumption is frequently violated in practice. In such cases, over/underdispersed models must be used to capture this extra (or lower) variability. Another problem frequently encountered is that the probability of zero in most real-life data is considerably underestimated by the standard Poisson model. For this reason, zero inflated models have been developed and used in practice.

Here we focus on two overdispersed distributions used to model count data. We further present zero inflated versions of such models. Finally we close this section with a discussion focused on other distributions used to model count data such as the bivariate Poisson distribution [see, e.g., Karlis and Ntzoufras (2003a)], and the Skellam distribution used to model differences of count data or, more generally, discrete integer-valued variables (Karlis and Ntzoufras, 2006).

8.3.1   The negative binomial model

A discrete random variable Y follows the negative binomial (NB) distribution $(Y \sim \mathrm{NB}(\pi, r))$ if the probability function is given by

$$f_{\mathrm{NB}}(y; \pi, r) = \frac{\Gamma(y+r)}{y!\Gamma(r)} \pi^r (1-\pi)^y$$

where $y = 0, 1, 2, \ldots, r$ and $r > 0$ a positive parameters. The mean and the variance of this distribution are equal to $r(1-\pi)/\pi$ and $r(1-\pi)/\pi^2$.

Frequently $r = N$ is an integer valued parameter. In such cases, this distribution is used to describe the number of times (equal to $N + y$) that we need to to repeat a Bernoulli experiment with success probability π until N successes are reached, for example, how many free throws a basketball player must attempt before scoring 10 times. In the more general case where $r \in \mathbf{R}, r > 0$, also referred to as *Polya distribution*, it is used to model overdispersed count data since the variance over mean ratio [i.e. the *dispersion index* (DI)] is given by

$$\mathrm{DI} = \frac{\mathrm{Var}(Y)}{E(Y)} = \frac{1}{\pi} .$$

When modeling count data, the parametrization $\lambda = r(1-\pi)/\pi$, or equivalently $\pi = r/(r+\lambda)$, is frequently used instead. Under this parametrization $E(Y) = \lambda$ and $V(Y) = \lambda(\lambda+r)$, while the dispersion index is equal to DI$= 1 + \lambda/r$. The negative binomial model with this parametrization can be also derived by the following Poisson-gamma mixture model

$$Y|u \sim \mathrm{Poisson}(\lambda u) \quad \text{and} \quad u \sim \mathrm{gamma}(r, r).$$

The resulting marginal distribution of Y is given by

$$f(y) = \int_0^\infty f(y|u)f(u)du = \frac{\Gamma(y+r)}{y!\Gamma(r)} \left(\frac{r}{r+\lambda}\right)^r \left(\frac{\lambda}{r+\lambda}\right)^y ,$$

which is a negative binomial distribution with parameters $r/(r+\lambda)$ and r.

The Poisson distribution with mean $\lambda = r(1-\pi)/\pi$ is the limiting case of the negative binomial distribution for $r \to \infty$. Although the canonical link [complementary log (i.e. clog) on π, $\log(1-\pi) = \eta$] link can be used, we recommend, especially when modeling count data, using the log-link on λ [which is equivalent to setting $\mathrm{logit}(\pi) = -r - \eta$] in order to ensure a straightforward parameter interpretation and a direct comparison to the corresponding parameters of the Poisson model.

This distribution is available in the current version of WinBUGS and can be modeled using the syntax

```
for (i in 1:n){
    y[i] ~ dnegbin( p[i], r )
    p[i] <- r/(r+lambda[i])
```

```
    log(lambda[i]) <- eta[i]
    eta[i] <- beta0 + beta1*x1[i] + ... + betap*xp[i]
}
```

The usual normal prior distributions can be used on model coefficients $\beta_j, j = 0, 1, 2, \ldots, p$. For parameter r, a gamma prior can used in a manner similar to that for the dispersion parameters in the normal and the inverse Gaussian cases. Moreover, the model can be easily extended by adding structure on the dispersion parameter r. Such extensions can be accommodated in WinBUGS without any complications.

Example 8.3. 1990 USA general social survey: Number of monthly sexual intercourses. In this example we consider the 1990 USA general social survey data concerning the number of sexual intercourses of each participant within the previous month. Accumulated data by gender are provided by Agresti (2002, pp. 569–570) and are reproduced in the book's Website with permission of John Wiley and Sons, Inc. Note that the sample means are 5.9 and 4.3 for males and females, respectively, while the variances are much higher (54.8 and 34.4, respectively), indicating that the data are clearly overdispersed.

Here we use the negative binomial distribution to model these data and compare the model with the standard Poisson model using the DIC measure.

The data consist of 550 observations. The data were inserted using ungrouped data and two variables: G_i, which is the gender for i subject (zero for males and one for females); and Y_i, which is the responded number of sexual intercourses during the previous month.

In the following we briefly present two approaches: (1) a simple approach using one separate negative binomial distribution for the data of each gender and (2) an approach structured as a generalized linear model with the gender as a covariate.

Simple approach. The model can be easily summarized by

$$Y_i \sim NB(\pi_{G_i+1}, r_{G_i+1}) \text{ for } i = 1, 2, \ldots, 550$$

with prior distributions

$$\pi_j \sim U(0, 1) \text{ and } r_j \sim \text{gamma}(0.001, 0.001)$$

for $j = 1, 2$ (males, females respectively).

The corresponding WinBUGS code is given by

```
model{
    for(i in 1:n){
        y[i] ~ dnegbin( p[ gender[i]+1], r[ gender[i]+1 ] )
    }
    for (j in 1:2){
        p[j]~dbeta(1,1)
        r[j]~dgamma(0.001,0.001)
    }
}
```

Finally, the mean, variance, and dispersion index for each gender are estimated using simple logical (deterministic) nodes defined by the following syntax

```
for (j in 1:2){
    lambda[j] <- r[j]*(1-p[j])/p[j]
    di[j] <- 1/p[j]
    var[j] <- di[j] * lambda[j]
}
```

Model-based approach. If we use the model-based approach with gender as a covariate, then we can express the model as

$$Y_i \sim NB(\pi_i^*, r_i^*)$$
$$\pi_i^* = \frac{r_i^*}{r_i^* + \lambda_i^*}, \quad r_i^* = r_{G_i+1}$$
$$\log(\lambda_i^*) = \beta_1 + \beta_2 G_i \text{ for } i = 1, 2, \ldots 550,$$

where π_i^*, r_i^*, and λ_i^* are the parameters for each individual i ($i = 1, \ldots, 550$) while π_j, r_j, and λ_j are the parameters for each gender j ($j = 1, 2$) and are given by

$$\lambda_1 = e^{\beta_0}, \quad \lambda_2 = e^{\beta_0 + \beta_1}, \pi_j = \frac{r_j}{r_j + \lambda_j} \text{ for } j = 1, 2.$$

Finally, for β_0 and β_1 we use the prior distributions

$$\beta_j \sim N(0, 1000) \text{ and } r_j \sim \text{gamma}(0.001, 0.001)$$

for $j = 1, 2$ (males, females, respectively).

The corresponding WinBUGS code is given by

```
model{
    for(i in 1:n){ y[i] ~ dnegbin( p.ind[i], r.ind[ i ] )
    p.ind[i] <- r.ind[i]/( r.ind[i]+lambda.ind[i] )
    log(lambda.ind[i]) <- beta[1] + beta[2] * gender[i]
    r.ind[i] <- r[ gender[i] + 1 ]
    }
    lambda[1] <- exp(beta[1])
    lambda[2] <- exp(beta[1]+beta[2])
    for (j in 1:2){
        r[j] ~ dgamma( 0.001, 0.001 )
        beta[j] ~ dnorm( 0.0, 0.001 )
    }
}
```

where p.ind, r.ind, and lambda.ind are the parameters π_i^*, r_i^*, and λ_i^*, respectively, of individual i. Finally, the dispersion index and the variances are specified as in the simple case described above.

Results. Table 8.5 summarizes posterior results from the two approaches above and the simple Poisson model. Both negative binomial (NB) and Poisson models indicate a significant difference in the number of monthly sexual intercourses declared by people of different genders. Female sexual activity is lower by 25% than the corresponding activity of males.

Dispersion indices of NB as well as DIC values indicate that the NB model is much better than the Poisson one. Concerning DIs, these range from 7.6 to 12.7 and from 9.3 to 15.8 for males and females, respectively, indicating a clear overdispersion. Moreover, DIC values for the NB model are much lower than the corresponding value for the Poisson model.

Note that λ_j (i.e. the expected counts) have similar posterior means under both Poisson and negative binomial models, but have different posterior variance since the dispersion of the sampling distribution is larger for the NB model. This leads to posterior distributions with wider 95% posterior intervals for the λ_j of the NB model.

Table 8.5   Posterior means (95% posterior intervals) for model parameters of negative binomial and Poisson model in Example 8.3

	NB — GLM	NB — simple	Poisson
β_1	1.76 (1.61, 1.93)		1.77 (1.71, 1.82)
β_2	-0.29 (-0.53, -0.05)		-0.31 (-0.38,-0.23)
λ_1	5.84 (5.02, 6.88)	5.86 (4.98, 6.86)	5.86 (5.55, 6.17)
λ_2	4.38 (3.61, 5.26)	4.31 (3.56, 5.17)	4.30 (4.07, 4.53)
e^{β_2}	0.75 (0.59, 0.95)		0.74 (0.68, 0.79)
r_1	0.67 (0.53, 0.83)	0.67 (0.54, 0.84)	
r_2	0.40 (0.32, 0.48)	0.40 (0.32, 0.49)	
π_1	0.10 (0.08, 0.13)	0.10 (0.08, 0.13)	
π_2	0.08 (0.06, 0.11)	0.09 (0.07, 0.11)	
DI_1	9.85 (7.79, 12.67)	9.80 (7.65, 12.47)	
DI_2	12.16 (9.42, 15.80)	11.89 (9.30, 15.32)	
DIC	2873.85	2873.46	5271.35

8.3.2 The generalized Poisson model

Another distribution frequently used to model overdispersed insurance claim counts is the generalized Poisson distribution introduced by Consul (1989); see also Consul and Famoye (1992) for a thorough description and in Ter Berg (1996) and Scollnik (1998) for actuarial applications of this model.

The *generalized Poisson* (GP) distribution is also known as *Lagrangian Poisson distribution*, and its probability function is given by

$$f(y|\zeta, \omega) = \frac{\zeta(\zeta + \omega y)^{y-1}}{y!} e^{-(\zeta + \omega y)}, \tag{8.1}$$

where ζ and ω are the distribution parameters. In this parametrization, the mean and the variance is given by $E(Y) = \zeta(1 - \omega)^{-1}$ and by $V(Y) = \zeta(1 - \omega)^{-3}$, respectively. To make the parametrization equivalent to the corresponding Poisson model, Ntzoufras et al. (2005) have recommended an alternative parametrization by setting $\lambda = \zeta(1 - \omega)^{-1}$. Then the corresponding probability function is given by

$$f(y|\lambda, \omega) = (1 - \omega)\lambda \frac{\{(1 - \omega)\lambda + \omega y\}^{y-1}}{y!} e^{-\{(1 - \omega)\lambda + \omega y\}}. \tag{8.2}$$

According to Ter Berg (1996), valid values for ω are within the interval $[0, 1)$. Typically, the distribution can be defined for $y = 0, 1, 2, \ldots, N$ with $\max\{-1, -\zeta/N\} \le \omega < 1$ with negative values leading to underdispersion; see Scollnik (1998) for details. Note that for positive counts we set $N = \infty$, leading to $\omega \in [0, 1)$. For $\omega = 0$, the GP also reduces to the simple Poisson model with mean λ. The mean of Y is equal to $E(Y) = \lambda$, while the variance and the dispersion index are equal to $V(Y) = \lambda(1 - \omega)^{-2}$ and DI $= (1 - \omega)^{-2}$, respectively.

The generalized Poisson model can be fitted using the zeros–ones trick described in Section 8.1.1. Hence we may specify the log-likelihood components l_i by

```
lambda.star[i] <- (1-omega)*lambda[i] + omega*y[i]
l[i] <-   log(1-omega) + log(lambda[i])
       + (y[i]-1)*log(lambda.star[i] ) - loggam(y[i]+1) - lambda.
           star[i]
```

where `lambda.star[i]` is the WinBUGS node used to define $\lambda_i^* = (1 - \omega)\lambda_i + \omega y_i$ and in this way simplify the log-likelihood component l_i.

In the syntax displayed above, ω is assumed constant across all parameters. Structure on ω parameter can be added by substituting ω by ω_i (`omega[i]` in WinBUGS) and then specifying its structure in a separate line similarly to the specification of the mean. The log-link on λ can be naturally used for an interpretation similar to that in the Poisson log-linear models.

Concerning the prior distributions, the usual normal prior distributions can be used for β_j parameters while a beta prior can be used for ω. When no information is available, a uniform defined on the interval $(0, 1)$ or a beta$(\frac{1}{2}, \frac{1}{2})$ prior can be used. Alternatively, if negative values of ω are allowed (allowing also for underdispersion), then a rescaled beta distribution can be adopted.

Here we use the GSS 1990 dataset (Example 8.3) to briefly illustrate results. The code can be found in this book's Webpage. More details and actuarial oriented examples can be found in Katsis and Ntzoufras (2005).

Example 8.3 (revisited—1): Fitting the generalized Poisson model on GSS 1990 data.
The generalized Poisson model has been fitted using either the simple or the model-based approach as described in the illustration of negative binomial model. The constant parameter was set equal to zero since e^{l_i} is the probability function of the GP distribution, resulting in $-l_i > 0$ for all $i = 1, \ldots, n$. With the simple approach, the DIC value, as reported by WinBUGS, is equal to 2926.44, which is lower than the corresponding value for the Poisson model but higher than the DIC value for the negative binomial one. With the posterior means of the dispersion indices, we observe that the GP model is far away from the value of one assumed by the Poisson distribution. Moreover, the GP model also indicates a considerable increase in variance with the 95% posterior intervals of DIs ranging from 8.5 to 17.7 and from 16.2 to 26.9 for males and females, respectively; see also the last column Table 8.6 for detailed posterior summaries.

Example 8.3 (continued): Using grouped data. In the previous example, the actual raw data were used with 550 observations. Aggregated data can be used instead in order to accelerate the MCMC algorithm. This can be easily achieved using the zeros–ones trick; see the following computational note for details. Using the aggregated data, only 3 seconds were required to generate 5000 iterations instead of 41 seconds needed when the raw data were used, increasing the speed of the algorithm by 13 times. Results are the same as in the case of using raw data since both the data and the model are the same (only rearranged in a more compact format).

COMPUTATIONAL NOTE (using grouped/aggregated data in WinBUGS)

In order to use aggregated data, we need two variables: one for all observed values y_k of Y (denoted by y[k]) and one of the corresponding observed frequencies/counts n_k (denoted by n[k]) for $k = 1, 2, \ldots, K$, where K are the distinct observed values. The corresponding WinBUGS code is

```
C<-1000
N <- sum( n[1:K] )
for (k in 1:K){
    zeros[k] <- 0
    zeros[k]    dpoiss( zeros.means[k] )
    log(zeros.means[k]) <- -n[k]*l[k] + C
    l[k]<- ...  # set here the log-density/probability function
}
```

8.3.3 Zero inflated models

Zero inflated models for count data have been widely used in statistical science to model a wide variety of real life count data such as manufacturing defects (Lambert, 1992); sexual behavior (Heibron, 1994); medical (Bohning, 1998; Cheung, 2002); dental (Bohning et al., 1999; Mwalili et al., 2008; Karlis and Ntzoufras, 2006); spatial (Agarwal et al., 2002); violence-related (Famoye and Singh, 2006), and sports data (Karlis and Ntzoufras, 2003a; Karlis and Ntzoufras, 2006); see also Ridout et al. (1998) and Gan (2000) and references cited therein. Such models introduce an extra probability parameter to capture an excess of zero values that cannot be estimated sufficiently by the assumed distribution or model.

The zero inflated version of a distribution D of random variable $Y \sim \text{ZID}(\pi_0, \boldsymbol{\theta})$ (where ZID = zero inflated distribution) has a probability function of the form

$$f_{\text{ZID}}(y) = \pi_0 I(y = 0) + (1 - \pi_0)f_D(y; \boldsymbol{\theta})$$

where $f_D(y|\boldsymbol{\theta})$ is the probability function of distribution D with parameters $\boldsymbol{\theta}$ and $f_{\text{ZID}}(y)$ is the zero inflated version of D with an additional parameter π_0 as the proportion of additional zeros. From the equation above, the probability of zero is equal to $\pi_0 + (1 - \pi_0)f_D(0|\boldsymbol{\theta})$, while the probability of $y > 0$ is given by $(1 - \pi_0)f_D(y|\boldsymbol{\theta})$. Moreover, the mean and the variance of this distribution are equal to

$$E(Y) = (1 - \pi_0)E(Y_D)$$
$$V(Y) = (1 - \pi_0)\Big(V(Y_D) + \pi_0 E(Y_D)\Big),$$

respectively, and the dispersion index is equal to $\text{DI} = V(Y_D)/E(Y_D) + \pi_0$, where $Y_D \sim D(\boldsymbol{\theta})$.

The zero inflated Poisson (ZIP) distribution is the simplest ZID and was introduced in the early 1960s by Cohen (1963), but a ZIP model was introduced in the statistical literature in the early 1990s by Lambert (1992) using covariates on both the Poisson rate λ and on the zero excess probability π_0. Alternative terms used by some researchers developing independently similar models is the *zero altered* (Heibron, 1994) or *zero adjusted* (Gupta et al., 1996) Poisson models. The full ZIP model has the following representation:

$$Y_i \sim \text{ZIP}(\pi_{0i}, \lambda_i) \text{ and } \log(\lambda_i) = \boldsymbol{X}_{(i)}\boldsymbol{\beta} .$$

The excessive proportion of zeros π_0 is usually assumed constant across all observations $i = 1, 2, \ldots, n$, but covariates can be also incorporated here with no difficulty. Hence the following structure

$$\log\left(\frac{\pi_{0i}}{1 - \pi_{0i}}\right) = \boldsymbol{X}_{(i)}^Z\boldsymbol{\beta}^Z,$$

where $\boldsymbol{X}_{(i)}^Z$ and $\boldsymbol{\beta}^Z$ are the values of the covariates and the corresponding coefficients involved in the linear predictor for π_0, can be also incorporated in the model with no difficultly.

Zero inflated models also exist for other distributions, such as the gamma distribution (Feuewerger, 1979). Nevertheless, interest mainly lies in discrete count data, resulting in a variety of zero inflated versions of the binomial (ZIB) (Hall, 2000), the negative binomial (ZINB) (Heibron, 1994; Mwalili et al., 2008), the generalized Poisson (ZIGP) (Gupta et al., 1996; Famoye and Singh, 2006), the bivariate Poisson (ZIBP) (Wahlin, 2001), and the multivariate Poisson model (MZIP) (Li et al., 1999). More recently, Karlis and Ntzoufras (2003a, 2005) have introduced diagonal inflated bivariate Poisson (DIBP) models emphasizing application in sports and the zero inflated version of Skellam's (1946) distribution [or *Poisson difference distribution* (ZIPD)] implemented in dental epidemiology (Karlis and Ntzoufras, 2006) and association football (Karlis and Ntzoufras, 2008) data. More details on zero inflated models can be found in Gan (2000).

Bayesian analysis of zero inflated models has been described in detail by Ghosh et al. (2006), who also provide an easy-to-use WinBUGS code for fitting the ZIP model. Detailed MCMC algorithms for estimation of the posterior distributions of the parameters in the ZIPD model were provided by Karlis and Ntzoufras (2006), while the same authors (Karlis and Ntzoufras, 2008) used a generalized ZIPD model for association football data implemented using WinBUGS.

Implementation of zero inflated distribution can be directly specified in WinBUGS using the zeros–ones trick using the syntax

```
l[i] <- log( p0[i] * equals( y[i], 0 ) + (1-p0[i])*fd[i] )
# set the probability function of each distribution
fd[i] <-  exp( -lambda[i] + y[i]*log(lambda[i]) - loggam(y[i]+1) )
```

for the ZIP. For other distributions we need to specify appropriately $\texttt{fd[i]}$.

Alternatively, the WinBUGS code can be based on the mixed Poisson representation

$$Y|U \sim \text{Poisson}\big(\lambda(1 - U)\big) \text{ with } U \sim \text{Bernoulli}(\pi_0)$$

resulting in

$$f_{\text{ZIP}}(y) = \pi_0 f_P(y; 0) + (1 - \pi_0)f_P(y; \lambda) = \pi_0 I(y = 0) + (1 - \pi_0)f_P(y; \lambda)$$

since the Poisson with mean equal to zero degenerates to $I(y = 0)$, that is, to a distribution with probability of zero value equal to one and all other values having zero probability. Hence the corresponding model is defined using the commands

```
for (i in 1:n){
    y[i] ~ dpois( mu[i] )
    mu[i] <- (1-u[i])*lambda[i]
    u[i] ~ dbern( p0 )
    log(lambda[i]) <- eta[i] # define here the linear predictor
}
```

This is the approach proposed by Ghosh et al. (2006).

A beta prior for π_0 (or a uniform prior when no information is available) can be used. The usual normal prior distributions can be facilitated for β and β^Z when linear structure is assumed for the logit of π_0.

The zero inflated version of the negative binomial can be fitted in a similar manner in the case of ZIP. Using the first approach, we need to define the success probability π of the negative binomial by

$$\pi_i = \frac{r_i}{r_i + \lambda_i(1 - U_i)} \quad \text{with} \quad U_i \sim \text{Bernoulli}(\pi_0),$$

which can be defined in WinBUGS by

```
r.ind[i]/( r.ind[i]+lambda.ind[i]*(1-u[i]) )
u[i] ~ dbern( p0 )
```

for $i = 1, \ldots, n$. Finally, the zero inflated generalized Poisson distribution (or any other distribution not in included in the list of the WinBUGS predefined distributions) can be fitted using the zeros–ones trick as described above.

Example 8.3 (revisited—2): Fitting the zero inflated models on GSS 1990 data. Zero inflated versions for the Poisson, negative binomial, and generalized Poisson distributions can be fitted using either the zeros–ones trick or the mixed Poisson model approach, with both approaches giving similar results. WinBUGS internal node of deviance (`deviance`) yields different results under both approaches since, in the mixed Poisson representation, the conditional deviance for each given u_i is calculated. Hence, the deviance calculated by WinBUGS in this case is given by

$$\text{Deviance}_c = -2\sum_{i=1}^{n} \log f(y_i|u_i, \pi_0, \lambda) = -2\sum_{i=1}^{n} I(u_i = 0) \log f_P(y_i; \pi_0, \lambda)$$

for each iteration instead of the actual deviance

$$
\begin{aligned}
\text{Deviance} &= -2\sum_{i=1}^{n} \log f(y_i|\pi_0, \lambda) \\
&= -2\sum_{i=1}^{n} \log \Big(\pi_0 I(u_i = 1) + (1 - \pi_0) f_P(y_i; \pi_0, \lambda) \Big),
\end{aligned}
$$

which marginalizes out the augmented data u_i. Hence we recommend specifying the latter as a deterministic/logical node in WinBUGS and using it directly.

Posterior summaries of model parameters for the zero inflated models for each gender are provided in Table 8.6. Results of all ZI models indicate that

- The zero inflated component considerably improves the model fit according to the DIC value, while ZIGP seems to fit the data better than do the other models under consideration here.

- The proportion of excessive zeros is different for the two genders: about 22–27% and 35–41% (depending on the model) of males and females, respectively, remaining inactive during the last month in addition to what was expected from the main distributional assumption.

- Overdispersion parameters are important for the model since both ZINB and ZIGP have much lower DIC values than do the corresponding Poisson models.

- Minor differences between the posterior distributions of r and ω parameters of the ZINB and ZIGP are observed between the two genders.

Table 8.6 Posterior means (95% posterior intervals) for model parameters of generalized Poisson and zero inflated versions of Poisson, negative binomial, and generalized Poisson models for Example 8.3

Parameter[a]	ZIP		ZINB		ZIGP		Generalized Poisson	
π_{01}	0.272	(0.219, 0.329)	0.226	(0.158, 0.291)	0.242	(0.184, 0.304)	—	
π_{02}	0.413	(0.360, 0.469)	0.355	(0.283, 0.423)	0.377	(0.312, 0.438)	—	
β_1	2.085	(2.029, 2.140)	2.026	(1.885, 2.162)	2.043	(1.907, 2.168)	1.767	(1.592, 1.966)
β_2	-0.094	(-0.183, -0.016)	-0.129	(-0.342, 0.078)	-0.110	(-0.297, 0.073)	-0.298	(-0.624, 0.008)
e^{β_2}	0.910	(0.833, 0.984)	0.883	(0.710, 1.081)	0.900	(0.743, 1.076)	0.751	(0.536, 1.008)
r_1 / ω_1	—		1.65	(1.15, 2.26)	0.58	(0.52, 0.64)	0.71	(0.65, 0.76)
r_2 / ω_2	—		1.42	(0.96, 2.00)	0.59	(0.53, 0.65)	0.75	(0.70, 0.81)
π_1 / ζ_1	—		0.18	(0.13, 0.23)	3.23	(2.75, 3.74)	1.71	(1.47, 1.96)
π_2 / ζ_2	—		0.17	(0.13, 0.23)	2.84	(2.36, 3.34)	1.07	(0.93, 1.23)
$E(Y_1)$	5.86	(5.31, 6.39)	5.88	(5.08, 6.77)	5.85	(5.08, 6.74)	5.88	(4.91, 7.14)
$E(Y_2)$	4.30	(3.84, 4.77)	4.30	(3.69, 5.04)	4.32	(3.67, 5.02)	4.38	(3.51, 5.56)
$\mathrm{Var}(Y_1)$	7.45	(6.98, 7.93)	55.98	(42.70, 71.37)	35.45	(24.82, 51.84)	71.47	(43.65, 122.4)
$\mathrm{Var}(Y_2)$	6.07	(5.59, 6.53)	36.54	(27.47, 47.48)	27.84	(19.00, 41.45)	75.44	(40.71, 144.0)
DI_1	1.27	(1.22, 1.33)	9.48	(8.15, 10.83)	6.03	(4.58, 8.05)	12.00	(8.38, 17.52)
DI_2	1.41	(1.36, 1.47)	8.46	(7.15, 9.70)	6.41	(4.82, 8.70)	16.87	(11.07, 26.79)
Deviance	3663	(3659, 3670)	2815	(2810, 2824)	2809	(2804, 2817)	2922	(2919, 2930)
DIC^b	3667.412		2821.139		2814.802		2926.440	

[a] Parameter indices refer to gender (1=males, 2=females).
[b] DIC values for the zero inflated models were calculated outside WinBUGS using the posterior means of the model parameters; DICs for Poisson and negative binomial models are equal to 5271.35 and 2873.46, respectively.

8.3.4 The bivariate Poisson model

Karlis and Ntzoufras (2003a) have constructed log-linear models based on the bivariate Poisson distribution [for details, see Kocherlakota and Kocherlakota (1992) and references cited therein] for modeling sports outcomes; see also Karlis and Tsiamyrtzis (2008) for conjugate Bayesian analysis.

Let us consider two discrete random variables Y_1 and Y_2, which jointly follow a bivariate Poisson distribution:

$$(Y_1, Y_2) \sim BP(\lambda_1, \lambda_2, \lambda_3).$$

Then, their joint probability function is given by

$$f_{BP}(y_1, y_2) = P_{Y_1, Y_2}(y_1, y_2)$$

$$= e^{-(\lambda_1 + \lambda_2 + \lambda_3)} \frac{\lambda_1^{y_1}}{y_1!} \frac{\lambda_2^{y_2}}{y_2!} \sum_{k=0}^{\min(y_1, y_2)} \binom{y_1}{k} \binom{y_2}{k} k! \left(\frac{\lambda_3}{\lambda_1 \lambda_2}\right)^k. \quad (8.3)$$

This bivariate distribution is derived by the following convenient latent variable representation

$$Y_1 = Z_1 + Z_3 \quad \& \quad Y_2 = Z_2 + Z_3 \text{ with } Z_k \sim \text{Poisson}(\lambda_k)$$

where Z_k are independent. From this representation, we can easily calculate that $E(Y_k) = \lambda_k + \lambda_3$ and that $V(Y_k) = \lambda_k + \lambda_3$ while the covariance $\text{Cov}(Y_1, Y_2) = \lambda_3$. Hence the distribution allows for a positive dependence between the two random variables Y_1 and Y_2. Note that for $\lambda_3 = 0$, the two variables are independent and the bivariate Poisson distribution reduces to the product of two independent Poisson distributions (also referred as a *double-Poisson model*) and is equivalent to the usual Poisson log-linear model; more details concerning the BP distribution can be found in Kocherlakota and Kocherlakota (1992) and Johnson et al. (1997).

Covariates can be linked with the logarithms of λ_k [see, e.g., Karlis and Ntzoufras (2003*a*)] or directly with the marginal means $\lambda_1 + \lambda_3$ and $\lambda_2 + \lambda_3$.

To specify the BP model in WinBUGS, we can use the latent variable representation and define it using the zeros trick. Hence when Y_1 and Y_2 are observed, we can rewrite the probability function (8.3) as

$$f_{BP}(y_1, y_2) = \sum_{k=0}^{\min(y_1,y_2)} \frac{e^{-\lambda_1} \lambda_1^{y_1-k}}{(y_1 - k)!} \frac{e^{-\lambda_2} \lambda_2^{y_2-k}}{(y_2 - k)!} \frac{e^{-\lambda_3} \lambda_3^k}{k!}$$

$$= \sum_{k=0}^{\min(y_1,y_2)} f_P(y_1 - k; \lambda_1) f_P(y_2 - k; \lambda_2) f_P(k; \lambda_3),$$

where $Z_1 = y_1 - k$, $Z_2 = y_2 - k$ and $Z_3 = k$. To reproduce this function in WinBUGS, we can generate z_3 latent data by considering them as Poisson distributed interval censored data at $[0, \min(y_1, y_2)]$, then set $z_1 = y_1 - z_3$ and $z_2 = y_2 - z_3$ and define the remaining likelihood using the zeros trick. Hence the WinBUGS code for this implementation is given by

```
C<-0
for (i in 1:n){
    # define the minimum of y1,y2
    miny[i] <- min ( y1[i], y2[i] )
    # generate z3 latent data
    z3[i] ~ dpois( lambda[3,i] ) I( 0, miny[i] )
    # calculate z1 and z2 latent data
    z1[1] <- y1[i] - x3[i]
    z2[1] <- y2[i] - x3[i]
    # set the remaining likelihood
    zeros[i] <- 0
    zeros[i] ~ dpois( zeros.mean[i] )
    zeros.mean[i] <- -l[i] + C
    l[i] <- -lambda[1,i] +z1[i]*log( lambda[1,i] ) -loggam( z1[i]+1 )
            -lambda[2,i] +z2[i]*log( lambda[2,i] ) -loggam( z2[i]+1 )
    # specify the linear predictor of lambdas
    for (k in 1:3){ log( lambda[i,k] ) <- eta[k,i] }
}
```

Note that initial values must be also defined for z_3. We recommend avoiding random generation of initial values in this case because noninteger values might be simulated resulting in logical errors within WinBUGS. A further problem that may appear with the addition of covariate structure on λ_3 is that the MCMC algorithm may adhere to very large values of λ_3 indicating zero covariance between the two response variables for specific individuals.

Hence, the user must monitor convergence and try alternative methods when suspicious behavior appears in the coefficients of the covariance term λ_3.

The zeros trick can also be used to implement the BP model. Additionally, a discrete uniform "pseudoprior" must be imposed on Z_3 latent data. An advantage of this approach is that we do not specify initial values for Z_3. Using this approach, z_3 is now defined by

```
miny[i]<-min(y1[i], y2[i])+1
# data generated by the continuous uniform distribution
u[i]~dunif(0, miny[i]);
# data truncated in order z3 to follow the appropriate
# discrete uniform distribution
z3[i]<-trunc( u[i] )
```

while the zeros part (which specifies the augmented likelihood) is now given by

```
zeros[i] <- 0
zeros[i] ~ dpois( zeros.mean[i] )
zeros.mean[i] <-     -l[i] + C
# log-likelihood for i observation
l[i]<- -lambda[i,1]+z1[i]*log(lambda[i,1])-loggam(z1[i]+1)
       -lambda[i,2]+z2[i]*log(lambda[i,2])-loggam(z2[i]+1)
       -lambda[i,3]+z3[i]*log(lambda[i,3])-loggam(z3[i]+1);
```

A short example using simulated data follows. Implementation in the association football data of Example 7.2 can be implemented in a straightforward manner; see Karlis and Ntzoufras (2003b) for more details.

Example 8.4. Bivariate simulated count data. To illustrate the bivariate Poisson model, we have generated 100 observations from the following model

$$
\begin{aligned}
(Y_{i1}, Y_{i2}) &\sim BP(\lambda_{1i}, \lambda_{2i}, \lambda_{3i}) \\
\log(\lambda_{1i}) &= 0.3 + 0.2x_{i1} + 0.1x_{i3} \\
\log(\lambda_{2i}) &= 0.5 - 0.3x_{i2} - 0.5x_{i5} \\
\log(\lambda_{3i}) &= x_{i4} \text{ for } i = 1, 2, \ldots, 100, \ k = 1, 2, 3,
\end{aligned}
$$

where x_{ij} (j=1,2,3,4,5) are random samples from the standardized normal distribution.

Posterior summaries of the full model and the true model are provided in Table 8.7. Normal prior distributions were used with zero mean and variance equal to 100. Results using both approaches described above give equivalent results. As we can see, estimated values are close to the actual ones. For this illustration normal prior distributions with zero mean and large variance were used. Model code of for this example is available in this book's Webpage.

8.3.5 The Poisson difference model

Karlis and Ntzoufras (2006) have recast interest in Skellam's (1946) distribution implementing Bayesian inference for dental epidemiology data. A model based on this distribution was used for association football data by Karlis and Ntzoufras (2008). The zero inflated version of this distribution was also considered in both of these publications. Skellam's (1946) distribution is not directly connected to the distribution of counts, but to their difference. For this reason it is also called the *Poisson difference* (PD) distribution. It can be used for modeling discrete response variables defined in $\mathbf{Z} = \{\ldots, -2, -1, 0, 1, 2, \ldots\}$

Table 8.7   Posterior summaries for model parameters of full and true bivariate Poisson models for simulated data of Example 8.4

| | | Full model | | | | True model | | | |
| | Actual | | | Percentiles | | | | Percentiles | |
Node	value	Mean	SD	2.5%	97.5%	Mean	SD	2.5%	97.5%
β_{10}	0.3	0.330	0.106	0.117	0.532	0.369	0.077	0.215	0.519
β_{11}	0.2	0.102	0.085	-0.070	0.266	0.141	0.073	-0.007	0.285
β_{12}	0.0	-0.081	0.072	-0.224	0.053	-0.079	0.068	-0.208	0.063
β_{13}	0.1	0.118	0.079	-0.036	0.274	—	—	—	—
β_{14}	0.0	-0.085	0.075	-0.226	0.061	—	—	—	—
β_{15}	0.0	-0.037	0.070	-0.178	0.101	—	—	—	—
β_{20}	0.5	0.446	0.102	0.245	0.641	0.502	0.072	0.356	0.636
β_{21}	0.0	-0.087	0.063	-0.205	0.041	—	—	—	—
β_{22}	-0.3	-0.416	0.057	-0.534	-0.307	-0.410	0.056	-0.521	-0.297
β_{23}	0.0	0.132	0.059	0.020	0.251	—	—	—	—
β_{24}	0.0	-0.010	0.058	-0.119	0.109	-0.533	0.049	-0.627	-0.437
β_{25}	-0.5	-0.575	0.060	-0.694	-0.453	—	—	—	—
β_{30}	0.0	-0.017	0.146	-0.325	0.243	—	—	—	—
β_{31}	0.0	0.016	0.073	-0.128	0.161	—	—	—	—
β_{32}	0.0	0.033	0.061	-0.088	0.152	—	—	—	—
β_{33}	0.0	-0.104	0.073	-0.249	0.038	—	—	—	—
β_{34}	1.0	0.979	0.072	0.842	1.129	0.962	0.035	0.892	1.030
β_{35}	0.0	0.056	0.076	-0.092	0.210	—	—	—	—

Note: Linear predictor of full model: $\log \lambda_{ki} = \beta_0 + \sum_{j=1}^{5} \beta_{kj} X_{ij}$, k=1,2,3.

and differences in correlated count data. The advantage in the latter case is that we re-move additive correlation in much the same manner as in the usual paired t test in classical inference. In this way, the problem can now be solved using a univariate response model.

The PD is defined as the distribution of a random variable Y with probability function

$$f_{\mathrm{PD}}(y|\lambda_1, \lambda_2) = P(Y = y|\lambda_1, \lambda_2) = e^{-(\lambda_1+\lambda_2)} \left(\frac{\lambda_1}{\lambda_2}\right)^{y/2} I_{|y|}\left(2\sqrt{\lambda_1\lambda_2}\right) \qquad (8.4)$$

for all $y \in \mathbf{Z}$, $\lambda_1, \lambda_2 > 0$, where $I_r(y)$ is the modified Bessel function of order r (Abramowitz and Stegun, 1974, p. 375) given by

$$I_r(x) = \left(\frac{x}{2}\right)^r \sum_{k=0}^{\infty} \frac{\left(x^2/4\right)^k}{k!\Gamma(r+k+1)}.$$

Generally, this can be expressed as the difference of two Poisson latent variables with parameters λ_1 and λ_2:

$$Z_k \sim \mathrm{Poisson}(\lambda_k), \text{ for } k = 1, 2 \Leftrightarrow Y = Z_1 - Z_2 \sim \mathrm{PD}(\lambda_1, \lambda_2).$$

Although the Skellam's (1946) distribution was originally derived as the difference be-tween two independent Poisson random variables, it can be also derived as the difference between distributions that have the following trivariate latent variable structure:

$$Y_k = Z_k + Z_3 \text{ with } Z_3 \sim D(\boldsymbol{\theta}_3) \ \& \ Z_k \sim \mathrm{Poisson}(\lambda_k) \text{ for } k = 1, 2. \qquad (8.5)$$

In other words Z_3 may follow any distribution with parameter vector $\boldsymbol{\theta}_3$. An interesting property is that the joint distribution of Y_1, Y_2 is a bivariate distribution with correlation induced by the common stochastic component in both variables Z_3. For example, if Z_3 follows a Poisson distribution, then the joint distribution is the bivariate Poisson distribution described in previous Section 8.3.4. Another interesting property is that the marginal distributions for Y_1 and Y_2 will be Poisson only when Z_3 is a Poisson distributed random variable (or zero with probability one). Therefore, in the general formulation, the marginal distributions of Y_1 and Y_2 do not need to be Poisson distributed, but they can be any convolution of a Poisson and another discrete random variable. This assumption is less restrictive, allowing also for modeling over/under-dispersed Y_1 and Y_2, in contrast to the assumption of the Poisson marginals of Y_1 and Y_2 imposed by the BP model.

The expected value of the $PD(\lambda_1, \lambda_2)$ distribution is equal to $E(Y) = \lambda_1 - \lambda_2$, while the variance is $Var(Y) = \lambda_1 + \lambda_2$. For large values of the $\lambda_1 + \lambda_2$, the distribution can be sufficiently approximated by the normal distribution. If λ_2 is very close to zero, then the distribution tends to a Poisson one; when λ_1 approaches zero, then the distribution is a negative mirror of the Poisson distribution with mean λ_2 (i.e., a Poisson distribution in the negative axis). Additional properties of the distribution are described by Karlis and Ntzoufras (2006).

Finally, the trivariate reduction scheme (8.5) of PD provides a convenient data augmentation scheme that can be efficiently used for the construction of the MCMC algorithm for estimation of the posterior distribution; see Karlis and Ntzoufras (2006) for details. This latent variable representation can be used to specify the PD model in WinBUGS.

Following the approach of Karlis and Ntzoufras (2006), for $y \geq 0$ we can write

$$
\begin{aligned}
f_{PD}(y; \lambda_1, \lambda_2) &= P(Z_1 - Z_2 = y | \lambda_1, \lambda_2) \\
&= P(Z_1 = y + Z_2 | \lambda_1, \lambda_2) \\
&= \sum_{w=0}^{\infty} P(Z_1 = y + Z_2, Z_2 = w | \lambda_1, \lambda_2) \\
&= \sum_{w=0}^{\infty} f_P(y + w; \lambda_1) f_P(w; \lambda_2).
\end{aligned}
$$

Similarly, for $y < 0$ we can write

$$
\begin{aligned}
f_{PD}(y; \lambda_1, \lambda_2) &= P(Z_2 = Z_1 - y | \lambda_1, \lambda_2) \\
&= \sum_{w=0}^{\infty} f_P(w; \lambda_1) f_P(w - y; \lambda_2).
\end{aligned}
$$

We can use these expressions to generate latent data w_i and then use the Poisson augmented likelihood. In WinBUGS, we can specify the PD model using using the following syntax:

```
C <- 0
UNIFLIM <- 100
# Likelihood
for (i in 1:n){
    w.cont[i]~dunif(0, UNIFLIM)
    w[i]<-trunc(w.cont[i])
    g[i]<-step( y[i] )
    z1[i]<- g[i] * ( y[i]+w[i]) + (1-g[i]) * w[i]
    z2[i]<- g[i] * w[i]          + (1-g[i]) * (-y[i]+w[i])
    zeros[i] <- 0
```

```
        zeros[i] ~ dpois( zeros.mean[i] )
        zeros.mean[i] <-  -l[i]+C
        l[i]<-  -lambda[i,1] + z1[i]*log(lambda[i,1]) - logfact( z1[i] )
                -lambda[i,2] + z2[i]*log(lambda[i,2]) - logfact( z2[i] )
        for (k in 1:2){ log( lambda[i,k] ) <- eta[i,k] }
    }
#
#   Priors
    for (k in 1:2){ for (j in 1:p) { beta[k,j]~dnorm(0.0, 0.01) } }
```

In order to use this approach, we have imposed a discrete uniform distribution on the latent data w_i. This is achieved indirectly by truncating the continuous uniform distribution. Note that, since the uniform distribution is used, an upper limit must be adopted. In this way, we essentially used a truncated version of the PD distribution. However, if this limit is high, the effect of this truncation will be minimal.

Example 8.5. Bivariate simulated count data of Example 8.4. Here we use the differences in the simulated data to fit the PD model on the difference in the simulated data of Example 8.4.

Posterior summaries of the full model and the true model are provided in Table 8.8. Results are quite close to the actual ones for λ_1 and λ_2. Nevertheless, the structure of the covariance term λ_3 cannot be estimated now, since differences are considered.

Table 8.8 Posterior summaries for model parameters of full and true Poisson difference models for simulated data of Example 8.4

| | Actual | Full model | | | | True model | | | |
| | | | | Percentiles | | | | Percentiles | |
Node	value	Mean	SD	2.5%	97.5%	Mean	SD	2.5%	97.5%
β_{10}	0.3	0.349	0.135	0.103	0.617	0.356	0.118	0.123	0.586
β_{11}	0.2	0.051	0.116	-0.195	0.291	0.166	0.082	0.013	0.337
β_{12}	0.1	-0.138	0.102	-0.344	0.069	-0.121	0.110	-0.358	0.088
β_{13}	0.0	0.194	0.109	-0.017	0.401	—	—	—	—
β_{14}	0.0	-0.022	0.096	-0.199	0.161	—	—	—	—
β_{15}	0.0	-0.092	0.104	-0.311	0.089	—	—	—	—
β_{20}	0.5	0.437	0.126	0.204	0.671	0.484	0.110	0.269	0.696
β_{21}	0.0	-0.116	0.083	-0.278	0.050	—	—	—	—
β_{22}	-0.3	-0.438	0.079	-0.600	-0.283	-0.444	0.076	-0.594	-0.300
β_{23}	0.0	0.189	0.079	0.029	0.342	—	—	—	—
β_{24}	-0.5	0.034	0.075	-0.117	0.172	-0.53	-0.535	0.061	-0.653
β_{25}	0.0	-0.607	0.072	-0.740	-0.470	—	—	—	—

Note: Linear predictor in full model: $\log \lambda_{ki} = \beta_0 + \sum_{j=1}^{5} \beta_{kj} X_{ij}$, k=1,2,3.

8.4 FURTHER GLM-BASED MODELS AND EXTENSIONS

Generalized linear models can be regarded as the general foundation for building more advanced models. In this section we briefly discuss some GLM based extensions.

The following sections focus on survival analysis models (which can be thought of as extensions of the models used for positive defined continuous response variables) and the multinomial regression models (which are natural extensions of the simple logistic regression models used for binomial responses). The section closes with a brief discussion of other related models, how to treat ordinal variables, and related issues.

8.4.1   Survival analysis models

Survival analysis refers to a family of statistical methods used to analyze duration of time until an event of interest (such as death) occurs. All models described in Section 8.2.2 can be used to analyze such response variables since they are positively defined continuous random variables. The difference in such data is that we now focus on the *hazard function* instead of the the mean duration time. Moreover, we have to deal with incomplete duration time data, termed as *censored*, which must be incorporated in the model.

In order to proceed, we need to specify the survival function $S(y) = 1 - F(y) = P(Y \geq y)$, which provides the probability of surviving until timepoint y. Moreover, the hazard function denoted by $h(y)$ is defined as

$$h(y) = \frac{f(y)}{S(y)} = -\frac{S'(y)}{S(y)} .$$

and can be interpreted as the death (or event) rate of an individual, provided that this person survived until time y:

$$h(y) = \lim_{\delta y \to 0} \frac{P(y \leq Y < y + \delta y | Y > y)}{\delta y} .$$

As we have already mentioned, focus is on modeling this hazard rate.

There are three types of censored data:

- *Right censoring*, referring to individuals who are followed from the beginning of the study until a time point where they are lost during the follow-up. The exact time of death is unavailable, but the know that the person survived until time y. This is the most frequent type of censoring encountered in survival analysis data.

- *Left censoring*, referring to cases where the exact time when the subject entered the study is unknown but the exact time of death is available.

- *Interval censoring*, referring to cases where both the exact times of death and the entry into the study are unknown. This type of censored data inform us that the individual was alive at specific timepoints, so we know that the survival time was greater than a specific value y.

The data of censored survival time consist of the response duration times y_i and the censoring indicators ξ_i which take values of zero and one depending on whether the corresponding time was death or censoring time. The likelihood is now given by

$$f(\boldsymbol{y}|\boldsymbol{\xi}, \boldsymbol{\theta}) = \prod_{i=1}^{n} f(\boldsymbol{y}|\boldsymbol{\theta})^{1-\xi_i} S(\boldsymbol{y}|\boldsymbol{\theta})^{\xi_i}$$

for right censored data. Hence the incomplete information is modeled via the survival function, which reflects the probability that the patient was alive for a duration greater than y_i.

In WinBUGS , censoring can be modeled using the commands I(a,), I(,b), and I(a,b) for right, left, and interval censoring, respectively. For right censored data, we need two variables in order to set up the model. The actual survival times using NA values when censored data first appear and the censoring times, which take zero values when actual survival times, are observed. Then, assuming, for example, a Weibull distribution, the likelihood can be expressed as

```
y[i] ~ dweib( v, lambda[i] )I( cens.time[i], )
```

which results in (1) the uncensored Weibull distribution when actual survival times are observed and hence cens.time[i]=0 and (2) the appropriate censored Weibull distribution with $y_i >$cens.time$_i$ when cens.time$_i > 0$. A detailed example is provided by Spiegelhalter et al. (2003a); see the mice data example. Similarly, we can treat other types of censored data, or we might build other survival analysis models.

Another popular model used in survival analysis is Cox's (1972) proportional hazards model. The model is essentially semiparametric since no assumptions concerning the baseline hazard functions are made. In order to implement Cox's model, we need to consider the formulation in the papers by Andersen and Gill (1982) and Clayton (1991) where the Cox model is represented as a counting process that was illustrated by Spiegelhalter et al. (2003a) using a detailed example (leuk data example).

Finally, survival models can be divided in two major categories: proportional hazards models and accelerated failure time models. For the first ones we assume that $h(y_i|\boldsymbol{x}_{(i)}) = \lambda(\boldsymbol{x}_{(i)})h_0(y_i)$; that is, the hazard function of an individual i with survival time y_i and covariate values $\boldsymbol{x}_{(i)}$ is proportional to the baseline hazard function, which does not depend on covariates.

On the other hand, accelerated failure time models assume that the hazard function is given by $h(y_i|\boldsymbol{x}_{(i)}) = \lambda(\boldsymbol{x}_{(i)})h_0(\lambda(\boldsymbol{x}_{(i)})y_i)$; that is, the baseline hazard function depends on the covariate values. Moreover, accelerated failure time models can be written using the following representation

$$\log Y = E(\log Y_0) - \log \lambda(\boldsymbol{x}_{(i)}) + \log \epsilon$$

where Y_0 be the random variable when all covariates are zero (baseline survival time) and ϵ is an error function with expected value equal to one (possible values are exponential, log-normal, gamma, and Weibull). Generally, for accelerated failure time models we have

$$S(y|\boldsymbol{x}) = S(ye^{-\eta}|\boldsymbol{x} = 0),$$

where η is the usual linear predictor. Hence the survival function for covariate values $\boldsymbol{x}$ is equal to the corresponding baseline survival function (i.e., for $\boldsymbol{x} = 0$) accelerated by $e^{-\eta_i}$.

For more details, the reader is referred to specialized survival analysis textbooks, such as the classic book by Cox and Oakes (1984) or the more recent one by Hosmer et al. (2008).

8.4.2 Multinomial models

Multinomial logistic regression models are natural extensions of the binomial logistic regression models and are used when the response of interest are multicategorical variables. For example, such models can be used to model the final outcome (win/draw/loss) of a football game instead of the actual score.

Assuming that the response variable $\boldsymbol{Y}_i = (Y_{i1}, \dots, Y_{iK})$ has K levels, where Y_{ik} denotes the frequency of the kth level, the multinomial logistic regression model can be written as

$$\boldsymbol{Y}_i \sim \text{multinomial}(\boldsymbol{\pi}_i, N_i)$$

and

$$\log \frac{\pi_{ik}}{\pi_{i1}} = \eta_{ik} = \beta_{0k} + \sum_{i=1}^{k} \beta_{jk}\gamma_{jk}x_{ij} \qquad (8.6)$$

for $k = 2, \ldots, K$, where $\boldsymbol{\pi}_i = (\pi_{i1}, \pi_{i2}, \ldots, \pi_{iK})^T$ is the vector of the probabilities for each level of variable Y for individual i with $\pi_{i1} = 1 - \sum_{k=2}^{K} \pi_{ik}$ and γ_{jk} are the usual binary indicators identifying the structure of the model and which variables specify or affect each odds. It is common practice to use similar structure for all odds π_{ik}/π_{i1}; see, for example, in Agresti (2002, chap. 7) for a comprehensive treatment of the subject.

Solving (8.6) is terms of response probabilities results in

$$\pi_{i1} = \frac{1}{1 + \sum_{k=2}^{K} e^{\eta_{ik}}} \quad \text{and} \quad \pi_{ik} = \frac{e^{\eta_{ik}}}{1 + \sum_{k=2}^{K} e^{\eta_{ik}}} \quad \text{for } k = 2, \ldots, K.$$

This can be summarized by

$$\pi_{ik} = \frac{e^{\eta_{ik}}}{\sum_{k=1}^{K} e^{\eta_{ik}}} \quad \text{with } \eta_{i1} = 0 \text{ for } i = 1, 2, \ldots, n.$$

The restriction $\eta_{i1} = 0$ can be indirectly imposed by setting all coefficients of the first linear predictor β_{j1} equal to zero. This can be implemented in WinBUGS using the commands

```
for (i in 1:n){
   y[i,1:K] ~ dmulti( p[i,1:K], N[i] )
   for (k in 1:K){
      eta[i,k] <- ....
      pi[i,k] <- expeta[i,k]/sum(expeta[i,1:K])
      expeta[i,k] <- exp( eta[i,k] )
   }
}
# Coefficient of the first/baseline category is constrained to zero
for (j in 1:P){ beta[j,1] <- 0.0 }
```

Usually individual data are observed, implying that $N_i = 1$. In such a case, the response variable will be given as a categorical variable with codes denoting each level. To specify this in terms of the multinomial distribution, it must be recoded using K dummy variable indicators that will identify which level appears in each case in a representation compatible with the multinomial notation. Alternatively, in WinBUGS, the categorical distribution (command dcat) can be used instead. In this way, the original variable (without transformation or dummy variables) can be used. The corresponding expression in WinBUGS is given by

```
y[i] ~ dcat( p[i,1:K] )
```

A simple approach that is frequently used express the above mentioned model by separate $K - 1$ simple logistic regression models comparing each model with a baseline category. Although, this might be problematic in classical inference when certain constraints on specific parameters need to be incorporated to the different logistic models, this can be easily incorporated in the WinBUGS model specification. Moreover, Agresti (1990, pp. 310–312, 2002, pp. 273–274) cautions that estimates may be inefficient when the baseline category does not have sufficient observations. Hence it is advisable to select as baseline the category with the highest number of observations and then transform estimates as desired. To implement the separate regression models approach in WinBUGS, we need only to define $Y_{ik} \sim \text{binomial}(p_{ik}^*, N_{ik})$ with $N_{ik} = Y_{ik} + Y_{i1}$. Hence the likelihood will be given by

```
for (i in 1:n){
    for (k in 2:K){
        N.star[i] <- y[i,1]+y[i,k]
        y[i,k] ~ dbin( p.star[i,k], N.star[i] )
        for (k in 2:K){
            logit(pi.star[i,k]) <- eta[i,k]
            eta[i,k] <- ....
        }
        # these quantities are not used in the model
        pi.star[i,1] <-0.0
        eta[i,1] <- 0.0
    }
}
```

Note that π_{ik}^* does not have the same meaning as in the original multinomial model since now it refers to the conditional probabilities given that only the category k and the baseline one are observed. Nevertheless, the ratio π_{ik}/π_{i1} estimates the same odds as in the multinomial logistic model.

Finally, when high-dimensional contingency tables are considered, we may use the connection between multinomial and log-linear models to indirectly fit the first one using the latter; see, for example, Agresti (2002) for more details. This was indeed the practice used for many years in conventional statistical packages. Nevertheless, it is not recommended in WinBUGS since direct fit of the multinomial model can be easily achieved.

A detailed multinomial example is provided by the second volume of WinBUGS examples [see Spiegelhalter et al. (2003b), aligators data example; see also Spiegelhalter et al. (1996c, table 7.16, pp. 51–54)]. WinBUGS code for a similar alligators dataset (Agresti, 2002, p. 304), with the alligators' size as quantitative covariate, as well as Agresti's (2002, table 7.15) political party data are provided in this book's Website.

Other related examples are the biopsies and the endometrial (endo) data examples, which are also provided by Spiegelhalter et al. (2003b). The latter illustrates also the association between multinomial and Poisson log-linear models in a simple 2×2 contingency table; see also Spiegelhalter et al. (1996c, pp. 55–57). Details concerning multinomial regression models and extensive examples can be also found in the recent books of Congdon; see Congdon (2006b, chap. 7) and Congdon (2005a, pp. 198–210).

Finally, other link functions such as the probit can be easily accommodated in the multinomial models, but here we restrict ourselves in the logistic case, which is the most frequent one; see, for example, Congdon (2005a, pp. 210–231) for details concerning the probit model.

8.4.3   Additional models and further reading

A wide variety of GLM-based models exist in the statistical bibliography. Unfortunately, not all of these models can be described in this introductory book. Hence, we focus on some basic models and their extensions. In this closing section, we briefly describe some additional models and provide some annotated references for further reading concerning GLM-based models. We also briefly review models for ordinal responses.

In the previous sections we have described how to model continuous variables defined in the whole real line (normal models), positive continuous variables (log-normal, gamma, inverse Gaussian, etc.), binary and multinomial models (using logit and probit models), and discrete random variables (based on Poisson and negative binomial models). Multivariate extensions of such models have been reported in the literature that account for correlations between observations; see Fahrmeir and Tutz (2001) for a good overview of the topic.

Concerning continuous variables, the multivariate normal or t distributions are frequently used [see, e.g., Tiao and Zellner (1964), Zellner (1976) and Box and Tiao (1973, sec. 8.2, 8.3)] while other skewed multivariate distributions can be adopted as proposed by Sahu et al. (2003). Concerning discrete count data, the multivariate Poisson models have been introduced in the statistical literature (Tsionas, 2001; Karlis and Melikotzidou, 2005), while multivariate versions of logistic regression models also exist [see, e.g., Glonek (1996), Movellan (2006) for a tutorial, and O'Brien and Dunson (2004) for the Bayesian approach].

Usually correlated repeated measurement data are modeled using random effects or hierarchical models, which are described in the following chapter. An alternative approach is modeling the marginals assuming a type of covariance between the variables. This is traditionally treated using generalized estimating equations (GEEs), which are not discussed extensively within the Bayesian framework. For a smooth introduction to these two topics, see Agresti (2002, chaps. 11,12). More recently, copulas have been considered for the joint modeling of multivariate data; see Kolev et al. (2006) for a review of the topic and Huard et al. (2006), Pitt et al. (2006) for Bayesian implementation of such models.

Another interesting topic is incorporation of ordinal variables in our model. Ordinal response variables have been extensively modeled using cumulative versions of multinomial logit or probit regression models, cumulative link models, or other variants of such models (e.g., continuation ratio, adjacent categories, or proportional odds models) [Agresti (2002, chap. 7) and Dobson (2002, sec. 8.4)]. Moreover, a full review of the topic is provided by Liu and Agresti (2005). In two-way contingency tables, Goodman's (1979, 1981) association models can be used to estimate the column and row scores of ordinal variables using order restrictions; for the Bayesian implementation of such models, see Iliopoulos et al. (2007a, b). Other models that can be used to reveal the ordinal structure of two cross-classified variables are the RC(M) models (Goodman, 1985), which are generalizations of the simple association models and are also related with the correlation models (i.e., the model-based analog of correspondence analysis); for more details, see Agresti (2002, secs. 9.5, 9.6) and Kateri et al. (2005) for the Bayesian implementation of RC(M) models.

To summarize, the reader can refer to Lindsey (1997) and Fahrmeir and Tutz (2001) for further variations and GLM extensions. More details on Bayesian inference of GLM-based models and their extensions can be found in the books by Congdon (2003; 2005a; 2006b), while more advanced GLM based models are extensively presented and described by Dey et al. (2000). Finally, a wide collection of the available Bayesian models and approaches used for categorical data analysis is described in the excellent review provided by Agresti and Hitchcock (2005).

Problems

8.1 Consider the following data:

```
1.3 2.1 2.1 2.3 1.0 1.0 2.8 2.1 2.7 2.3 3.2 4.2 2.3 1.1 2.1
2.4 1.7 1.7 2.1 1.8 2.0 3.0 2.0 2.3 1.3 1.0 2.2 2.1 1.9 1.8
```

a) Use the inverse Gaussian, gamma, and log-normal distributions for these data.

b) Compare the fitted distributions with the actual data. Which distribution do you think fits the data best?

c) Use the DIC to identify which distribution fits the data best. For the inverse Gaussian distribution, calculate DIC using a software other than WinBUGS.

    d) Use a normal distribution for the logarithms of the data above. Compare the posterior distributions of the parameters with those obtained by assuming the log-normal distribution for the data.

    e) Calculate DIC for the model of (d) and compare it with the corresponding one estimated using the log-normal distribution. How are they connected?

8.2 Return to the data of Problem 6.9.

    a) Use the inverse Gaussian, gamma, and log-normal distributions to determine which variables influence the total money spent according to the receipt total.

    b) Use DIC to finally determine which variables are appropriate for each model.

    c) Interpret the parameters of each model on the basis of their posterior distributions.

    d) Compare the models assuming different distributions using DIC. Also compare your results with the corresponding ones assuming the normal distribution for our data.

8.3 Consider the data of example `mice` of WinBUGS examples volume 1 (Spiegelhalter et al., 2003*a*).

    a) Use the Weibull, gamma, and log-normal distributions to model the survival times. For the incomplete censored times for predefined distributions, use the command `I(,)`; see Spiegelhalter et al. (2003*a*, *d*) for more details.

    b) Use DIC to identify which variables are appropriate for each model.

    c) Interpret the parameters of each model on the basis of their posterior distributions.

    d) Compare the models assuming different distributions using DIC.

8.4 Consider the data of the example `kidney` of WinBUGS examples volume 1 (Spiegelhalter et al., 2003*a*).

    a) Use the Weibull, gamma, and log-normal distributions to model the survival times for the first instance only (ignore random effects). For the incomplete censored times, use the command `I(,)`; see Spiegelhalter et al. (2003*a*, *d*) for more details.

    b) Use DIC to determine which variables are appropriate for each model.

    c) Interpret the parameters of each model according to their posterior distributions.

    d) Compare the models assuming different distributions using DIC.

8.5 Use the inverse Gaussian distribution to model the survival data `mice` and `kidney` analyzed in Problems 8.3 and 8.4. For the censored observation, use the survival function of the inverse Gaussian distribution given by

$$S(y) = P(Y \geq y) = 1 - \Phi\left(\sqrt{\frac{\lambda}{y}} \frac{y - \mu}{\mu}\right) - e^{2\lambda/\mu} \Phi\left(\sqrt{\frac{\lambda}{y}} \frac{-y - \mu}{\mu}\right).$$

    a) Compare the results with the corresponding ones from the models using other distributions.

    b) Calculate DIC outside WinBUGS and compare it with the corresponding DIC values for the models based on other distributions.

8.6 Consider the following data

Y	0	1	2	3	4	5	6	7	8	9	10	11	12	13	14	15	20
Frequencies	31	37	37	35	23	30	31	8	25	13	8	2	5	6	3	5	1

a) Use the Poisson, negative binomial, and generalized Poisson distributions to model these data.

b) Compare the fitted distributions with the actual data. Which distribution do you think fits the data best?

c) Use the DIC to identify which distribution fits the data best.

8.7 Consider the following data

Y	-7	-5	-4	-3	-2	-1	0	1	2	3	4	5	6	7	8	9	12
Frequencies	1	2	4	6	7	14	21	28	30	29	20	10	14	5	4	4	1

a) Use the Poisson difference distribution to model these data.

b) Compare the fitted distributions with the actual data. Which distribution do you think fits the data best?

c) Use the DIC to identify which distribution fits the data best.

8.8 Consider the `Epil` data from WinBUGS examples volume I (Spiegelhalter et al., 2003a).

a) Construct a bivariate Poisson model comparing the first and the second instances.

b) Interpret the model coefficients.

c) Determine whether the covariance term can be assumed constant or whether covariates are needed to appropriately estimate this term.

8.9 Consider the `Epil` data from WinBUGS examples volume 1 (Spiegelhalter et al., 2003a).

a) For the differences $Y_2 - Y_1$, $Y_3 - Y_1$, and $Y_4 - Y_1$, construct models on the basis of the Poisson difference distribution.

b) Interpret the model coefficients.

8.10 Use the alligators' primary food choice from the data of WinBUGS example `alligators` (Spiegelhalter et al., 2003b) as the response variable in the following models.

a) Fit binomial regression models comparing each type of food with fish (first level of the preferred choice of food). Interpret the model parameters.

b) Fit a model for the food preference according to the multinomial distribution. Compare the results with the ones from simple binomial-based models.

8.11 Use the zero–one tricks to fit the folded normal distribution to the data of Problem 8.1. The density function of the folded normal distribution is given by

$$f(y; \mu, b) = \frac{1}{b\sqrt{2\pi}}(e^{ay/b^2} + e^{-ay/b^2})e^{-(y^2+a^2)/(2b^2)},$$

while the mean is given by

$$E(Y) = b\sqrt{\frac{2}{\pi}}e^{-a^2/(2b^2)} - a\left\{1 - 2\Phi\left(\frac{a}{b}\right)\right\}.$$

a) Estimate the posterior distributions of $E(Y)$, a, and b.

b) Compare the fitted distribution with the observed data and the corresponding fitted distributions of Problem 8.1.

8.12 Use a multinomial model for the data of Italian championship 2000 described in Problem 7.4 (consider only the final outcome of each game; win/draw/loss, and use structure similar to that in the Poisson regression model described in Section 7.4.3).

  a) Evaluate the performance of each team on the basis of the posterior distributions of the parameters.

  b) Reproduce the league. Compare the results with the corresponding ones from the Poisson model.

8.13   Use a Poisson difference distribution to model the goal differences in the water polo data described in Problem 7.3; see also Karlis and Ntzoufras (2008) for details.

  a) Evaluate the performance of each team on the basis of the posterior distributions of the parameters.

  b) Reproduce the league using simulated values from the fitted model. Compare the results with the corresponding ones from the Poisson model.

CHAPTER 9

BAYESIAN HIERARCHICAL MODELS

9.1 INTRODUCTION

Bayesian models have an inherently hierarchical structure. The prior distribution $f(\boldsymbol{\theta}|\boldsymbol{a})$ of the model parameters $\boldsymbol{\theta}$ with prior parameters $\boldsymbol{a}$ can be considered as one level of hierarchy, with the likelihood as the final stage of a Bayesian model resulting in the posterior distribution $f(\boldsymbol{\theta}|\boldsymbol{y}) \propto f(\boldsymbol{y}|\boldsymbol{\theta})f(\boldsymbol{\theta}\,;\,\boldsymbol{a})$ via the Bayes theorem; see Figure 9.1 for a graphical representation of the hierarchical structure of a typical Bayesian model.

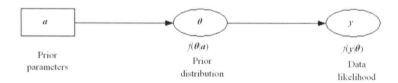

Figure 9.1 Graphical representation of standard Bayesian model. Squared nodes refer to constant parameters, oval nodes refer to stochastic components of the model.

To capture the complicated structure of some data, the prior is frequently structured using a series of conditional distributions called *hierarchical stages* of the prior distribution. Hence, a Bayesian hierarchical model is defined when a prior distribution is also assigned on the prior parameters $\boldsymbol{a}$ associated with the likelihood parameters $\boldsymbol{\theta}$. The posterior distribution can be written as

$$f(\boldsymbol{\theta}|\boldsymbol{y}) \quad \propto \quad f(\boldsymbol{y}|\boldsymbol{\theta})f(\boldsymbol{\theta}\,;\,\boldsymbol{a})f(\boldsymbol{a}\,;\,\boldsymbol{b})$$
$$\propto \quad f(\boldsymbol{y}|\boldsymbol{\theta})f(\boldsymbol{\theta}|\boldsymbol{a})f(\boldsymbol{a}|\boldsymbol{b})\,.$$

The prior distribution in this model formulation is characterized by two levels of hierarchy: $f(\boldsymbol{\theta}|\boldsymbol{a})$ (first level) and $f(\boldsymbol{a}|\boldsymbol{b})$ (second level). Prior distributions of the upper levels of a hierarchical prior are called *hyperpriors* and the corresponding parameters, *hyperparameters*; in the example above, $f(\boldsymbol{a}|\boldsymbol{b})$ is the hyperprior and $\boldsymbol{b}$ are the hyperparameters of the prior parameters $\boldsymbol{a}$. The structure can be extended by adding more levels of hierarchy if needed.

This structure is easily depicted using directed acyclic graphs and can be designed within WinBUGS using the DOODLE tool; see Appendix A for details. Figure 9.2 depicts the simple two-stage hierarchical model described above.

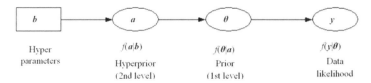

Figure 9.2 Graphical representation of a two-stage Bayesian hierarchical model. Squared nodes refer to constant parameters, oval nodes refer to stochastic components of the model.

Generally, any Bayesian model with parameters θ and ϕ and prior distribution $f(\theta, \phi)$ can be written in a hierarchical structure if the joint prior distribution is decomposed to a series of conditional distributions such as $f(\theta, \phi) = f(\theta|\phi)f(\phi)$. In hierarchical models, hyperparameters ϕ are rarely involved in the model likelihood.

Hierarchical models can be considered as a large set of stochastic formulations that include many popular models such as the random effects, the variance components, the multilevel and the generalized linear mixed models (GLMM). Latent variable models can also be viewed as hierarchical models since random effects can also have a latent variable interpretation.

Details concerning Bayesian hierarchical models can be found in many standard textbooks; see, for example, Gelman et al. (2004, chaps. 5, 15) and references cited therein, Robert (2007, chap. 10), Lawson et al. (2003, chap. 2), and Woodworth (2004, chap. 11). An extensive treatment of the subject can also be found in Dey et al. (2000) and Gelman and Hill (2006).

9.1.1   A simple motivating example

A motivation for introducing hierarchical structure in a simple model is provided by Spiegelhalter et al. (2004, pp. 91–94). Following their example, we assume the case of having K groups or units under investigation from which we collect a sample of responses $Y_{ik} \sim D(\boldsymbol{\theta})$; $i = 1, 2, \ldots, n_k$ and $k = 1, \ldots, K$. Initially, we can assume two alternative models:

1. One model that estimates a common mean effect μ (pooled effect). With this approach, when, for example, a normal distribution is assumed, we express the model as

$$Y_{ik} \sim N(\mu, \sigma^2) \text{ for } i = 1, \ldots, n_i \text{ and } k = 1, 2, \ldots, K.$$

2. One model that estimates different independent mean effects μ_k for each group or unit (fixed effects). In this approach, when a normal distribution is assumed, we express the model as

$$Y_{ik} \sim N(\mu_k, \sigma^2) \text{ for } i = 1, \ldots, n_i \text{ and } k = 1, 2, \ldots, K.$$

Usually, we are interested in model 2, which estimates the expected performance of each unit. A disadvantage of this approach is that each mean effect μ_k is estimated independently from the other groups. Hence, in a group with small sample size, the posterior uncertainty will be large.

In such cases it is logical to assume that all expected performances μ_k are observables from a population distribution with mean μ, namely, an overall population average effect. Therefore, a two stage prior distribution may be adopted with hyperpriors:

$$\mu_k \sim N(\mu, w^2).$$

In this way, all mean effects are associated, allowing for "borrowing strength" between groups or units and provide more accurate estimations, especially when the sample size of specific groups is small. The posterior mean of each μ_k is a weighted mean of the corresponding sample mean of the kth group and the overall mean effect μ.

9.1.2 Why use a hierarchical model?

Hierarchical models are inherently implied in population-based problems such the one described in Section 9.1.1. Within this context, hierarchical models are widely used in meta-analysis in medical research, where information from different sources or studies on the same topic is available; see Woodworth (2004, pp. 223–225) for an example.

More generally, hierarchical models describe efficiently complex datasets incorporating correlation or including other properties in our model. Hence, when multivariate or repeated responses are observed, correlation can be incorporated in the model via a common "random" effect for all measurements referring to the same individual. This introduces a marginal correlation between repeated data, while interpretation is based on the conditional means. Therefore, given the random effects, the structure and the interpretation are similar to common generalized linear models. Accordingly, hierarchical models naturally appear, for example, when modeling spatiotemporal data in which correlation between time and space can be added by using common random effects on adjacent (in time or space) responses.

Hierarchical models can also be used to imply a complicated marginal distribution but (at the same time) keep the conditional structure as simple as possible. Such a characteristic example is the use of multiplicative gamma distributed random effects in the Poisson model, which implies a negative binomial marginal sampling distribution. In this way, in the Poisson case, we account for overdispersion, which may be present in our response data. Nevertheless, the conditional structure still consists of a series of Poisson–gamma models. Using random effects and the corresponding hierarchical structure to appropriately specify the marginal sampling distribution is frequently referred to as *data augmentation*. Such an approach considerably simplifies the MCMC scheme that can be used to estimate the posterior distributions of interest.

9.1.3   Other advantages and characteristics

An important characteristic of hierarchical models is that each parameter referring to a specific group or unit borrows strength from the corresponding parameters of other groups or units with similar characteristics. In other words, a shrinkage effects towards the population mean is present with use of hierarchical models. The volume of shrinkage depends on the variance between the random parameters. This can be quite beneficial, especially when a small number of individuals is observed in some groups. In such cases, the reduction of the uncertainty is large since information from other groups or units with smaller variability is incorporated in the posterior estimates. The so-called exchangeability assumption plays a central role in this behavior. The exchangeability assumption is used when no information concerning the structure of the parameters of interest is available. With this approach, we assume that all random parameters come from a common population distribution and their ordering does not affect the model and the results, that is, the hyperprior is invariant to permutations of the random parameters. In general we can say that θ_j for $j = 1, \ldots, K$ are *exchangeable* if $\theta_j | \phi \sim D(\phi)$ for all j. For more details and an interesting discussion, see Gelman et al. (2004, sec. 5.2).

Generally, hierarchical models are more flexible than are the typical nonhierarchical (or fixed effects) models since a more complicated structure is accommodated in the model. For this reason, they describe the data better, especially when the sample size is large. On the other hand, a complicated hierarchical formulation may lead to a model that overfits the data; that is, it may describes the current dataset better but might not allow for enough uncertainty in order to predict sufficiently future observations.

Finally, Robert (2007, sec. 10.2.2) provides a series of justifications and advantages for using hierarchical models, including the fact that the prior is decomposed into two main parts: one referring to structural information or assumptions concerning the model and one referring to the actual subjective information of the model parameters. Another advantage, according to the same author, is that hierarchical structure leads to a more robust analysis, reducing subjectivism since posterior results are averaged across different prior choices of parameters of interest. Finally, the hierarchical structure simplifies both the interpretation and the computation of the model since the corresponding posterior distribution is simplified, resulting in conditional distributions of simpler form. This allows for the implementation of simpler Gibbs-based sampling schemes; for examples, see Sections 2.3.3 (Gibbs sampler) and 2.3.5 (slice sampler).

9.2   SOME SIMPLE EXAMPLES

9.2.1   Repeated measures data

> **Example 9.1 . Repeated measurements of blood pressure.** Let us consider blood pressure measurements, which are well known for their variability. For illustration we consider the repeated measurements of blood pressure from 20 healthy individuals (see Table 9.1). The aim here is to estimate within-individual variability and between-individual variability.

9.2.1.1   Model formulation.   In this example, we can split the overall variability into two sources: (1) between-subject variability and (2) within-subject variability. Moreover, there is an obvious correlation between the measurements of the same individual. To

Table 9.1   Blood pressure measurements of 20 healthy individuals

Individual	1	2	3	4	5	6	7	8	9	10	11	12	13	14	15	16	17	18	19	20
Measurement																				
1st	108	91	93	104	99	95	93	99	90	92	101	97	97	96	106	100	90	88	92	100
2nd	98	94	96	99	97	98	97	96	100	95	89	97	100	95	100	98	99	98	92	101

account for the within-persons variability (source 2) and the correlation, we may introduce a random effect a_i for each individual. Hence we can formulate the model as

$$Y_{ij} = \mu + a_i + \epsilon_{ij} \text{ with } \epsilon_{ij} \sim N(0, \sigma^2) \text{ and } a_i \sim N(0, \sigma_a^2)$$

or equivalently

$$Y_{ij} \sim N(\mu_{ij}, \sigma^2) \text{ with } \mu_{ij} = \mu + a_i \text{ and } a_i \sim N(0, \sigma_a^2)$$

for $i = 1, 2, \ldots, n$ and $j = 1, 2$. Variance component σ_a^2 (the random effects variance) measures the between-subject variability (source 1), while σ^2 accounts for the remaining within-subject variability.

The usual prior distributions with large variances can be used when no information is available. Hence we can adopt

$$\mu \sim N(0, 1000), \quad \sigma^2 \sim IG(0.001, 0.001), \text{ and } \sigma_a^2 \sim IG(0.001, 0.001)$$

to express our prior ignorance.

This simple model is equivalent to assuming that

$$\boldsymbol{y}_i | \mu, \sigma^2, \sigma_a^2 \sim N_2 \left(\mu \mathbf{1}_2, \begin{bmatrix} \sigma^2 + \sigma_a^2 & \sigma_a^2 \\ \sigma_a^2 & \sigma^2 + \sigma_a^2 \end{bmatrix} \right)$$

where $\boldsymbol{y}_i = (Y_{i1}, Y_{i2})^T$ and $\mathbf{1}_2 = (1, 1)^T$ since Y_{ij} is expressed as the sum of two independent normal distributions and the parameters are given by

$$
\begin{aligned}
E(Y_{ij}) &= \mu, \\
\text{Var}(Y_{ij}) &= \text{Var}(a_i) + \text{Var}(\epsilon_{ij}) = \sigma_a^2 + \sigma^2 \\
\text{Cov}(Y_{i1}, Y_{i2}) &= \text{Cov}(\mu + a_i + \epsilon_{i1}, \mu + a_i + \epsilon_{i2}) \\
&= \text{Var}(a_i) + \text{Cov}(a_i, \epsilon_{i2}) + \text{Cov}(\epsilon_{i1}, a_i) + \text{Cov}(\epsilon_{i1}, \epsilon_{i2}) = \sigma_a^2.
\end{aligned}
$$

Thus, the total variability of the response Y_{ij} is equal to $\sigma_a^2 + \sigma^2$, while the covariance between two measurements of subject i is equal to the between-subject variability (the random effects variance). Thus, we can calculate the within-subject correlation by

$$r_{12} = \text{Cor}(Y_{i1}, Y_{i2}) = \frac{\sigma_a^2}{\sigma_a^2 + \sigma^2}. \tag{9.1}$$

Values close to one imply large within-subject correlation, indicating the importance of hierarchical/mixed models, while values close to zero imply low within-subject correlation, indicating that random effects do not improve our model.

We can generalize these calculations for K (exchangeable) repeated measurement. In such case

$$\boldsymbol{y}_i | \mu, \sigma^2, \sigma_a^2 \sim N_K \left(\mu \mathbf{1}_K, \sigma_a^2 \mathbf{1}_{[K \times K]} + \sigma \boldsymbol{I}_K \right),$$

where $\boldsymbol{y}_i = (Y_{i1}, Y_{i2}, \ldots, Y_{iK})^T$, and $\mathbf{1}_K$ and $\mathbf{1}_{[I \times J]}$ are a vector of length K and a matrix of dimension $I \times J$, respectively, with all elements equal to one.

Although the two model formulations are equivalent, the hierarchical structure facilitates the model and also provides estimates of individual effects. This simpler structure considerably simplifies interpretation of the parameters and the required MCMC algorithm by augmenting the parameter space using the random effects.

9.2.1.2 WinBUGS code.   To fit the hierarchical model described above, we need to set up a double nested loop (one for individuals and one for repeated measures). The following code can be used to define the model:

```
model {
   for   (i  in 1:n) {
         for (j in 1:K){
            y[i,j]  ~ dnorm ( mu[i,j], tau )
            mu[i,j] <- m + a[i]
         }
         a[i]~dnorm ( 0, tau.a)
   }
   # prior distributions
   m  ~ dnorm ( 0.0, 0.001)
   tau~dgamma (0.001, 0.001)
   tau.a~dgamma (0.001, 0.001)
}
```

Correlations, variances, and standard deviations can be calculated using simple logical/deterministic nodes.

The corresponding multivariate model can be directly fitted using the command `dmnorm`. The corresponding code for fitting the equivalent multivariate normal model directly is provided in the book's Webpage (`www.stat-athens.aueb.gr/~jbn/winbugs_book`). Note that generation was slower when using the multivariate normal approach.

9.2.1.3 Results.   Posterior summaries of the hierarchical model used for the above-mentioned data are presented in Table 9.2. The posterior mean of the total variation is found equal to 20.72, from which 1.25 is attributed to between-individual variabilitity. Correlation between measurements is relatively low (posterior mean ~ 0.06), which can be interpreted that 6% of the total variation due to between-subject variability. Posterior intervals of individual random effects are depicted in Figure 9.3. The same results were obtained by direct application of the multivariate normal approach described in Section 9.2.1.1.

9.2.1.4 Handling missing data.   Hierarchical models can also contribute to the estimation of missing values. If no random effects are assumed, then may use one of the following approaches. We can assume a simple model with $Y_{ij} \sim N(\mu, \sigma^2)$ ignoring individual effects or a "fixed" effects model with $Y_{ij} \sim N(\mu + a_i, \sigma^2)$ and a simple prior distribution assigned directly on a_i. In the first case all estimates of the missing values will be based on the posterior density of μ ignoring individual differences. In the second case, an estimate of the individual effect will be taken directly from the prior when all individual values are missing, while in the hierarchical model presented in the Section 9.2.1.1, an estimate will be produced by borrowing information from observations in the remaining groups. For illustration, let as assume the dataset given in Table 9.3.

Table 9.2 Posterior summaries for Example 9.1

Node	Posterior Mean	SD	Posterior percentiles 2.5%	97.5%
μ	96.70	0.74	95.21	98.14
τ	0.05	0.01	0.03	0.09
τ_a	109.50	265.40	0.11	876.20
σ^2	19.47	5.02	11.42	31.20
σ_a^2	1.25	2.63	0.00	9.16
$\sigma^2 + \sigma_a^2$	20.72	4.99	13.11	32.29
Correlation	0.06	0.11	0.00	0.40
σ	4.38	0.56	3.38	5.59
σ_a	0.72	0.86	0.03	3.03

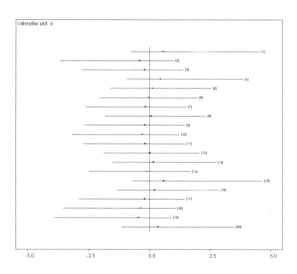

Figure 9.3 95% posterior intervals of random effects for Example 9.1.

Table 9.3 Blood pressure measurements of 20 healthy individuals with missing values

Individual:	1	2	3	4	5	6	7	8	9	10	11	12	13	14	15	16	17	18	19	20
Measurement																				
1st	108	91	—	104	99	95	93	99	90	92	101	97	97	96	106	100	90	88	92	100
2nd	—	—	—	—	97	98	97	96	100	95	89	97	100	95	100	98	99	98	92	101

The posterior distributions of between-subject and within-subject variabilities (σ_a^2 and σ^2, respectively) are similar to those for the full dataset (posterior means equal to 1.39 and 21.73, respectively). Concerning the estimates, we observe that for individuals 2 and 3, posterior estimates of blood pressure are close to 96.4 and 96.8 while for individuals 1 and 4 are slightly higher (97.4 and 97.2, respectively).

For the simpler model with a common mean, all posterior means of the missing values are equal (as expected) to 96.7. Finally, for the fixed effects model, estimates are significantly affected by the observed measurements. Hence for individuals 1 and 4, estimates of posterior means are quite high and close to the observed values of the first measurement, while for individual 2, the estimated measure is quite low. For individuals with no measurements, the posterior means are equal to 93.8, but the posterior variance is quite high, corresponding to the large prior variance (equal to 1000); see Table 9.4 for details. On the other hand, results from the random effects (hierarchical) model provides accurate estimates of the missing values in all cases with posterior intervals of lower range.

Table 9.4   Posterior summaries for missing data for incomplete dataset of Table 9.3

Individual	Measurement	Model	Posterior Mean	SD	Posterior percentiles 2.5%	97.5%	Actual values[a]
1	2	RE	97.4	5.0	87.6	107.5	(98)
		FE	107.7	6.1	95.6	119.7	108
2	2	RE	96.4	4.9	86.7	105.9	(94)
		FE	91.1	6.1	79.0	103.4	91
3	1,2	RE	96.8	4.9	87.2	106.4	93
		FE	93.8	32.6	28.2	156.8	96
4	2	RE	97.3	4.9	87.7	107.0	(99)
		FE	103.8	6.0	91.9	115.9	104
All		CM	96.7	4.8	87.2	106.3	—

[a] Omitted values are denoted in parentheses.
Abbreviations: RE=hierarchical models (model using individual random effects); FE=nonhierarchical (model using individual fixed effects); CM=common mean (nonhierarchical model with common mean).

Hence the hierarchical model provides a reasonable compromise between the two approaches. It shrinks estimates toward the overall mean by borrowing information from other individuals, but at the same time it accounts for individual differences by providing more reliable and accurate estimates.

9.2.2   Introducing random effects in performance parameters

Example 9.2. Kobe Bryant's field goals in NBA (Example 1.4 revisited). Here we reconsider Example 1.4, in which Kobe Bryant's field goals in the NBA were examined. A simple logit model for the success probability of Kobe Bryant's field goal attempts is adopted with log-odds (or logits) from the same population mean. Hence we assume that the performance of the player is similar (exchangeable) from season to season.

We use the following simple hierarchical model

$$Y_t \quad \sim \quad \text{binomial}(\pi_t, N)$$
$$\text{logit}(\pi_t) \quad = \quad \log\left(\frac{\pi_t}{1 - \pi_t}\right) = \theta_t$$
$$\theta_t \quad \sim \quad N(\mu_\theta, \sigma_\theta^2) \text{ for } t \in \{1999, 2000, \dots, 2006\}$$
$$\mu_\theta \quad \sim \quad N(0, 100) \text{ and } \sigma_\theta^2 \sim \text{IG}(0.01, 0.01),$$

which can be compiled in WinBUGS using the following syntax:

```
model {
    for (t in 1:YEARS){
        # stochastic part of the likelihood
        y[t]    ~ dbin( pi[t], N[t] )
        # link function
        logit(pi[t]) <- theta[t]
        # 1st stage of the prior
        theta[t] ~ dnorm(  mu.theta ,   tau.theta)

    }
    # hyperprior
    mu.theta~dnorm( 0, 0.001)
    tau.theta~dgamma( 0.01, 0.01)
    s.theta <- sqrt(1/tau.theta)
    # probability estimated for mu.theta (overall success odds)
    p.theta <- 1/(1+exp(-mu.theta))
}
```

Table 9.5 provides estimates from the hierarchical and the corresponding fixed effects logit models. All success probabilities and log-odds have smaller variability in the hierarchical model as we have already described; see also Figure 9.4. Finally, DIC indicates that the hierarchical model is better in terms of prediction by the corresponding fixed effects model (74.3 vs. 77.5).

9.2.2.1 *State space model.* The model of Section 9.2.2 assumes that all performance parameters are exchangeable, that is, the performance remains constant across time and the observable seasonal success rate are simply a sample from a constant latent performance random variable.

A model that is frequently used when data occur within a specific time duration, is the state space model, where the random coefficients are assumed to follow a normal distribution with the mean of the previous value of the performance parameter. Hence the model can be written as

$$Y_t \quad \sim \quad \text{binomial}(\pi_t, N)$$
$$\text{logit}(\pi_t) \quad = \quad \log\left(\frac{\pi_t}{1 - \pi_t}\right) = \theta_t$$

Table 9.5   Posterior summaries for model parameters and Kobe Bryant's field goal success probability (%) using fixed effects and hierarchical logit models (Example 9.2)

Node	Fixed effects model				Hierarchical model			
	Mean	SD	2.5%	97.5%	Mean	SD	2.5%	97.5%
μ_θ^a	-0.178	0.019	-0.216	-0.140	-0.179	0.039	-0.257	-0.102
σ_θ^b	0.079	0.019	0.0440	0.118	0.090	0.032	0.048	0.172
$\pi_{\mu_\theta}^c$	45.6	0.5	44.6	46.5	45.5	1.0	43.6	47.5
π_{1999}	46.9	1.4	44.0	49.7	46.4	1.2	44.0	48.8
π_{2000}	46.4	1.3	43.9	48.9	46.2	1.1	44.0	48.4
π_{2001}	46.9	1.3	44.4	49.3	46.5	1.1	44.4	48.8
π_{2002}	45.1	1.1	42.9	47.4	45.2	1.0	43.3	47.2
π_{2003}	43.8	1.4	41.0	46.6	44.4	1.2	41.9	46.8
π_{2004}	43.3	1.4	40.6	46.0	44.0	1.2	41.6	46.3
π_{2005}	45.0	1.1	42.9	47.1	45.1	1.0	43.2	47.0
π_{2006}	47.2	1.7	43.9	50.6	46.5	1.4	43.9	49.4

[a] μ_θ in fixed effects model was calculated as the arithmetic mean of all θ_i in each iteration.
[b] σ_θ in fixed effects model was calculated as the standard deviation of all θ_i in each iteration.
[c] $\pi_{\mu_\theta} = 1/(1 + e^{\mu_\theta})$ is each iteration.

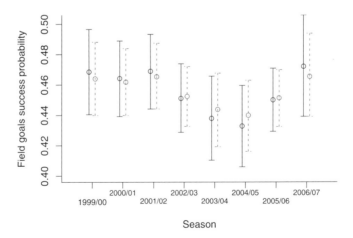

Figure 9.4   95% posterior intervals of model parameters for fixed effects (solid lines) and hierarchical (dashed lines) models for Kobe Bryant's field goals (Example 9.2).

$$\theta_t \quad \sim \quad N(\theta_{t-1}, \sigma_\theta^2) \text{ for } t \in \{2000, \dots, 2006\}$$
$$\theta_{1999} \quad \sim \quad N(0, 100) \ \& \ \sigma_\theta^2 \sim \text{IG}(0.01, 0.01)$$

The model defined above is more sensible in terms of interpretation, since the perfor-mance of the player depends on the previous season's performance, incorporating in this way both physical and psychological effects, while the model defined in the previous section assumes exchangeable performances and hence stability across time.

As expected, this model describes better the performance of the player. The DIC value was estimated at 67.8, which indicates a better fitted model than the exchangeable hier-archical model of the previous section. Estimates of the posterior success probabilities are presented in Figure 9.5 for comparison with the hierarchical model with exchangeable random effects.

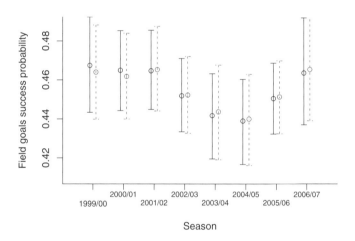

Figure 9.5   95% posterior intervals of model parameters for parameters of state space (solid lines) and hierarchical model (dashed lines) models for Kobe Bryant's field goals (Example 9.2).

9.2.3   Poisson mixture models for count data

Example 9.3. 1990 USA general social survey (Example 8.3 revisited). In this section we consider two hierarchical models, which are frequently used for count data, for the data of Example 8.3: (1) the Poisson–gamma model and (2) the Poisson–log-normal model. Both models are simply finite mixtures of the Poisson distribution with different assumptions concerning the mixing distribution, resulting in a marginal sampling distribution that differs from the Poisson distribution, which is the standard assumed model for count data.

9.2.3.1 *The Poisson–gamma model.*   The Poisson–gamma model is frequently used to model count data with overdispersion. The model is given by

$$Y_i \quad \sim \quad \text{Poisson}(\lambda_i u_i)$$

$$u_i \quad \sim \quad \text{gamma}(r_i, r_i)$$

for $i = 1, 2, \ldots, n$. Covariates can be used to define both λ_i and r_i. Here we consider only the gender as a possible covariate; hence we may simply replace λ_i by λ_{G_i}, where G_i is a gender index taking values 1 or 2 when the ith individual is male or female, respectively. Similarly, we may replace r_i by r_{G_i} if we wish to assume different gender parameters for the mixing distribution.

The marginal likelihood of this simple hierarchical model is given by integrating the random effects u_i as described in Section 8.3.1, resulting in a negative binomial distribution with probability function

$$f(y|\lambda, r) \quad = \quad \frac{\Gamma(y+r)}{\Gamma(r)y!} \left(\frac{\lambda}{\lambda+r} \right)^y \left(\frac{r}{\lambda+r} \right)^r.$$

The code for fitting this model is given by the following syntax:

```
model{
    for(i in 1:n){
        # Poisson part
        y[i] ~ dpois( mu[i] )
        # defining the mean of the Poisson
        mu[i] <- lambda[ gender[i]+1 ] * u[i]
        # mixing distribution
        u[i] ~ dgamma( r[ gender[i]+1 ] , r[ gender[i]+1 ] )
    }
    for (j in 1:2){
        # prior distributions
        lambda[j]~dgamma( 0.001, 0.001 )
        r[j]~dgamma( 0.001, 0.001 )
    }
}
```

Results are the same as in the negative binomial illustration; see Section 8.3.1 for details. The only difference we observe is the DIC value (2122.03 instead of 2873.4) since it is now based on the conditional likelihood $f(y|\lambda_1, \lambda_1, r_1, r_2, u)$ instead of the marginal likelihood $f(y|\lambda_1, \lambda_1, r_1, r_2) = \int f(y|\lambda_1, \lambda_1, r_1, r_2, u) f(u|r_1, r_2) du$,, where $u = (u_1, \ldots, u_n)^T$; for details, see the computational note at the end of Section 9.2.3.

9.2.3.2 The Poisson–log-normal model.

A usual practice in building hierarchical models is to express the model as a typical GLM and then add random errors (or effects) at the linear predictor. Following this practice, we can formulate a log-linear Poisson model and add a random effect that is usually assumed to be normally distributed. Hence the model can be written using the following structure

$$\begin{aligned} Y_i &\sim \text{Poisson}(\mu_i) \\ \log(\mu_i) &= \beta_0 + \beta_1 X_{i1} + \cdots + X_{ip} + \epsilon_i \\ \epsilon_i &\sim N(0, \sigma_\epsilon^2). \end{aligned}$$

In the current example only one covariate is assumed, which is the gender. A single dummy variable for the corner parametrization with "male" as the reference category can be defined by setting $D_i = G_i - 1$.

The model formulation is equivalent to assuming

$$\begin{aligned} Y_i &\sim \text{Poisson}(\lambda_i u_i) \\ u_i = \exp(\epsilon_i) &\sim \text{LN}(0, \sigma_\epsilon^2) \\ \log(\lambda_i) &= \beta_0 + \beta_1 X_{i1} + \cdots + X_{ip}. \end{aligned}$$

This structure is identical to the one used for the Poisson–gamma model, but now the mixing distribution is no longer gamma but log-normal. Hence the only difference between the two models lies in the assumption of the distribution of the random effects and the corresponding assumed data distribution, which is now Poisson–log-normal instead of negative binomial. Unfortunately, the probability function of the Poisson–log-normal distribution is not available analytically, making interpretation of this model more complicated than that of the negative binomial one. Nevertheless, the mean and the variance can be easily calculated and are given by

$$E(Y|\lambda, \sigma_\epsilon^2) = \lambda e^{\sigma_\epsilon^2/2} \text{ and } V(Y|\lambda, \sigma_\epsilon^2) = \lambda e^{\sigma_\epsilon^2/2} + \lambda^2 e^{2\sigma_\epsilon^2} - \lambda^2 e^{\sigma_\epsilon^2}.$$

In these data we have used a different variance for each gender.

The code for the Poisson–log-normal model for this dataset is straightforward:

```
model{
    for(i in 1:n){
        # Poisson part
        y[i] ~ dpois( mu[i] )
        # defining the mean of the Poisson
        log(mu[i]) <- beta[ gender[i]+1] + epsilon[i]
        # mixing distribution
        epsilon[i] ~ dnorm( 0, tau[ gender[i]+1] )
    }
    for (j in 1:2){
    # prior distributions
    beta[j]~dnorm( 0, 0.001)
    tau[j]~dgamma(0.001,0.001)
    s[j] <- sqrt( 1/tau[j] )
    s2[j]<-1/tau[j]
    #
    lambda[j] <- exp(beta[j])
    mean.gender[j] <- exp(beta[j] + 0.5*s2[j] )
    var[j] <-    lambda[j]*exp( s2[j]/2 )
                + lambda[j]*lambda[j]*exp( 2*s2[j] )
                - lambda[j]*lambda[j]*exp( s2[j] )
    DI[j] <- var[j]/mean.gender[j]
    }
}
```

The DIC value for this model is calculated to be equal to 2295.48, which is much higher than the corresponding DIC value for the hierarchical Poisson–gamma model (2122.03), indicating a better fit for the latter. Note that we can compare DIC values resulting only from hierarchical models since their values are based on the conditional likelihood as described in the computational note at the end of this section.

The insufficient fit of the Poisson–log-normal model is also evident when we examine the posterior distributions of $E(Y)$ and $V(Y)$ for each gender. For example, the posterior means of $E(Y)$ are equal to 6.9 and 6.6, respectively, which are far away from the corresponding sample estimates (equal to 5.85 and 4.30, respectively), while similar differences are also observed for the corresponding variances.

COMPUTATIONAL NOTE (DIC for mixture models WinBUGS)

For a hierarchical model with the structure

$$Y_i|u_i \quad \sim \quad f(y_i|u_i, \boldsymbol{\theta})$$
$$u_i \quad \sim \quad f(u_i|\boldsymbol{\theta}_u),$$

DIC (in WinBUGS) is calculated using deviance measure

$$D_c(\boldsymbol{u}, \boldsymbol{\theta}) = -2\log f(\boldsymbol{y}|\boldsymbol{u}, \boldsymbol{\theta})$$

based on the conditional likelihood

$$f(\boldsymbol{y}|\boldsymbol{u}, \boldsymbol{\theta}) = \prod_{i=1}^{n} f(y_i|u_i, \boldsymbol{\theta}).$$

Hence DIC in this case is given by

$$\text{DIC} = 2\overline{D_c(\boldsymbol{u}, \boldsymbol{\theta})} - D_c(\overline{\boldsymbol{u}}, \overline{\boldsymbol{\theta}}),$$

while in the case where we fit the model directly

$$Y_i \sim f(y_i|\boldsymbol{\theta}, \boldsymbol{\theta}_u),$$

based on the marginal distribution

$$f(Y_i|\boldsymbol{\theta}, \boldsymbol{\theta}_u) = \int f(Y_i|\boldsymbol{\theta}, u_i) f(u_i|\boldsymbol{\theta}_u) du_i,$$

the DIC is given by

$$\text{DIC} = 2\overline{D(\boldsymbol{\theta}, \boldsymbol{\theta}_u)} - D(\overline{\boldsymbol{\theta}}, \overline{\boldsymbol{\theta}}_u)$$

with

$$D(\boldsymbol{\theta}, \boldsymbol{\theta}_u) = -2\sum_{i=1}^{n} \log f(Y_i|\boldsymbol{\theta}, \boldsymbol{\theta}_u).$$

9.2.4   The use of hierarchical models in meta-analysis

The hierarchical structure naturally arises in meta-analysis, where similar parameters are estimated in different studies under different circumstances. In the following paragraph we present a simple example where the odds ratios from different studies were estimated and the aim is to combine results from all studies and produce a common estimate of the risk factor for the disease under study.

Example 9.4. Analysis of odds ratios from various studies. Suppose the odds ratios measuring the effect of smoking on lung cancer presented in Table 9.6. In the seven studies listed in Table 9.6, only the 95% confidence intervals are available. For three additional studies, the full 2×2 contingency tables were also available given in Table 9.7. The aim in this illustrative example is to obtain an overall estimate of the odds ratio using information available from all 10 studies of Tables 9.6 and 9.7.

Table 9.6   Odds ratios of lung cancer for smokers versus nonsmokers for studies 1–7

Study	Odds ratio	95% CI
1	3.89	0.92 – 16.30
2	3.97	2.20 – 7.16
3	3.88	2.47 – 6.08
4	17.47	14.24 – 21.43
5	5.35	2.44 – 11.74
6	9.10	5.57 – 14.86
7	12.41	2.94 – 3.96

Table 9.7 2×2 contingency tables of lung cancer and smoking for studies 8–10

Study	Odds Ratio		Cases	Controls
8	3.48	Smokers	49	29958
		Nonsmokers	33	70186
9	33.10	Smokers	12	0
		Nonsmokers	89	118
10	3.43	Smokers	29	4
		Nonsmokers	171	81

Bear in mind that the sample estimates of the log-odds ratios are asymptotically normal, we may use the normal distribution to model the available information of Table 9.6. Hence we may use the following simple hierarchical model

$$\log \widehat{\mathrm{OR}}_k \sim N(\theta_k, \hat{\sigma}_k^2)$$
$$\theta_k \sim N(\theta, \sigma_\theta^2) \text{ for } k = 1, 2, \ldots, 7, \ i = 1, 2,$$

where $\widehat{\mathrm{OR}}_k$ are the estimated odds ratio for the kth study as provided in Table 9.7, while $\hat{\sigma}_k^2$ is the standard error of the corresponding $\log \widehat{\mathrm{OR}}_k$ calculated by $\hat{\sigma}_k = \log(U/L)/(2 \times 1.96)$, with U and L, respectively, denoting the upper and the lower limits of the 95% confidence interval of the odds ratio of Table 9.6. The first equation approximates the likelihood of each study since the original data of the corresponding 2×2 contingency tables are not available. For the last three studies the full data are available, and therefore the following model can be used

$$Y_{i1k} \sim \text{binomial}(\pi_{i1k}, Y_{i1k} + Y_{i2k}),$$

$$\log\left(\frac{\pi_{i1k}}{1 - \pi_{i1k}}\right) = a_k + \theta_k I(i = 1)$$

$$\theta_k \sim N(\theta, \sigma_\theta^2) \text{ for } k = 8, 9, 10, \ i = 1, 2,$$

where Y_{ijk} refers to the number of observations in the kth study with smoking and cancer status i (1=smokers, 2=nonsmokers) and j (1=case, 2=control), respectively. Parameter θ_k is the corresponding odds ratio, while a is the odds of the disease for the nonsmoking group. Usual noninformative prior distributions can be used for a_k, θ, and σ_θ^2.

The code for the first part (seven studies) of the hierarchical model can be specified in WinBUGS using the following syntax

$$\left(\text{precision.logor}[K] \leftarrow \frac{1}{\text{pow}(\text{selogor}[K], 2)}\right)$$
$$\left(\text{logor}[K] \sim \text{dnorm}(\text{theta}[K], \text{precision.logor}[K])\right)$$

```
for (k in 1:K1){
    logor[k] <- log(or[k])
    selogor[k] <- (U[k]/L[k])/(2*1.96)
    logor[k] ~ dnorm( theta[k], selogor[k] )
    theta[k]~dnorm( mu.theta, tau.theta )
    OR[k] <- exp(theta[k])
}
```

while for the second part (studies 8–10) the model is specified using the following syntax:

```
for (k in 1:K2){
    for (i in 1:2){
        N[i,k] <- Y[i,1,k]+Y[i,2,k]
        Y[i,1,k] ~ dbin( p[i,k], N[i,k] )
        logit(p[i,k]) <- a[k] + theta[K1+k] * equals(i,1)
    }
    theta[K1+k] ~ dnorm( mu.theta, tau.theta )
    OR[K1+k] <- exp(theta[K1+k])
}
```

Finally, the prior distributions are defined as usual by

```
for( k in 1:3) { a[k] ~ dnorm( 0.0, 0.001) }
mu.theta    ~ dnorm( 0.0, 0.001)
tau.theta   ~ dgamma( 0.001, 0.001)
```

while estimated odds ratios for each study using the preceding model are given by e^{θ_k}. The overall estimate of the odds ratio is given by e^θ.

The posterior mean of the overall odds ratio is found equal to 5.28, with 95% of the posterior values ranging from 3.33 to 9.05. Error bars of the estimated odds ratios of each study using the hierarchical model presented above are depicted in Figure 9.6.

Additional details concerning hierarchical models and meta-analysis can be found in Woodworth (2004, chap. 11).

9.3   THE GENERALIZED LINEAR MIXED MODEL FORMULATION

A popular hierarchical model formulation is that for the so called *generalized linear mixed models*, which is based on the GLM formulation also having a hierarchical structure by including random coefficients/effects in the usual linear predictor.

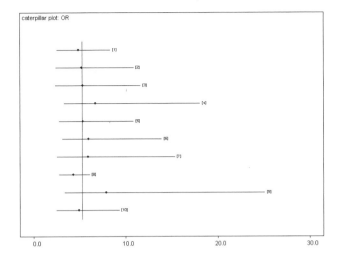

Figure 9.6 95% posterior intervals of adjusted odds ratios of lung cancer for smokers versus nonsmokers for each study in Example 9.4.

Hence the model can be formulated as

$$Y \sim \mathcal{D}(\boldsymbol{\theta}) \tag{9.2}$$
$$E(Y) = X\boldsymbol{\beta} + Z\boldsymbol{b} \tag{9.3}$$
$$\boldsymbol{b} \sim N(\boldsymbol{0}, \boldsymbol{G}), \tag{9.4}$$

where Y is a random response vector that may include repeated measures of the same variable or measurements of correlated variables and X, Z are the data or design matrices for fixed and random effects $\boldsymbol{\beta}$ and $\boldsymbol{b}$, respectively.

The assumption of normal random effects [Eq. (9.4) in the formulation above] can be easily substituted by another distribution without any difficulty within Bayesian inference, resulting in a different marginal sampling distribution for Y assumed by the adopted model.

The normal model is a special case of the preceding formulation for $\mathcal{D}$ = normal and $\boldsymbol{\theta} = (\boldsymbol{\mu}, \boldsymbol{\Sigma})$. Note that some structure must be imposed on $\boldsymbol{\Sigma}$ in order to avoid an over-parametrized model. The simplest case is to set $\boldsymbol{\Sigma} = \sigma^2 \boldsymbol{I}$. Variance components in normal models are calculated using logic similar to that followed in Section 9.2.1.1. The total variability is calculated via $\mathrm{Var}(Y_{ij})$, while partial variances are provided by the corresponding random effects variances. Correlations within each level of hierarchy also provide useful decomposition and interpretation concerning the source of variability and the necessity of the corresponding random effects and levels of hierarchy.

9.3.1 A hierarchical normal model: A simple crossover trial

A frequent implementation of hierarchical models is within the context of crossover trials, in which different treatments are given with different sequences in groups of patients. Here we illustrate a simple two treatment two-period crossover trial in which patients are divided into two groups. The first group receives treatment A for the first period and treatment B for the second period of the study, while the other group receives the same treatments in the reverse order. Random effects are used to capture the correlation that results from the within-patient variability. An additional feature that arises from this crossover design is that the treatment implemented in the first period might possibly influence the effectiveness of the treatment implemented in the second period. This effect is referred to as a *carryover effect*. This effect is sometimes eliminated if a treatment-free interval is introduced between the two treatment periods (called the *washout period*). The carryover effect can be estimated and tested using the interaction between the period and the treatment effect in hierarchical models.

> **Example 9.5. An AB/BA crossover trial.** The data in this example are results
> from a study presented by Brown and Prescott (2006, pp. 275–279) comparing two
> diuretics in the treatment of mild and moderate heart failure. Baseline observations
> were also taken just before the first treatment period. The duration of each treat-
> ment period was 5 days without any washout period. To avoid carryover effects,
> measurements of the first 2 days were ignored. Although the primary outcome
> measurement was micturition, we analyze two secondary response variables that
> were available: edema status and diastolic blood pressure (DBP). The first variable
> was calculated as the sum of left and right ankle diameters, while the second vari-
> able was the sum of three DBP readings (here we will consider the corresponding
> means). These response variables were measured before the first treatment period
> and after the end of each treatment period. In total, 94 patients participated in the
> study. For 20 patients only one measurement is available in this dataset. The aim
> here is to compare the effectiveness of the two treatments. Data of this example are
> available in the book's Website and are reproduced with permission of John Wiley
> and Sons, Inc.

Following Brown and Prescott (2006), four models were fitted for each response (edema and DBP) and compared using DIC. In all four models, the treatment and the period effect were included in the analysis as fixed effects. The patient effect was also used in all models as either fixed or random effect. The baseline measurement was also introduced in two of the models to assess their importance for the model. Finally, an additional model was introduced in which the interaction effect between the period and the treatment was included in the linear predictor in order to account for possible carryover effects.

We can incorporate all models using the usual γ indicators in the model formulation. Hence all four models can be described by the following expression

$$
\begin{aligned}
Y_i &\sim N(\mu_i, \sigma^2) \\
\mu_i &= \beta_1 + \beta_2 \, \text{period}_i + \beta_3 T_i + \gamma_1 a_{P_i}^{\text{random}} + (1 - \gamma_1) a_{P_i}^{\text{fixed}} + \gamma_2 \beta_4 B_i \\
a_k^{\text{random}} &\sim N(0, \sigma_{\text{patients}}^2) \\
a_k^{\text{fixed}} &\sim N(0, 10^{-3}),
\end{aligned}
$$

for $i = 1, 2, \dots, N$ and $k = 1, 2, \dots, n$, where N now denotes the total number of obser- vations available, n is the number of individuals/patients participating in the study, period$_i$ and T_i are dummy variables indicating the period and the treatment of observation i (or indi- vidual P_i) (corner constraints and zero/one indicator variables are used here), and a_k^{random} and a_k^{fixed} are the random and fixed effects for individual k ($k = 1, 2, \dots, n$). The variable B_i refers to the baseline measurement of Y for observation i (or individual P_i). Since we consider two different responses (edema and DBP) here, an additional indicator must be added for the response variable used. The model structure is defined using binary indicators

γ_1 (0=fixed effects, 1=random effects) and γ_2 (1 or 0 depending on whether the baseline measurement is included in the linear predictor). The usual noninformative priors can be used for the remaining parameters. The WinBUGS code for this model, which encompasses all four different formulations, is given in Table 9.8.

Table 9.8   WinBUGS code for models with edema and DBP responses in Example 9.5

```
model{
    for (i in 1:N){
        # model for oed
        oed[i]~dnorm( mu.oed[i], tau.oed[1])
        mu.oed[i]<- b.oed[1]+ b.oed[2] *(period[i]-1)
                            + b.oed[3] *(treatment[i]-1)
                            + g[1]*a.oed[ patient[i] ]
                            + (1-g[1])*a.oed.fixed[ patient[i] ]
                            + g[2]*b.oed[4]*oedbase[i]
        # model for dbp
        dbp[i]~dnorm( mu.dbp[i], tau.dbp[1])
        mu.dbp[i]<- b.dbp[1]+ b.dbp[2] *(period[i]-1)
                            + b.dbp[3] *(treatment[i]-1)
                            + g[1]*a.dbp[ patient[i] ]
                            + (1-g[1])*a.dbp.fixed[ patient[i] ]
                            + g[2]*b.dbp[4]*dbpbase[i]
    }
    for (i in 1:n){
        # Hyper priors for individual/patients random effects
        a.oed[i]~dnorm( 0.0, tau.oed[2])
        a.dbp[i]~dnorm( 0.0, tau.dbp[2])
        # noninformative priors for individual/patients fixed effects
        a.oed.fixed[i]~dnorm( 0.0, 0.001)
        a.dbp.fixed[i]~dnorm( 0.0, 0.001)
    }
    for (i in 1:p){
        b.oed[i]~dnorm( 0.0, 0.001)
        b.dbp[i]~dnorm( 0.0, 0.001)
    }
    tau.oed[1]~dgamma( 0.001,0.001)
    tau.oed[2]~dgamma( a, a)
    tau.dbp[1]~dgamma( 0.001,0.001)
    tau.dbp[2]~dgamma( a, a)
    a <- g[1]*0.001 + (1-g[1])*1
    #
    s2[1]<-1/tau.oed[1]
    s2[2]<-1/tau.oed[2]
    s2[3]<-1/tau.dbp[1]
    s2[4]<-1/tau.dbp[2]
    r[1] <- s2[2]/(s2[2]+s2[1])
    r[2] <- s2[4]/(s2[4]+s2[3])
    for( i in 1:N ){
        res1[i] <- oed[i] - mu.oed[i]
        res2[i] <- dbp[i] - mu.dbp[i]
    }
    R[1] <- 1 - pow( sd(res1[1:N])/sd(oed[1:N]), 2)
    R[2] <- 1 - pow( sd(res2[1:N])/sd(dbp[1:N]), 2)
}
```

Key: $\gamma = (0,0)$ — fixed patient effects and no baseline measurement; $\gamma = (1,0)$ — fixed patient effects and baseline measurement, $\gamma = (0,1)$ — random patient effects and no baseline measurement, $\gamma = (1,1)$ — random patient effects and baseline measurement.

DIC values of all models are presented in Table 9.9. All DIC values were calculated using 50,000 iterations since instability was observed for chains of length equal to 10,000 iterations (especially for hierarchical models). From this table we observe that the fourth model (with random effects and the baseline measurement as covariate) has the lower DIC values. According to DIC, improvement of the fit is higher for edema than the corresponding one for DBP. The log-normal distribution was additionally used for the fourth model of Table 9.9 in order to check for deviations from the normality assumption, resulting in considerably higher DIC values (1095.5 and 494.1 for edema and DBP, respectively) than the corresponding ones under the normal assumption. Finally, carryover effects were also checked by including the interaction terms between the period and the treatment. Posterior distributions of the corresponding carry over effects lie around zero while DIC values are close to the ones for models with no carryover effects, indicating that these effects can be omitted from the model.

Table 9.9   Fitted models and DIC values for Example 9.5

				DIC	
Model[a]	γ	Patient effect	Baseline	Edema	DBP
1	$(0,0)$	Fixed	No	1094.1	471.3
2	$(1,0)$	Random	No	1084.8	470.6
3	$(0,1)$	Fixed	Yes	1093.9	471.2
4	$(1,1)$	Random	Yes	1065.4	462.4

[a] In all models the treatment and the period effect were included as fixed effects.

Posterior summaries of the finally selected model are presented in Table 9.10. For DBP, both the period and the treatment effects are important since zero lies at the tails of the corresponding posterior distributions. On the other hand, the period and the treatment effects on edema are minor since their posterior distributions lie around the zero value. For both models the baseline measurement is important for determination of the corresponding values in each stage of the study.

Table 9.10   Posterior summaries of parameters of model 4 in Table 9.9 for Example 9.5

		Edema					Diastolic blood pressure				
		Posterior			Posterior percentiles		Posterior			Posterior percentiles	
		Mean	Median	SD	2.5%	97.5%	Mean	Median	SD	2.5%	97.5%
Constant	β_1	24.57	24.59	5.29	14.2	34.97	-0.25	-0.25	1.51	-3.19	2.75
Period	β_2	-1.03	-1.03	0.77	-2.54	0.48	-0.24	-0.24	0.12	-0.47	0.0016
Treatment	β_3	-1.01	-1.01	0.76	-2.53	0.47	-0.31	-0.31	0.12	-0.54	-0.07
Baseline	β_4	0.69	0.69	0.06	0.58	0.81	0.99	0.99	0.03	0.93	1.04
Variance components											
Within-patient	σ^2	0.543	0.532	0.093	0.391	0.754	23.12	22.43	4.88	16.38	33.43
Between-patient	$\sigma^2_{patients}$	3.833	3.773	0.625	2.787	5.229	25.30	25.12	6.80	13.36	39.11
Correlation											
Within-patient	r_{12}	0.874	0.876	0.027	0.814	0.918	0.523	0.529	0.085	0.341	0.673

Concerning the variance components, for both response variables the between-patient variability is large in comparison to the remaining within-patient variability. To be more specific for edema, the between-patient variability is about 7 times the within-patient variability when comparing their posterior means, while for DBP, the two variance components are about the same. The within-subject correlation r_{12} [see Eq. (9.1)] ranges from 0.81 to 0.92 with 95% posterior probability for edema measurements, while for DBP the corresponding correlation is around 0.53, ranging from 0.34 to 0.67 with the same posterior probability. The large within-subject correlation indicates that using the crossover design is beneficial and allows us to successfully quantify and remove the corresponding variability.

9.3.2   Logit GLMM for correlated binary responses

When a series of repeated binary responses Y_{it} for $t = 1, 2, \ldots, T$ over the same individuals $i = 1, 2, \ldots n$ are considered, the following model can be adopted

$$Y_{it} \sim \text{Bernoulli}(\pi_{it})$$

$$\log \left(\frac{\pi_{it}}{1 - \pi_{it}} \right) = \beta_{0t} + \sum_{j=1}^{p} \beta_{jt} X_{ijt} + b_i \tag{9.5}$$

$$b_i \sim N(0, \sigma^2),$$

where β_{0t} are fixed constants depending on time sequence t, X_{ijt} are time-varying covariates for individual i and time sequence j, β_{jt} are their corresponding coefficients, and b_i are individual random effects capturing within-patient correlation. The structure of the model formulated above can be slightly modified depending on the data and the problem at hand; for example, covariates or their effects may not depend on the time index t.

A much simpler version of this model can be considered if no covariates are used. In such a case, the linear predictor can be simplified to

$$\log \left(\frac{\pi_{it}}{1 - \pi_{it}} \right) = b_i + \beta_t. \tag{9.6}$$

This model is frequently used in psychometrics and social sciences to model correlated questionnaire responses (items). The random effect b_i depicts the individual's ability to reply positively to each question, while β_t reflects the frequency rate of a positive reply to the question t (and therefore easiness of the question in some cases). This type of model was initially introduced by Rasch (1961) and is referred in the relevant literature as *Rasch* or *item response models* (Agresti, 2002, p. 495). The simplest case of this model is the one for data with two binary measurements ($T = 2$). In such a case, the model is the analog of the one used by Cox (1958) to derive the conditional likelihood estimate of odds ratio for two correlated binary responses. Other link functions such as the probit or the clog–log can be alternatively used without any problem.

Decomposition of the variance components is more complicated for models with binomial responses. One simple approach is to adopt the latent variable approach and express everything in terms of the variability of this latent variable. Hence, for the logit model, we assume that a latent variable Z_i exists following the logistic distribution with mean μ_i and dispersion parameter equal to one. The model is summarized as follows:

$$Z_{it} \sim \text{Logistic}(\mu_{it}, 1)$$

$$\mu_{it} = \beta_{0t} + \sum_{j=1}^{p}\beta_{jt}X_{ijt} + b_i$$

$$Y_{it} = 1 \text{ if } Z_{it} > 0 \text{ and } Y_{it} = 0 \text{ otherwise.}$$

In this formulation, the variance of the latent variable given the random effects is equal to $\text{Var}(Z_{it}|\mu_{it}) = \text{Var}(Z_{it}|b_i) = \pi^2/3$ from the logistic regression assumption. Moreover, the total variability of the latent variable is given by

$$
\begin{aligned}
\text{Var}(Z_{it}) &= \text{Var}_{\mu_{it}}\Big(E(Z_{it}|\mu_{it})\Big) + E_{\mu_{it}}\Big(V(Z_{it}|\mu_{it})\Big) \\
&= \text{Var}_{\mu_{it}}\Big(\beta_{0t} + \sum_{j=1}^{p}\beta_{jt}X_{ijt} + b_i\Big) + E_{\mu_{it}}\left(\frac{\pi^2}{3}\right) = \text{Var}_{\mu_{it}}\left(b_i\right) + \frac{\pi^2}{3} \\
&= \sigma^2 + \frac{\pi^2}{3},
\end{aligned}
$$

resulting in within-subject correlation of the latent measurement equal to

$$r_z = \frac{\sigma^2}{\sigma^2 + \pi^2/3}.$$

Similar logic can be used for the rest of the link function. For example, for the probit model the latent variable is assumed to be a standardized normal distribution with $\text{Var}(Z_{it}|b_i) = 1$.

Usually, we wish to express the variability in terms of the original response variable. For this reason alternative approaches have been proposed in the literature using either first-order Taylor-based approximations (model linearization approach) or simulation (Goldstein et al., 2002; Browne et al., 2005). Here we do not pursue this issue further but use a much simpler approach to estimate the changes in the variability of the original measurement by considering a simple R^2 type measure and the corresponding modification of the measure with the addition of hierarchical levels to model structure. Here we define this measure as $R^2_{\text{bin}} = 1 - \sigma^2_{\text{res}}/S_Y^2$, where S_Y^2 denotes the sample variance of the response Y and σ^2_{res} denotes the variance measure of the unstandardized residuals, given by

$$\sigma^2_{\text{res}} = \frac{1}{N-1}\sum_{i=1}^{n}\sum_{t=1}^{T}(Y_{it} - N_{it}\pi_{it})^2,$$

where N_{it} are the number of Bernoulli replications for the i, t combination and $N = \sum_{i=1}^{n}\sum_{t=1}^{T}N_{it}$ is the total number of Bernoulli observations over all available data. When binary data are used (as in the example here), then $N_{it} = 1$ for all i, t.

In the following paragraphs, we present a simple example in a matched-pair clinical trial using two binary responses (Example 9.6) and a social sciences example using three correlated questionnaire items with gender as a covariate (Example 9.7).

9.3.2.1 *The logit model in* 2×2 *tables of dependent binary data.* In this section we present a simple hierarchical model used when a simple 2×2 contingency table arises from dependent binary data. In the illustrative example that follows we compare two treatments given in a matched-pair clinical trial.

Example 9.6. **Treatment comparison in matched-pair clinical trials.** Let us consider a matched-pair clinical trial in which 60 pairs of patients were considered. Each patient in the first treatment group was matched to one individual in the other treatment group according to age, gender, and severity of the disease. After 2 weeks of treatment, the presence or absence of the disease was measured and denoted by Y_{i1} and Y_{i2} for patients receiving treatments A and B, respectively. Table 9.11 summarizes the data. Each entry of this table represents observed frequencies for each combination of the responses (Y_{i1}, Y_{i2}).

The aim here is to estimate the efficiency of the new treatment (B) in comparison to the standard treatment (A). Traditional Poisson log-linear models or logit models presented in Chapter 7 cannot be used since measurements for each pair of patients are correlated and this must be accounted for in the model structure.

Table 9.11 Data of Example 9.6

	Treatment B	
Treatment A	Not cured	Cured
Not cured	7	14
Cured	2	37

Here we the simplified version of model (9.5) using the linear predictor of the type (9.6) with $T = 2$ binary responses. Note that instead of using β_t parameters for each individual in the matched pair, we use the parameterization β_1 and $\beta_1 + \beta_2$ for each individual in order to directly estimate the log-odds ratio using β_2 parameter.

In order to import the aggregated data of Table 9.11 in WinBUGS , we have used vectors

$$
\begin{aligned}
\boldsymbol{f} &= (n_{11}, n_{21}, n_{12}, n_{22})^T = (7, 2, 14, 37)^T \\
\boldsymbol{Y}_1 &= (0, 1, 0, 1)^T \\
\boldsymbol{Y}_2 &= (0, 0, 1, 1)^T
\end{aligned}
$$

where n_{ij} are the frequencies of the contingency table (Table 9.11) and $\boldsymbol{Y}_t$, $t = 1, 2$, are the values of the binary variables in the instances (here, each individual in the same matched pair) that correspond to each element of of frequency vector $\boldsymbol{f}$.

In WinBUGS we need to reconstruct individual data. One problem that arises is that the range of loop indices cannot be defined within WinBUGS code. Thus we construct a vector $\boldsymbol{F}$ with elements of the cumulative counts of vector $\boldsymbol{f}$. Hence each element of $\boldsymbol{F}$ will be given by

$$
F_i = \sum_{k=1}^{i} f_{k-1} \text{ for } i = 1, 2, 3, 4, 5
$$

with f_0 set equal to zero. In this specific case

$$
\boldsymbol{F} = (0, 7, 9, 23, 60)^T .
$$

Then we need to specify individual data y_{i1} and y_{i2} using the following loop:

$$
y_{it} = Y_{tk} \text{ for } t = 1, 2; \ i = F_k + 1, \dots, F_{k+1}; \ k = 1, 2, 3, 4 .
$$

The model definition is completed using the following expressions

$$y_{it} \quad \sim \quad \text{Bernoulli}(\pi_{it})$$

$$\log\left(\frac{\pi_{it}}{1-\pi_{it}}\right) \quad = \quad b_i + \beta_1 + \beta_2 I(y_{i2}=1)$$

for $t = 1,2$ and $i = 1,\dots,n$, where n is the total number of matched pairs in the study.

We further need to define random effects (i.e., the mixing distribution) and the prior distributions. For the random effects we use a normal mixing distribution with $b_i \sim N(0, \tau^{-1})$, while the usual low-information prior distributions are used for the model parameters β_1, β_2, and τ: $\beta_t \sim N(0, 1000)$ for $t = 1,2$ and $\tau \sim \text{gamma}(10^{-4}, 10^{-4})$. Finally, we use logical nodes to define the odds ratio as $OR = e^{\beta_2}$ and the standard deviation of the random effects as $\sigma = \sqrt{1/\tau}$. The full WinBUGS code and the data specification are provided in Table 9.12.

Table 9.12   WinBUGS code for modeling the two correlated binary responses in matched-pair clinical trial data of Example 9.6

```
model{
    for (k in 1:4){
        for (i in (F[k]+1):F[k+1]){
        # model for the 1st individual (Treatment A) of pair i
        y1[i]<-Y1[ k ]                  # definition of y1
        y1[i]~dbern( p1[i] )            # stochastic component
        logit(p1[i])<- b[i] +beta[1] # linear predictor and link
        # model for the 2nd individual (Treatment B) of pair i
        y2[i]<-Y2[ k]                   # definition of y2
        y2[i]~dbern( p2[i] )              # stochastic component
        logit(p2[i])<- b[i] + beta[1]+beta[2]# linear predictor & link
        # random effects for individuals belonging in i pair
        b[i]~dnorm( 0, tau)
    }}
    # PRIORS
    tau~dgamma( 0.001, 0.001)# prior for the precision of random effects
    beta[1]~dnorm(0,0.001)    # prior for beta[1] (fixed effect)
    beta[2]~dnorm(0,0.001)    # prior for beta[2] (fixed effect)
    # LOGICAL NODES
    or<-exp(beta[2])   # definition of odds ratio
    sigma <- sqrt(1/tau)        # standard deviation of the random effects
    # estimating variance components
    s2<-1/tau
    r <- s2/(s2 + pow(pi,2)/3)
    ss.res <-    sum(sq.res1[1:n]) + sum(sq.res2[1:n])
    overallp <- (sum(y1[1:n]) + sum(y2[1:n]))/(2*n)
    for (i in 1:n){
        ssy1[i] <- pow( y1[i]-overallp, 2)
        ssy2[i] <- pow( y2[i]-overallp, 2)
    }
    ss.y   <- sum( ssy1[1:n] ) + sum( ssy2[1:n] )
    R2bin <- 1 - ss.res/ss.y
}
DATA
list( F=c(0,7, 9,23,60), Y1=c(0,1,0,1), Y2=c(0,0,1,1) )
# F are accumulated counts
# actual frequency data are f=c(7,2,14,37)
```

Posterior summaries of the estimated model parameters, using a sample of 3000 iterations and 1000 as burnin period are provided in Table 9.13. The log-odds ratio (β_2) is equal to $\sim$2.0 (mean equal to 2.1, median equal to 2.0, and rough estimate of the posterior mode equal to 1.87), which is very close to the sample estimate $\log\left(\frac{14}{2}\right) = 1.95$. Since the posterior distribution of the odds ratio is highly skewed, we can use the posterior geometric mean, the median as point estimates, or the mode as point estimates giving values equal to $e^{2.1} = 8.17, 7.46$ and $e^{1.87} = 6.49$, respectively. The corresponding 95% posterior interval ranges from 1.95 to 54.04, indicating better performance of the second treatment, whereas using the posterior median, the second treatment is at least 6.5 times more effective than the first one with probability 0.5.

Table 9.13 Posterior summaries for parameters of model used for the two correlated binary responses in matched-pair clinical trial data of Example 9.6

node	mean	sd	MC error	2.5%	median	97.5%	start	sample
beta[1]	1.25	0.65	0.043	0.16	1.17	2.67	1001	3000
beta[2]	2.10	0.85	0.062	0.67	2.00	3.99	1001	3000
or	12.77	21.77	1.260	1.95	7.46	54.04	1001	3000
sigma	2.66	1.16	0.115	0.81	2.53	5.48	1001	3000
r	0.62	0.19	0.021	0.17	0.66	0.90	1001	3000
R2bin	0.45	0.13	0.012	0.17	0.46	0.66	1001	3000

The posterior mean of the matched-pair random effect σ is equal to 2.66. Using the latent variance approach, the within-patient correlation of the latent "sickness" variable is equal to 0.62, indicating an important correlation between-subject that cannot be ignored. The R^2-type statistic has posterior mean equal to 45%, which suggests a large improvement over the corresponding model with no random effects (posterior mean R^2-type statistic equal to 3.8%).

Example 9.7. Modeling correlation in questionnaire binary responses: Schizotypal personality questionnaire. According to the DSM-III-R diagnostic and statistical manual of mental disorders edited by the American Psychiatric Association (1987), a "schizotypal" person suffers from minor episodes of "pseudoneurotic" problems. The prevalence rate of schizotypy is about 10% in the general population. Study of the schizotypal personality in psychiatric research is of prominent importance for two reasons: (1) shizotypal subjects have increased risk of developing schizophrenia during their lifetimes, and (2) since they are healthy persons, they can participate in psychiatric/psychological research studies (by completing questionnaires — psychometric instruments), which schizophrenics are unable to do. Although several scales have been proposed in the literature, Raine's (1991) SPQ scale, a 74-item self-administered questionnaire, is the most popular questionnaire used to measure the concepts of schizotypal personality. The questionnaire consists of binary zero/one (yes/no) items. It provides subscales for nine schizotypal features as well as an overall scale for schizotypy. Each subscale is calculated as the sum of the questionnaire items that refer to each schizotypal subscale.

Here we analyze the original binary responses of the SPQ scale administered within a small student survey completed in Greece examining the association between schizotypal traits and impulsive and compulsive buying behavior of university students (Iliopoulou, 2004). A total number of 167 students were finally considered here, which is subset of the original full dataset. Here we use the logit mixed model to quantify the within-subject variability. In a second stage, we use information on subscales to improve the model and estimate partial estimates of variability.

We adopt model (9.5) with linear predictor (9.6) with random individual effects b_i and fixed item coefficients β_j. We estimate within-subject variability and correlation using the simple latent variable interpretation. The WinBUGS code for this model is provided in Table 9.14.

Table 9.14   WinBUGS code for simple individual random effects model for SPQ data of Example 9.7

```
model{
    pi<-3.14
    for (i in 1:n){
        for (j in 1:P){
            spq[i,j]~dbern( p[i,j] )
            logit(p[i,j]) <- b[i] + beta[j]

            res[i,j]<- spq[i,j] - p[i,j]
        }
        b[i]~dnorm(0, tau)
    }
    for (j in 1:P){scales[j]~dnorm(0,1)}
    for (j in 1:P){ beta[j]~dnorm( 0.0,  0.001) }
    tau ~dgamma( 0.001, 0.001)

    s2<-1/tau
    s<-sqrt(1/tau)
    r <- s2/(s2+pi*pi/3)

    R2bin <- 1 - pow( sd(res[1:n,1:P])/sd(spq[1:n,1:P]) ,2)
}
```

According to our results, the within-subject correlation of the latent variable is equal to 0.19. Nevertheless, the DIC value of the adopted model is found equal to $12{,}343.3$, which is lower than the corresponding values for models with individual fixed or any individual effects ($12{,}355.400$ and $13{,}553.9$, respectively), indicating that nonzero correlation within-subject measurements exist.

A similar analysis for the examination of random item effects lead to a nondecreasing DIC, indicating a minor correlation between individuals as expected. Adding individual effects increases the R^2-type measure to 0.24 from 0.14, which was the corresponding value for the model with no individual effects.

We can further introduce random effects clustering due to the different subscales which measure different characteristics of the schizotypal personality. These characteristics are: ideas of reference (9), odd beliefs (7), unusual perceptual experiences (9), odd speech (9), suspiciousness (8), constricted affect (8), odd behavior (7), no close friends (9), and social anxiety (8); the number of items of each corresponding subscale is denoted within parentheses. The inclusion of each item within one of the these subscales is coded with values ranging from 1 to 9, respectively. The linear predictor of the model now accommodates individual random effects v_{is_j}, where $s_j \in \{1, 2, \ldots, 9\}$ denotes the subscale to which item j belongs. Hence the linear predictor (9.6) is now formulated as

$$\log\left(\frac{\pi_{it}}{1 - \pi_{it}}\right) = b_i + \beta_t + v_{is_j},$$

with

$$v_{is} \sim N(0, \sigma_{\text{scales},s}^2) \text{ for } i = 1, \ldots, n = 167; \ s = 1, 2, \ldots, 9.$$

This model assumes (under the simple latent variable interpretation) that within-subject variability for items within the same subscales is given by

$$\text{Cor}(Z_{ij}, Z_{ij'}) = \frac{\sigma^2 + \sigma^2_{\text{scales},s}}{\sigma^2 + \sigma^2_{\text{scales},s} + \pi^2/3}, \quad \text{where } s_j = s_{j'} = s,$$

and within-subject variability for items within different subscales is given by

$$\text{Cor}(Z_{ij}, Z_{ij'}) = \frac{\sigma^2}{\sqrt{\sigma^2 + \sigma^2_{\text{scales},s_j} + \pi^2/3}\sqrt{\sigma^2 + \sigma^2_{\text{scales},s_{j'}} + \pi^2/3}}, \quad \text{where } s_j \neq s_{j'}.$$

The latter is formulated in WinBUGS by considering a matrix with ellements r_{k_1,k_2} for $k_1, k_2 \in \{1, 2, \ldots, 9\}$ using the formula

$$r_{k_1,k_2} = \frac{\sigma^2 + \sigma^2_{\text{scales},k_1} I(k_1 = k_2)}{\sqrt{\sigma^2 + \sigma^2_{\text{scales},k_1} + \pi^2/3}\sqrt{\sigma^2 + \sigma^2_{\text{scales},k_2} + \pi^2/3}}$$

and the corresponding WinBUGS syntax

```
for (k in 1:9){ tot.s2.scales[k] <- s2+s2.scales[k]+pi*pi/3   }
for (k1 in 1:9){ for (k2 in 1:9){
    r[k1,k2]    <- (s2+s2.scales[k1]*equals(k1,k2))/
                    sqrt( tot.s2.scales[k1]*tot.s2.scales[k2])
}}
```

where `tot.s2.scales[k]` is the total variability for each latent variable Z_{ij}. The full code for this model is provided in Table 9.15.

DIC $(11, 356.2)$ is much lower for the model with clustering random effects of items because of the underlying subscales. Moreover, the posterior mean of the R^2-type statistic is equal to 0.36, indicating a decrease in residual variability of about 50% and 155% for the models with or without individual random effects, respectively. Posterior means of within-subject correlation is presented in Table 9.16. Within-subject correlations for items of different subscales range from 0.13 to 0.19, while the corresponding correlation of items in the same subscale range from 0.33 to 0.54. Items of odd behavior and social anxiety demonstrate the highest within-subject correlations with posterior means equal to 0.54 and 0.52, respectively, followed by the corresponding correlations for items of odd beliefs and constricted affect with values around 0.42. Results of this model indicate that within-subject correlation of the items belonging in the same subscale changes from item to item and that subscales explain an important percent of the total SPQ variability.

Table 9.15 WinBUGS code for random effects model, including intrasubscale variability for SPQ data of Example 9.7

```
model{
  pi<-3.14
  for (i in 1:n){
      for (j in 1:P){
          spq[i,j]~dbern( p[i,j] )
          logit(p[i,j]) <- b[i] + beta[j] + v[i, scales[j] ]
          res[i,j]<- spq[i,j] - p[i,j]
      }
      b[i]~dnorm(0, tau.ind)
      for (k in 1:9){ v[i,k]~dnorm( 0.0, tau[k] )}
  }
  for (j in 1:P){ beta[j]~dnorm( 0.0, 0.001)   }
  tau.ind ~dgamma( 0.001, 0.001)
  for (k in 1:9){
    tau[k] ~dgamma( 0.001, 0.001)
    s2.scales[k]<-1/tau[k]
    s.scales[k]<-sqrt(1/tau[k])
  }
  s2<-1/tau.ind
  s<-sqrt(1/tau.ind)
  for (k in 1:9){ tot.s2.scales[k] <- s2+s2.scales[k]+pi*pi/3   }
  for (k1 in 1:9){
      for (k2 in 1:9){ r[k1,k2]     <- (s2+s2.scales[k1]*equals(k1,k2))/
                                       sqrt( tot.s2.scales[k1]*tot.s2.
                                             scales[k2]) }
  }
  R2bin <- 1 - pow( sd(res[1:n,1:P])/sd(spq[1:n,1:P]) ,2)
}
```

Table 9.16 Posterior means of within-subject correlations for random effects model, including intrasubscale variability for SPQ data of Example 9.7

	1 Ideas of reference	2 Odd beliefs	3 Perceptual experiences	4 Odd speech	5 Suspic.[a]	6 Constricted affect	7 Odd behavior	8 No close friends	9 Social anxiety
1 Ideas of reference	0.359								
2 Odd beliefs	0.175	0.426							
3 Perceptual experiences	0.188	0.178	0.341						
4 Odd speech	0.188	0.178	0.191	0.338					
5 Suspiciousness	0.181	0.171	0.184	0.184	0.389				
6 Constricted affect	0.177	0.167	0.179	0.179	0.172	0.417			
7 Odd behavior	0.157	0.148	0.159	0.159	0.153	0.149	0.540		
8 No close friends	0.185	0.175	0.188	0.188	0.181	0.176	0.157	0.359	
9 Social anxiety	0.161	0.152	0.163	0.163	0.157	0.153	0.136	0.161	0.518

[a] Suspicoiousness.

Key: Diagonal correlations (r_{kk}) = within-subject correlations for items of same subscale k; $r_{k_1 k_2}$ with $k_1 \neq k_2$ = within-subject correlations for items of the different subscales k_1 and k_2.

9.3.3 Poisson log-linear GLMMs for correlated count data

In this section we consider repeated count/discrete responses Y_{it} for $t = 1, 2, \ldots, T$ over the same individuals (or more generally conditions) $i = 1, 2, \ldots, n$. As we have already mentioned in Section 9.2.3.2, random effects can be simply used to model overdispersion. Hence when we simply use the model

$$
\begin{aligned}
Y_{it} &\sim \text{Poisson}(\lambda_{it}) \\
\log \lambda_{it} &= \mu_{it} + u_{it} \\
\mu_{it} &= \beta_{0t} + \sum_{j=1}^{p} \beta_{jt} X_{ijt} \\
u_{it} &\sim N(0, \sigma^2),
\end{aligned}
\tag{9.7}
$$

where μ_{it} is the part of the linear predictor that involved only the usual fixed effects. This model assumes variance

$$
\begin{aligned}
V(Y_{it}) &= V_u\Big[E(Y_{it}|u_{it})\Big] + E_u\Big[V(Y_{it}|u_{it})\Big] \\
&= e^{\mu_{it}}\Big\{e^{\mu_{it}} V(e^{u_{it}}) + E(e^{u_{it}})\Big\},
\end{aligned}
$$

which simplifies to

$$
V(Y_{it}) = e^{\mu_{it}}\Big\{e^{\mu_{it}}(e^{\sigma^2} - 1)e^{\sigma^2} + e^{\sigma^2/2}\Big\}
\tag{9.8}
$$

when the usual log-normal random effects are used. The model is overdispersed since

$$
E(Y_{it}) = e^{\mu_{it}} E(e^{u_{it}}) = e^{\mu_{it}} e^{\sigma^2/2}
$$

with dispersion index

$$
\text{DI}(Y_{it}) = 1 + e^{\mu_{it}}\text{DI}(e^{u_{it}}) = 1 + e^{\mu_{it}}(e^{\sigma^2} - 1)e^{\sigma^2/2}.
\tag{9.9}
$$

Although the model introduces overdispersion over the response variable, it assumes zero correlation between consequent measurements for individual observations i. For this reason, we may rewrite the linear predictor of the model described above as

$$
\begin{aligned}
\log \lambda_{it} &= \mu_{it} + b_i \tag{9.10} \\
b_i &\sim N(0, \sigma^2), \tag{9.11}
\end{aligned}
$$

where b_i are the individual random effects of observations Y_{it} for $t = 1, 2, \ldots, T$. This model still induces overdispersion, resulting in the same variance and dispersion index as above [see Eqs. (9.8) and (9.9), respectively], but at the same time it introduces correlation between measurements Y_{it_1} and Y_{it_2} of the same individual/observation i since

$$
\begin{aligned}
\text{Cov}(Y_{it_1}, Y_{it_2}) &= E_{b_i}\Big[E(Y_{it_1} Y_{it_2}|b_i)\Big] - E_{b_i}\Big[E(Y_{it_1}|b_i)\Big] E_{b_i}\Big[E(Y_{it_2}|b_i)\Big] \\
&= e^{\mu_{it_1}+\mu_{it_2}} E\big(e^{2b_i}\big) - e^{\mu_{it_1}+\mu_{it_2}}\Big[E\big(e^{b_i}\big)\Big]^2 \\
&= e^{\mu_{it_1}+\mu_{it_2}}(e^{\sigma^2} - 1)e^{\sigma^2},
\end{aligned}
$$

resulting in

$$\text{Cor}(Y_{it_1}, Y_{it_2}) = \frac{e^{\mu_{it_1} + \mu_{it_2}} (e^{\sigma^2} - 1)e^{\sigma^2}}{\sqrt{e^{\mu_{it_1}} \left\{ e^{\mu_{it_1}} (e^{\sigma^2} - 1)e^{\sigma^2} + e^{\sigma^2/2} \right\}}}$$

$$\times \frac{1}{\sqrt{e^{\mu_{it_2}} \left\{ e^{\mu_{it_2}} (e^{\sigma^2} - 1)e^{\sigma^2} + e^{\sigma^2/2} \right\}}}$$

$$= \left\{ \left(1 + \omega e^{-\mu_{it_1}} \right) \left(1 + \omega e^{-\mu_{it_2}} \right) \right\}^{-1/2} \text{ with } \omega = (e^{\sigma^2} - 1)^{-1} e^{-\sigma^2/2}.$$

From this expression, the within-subject correlation depends on the fixed linear predictor μ_{it} and therefore on the covariate values. Indicative values will be reported on the basis of sample means, sample minimum and maximum values of the observed covariates.

Additional random effects can be added depending on the problem at hand. Although variance can be decomposed using arguments similar to those used above, more complicated hierarchical structure introduced in the model results in more complicated computations concerning the variance components.

We now present a simple example with data from water polo and the corresponding scores from the World Cup of year 2000. We use the hierarchical formulations above to model overdispersion and correlation between the two competitors.

Example 9.8. Modeling water polo World Cup 2000 data. Let us consider results of the water polo tournament held at the city of Fukuoka, Japan during July 2001 for the "9th World Swimming Championships." A total of 16 national water polo teams competed with each other. Initially four round-robin groups with four teams in each group were formed. The best two teams from each group qualified for a second round robin, and best two teams were qualified for the semifinals, and so on. The score of each team refers to the number of goals scored by each team within a game. Poisson seems to be a realistic assumption for such data. Usual scores are around eight goals for each team, with a strong correlation between the scores of the competing teams. Here we consider the following formulation:

$$\begin{aligned}
Y_{it} &\sim \text{Poisson}(\lambda_{it}) \text{ for } t = 1, 2 \\
\log \lambda_{it} &= \mu_{it} + \gamma_1 u_{it} + \gamma_2 b_i \\
\mu_{i1} &= \mu + a_{\text{team}_{1i}} + d_{\text{team}_{2i}} \\
\mu_{i2} &= \mu + a_{\text{team}_{2i}} + d_{\text{team}_{1i}} \\
u_{it} &\sim N(0, \sigma_u^2) \\
b_i &\sim N(0, \sigma_b^2).
\end{aligned}$$

Here team_{1i} and team_{2i} indicate the two teams competing each other, the usual sum-to-zero constraints were imposed on both attacking and defensive parameters a_k and d_k, and parameters γ_1 and γ_2 are only binary indicators used to identify which model is fitted each time. Hence we consider four models: the fixed effects model with $(\gamma_1, \gamma_2) = (0, 0)$, the overdispersed Poisson model with no correlation with $(\gamma_1, \gamma_2) = (1, 0)$, the overdispersed correlated Poisson model $(\gamma_1, \gamma_2) = (0, 1)$, and model with both game and individual random effects $(\gamma_1, \gamma_2) = (1, 1)$.

The DIC results and R^2-type statistics based on the residuals of each model are presented in Table 9.17. The latter was calculated using the following equation within each MCMC

Table 9.17 Comparison of models for water polo data of Example 9.8

Model (γ_1, γ_2)		DIC	$E(R^2_{\text{Pois}}|\boldsymbol{y})$
1	Fixed effects $(0,0)$	590.3	0.291
2	Individual RE $(1,0)$	591.0	0.322
3	Game RE $(0,1)$	587.2	0.376
4	Game + individual RE $(1,1)$	587.9	0.385

Key: RE = random effects; $E(R^2_{\text{Pois}}|\boldsymbol{y})$ = posterior mean of R^2_{Pois}.

iteration:

$$R^2_{\text{Pois}} = 1 - \frac{\sum_{i=1}^{n}\sum_{t=1}^{2} (y_{it} - \lambda_{it})^2}{\sum_{i=1}^{n}\sum_{t=1}^{2} (y_{it} - \overline{y})^2}.$$

Thus we compare each model with the constant one. The third model, with game random effects, is indicated as the best-fitted model according to DIC. This implies that the goals scored in a game by the two competing teams demonstrate a nonzero correlation. The WinBUGS code for model with game specific random effects is given in Table 9.18.

Table 9.18 WinBUGS code for game-specific random effects model for water polo data of Example 9.8

```
model{
  for (i in 1:n){
       g[i,1] <- G1[i]
       g[i,2] <- G2[i]
       g[i,1]~dpois(  lambda[i,1]  )
       g[i,2]~dpois(  lambda[i,2]  )
       log( lambda[i,1] )<- mu + att[ team1[i] ] + def[ team2[i] ] +b[i]
       log( lambda[i,2] )<- mu + att[ team2[i] ] + def[ team1[i] ] +b[i]
       res2[i,1] <- pow( g[i,1] - lambda[i,1] , 2)
       res2[i,2] <- pow( g[i,2] - lambda[i,2] , 2)

       b[i]  ~ dnorm( 0, tau)
  }
  mu~dnorm( 0, 0.001)
  att[1]   <-  -sum(att[2:teams])
  def[1] <-  -sum(def[2:teams])
  for (k in 2:teams){
       att[k]~dnorm( 0, 0.001)
       def[k]~dnorm( 0, 0.001)
  }
  tau~dgamma( 0.001, 0.001)
  ss.res <- sum(res2[1:n,1:2])
  ss.y   <- pow( sd(g[1:n,1:2]), 2)*(2*n-1)
  R2pois <- 1 - ss.res/ss.y
  for(i in 1:n){
     game[i]~dnorm(0,1)
     phase[i]~dnorm(0,1)
  }
}
```

The posterior mean of the constant is equal to 1.84, which corresponds to about 6.3 expected goals when two teams of average attacking and defensive abilities compete with each other. Note that positive attacking and negative defensive parameters indicate teams with attacking and defensive skills better than those of an "average"-level team in the tournament (denoted by the zero value). Estimates of the attacking and defensive parameters are presented in Figures 9.7 and 9.8. From these figures we observe that Spain (the contest winner) had the best overall defensive parameter but only the fifth overall best attacking parameter. Russia (third in the competition) had the best attacking parameter but its defensive parameter is slightly worse than the average defensive level. Similarly, Italy (fourth in the competition) had the second best defensive parameter, but its attacking parameter was average. Yugoslavia (second in the league) had the fourth and the third best attacking and defensive parameters, respectively. We can compare the performance of the remaining teams in the same manner. Estimation of the full league under the fitted model can be performed using the predictive distribution; for mored details, see Section 7.4.3.5 and Chapter 10.

To understand the effect of the random coefficients on the imposed variance and correlation of the water polo data, we have calculated all individual correlations and dispersion indices using the following WinBUGS syntax:

```
es <- exp( 1/tau )
omega <- 1/( (es-1)*sqrt(es) )
for (i in 1:n){
    m[i,1] <- exp( mu + att[ team1[i] ] + def[ team2[i] ] )
    m[i,2] <- exp( mu + att[ team2[i] ] + def[ team1[i] ] )
    DI[i,1] <-    1 + m[i,1]* (es-1)*sqrt(es)
    DI[i,2] <- 1 +   m[i,2]* (es-1)*sqrt(es)
    cor[i] <- 1/sqrt( (1+omega/m[i,1])*(1+omega/m[i,2]) )
}
```

Finally, we have calculated the minimum, median, mean, maximum, and the quartiles over all games of the imposed dispersion index and the within-game correlations for each iteration. Their posterior summaries are presented in Table 9.19. The correlation averaged around 0.14, the maximum value is found equal to 0.20, while the variance is overdispersed by 19% on average or by 46% at the maximum value of the sample.

Table 9.19 Posterior summaries of selected within-game correlations and dispersion indices for water polo data of Example 9.8

	Posterior summaries					
	Correlation			Dispersion index		
		Percentiles			Percentiles	
Sample statistic	Mean	2.5%	97.5%	Mean	2.5%	97.5%
Minimum	0.078	0.003	0.217	1.051	1.002	1.171
2.5% percentile	0.116	0.005	0.307	1.132	1.005	1.413
Mean	0.136	0.006	0.353	1.188	1.007	1.590
Median	0.136	0.006	0.353	1.179	1.006	1.559
97.5% percentile	0.154	0.007	0.395	1.231	1.008	1.731
Maximum	0.198	0.010	0.486	1.458	1.015	2.528

Note: Sample statistics related to the corresponding values of correlation and variance over all calculated values within each MCMC iteration.

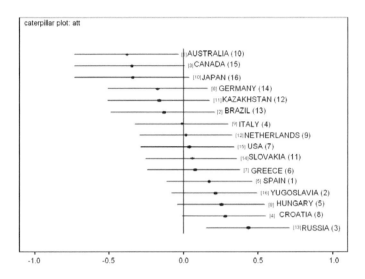

Figure 9.7   95% posterior intervals of attacking parameters for water polo data of Example 9.8

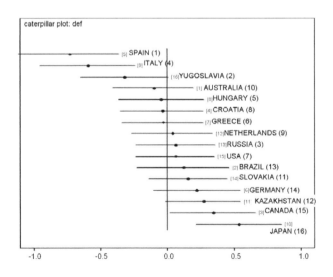

Figure 9.8   95% posterior intervals of defensive parameters for the water polo data of Example 9.8

9.4   DISCUSSION, CLOSING REMARKS, AND FURTHER READING

In this chapter we have briefly introduced the reader to the basic notions of hierarchical modeling. The term *hierarchical models* refers more to a general set of modeling principles than to a specific family of models. The basic idea of hierarchical modeling (also known as *multilevel, repeated measures, mixed,* or *longitudinal models*) is to organize the model using a set of sequential statements of conditional relationships. The levels of these models are usually specified by the existence of several nested units of observations such as, for example, patients, hospitals, and geographic areas of the hospitals and countries. Each unit in this set may contribute effectively in the total observed variability and can be modeled by introducing a set of random effects coming from a general population distribution. Bayesian theory and hierarchical modeling are closely related since even simple Bayesian models embed the idea of hierarchy because of the conditional structure of the posterior distribution, which is decomposed to the data likelihood multiplied by the prior distribution.

In this chapter we focused on simple hierarchical models that mainly included random intercept models. The use of hierarchical models (or multilevel, repeated measures, mixed, or longitudinal models) is valuable in statistics since they are based on the logic of simple generalized linear models by including random terms in the linear predictor that can be used to introduce dependence or overdispersion or to simply change the resulted marginal sampling distribution of the data (i.e., the likelihood). They are also used to combine information between different observations or studies or introduce a clustering effect on observations within the data.

Details concerning classical hierarchical models can be found in Agresti (2002, chap. 11) along with a very nice review at the end of the same chapter. Interesting books on the subject include the more recent ones by Hedeker and Gibbons (2006), Brown and Prescott (2006), and de Leeuw and Meijer (2008), while a variety of papers on the topic can be found in the related special issue of the *Journal of Educational and Behavioral Statistics* (vol. 20, issue no. 2, 1995), including the fruitful critical review by Professor D. Draper (Draper, 1995).

Concerning Bayesian hierarchical models, their basic notions have been introduced by Lindley and Smith (1972) within the framework of the normal linear model. Computational developments in the 1990s and MCMC algorithms have made Bayesian hierarchical models very popular; see Zeger and Karim (1991) for implementation of the Gibbs sampler in hierarchical models. Bayesian generalized linear mixed, hierarchical longitudinal, and nonlinear hierarchical models are presented by Clayton (1996), Carlin (1996), and Bennet et al. (1996) in Gilks et al. (1996). A nice and comprehensive review can be found in Seltzer et al. (1996). Details on hierarchical modeling are also provided by Gelman et al. (2004, chaps. 5, 13) and Carlin and Louis (2000, sec. 7.4). Extensive treatment of the subject is also presented by several authors in Dey et al. (2000), while more recently Gelman and Hill (2006) have published a very nice book on the topic. Special topics within the framework of hierarchical models have been examined by Professor Gelman and his associates, including the specification of the prior distribution for variance components (Gelman, 2006), measures for variance components models (Gelman and Pardoe, 2006), and the use of redundant parametrizations (Gelman et al., 2008).

In the wide class of the hierarchical model, we can also include the latent variables models [see Bartholomew and Knott (1999) for a general treatment of the topic and Dunson (2000), Dunson and Herring (2005) for details concerning the Bayesian implementation], errors in variable models [for a nice review of the Bayesian implementation, see Wakefield and

Stephens (2000)], and models for time series data where the parameters change dynamically [see Aguilar et al. (1999) for a nice review and West and Harrison (1997) for more details].

Finally, additional examples of hierarchical models implemented in WinBUGS can be found in the three volumes of WinBUGS examples:

1. In the first volume of examples (Spiegelhalter et al., 2003*a*) the following datasets are available

   - `rats` (normal hierarchical model)
   - `pump` (gamma–Poisson hierarchical model)
   - `seeds` (random effects logistic regression)
   - `surgical`, used for institutional ranking (the second model)
   - `salm`, used to model extra-Poisson variation in a dose–response study
   - `equiv`, used to model bioequivalence in a crossover trial data
   - `dyes` (variance components model)
   - `epil` (repeated measures on Poisson counts)
   - `blocker` (random effects meta-analysis of clinical trials)
   - `Oxford` (smooth fit to log-odds ratios in case control studies)
   - `LSAT` (latent variable models for item–response data)
   - `inhalers` (random effects model for ordinal responses from a crossover trial)
   - `kidney` (Weibull regression with random effects)
   - `leuk` (survival analysis using Cox regression with frailties)

2. In the second volume (Spiegelhalter et al., 2003*b*), the following examples refer to hierarchical models:

   - `orange trees` (a hierarchical, nonlinear model)
   - `air` (covariate measurement error)
   - `cervix` (case–control study with errors in covariates)
   - `jaw` (repeated measures analysis of variance)
   - `birats` (a bivariate normal hierarchical model)
   - `schools` (multivariate hierarchical model of examination results)

3. The example `hepatitis` (random effects model with measurement error) can be found in the third volume of examples (Spiegelhalter et al., 2003*c*).

Further examples of Bayesian hierarchical models for repeated ordinal data (implemented in WinBUGS) are also provided by Qiu et al. (2002). Lawson et al. (2003) present the basic notions of Bayesian hierarchical models and multilevel modeling (see chaps. 2 and 3 respectively), while implementation using WinBUGS and MLWin is also illustrated in detail. WinBUGS code for the fitting hierarchical models for meta-analysis can be found in Woodworth (2004, chap. 11). Numerous examples are also provided in the Congdon's books on Bayesian models Congdon (2003; 2005*a*; 2006*b*). Finally, Lynch (2007, chap. 9) provides examples of WinBUGS hierarchical models, including random coefficient and growth curve models.

Problems

9.1 Use a Poisson log-linear model with random effects
  a) For the water polo data of Problem 7.3.
  b) For the Italian soccer/football data of Problem 7.4.

9.2 Use a hierarchical model for the data of the WinBUGS example `salm` (Spiegelhalter et al., 2003*a*) used in Problem 7.5.
  a) Compare the results with the corresponding ones from a model without random effects.
  b) What do we achieve here by using random effects?

9.3 Use a hierarchical model for the data of the WinBUGS example `seeds` (Spiegelhalter et al., 2003*a*) used in Example 7.6.
  a) Compare the results with the corresponding ones from a model without the random effects.
  b) What do we achieve here by adopting random effects?

9.4 Use a hierarchical model for the data of the WinBUGS example `rats` (Spiegelhalter et al., 2003*a*).
  a) What do we achieve here by using random effects?
  b) Calculate the within-rat correlation.

9.5 Use a hierarchical model for the data of the WinBUGS example `surgical` (Spiegelhalter et al., 2003*a*).
  a) Compare the random and the fixed effects models.
  b) What do we achieve here by using the random effects?

9.6 Use the data of the WinBUGS example `kidney` to construct a hierarchical random effects model for the survival times. Estimate the within-subject variability and correlation in this case.

CHAPTER 10

THE PREDICTIVE DISTRIBUTION AND MODEL CHECKING

10.1 INTRODUCTION

10.1.1 Prediction within Bayesian framework

In Bayesian theory, predictions of future observables are based on *predictive distributions*, that is, the distribution of the data averaged over all possible parameter values. For this reason, when data y have not been observed yet, predictions are based on the marginal likelihood

$$f(y) = \int f(y|\theta)f(\theta)d\theta, \tag{10.1}$$

which is the likelihood averaged over all parameter values supported by our prior beliefs. Hence, $f(y)$ is also called *prior predictive distribution*.

Usually, after having observed data y, one finds the prediction of future data y' more interesting. Following this logic, we calculate the posterior predictive distribution

$$f(y'|y) = \int f(y'|\theta)f(\theta|y)d\theta, \tag{10.2}$$

which is the likelihood of the future data averaged over the posterior distribution $f(\theta|y)$. This distribution is termed as the *predictive distribution* since prediction is usually attempted only after observation of a set of data y.

Future observations y' can be alternatively viewed as additional parameters under estimation. From this perspective, the joint posterior distribution is now given by $f(y', \theta|y)$.

Bayesian Modeling Using WinBUGS, by Ioannis Ntzoufras
Copyright ©2009 John Wiley & Sons, Inc.

Inference on the future observations y' can be based on the marginal posterior distribution $f(y'|y)$ by integrating out all nuisance parameters, one of which in this case, is the parameter vector θ. Hence, the predictive distribution is given by

$$
\begin{aligned}
f(y'|y) &= \int f(y', \theta|y)d\theta \\
&= \int f(y'|\theta, y)f(\theta|y)d\theta,
\end{aligned}
$$

resulting in (10.2) since past and future observables (y and y', respectively) are conditionally independent given the parameter vector θ.

According to Press (1989, pp. 57–58), inference must be based on predictive distributions since they involve observables while the posterior distribution also involves parameters that are never observed. Hence, by using the predictive distribution, we can quantify our knowledge about future as well as measure the probability of reobserving each y_i assuming that the adopted model is true. For this reason, we may use the predictive distribution not only to predict future observations but also to construct goodness-of-fit diagnostics and perform model checks for each model's structural assumptions.

10.1.2 Using posterior predictive densities for model evaluation and checking

The posterior predictive density $f(y'|y, m)$ of a model m is frequently used for checking the assumptions of a model and its goodness-of-fit. The main reason is that we can easily generate replicated values y^{rep} from the posterior predictive distribution by adding a single simple step within any MCMC sampler using the likelihood function $f\left(y^{\text{rep}}|\theta^{(t)}\right)$ evaluated at parameter values $\theta^{(t)}$ of the current state of the algorithm; see Section 10.2. The predictive data y^{rep} reflect the expected observations after replicating our experiment in the future, having already observed y and assuming that the adopted model is true. If the adopted model is appropriate for describing the observed data, then vectors y and y^{rep} will be close. Hence, a comparison of these two vectors provides information concerning the fit of the model and possible outliers. Such a comparison can be facilitated by considering summary functions $D(y, \theta)$, which play the role of a test statistic for checking the assumption under investigation and measure discrepancies between the data and the model (Gelman et al., 1996). Assessment of the posterior distributions of $D(y^{\text{rep}}, \theta)$ and $D(y, \theta)$ provides individual as well as overall goodness-of-fit measures that can be summarized graphically or using tail area probabilities called *posterior predictive p-values* (Meng, 1994) given by

$$
\text{Posterior p-value} = P\left(D(y^{\text{rep}}, \theta) > D(y, \theta)\Big|y\right).
$$

Posterior p-values are posterior probabilities and can be directly interpreted as the probability of observing in the future samples with $D(y, \theta)$ higher than the one already observed. Values around 0.5 indicate that the distributions of the replicated and actual data are close, while values close to zero or one indicate differences between them (Gelman and Meng, 1996, p. 191); additional details can also be found in Bayarri and Berger (2000). Posterior p-values must not be used or interpreted as the "probability that the current model is true" (Gelman and Meng, 1996, p. 192). Although the scaling of p-values is more familiar, it is not clear which cutpoints must be used to identify inadequate models, and further calibration is needed (Sellke et al., 2001).

In the preceding model checks, data are used twice: first for estimation of the posterior predictive density and second for comparison between the predictive density and the data. Although such comparisons clearly violate the likelihood principle, Meng (1994) argues in favor of "posterior predictive checks" provided that they are used only as measures of discrepancy between the model and the data in order to identify poorly fitted models (model adequacy) and not for model comparison and inference (Carlin and Louis, 2000, p. 48).

We can divide model checks into individual and overall diagnostics. Individual checks are based on each y_i and y_i^{rep} separately, and their aim is to trace outliers or surprising observations under the assumed model. Overall predictive diagnostics aim to check more general assumptions of the model such as normality or the goodness-of-fit of the model.

Steps in a selection of the possible posterior predictive checks are listed here:

1. Plot and compare the frequency tabulations of y^{rep} and y (for discrete data).

2. Plot and compare the cumulative frequencies of y^{rep} and y (for continuous data).

3. Plot and compare ordered data $\left(y_{(1)}^{\text{rep}}, \ldots, y_{(n)}^{\text{rep}} \right)$ and $\left(y_{(1)}, \ldots, y_{(n)} \right)$ for continuous data.

4. Plot estimated posterior predictive ordinate $f(y_i|y)$ against y_i to trace surprising values. The posterior predictive ordinate

$$\text{PPO}_i = f(y_i|y) = \int f(y_i|\boldsymbol{\theta}) f(\boldsymbol{\theta}|y) d\boldsymbol{\theta} \qquad (10.3)$$

provides us the probability of again observing y_i after having observed y. Low values of $f(y_i|y)$ indicate observations originating from the tail areas of the assumed distribution, while extremely low values indicate possible outliers. A large amount of y_i with small PPO may indicate a poorly fitted model. The scaling of PPO depends on the structure of the assumed data distribution. Thus, no general rule for identifying surprising values using PPOs can be adopted. For this reason, each PPO value must be monitored relative to other PPO values such as, for example, their maximum values. A graph of PPOs versus y_i considerably simplifies their evaluation.

5. Use test statistics and tail area probabilities (posterior p-values) to quantify differences concerning:

    a. Outliers: We may identify outliers using individual test statistics on the basis of residual values; see Section 10.3.5 for more details.

    b. Certain structural assumptions of the model: We may use global test statistics to assess the plausibility of structural assumptions of the model. For example, we may compare the skewness and the kurtosis of y^{rep} with the corresponding observed measures in order to check for the plausibility of assuming a mesokurtic and symmetric distribution for y; see a similar example in Spiegelhalter et al. (1996a, pp. 43–47).

    c. Fitness of the model: Usual measures are the χ^2 type of statistic (Gelman et al., 1995, p. 172)

$$\chi^2(y, \boldsymbol{\theta}) = \sum_{i=1}^{n} \frac{\left[y_i - E(Y_i|\boldsymbol{\theta}) \right]^2}{Var E(Y_i|\boldsymbol{\theta})}$$

or the deviance measure

$$\text{Deviance}(\boldsymbol{y}, \boldsymbol{\theta}) = -2 \sum_{i=1}^{n} \log f(y_i|\boldsymbol{\theta}) \, .$$

For more details and further statistics used for checking the goodness-of-fit, see Section 10.3.7.

10.1.3   Cross-validation predictive densities

Although the full predictive distribution $f(\boldsymbol{y}'|\boldsymbol{y})$ is useful for prediction, its use for model checking is questionable because of their double use of the data. For this reason, several authors (Gelfand, Dey and Chang, 1992; Gelfand, 1996; Vehtari and Lampinen, 2003; Draper and Krnjajić, 2006) have proposed the use of cross-validatory predictive densities. In this approach, the full set of data $\boldsymbol{y}$ is divided to in two subsets $(\boldsymbol{y}_1, \boldsymbol{y}_2)$. The first set $\boldsymbol{y}_1$ is used to fit the model and estimate the posterior distribution of interest, while the remaining observations $\boldsymbol{y}_2$ are used for model evaluation and checking by calculating the cross-validatory predictive density:

$$f(\boldsymbol{y}_2|\boldsymbol{y}_1) = \int f(\boldsymbol{y}_2|\boldsymbol{\theta}) f(\boldsymbol{\theta}|\boldsymbol{y}_1) d\boldsymbol{\theta} \, .$$

One major difficulty in this approach is the selection of $\boldsymbol{y}_1$ and $\boldsymbol{y}_2$. Moreover, different splits of the data provide different results making the problem difficult to handle. A solution is proposed by Geisser and Eddy (1979), who proposed using the leave-one-out cross-validation (CV-1) predictive density

$$f(y_i|\boldsymbol{y}_{\setminus i}) = \int f(y_i|\boldsymbol{\theta}) f(\boldsymbol{\theta}|\boldsymbol{y}_{\setminus i}) d\boldsymbol{\theta},$$

where y_i is the ith observation of $\boldsymbol{y}$ and $\boldsymbol{y}_{\setminus i}$ is $\boldsymbol{y}$ after omitting y_i. This quantity is also known as the conditional predictive ordinate (CPO) (Gelfand, 1996). It can provide a quantitative measure for the effect of observation i on the overall prior predictive density $f(\boldsymbol{y})$ since

$$\text{CPO}_i = f(y_i|\boldsymbol{y}_{\setminus i}) = \frac{f(\boldsymbol{y})}{f(\boldsymbol{y}_{\setminus i})} \, .$$

This quantity is equivalent to the posterior predictive ordinate (PPO) given in (10.3), and it also provides a measure for tracing outliers. Small CPOs indicate observations that are not expected under the cross-validation predictive distribution of the current model.

An overall measure of fit can be constructed by by the product of CPOs (cross-validation predictive likelihood); for more details, see Chapter 11.

10.2   ESTIMATING THE PREDICTIVE DISTRIBUTION FOR FUTURE OR MISSING OBSERVATIONS USING MCMC

Let us consider a usual normal regression model as defined in Section 5.2 and a (unknown) future observation Y_{n+1} with (known) covariate values $\boldsymbol{x}_{(n+1)} = (x_{n+1,1}, x_{n+1,2}, \dots, x_{n+1,p})$.

Then we can estimate its expected value $E\left(Y_{n+1} | \boldsymbol{y}, \boldsymbol{x}_{(n+1)}\right)$ using the predictive distribution

$$f\left(y_{n+1} | \boldsymbol{y}, \boldsymbol{x}_{(n+1)}\right) = \int f\left(y_{n+1} | \boldsymbol{\beta}, \sigma^2, \boldsymbol{x}_{(n+1)}\right) f(\boldsymbol{\beta}, \sigma^2 | \boldsymbol{y}) \, d\boldsymbol{\beta} \, d\sigma^2 \ .$$

Quantity y_{n+1} can considered as an additional parameter under estimation. Thus it can be generated within an MCMC scheme from the conditional posterior distribution

$$f\left(y_{n+1} | \boldsymbol{\beta}, \sigma^2, \boldsymbol{y}, \boldsymbol{x}_{(n+1)}\right) = f\left(y_{n+1} | \boldsymbol{\beta}, \sigma^2, \boldsymbol{x}_{(n+1)}\right)$$

since Y_i are independent, identically distributed (i.i.d.) random variables. Hence, we only need to generate y_{n+1} from the distribution assumed by its model structure with the appropriate parameter values. For the usual normal regression model, we can generate y_{n+1} in the tth iteration of the algorithm by

$$y_{n+1}^{(t)} \sim N\left(\mu_{n+1}^{(t)}, \sigma^{2,(t)}\right) \ \text{with} \ \mu_{n+1}^{(t)} = E\left(Y_{n+1} | \mu_{n+1}^{(t)}, \boldsymbol{x}_{(n+1)}\right) = \beta_0^{(t)} + \sum_{j=1}^{p} \beta_j x_{n+1,j}^{(t)} \ .$$

In WinBUGS we need only to define an additional stochastic node ynew

```
ynew ~ dnorm( munew , tau )
munew <- beta0 + inprod( beta[] , xnew[] )
```

where xnew[] is the vector with elements of the explanatory values for the future (to-be-estimated) response. To complete the specification of the additional nodes, we need to specify xnew in the data of the WinBUGS model code. Moreover, in the data, we must specify that the value of ynew is not available by setting ynew=NA in the list data format. As we have already mentioned, ynew is treated in a manner similar to that used for parameters. A simple monitor of the posterior distribution of ynew produces a sample from the posterior predictive distribution, enabling us to calculate posterior summaries, density plots, and other properties.

Otherwise, we may incorporate additional to-be-predicted observations in the data vector $\boldsymbol{y}$ by substituting specific elements with NA values. Therefore, within the Bayesian framework (and WinBUGS), both future and missing observations are handled in a similar manner using the predictive distribution.

10.2.1 A simple example: Estimating missing observations

Let us consider again the schizotypal personality data of Example 5.3 and the model fitted using the WinBUGS code of Table 5.15. In order to estimate the missing observations, we substitute values 24 and 13 with NA values. Hence the data can now be written as

```
list( LA=3, LB=2, K=2,
     y = structure(
               .Data=c( 9, 14, 18, 29,
                       25, 26, 22, 25,
                       23, NA, 12, NA),
              .Dim = c( 3,2,2 )
              )
     )
```

in list format or as

```
list( LA=3, LB=2, K=2)
 y[,1,1] y[,1,2] y[,2,1] y[,2,2]
```

```
      9        14        18        29
     25        26        22        25
     23        NA        12        NA
END
```

in rectangular (data) format.

After compiling and running the model, posterior summaries of y provide values only for the missing (i.e., stochastic) components of this vector; see Table 10.1 and Figure 10.1 for posterior summaries and densities, respectively.

Table 10.1 Posterior summaries for missing values y_{312} and y_{322} in Example 5.3

node	mean	sd	MC error	2.5%	median	97.5%	start	sample
y[3,1,2]	22.76	8.131	0.2267	5.568	22.87	39.09	1001	2000
y[3,2,2]	12.21	8.2	0.187	-4.321	12.03	29.99	1001	2000

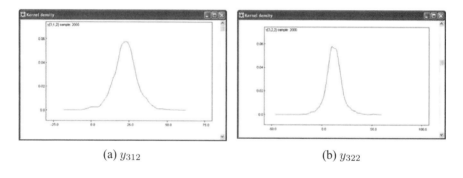

(a) y_{312} (b) y_{322}

Figure 10.1 Posterior predictive densities for y_{312} and y_{322} in Example 5.3.

According to this estimate, males and females living in villages with high income are a posteriori expected to score about 23 and 12 points in the SPQ scale, respectively. From the analysis presented above, a major deficiency of the proposed model is revealed. The normal distribution is not appropriate for such data since it assumes that the response variable also assumes negative values. Thus, especially for the second case, which corresponds to females, the model assigns to negative values a considerable posterior probability. For this reason, we may transform the response using the log-SPQ score to ensure that positive values for the predicted SPQ scores will be estimated; this is equivalent to using the log-normal distribution instead.

In order to fit the log-transformed data (with missing values), we have two alternatives to: either insert directly the log-transformed data with NAs in the missing observations or change the response distribution to log-normal using the command y[i]~dlnorm(mu[i],tau). Posterior summaries and densities of the predictive distributions of y_{312} and y_{322} using the log-transformed data are provided in Table 10.2 and Figure 10.2, respectively. According to the results, we expect that males and females leaving in villages with high income are a posteriori expected to score about 25 and 14 points in the SPQ scale respectively. The 95% posterior predictive intervals are much wider using this model, providing sensible values.

From Figure 10.2 we observe that a problem similar to that in the case of the simple normal model arises since values higher than 74 have a minor but positive probability of appearing. Hence, an even more realistic model can be based on the binomial distribution and the logistic regression model, resulting in a discrete predictive distribution; see Chapter 7.

Table 10.2 Posterior summaries for missing values y_{312} and y_{322} in Example 5.3 using log-transformed data

node	mean	sd	MC error	2.5%	median	97.5%	start	sample
y[3,1,2]	25.32	13.55	0.3768	8.704	22.89	56.77	1001	2000
y[3,2,2]	13.53	8.499	0.1969	4.652	12.08	31.72	1001	2000

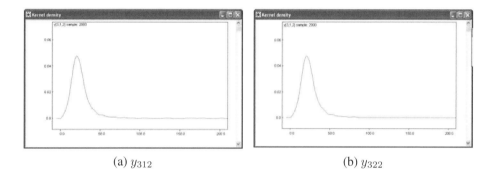

(a) y_{312} (b) y_{322}

Figure 10.2 Posterior predictive densities for y_{312} and y_{322} in Example 5.3 using log-transformed data.

10.2.2   An example of Bayesian prediction using a simple model

Example 10.1. Outstanding car insurance claim amounts. Table 10.3 lists the amounts (in thousand euros) paid per year for car accidents from an insurance company in Greece, while inflation factors are provided in Table 10.4 (Ntzoufras and Dellaportas, 2002). Original amounts were transformed to Euros to make their scale familiar to the current currency of the European Union. Data are provided in a tabular form, with rows indicating in the actual accident year of the claim amounts (factor A) and columns indicating years that the compensation was delayed (factor B). Thus, y_{ij} denotes the amount that the company paid in year i with delay of $j - 1$ years. Delaying pending claims is a common practice for insurance companies. Such delays usually occur for various administrative reasons. It is of central interest in actuarial practice to precisely estimate such quantities in order to reserve appropriate amounts and be able to pay its obligations to customers. For this reason, this problem has attracted the attention of many researchers.

Assuming that these data were available at the end of year 1995, the lower triangle in Table 10.3 displays future payments that were not available at the time when the data were collected. Hence, in this example, we focus on the estimation of amounts missing from the lower triangle of Table 10.3.

Table 10.3   Claim amounts data (in thousands of Euros) for Example 10.1

	Year	1	2	B 3	4	5	6	7
	1989	1546.6	647.5	382.2	246.8	211.8	63.6	146.3
	1990	2099.0	1001.8	487.2	293.0	318.8	269.9	
	1991	3422.2	1257.1	488.4	456.0	562.4		
A	1992	4948.8	1899.7	984.0	1253.3			
	1993	8161.3	2820.3	1304.8				
	1994	10622.0	3897.7					
	1995	11744.9						

Source: Ntzoufras and Dellaportas (2002). Copyright 2002 by the Society of Actuaries, Schaumburg, Illinois. Reprinted with permission.

Table 10.4   Inflation factors for Example 10.1

Year	1989[a]	1990	1991	1992	1993	1994	1995
Inflation $(\%)$	100.0	120.4	143.9	166.6	190.6	214.2	235.6

Year	1996	1997	1998	1999	2000	2001
Inflation $(\%)$	248.4	261.8	273.6	279.4	287.5	298.1

[a] Year 1989 was used as the baseline year.

Source: Ntzoufras and Dellaportas (2002). Copyright 2002 by the Society of Actuaries, Schaumburg, Illinois. Reprinted with permission.

10.2.2.1 *Model formulation.* We base our predictions on the simplest and most popular model used in the relevant literature. The proposed model is a simple two-way main effects ANOVA model with response the log-amounts data. For details concerning this model and other related approaches, see Ntzoufras and Dellaportas (2002) and references cited therein. Hence the model is given by the formulation

$$Z_{ab} = \log \frac{Y_{ab}}{\inf_{a+b-1}}, \quad Z_{ab} \sim N(\mu_{ab}, \sigma^2), \quad \mu_{ab} = \mu + \alpha_a + \beta_b, \quad a, b = 1, \ldots, L \quad (10.4)$$

where Z_{ab} are the log-deinflated claim amounts, L are the number of years considered and $\inf_{a+b-1}$ is the inflation corresponding to the payment year of cell ab. A detailed description of building the above model in WinBUGS can be found in Scollnik (2002).

The model is alternatively expressed by

$$Y_{ab} \sim LN(\mu_{ab}, \sigma^2), \quad \mu_{ab} = \log(\inf_{ab}) + \mu + \alpha_a + \beta_b, \quad a, b = 1, \ldots, L \quad (10.5)$$

where $LN(\mu, \sigma^2)$ are the log-normal distributions with parameters μ and σ^2 (which are the mean and the variance of $\log Y$). This expression is convenient for WinBUGS since the likelihood of the model can be directly expressed using the syntax

```
for (a in 1:L){
    for (b in 1:L){
        y[a,b] <- dlnorm( mu[a,b], tau )
        mu[a,b] <- log(inf[ a+b-1 ]) + mu0 + alpha[a] + beta[b]
}}
```

Monitoring of y directly provides us the predictive posterior distributions for the lower missing triangle of Table 10.3.

We additionally need to calculate total payments for each year denoted by T_i for $i = L+1, \ldots, 2L-1$ calculated by

$$T_i = \sum_{(a,b):a+b=i+1} Y_{ab} = \sum_{a=i-L+1}^{L} Y_{a,i+1-a} \cdot$$

Moreover, the total amount to be paid due to the outstanding claims is also of interest. It is calculated simply by the sum of the aggregated amounts

$$T = \sum_{i=L+1}^{2L-1} T_i \cdot$$

In WinBUGS these quantities are computed using the following syntax

```
for ( i in (L+1):(2*L-1) ){
    for ( a in 1:(i-L)    ){Y.T[i-L,a] <- 0.0}
    for ( a in (i-L+1):L ){Y.T[i-L,a] <- y[ a, i+1-a ] }
    T[i-L]<-sum( Y.T[i-L,1:L] )
}
```

in which the matrix Y.T of dimension $(L-1) \times L$ is first calculated by

$$Y.T_{i-L,a} = \begin{cases} 0 & \text{for } a = 1, 2, \ldots, i-L \\ Y_{a,i+1+a} & \text{for } a = i-L+1, \ldots, L \end{cases}$$

for $i = L+1, L+2, \ldots, 2L-1$. The annual sums are calculated by

$$\texttt{T[i-L]} = T_i = \sum_{a=1}^{L} Y.T_{i-L,a}$$

since

$$\sum_{a=1}^{L} Y.T_{i-L,a} = \sum_{a=1}^{i-L} Y.T_{i-L,a} + \sum_{a=i-L+1}^{L} Y.T_{i-L,a} = \sum_{a=i-L+1}^{L} Y_{a,i+1-a}$$

and $\sum_{a=1}^{i-L} Y.T_{i-L,a} = 0$.

To complete the model formulation, we adopt the usual corner constraints: $\alpha_1 = \beta_1 = 0$. Consequently, expression (10.4) assumes that the expected log-adjusted claim amount μ_{ab}, originated at year a and paid with delay of $b - 1$ years, is modeled via a linear predictor that consists of the expected log-adjusted claim amount μ originating and paid within the first accident notification year, a factor that reflects expected changes due to the origin year α_a, and a factor depending on the delay pattern β_b. These constraints imply that the parameters to be estimated are μ, σ^2, $\boldsymbol{\alpha} = (\alpha_2, \dots, \alpha_r)^T$, $\boldsymbol{\beta} = (\beta_2, \dots, \beta_r)^T$, and $\boldsymbol{Y}^L = \{Y_{ij} : i + j > r + 1\}$. Finally, the usual prior setup is adopted by

$$\mu \sim N(0, \sigma_\mu^2), \quad \alpha_a \sim N(0, \sigma_{\alpha_i}^2), \quad \beta_b \sim N(0, \sigma_{\beta_j}^2), \quad a, b = 2, \dots, L, \quad \tau = \sigma^{-2} \sim G(a_\tau, b_\tau).$$

For the type of problem we are interested in, low-information prior distributions can be produced by

$$\sigma_\mu^2 = 1000, \quad \sigma_{\alpha_a}^2 = 100, \ a = 2, \dots, L, \quad \sigma_{\beta_b}^2 = 100, \ b = 2, \dots, r, \quad a_\tau = b_\tau = 0.001.$$

Data in WinBUGS code can be provided in a list format, with $\boldsymbol{y}$ given in $L \times L$ tabular format with the lower triangle filled with NAs; see Table 10.5 for the list data format. The WinBUGS code for the full model of this example is presented in Table 10.6.

Table 10.5 Data in list format for Example 10.1

```
DATA (LIST)
list( L=7,
     inf=c(1.000 , 1.204, 1.439, 1.666, 1.906, 2.142, 2.356,
                   2.484, 2.618, 2.736, 2.794, 2.875, 2.981),
y = structure(
        .Data=c(1546.6,   647.5,   382.2,   246.8, 211.8,  63.6, 146.3,
                2099.0, 1001.8,   487.2,   293.0, 318.8, 269.9,    NA,
                3422.2, 1257.1,   488.4,   456.0, 562.4,    NA,    NA,
                4948.8, 1899.7,   984.0, 1253.3,    NA,    NA,    NA,
                8161.3, 2820.3, 1304.8,     NA,    NA,    NA,    NA,
               10622.0, 3897.7,     NA,     NA,    NA,    NA,    NA,
               11744.9,     NA,     NA,     NA,    NA,    NA,    NA),
        .Dim = c( 7,7 ) )
)
```

Table 10.6 WinBUGS code for Example 10.1[a]

```
model{
    # model's likelihood
    for (a in 1:L){
    for (b in 1:L){
            y[a,b] ~ dlnorm( mu[a,b], tau )
            mu[a,b] <- log(inf[a+b-1]) + mu0 + alpha[a] + beta[b]
        }}
    #### CR Constraints
    alpha[1] <- 0.0
    beta[1] <- 0.0
    # priors
    mu0~dnorm( 0.0, 1.0E-04)
    for (a in 2:L){
        alpha[a]~dnorm( 0.0, 1.0E-04)
        beta[a] ~dnorm( 0.0, 1.0E-04)
    }
    tau ~dgamma( 0.01, 0.01)
    s <- sqrt(1/tau) # precision

    for ( i in (L+1):(2*L-1) ){
        for ( a in 1:(i-L)   ){Y.T[i-L,a] <- 0.0}
        for ( a in (i-L+1):L ){Y.T[i-L,a] <- y[ a, i+1-a ] }
        T[i-L]<-sum( Y.T[i-L,1:L] )
    }
    Total <- sum(T[])
}

INITS
list( mu0=1.0, alpha=c(NA, 0,0,0,0,0,0), beta=c(NA, 0,0,0,0,0,0), tau
=1.0 )
```

[a]Data are given in Table 10.5.

Results.   Posterior summaries of model parameters using 10,000 iterations (after discarding the initial 1000 iterations) are given in Table 10.7. Posterior summaries are close to the ones reported by Ntzoufras and Dellaportas (2002) and Scollnik (2002). Differences are observed to the intercept μ_0 with posterior mean equal to 7.229. By adding the log-exchange rate between euros and drachmas to the constant term, we get $7.23 + \log(340.75) = 13.06$, which is the value reported by the authors cited above.

From the posterior distributions of the model parameters we observe

- A positive effect of accident year to the deinflated amounts which shows that claim amounts increase from year to year (possibly because of the increased number of contracts by the insurance company).

- A negative effect of the delay year to the outstanding claims, which is plausible since the amount of unsettled claims reduces with time.

Table 10.7   Posterior summaries for model parameters in Example 10.1[a]

node	mean	sd	MC error	2.5%	median	97.5%	start	sample
mu0	7.229	0.1687	0.006309	6.889	7.229	7.563	1001	10000
alpha[2]	0.3572	0.1655	0.004067	0.02533	0.3595	0.6804	1001	10000
alpha[3]	0.4682	0.1807	0.004547	0.1056	0.4669	0.8315	1001	10000
alpha[4]	0.917	0.1938	0.00479	0.5273	0.9197	1.307	1001	10000
alpha[5]	1.068	0.2145	0.004901	0.6363	1.069	1.491	1001	10000
alpha[6]	1.271	0.2471	0.005466	0.7894	1.273	1.758	1001	10000
alpha[7]	1.284	0.3317	0.006881	0.6192	1.282	1.938	1001	10000
beta[2]	-1.085	0.167	0.003729	-1.418	-1.084	-0.7527	1001	10000
beta[3]	-1.942	0.1795	0.004298	-2.303	-1.94	-1.586	1001	10000
beta[4]	-2.247	0.1931	0.004576	-2.633	-2.245	-1.862	1001	10000
beta[5]	-2.441	0.2139	0.004726	-2.866	-2.442	-2.016	1001	10000
beta[6]	-3.342	0.2501	0.005683	-3.833	-3.34	-2.852	1001	10000
beta[7]	-3.099	0.3293	0.007312	-3.758	-3.095	-2.455	1001	10000
s	0.282	0.0572	0.001141	0.1962	0.2739	0.4198	1001	10000

[a]Total of 10,000 iterations kept; 1000 burnin.

Returning to the estimated to-be-paid claim amounts, posterior predictive means for each future amount are presented in Table 10.8, while posterior predictive summaries for claim totals are provided in Table 10.9 and Figure 10.3. Therefore, the amounts that the company is expected to pay for car accidents that occurred during 1989–1995 are equal to 8883, 5143, 3767, 2430, 1282, and 769 thousand euros for years 1996, 1997, . . . , 2001, respectively. The total corresponding amounts are found a posteriori equal to 22.27 million euros. Finally, if the company wishes to ensure that reserved accounts will cover its obligations with high probability, then reserved accounts can be based on posterior percentiles (e.g., 95th, 97.5th or 99th percentiles). Using this conservative approach, if the insurance company reserved the total amount of 34.1 million euros, then it would be able to fully cover the future obligations due to pending claims with probability equal to 97.5%.

Table 10.8   Posterior predictive means for claim amounts (in thousands of euros) for Example 10.1

	Year	1	2	B 3	4	5	6	7
	1989							
	1990							242.2
	1991						208.4	287.0
A	1992					798.4	346.2	472.2
	1993				1126.0	979.9	421.1	562.1
	1994			1882.0	1461.0	1272.0	531.3	715.2
	1995		4626.0	2068.0	1602.0	1337.0	566.8	769.1

Table 10.9   Posterior predictive summaries for to-be-paid total claim amounts for Example 10.1[a]

```
node   mean     sd      MC error 2.5%    median 97.5%    start sample
T[1]   8883.0   2412.0 33.9    5488.0   8485.0 14760.0 1001  10000 (1996)
T[2]   5143.0   1285.0 21.28   3207.0   4961.0 8069.0  1001  10000 (1997)
T[3]   3767.0   1065.0 17.06   2218.0   3604.0 6283.0  1001  10000 (1998)
T[4]   2430.0   765.4  13.13   1326.0   2301.0 4251.0  1001  10000 (1999)
T[5]   1282.0   496.0  8.379   604.5    1204.0 2450.0  1001  10000 (2000)
T[6]   769.1    462.5  7.028   238.9    677.2  1835.0  1001  10000 (2001)
Total  22270.0 4911.0 90.62   14740.0  21600.0 34110.0 1001 10000
```

[a] Total of 10,000 iterations kept; 1000 burnin.

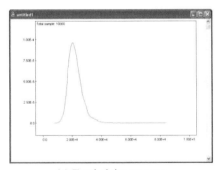

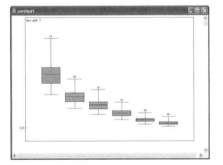

(a) Total claim amount   (b) Total to-be-paid claim amounts $(T_1, \ldots, T_6)$

Figure 10.3   Posterior predictive plots for total outstanding claim counts for Example 10.1 (total of 10,000 iterations kept, 1000 burnin).

10.3   USING THE PREDICTIVE DISTRIBUTION FOR MODEL CHECKING

10.3.1   Comparison of actual and predictive frequencies for discrete data

As we have already mentioned, we can use the predictive distribution to re-generate the distributions of future (or replicated) data y^{rep}. This distribution can be used to compare the predicted values under the assumed model with the actual data. Large deviations of the two distributions may indicate that the model is inappropriate for the data used. The comparison of the predictive distributions y^{rep} is relative easy and convenient when discrete variables are considered. In the following we illustrate this by using a simple example with data from association football (soccer), which is popular in Europe.

Example 10.2. The distribution of Manchester United's goals in home games for season 2006–2007. In this example we consider the number of goals scored and conceded by Manchester United in home games for Season 2006/07 in the English premiership of association football (soccer), presented in Table 10.10. The aim here is to estimate the expected number of goals using a simple Poisson model.

Table 10.10   Data for Example 10.2: Frequency of goals scored and conceded in home games of Manchester United for premiership season 2006–2007

	\multicolumn{6}{c}{Number of goals}	
	0	1	2	3	4	5	Total games
Frequencies for scored goals	2	3	4	6	3	1	19
Frequencies for conceded goals	8	10	1				

Model formulation and WinBUGS code.   A simple Poisson model with common mean $y_{ik} \sim \text{Poisson}(\lambda_k)$ is used for the scored ($k=1$) and conceded ($k=2$) goals. Although the usual (conjugate) gamma prior can be adopted, here we facilitate a log-normal prior with zero mean and large variance (10^3) in the log-scale, which is the default choice in the Poisson log-linear model presented in Chapter 7.

Since the data are given in tabulated format, we can express the likelihood by

$$f(y|\lambda) = \prod_{i=1}^{n_f}\left[e^{-\lambda}\frac{\lambda^{y_i}}{y_i!}\right]^{f_{y_i}} = \prod_{i=1}^{n_f}\prod_{j=1}^{f_{y_i}}\left[e^{-\lambda}\frac{\lambda^{y_i}}{y_i!}\right] = \prod_{i=1}^{n_f}\prod_{j=1}^{f_{y_i}}\left[e^{-\lambda}\frac{\lambda^{z_{ij}}}{y_i!}\right],$$

where $z_{ij} = y_i$, n_f is the number of observed frequencies and f_y is the frequency of $Y = y$. In WinBUGS it is possible to fit models for data given in a frequency table format using the syntax

```
for (i in 1:nf){
    for ( j in 1:fx[i] ){
        z[i,j] <- y[i]
        z[i,j] ~ dpois( lambda )
}}
```

when the data for the scored goals are imported in the following list format

```
list( nf=6, y=c(0,1,2,3,4,5), fx=c(2,3,4,6,3,1)  )
```

where nf is the number of available frequencies.

To generate replicated data y^{rep} of the same size, we simply add the command

```
y.rep[i,j]~dpois( lambda )
```

within the double loop above. More generally, we may generate a set of data of size n^{rep} (n.rep in WinBUGS) using a vector node by setting

```
for (i in 1:n.rep){ y.rep[i]~dpois( lambda ) }
```

outside the above mentioned likelihood double loop. In the first case, the distribution of the predictive and observed counts can be directly compared; in the second case, the distribution of the corresponding relative frequencies must be used instead. The frequency table of each replicated set can be stored and monitored using the syntax

```
for (k in 1:nf){
    for (i in 1:n.rep){ freq.bin[k,i]<-equals(y.rep[i], y[k]) }
        freq[k] <- sum( freq.bin[k,1:n.rep] )
        rel.freq[k] <- freq[k]/n.rep
}
```

In this syntax we first calculate matrix freq.bin of dimension $n_f \times n^{\text{rep}}$ (here 6×19) with elements freq.bin[k,i] taking value equal to one if $y_i^{\text{rep}} = y_k$ and zero otherwise. Then the frequencies are calculated as the sum of each row of this matrix. Concerning the conceded goals, the same model is used by substituting the data by the following list data format

```
list( nf=6, n.rep=19, y=c(0,1,2,3,4,5), fx=c(8,10,1,0,0,0) )
```

Results. Posterior summaries for the mean scored and conceded goals as well as the frequencies of the future (replicated) goals are presented in Tables 10.11 and 10.12. Error bars of the frequencies of the goals are given in Figure 10.4, produced in WinBUGS and

Table 10.11 Posterior summaries for λ and frequencies of replicated scored goals for Manchester United in Example 10.2[a]

node	mean	sd	MC error	2.5%	median	97.5%	start	sample
lambda	2.407	0.3432	0.007838	1.762	2.399	3.094	1001	2000
freq[1]	1.821	1.466	0.03602	0.0	2.0	5.0	1001	2000
freq[2]	4.19	1.99	0.04141	1.0	4.0	9.0	1001	2000
freq[3]	4.822	1.913	0.04489	1.0	5.0	9.0	1001	2000
freq[4]	3.933	1.804	0.03462	1.0	4.0	8.0	1001	2000
freq[5]	2.277	1.506	0.02857	0.0	2.0	6.0	1001	2000
freq[6]	1.182	1.103	0.02522	0.0	1.0	4.0	1001	2000
rel.freq[1]	0.09584	0.07717	0.001896	0.0	0.1053	0.2632	1001	2000
rel.freq[2]	0.2205	0.1047	0.00218	0.05263	0.2105	0.4737	1001	2000
rel.freq[3]	0.2538	0.1007	0.002362	0.05263	0.2632	0.4737	1001	2000
rel.freq[4]	0.207	0.09493	0.001822	0.05263	0.2105	0.4211	1001	2000
rel.freq[5]	0.1198	0.07928	0.001504	0.0	0.1053	0.3158	1001	2000
rel.freq[6]	0.06221	0.05804	0.001327	0.0	0.05263	0.2105	1001	2000

[a]Total of 1000 iterations kept; 1000 burnin.

compared to the original data superimposed (dashed lines). Although the model is simple, we observe that the Poisson distribution seems to describe the original scored goals sufficiently since the replicated ones are quite close to them (all 95% posterior intervals include the observed data, indicating that the orginal dataset can be regenerated using the proposed model). Concerning the conceded goals, we observe that all actual counts (except for $y = 0$) lie at the tails of each predictive distribution, which indicates that a more sophisticated distribution is needed for these data. Finally, combining the two datasets and models in one WinBUGS model will also enable us to produce estimates of the goal difference and the probability of win, draw, and loss for Manchester United.

Table 10.12   Posterior summaries for λ and frequencies of replicated conceded goals for Manchester United in Example 10.2

node	mean	sd	MC error	2.5%	median	97.5%	start	sample
lambda	0.6301	0.1789	0.00393	0.335	0.6082	1.033	1001	2000
freq[1]	10.26	2.721	0.06067	5.0	10.0	15.0	1001	2000
freq[2]	6.178	2.081	0.04327	2.0	6.0	10.0	1001	2000
freq[3]	1.974	1.458	0.03346	0.0	2.0	5.0	1001	2000
freq[4]	0.487	0.7535	0.01591	0.0	0.0	2.0	1001	2000
freq[5]	0.0825	0.3061	0.007382	0.0	0.0	1.0	1001	2000
freq[6]	0.015	0.1256	0.002836	0.0	0.0	0.0	1001	2000
rel.freq[1]	0.5401	0.1432	0.003193	0.2632	0.5263	0.7895	1001	2000
rel.freq[2]	0.3251	0.1095	0.002277	0.1053	0.3158	0.5263	1001	2000
rel.freq[3]	0.1039	0.07676	0.001761	0.0	0.1053	0.2632	1001	2000
rel.freq[4]	0.02563	0.03966	8.371E-4	0.0	0.0	0.1053	1001	2000
rel.freq[5]	0.004342	0.01611	3.885E-4	0.0	0.0	0.0526	1001	2000
rel.freq[6]	7.895E-4	0.00661	1.493E-4	0.0	0.0	0.0	1001	2000

[a] Total of 1000 iterations kept; 1000 burnin.

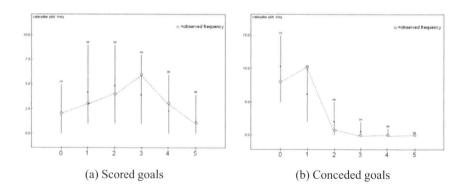

(a) Scored goals   (b) Conceded goals

Figure 10.4   Posterior predictive plots for frequencies of replicated and actual goals conceded for Manchester United in Example 10.2 (total of 1000 iterations kept, 1000 burnin).

10.3.2   Comparison of cumulative frequencies for predictive and actual values for continuous data

When the data vector $\boldsymbol{y}$ is continuous, then we need to consider a more sophisticated approach to compare the actual and replicated values. Such a comparison can be based on the predictive cumulative relative frequencies $F_{y_i}^{\text{pred}} = P(Y < y_i|\boldsymbol{y})$ compared with the observed cumulative relative frequencies F_{y_i}. The predictive predictive cumulative relative frequencies are given by

$$
F_{y'}^{\text{pred}} = P(Y < y'|\boldsymbol{y}) \;=\; \int_{-\infty}^{y'} \int f(z|\boldsymbol{\theta}) f(\boldsymbol{\theta}|\boldsymbol{y})\, d\boldsymbol{\theta}\, dz
$$
$$
\;=\; \int_{-\infty}^{\infty} \int I(z < y') f(z|\boldsymbol{\theta}) f(\boldsymbol{\theta}|\boldsymbol{y})\, d\boldsymbol{\theta}\, dz \;.
$$

This quantity can be estimated from an MCMC output by the expression

$$
\hat{F}_{y'}^{\text{pred}} = \frac{1}{n} \sum_{i=1}^{n} I(y_i^{\text{rep}} < y'),
$$

assuming that we obtain $\boldsymbol{y}^{\text{rep}}$ of size equal to the original sample n. Then a plot of the 95% posterior intervals of $\hat{F}_i^{\text{pred}}$ by F_i will provide us with a good visual representation of the predictive cumulative frequencies in comparison to the observed data.

Hence, in WinBUGS , if we use the syntax

```
for (i in 1:n){
    for (k in 1:n){
        pred.lower.yi[i,k] <- step( y[i] - y.rep[k] )
    }
    F.pred[i] <- sum( pred.lower.yi[i,1:n] )/n
}
```

we first identify which $y_i > y_k^{\text{rep}}$ using the command `step`. Each comparison is stored in the binary matrix `pred.lower.yi` of dimension $n \times n$. Then the cumulative frequencies $F_{y_i}^{\text{pred}}$ of the predictive values in each MCMC cycle are calculated by the sum of each row. An error bar plot of the 95% posterior intervals of $F_{y_i}^{\text{pred}}$ versus observed frequencies F_{y_i} can be produced using either the `compare` tool of WinBUGS or another statistical program.

Example 10.3. Simulated normal data. To demonstrate the approach described above, we have generated the following 19 values from the standardized normal distribution forming dataset 1; see Table 10.13. To further illustrate how the predictive measures and plots behave in the presence of outliers, we have added the values of 5 and 10 in the data of the originally simulated dataset 1, forming in this way datasets 2 and 3, respectively.

Table 10.13   Dataset 1 of Example 10.3: Simulated data from $N(0, 1)$

0.51, 0.10, -2.53, 1.00, 0.65, -0.95, 2.76, 1.33, 0.25, 1.48, -0.25, -0.45, 2.11, -0.76, -1.51, -0.35, 0.18, 1.35, -0.18

Results. We have run a simple normal model using WinBUGS and calculated the predictive cumulative frequencies for each observed value for the three datasets. Figure 10.5 provides posterior predictive error bars of cumulative relative frequencies for each y_i against its corresponding observed frequencies. If the model is appropriate, then these values must be close; hence the $y = x$ line is also plotted as a reference line. For the first dataset (Figure 10.5a), we observe that both predicted and observed frequencies are close with small deviations from the reference line. For the second dataset (Figure 10.5b), larger deviations are observed with the last parameter having a narrow error bar indicating that very few predictive values were higher than $y_{(20)} = 5$. Finally, for the last dataset (Figure 10.5c), deviations between predictive and observed frequencies are high (where the reference line is outside specific error bars), indicating poor fit of the model. Moreover, no value was generated from the predictive distribution with values higher than $y_{(20)} = 10$, indicating that this value has a very low probability of observing this value if the current model holds. Finally, in Table 10.14, the effect of the artificial observations added in datasets 2 and 3 on the posterior distribution of μ and σ is clear (both the posterior mean and the standard deviation were increased when a single outlier value was added in the dataset).

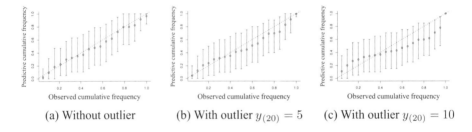

(a) Without outlier   (b) With outlier $y_{(20)} = 5$ (c) With outlier $y_{(20)} = 10$

Figure 10.5   Posterior predictive error bars of cumulative frequencies versus observed cumulative frequencies for Example 10.3.

Table 10.14   Actual values and posterior means (standard deviations) for model parameters (μ, σ) for Example 10.3

Parameter	True values	Dataset 1	Dataset 2	Dataset 3
μ	0.0	0.26 (0.30)	0.48 (0.39)	0.73 (0.59)
σ	1.0	1.33 (0.25)	1.71 (0.30)	2.63 (0.47)

10.3.3 Comparison of ordered predictive and actual values for continuous data

An alternative and efficient predictive check can be based on the ordered data $y_{(i)}$ and $y_{(i)}^{\text{rep}}$. For each ordered observation, we compare $y_{(i)}$ with the distribution $f\left(y_{(i)}^{\text{rep}} \middle| \boldsymbol{y}\right)$ estimated by the generated values $y_{(i)}^{\text{rep}}$.

The distribution of $f\left(y_{(i)}^{\text{rep}}\middle|\boldsymbol{y}\right)$ can be estimated within an MCMC cycle by adding the following steps:

1. Generate $y_i^{\text{rep},(t)}$ from $f\left(y_i\middle|\boldsymbol{\theta}^{(t)}\right)$.

2. Order $y_i^{\text{rep},(t)}$ to get $\left(y_{(1)}^{\text{rep},(t)},\dots,y_{(n)}^{\text{rep},(t)}\right)$.

After generating $\left(y_{(1)}^{\text{rep},(t)},\dots,y_{(n)}^{\text{rep},(t)}\right)$ we compare the distribution of the generated values $y_{(i)}^{\text{rep},(t)}$ with $y_{(i)}$. Graphical comparison of these two quantities can be produced by an error bar of $y_{(i)}^{\text{rep},(t)}$ for each value of $y_{(i)}$.

Example 10.3. Simulated normal data (continued). Error bars of the predicted ordered data against the corresponding observed ones for the simulated datasets of Example 10.3 are given in Figure 10.6. Bars crossed by the diagonal line $y = x$ indicate that observed and predicted ordered data are close. From the error bars, it is clear that for the first dataset (Figure 10.6a), the model is sufficient since all predicted and observed ordered data are very close. For the second dataset (Figure 10.6b), the picture is slightly worse with the outlier value $y_{(20)} = 5$ found in the upper tail of the corresponding error bar. The overall picture indicates that $y_{(20)} = 5$ might be an outlier, and not that the model must be rejected. Finally, for the third dataset, Figure 10.6c indicates that the model must be rejected since we observe high deviations between predicted and observed data, while observation $y_{(20)} = 10$ is flagged as an outlier since the corresponding posterior interval does not contain this value. The latter also implies that the probability of observing this value is rather low under the assumed model.

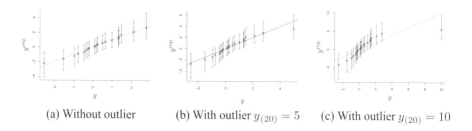

(a) Without outlier   (b) With outlier $y_{(20)} = 5$ (c) With outlier $y_{(20)} = 10$

Figure 10.6   Posterior predictive error bars of $y_{(i)}^{\text{rep}}$ versus observed data $y_{(i)}$ for Example 10.3.

10.3.4 Estimation of the posterior predictive ordinate

Instead of replicating the data, we can directly estimate the ith posterior predictive ordinate $\text{PPO}_i = f(y_i|\boldsymbol{y})$ by

$$\hat{f}(y_i|\boldsymbol{y}) = \frac{1}{T'}\sum_{t=1}^{T'} f\left(y_i\middle|\boldsymbol{\theta}^{(t)}\right),$$

where $\boldsymbol{\theta}^{(t)}$ is the vector of parameter values generated in the tth iteration of the MCMC algorithm. Thus, we can directly calculate $\hat{f}(y_i|\boldsymbol{y})$ in WinBUGS by considering a deterministic node set equal to the likelihood evaluated at the current values of $\boldsymbol{\theta}$. For discrete distributions, we can compare $\hat{f}(y_i|\boldsymbol{y})$ with the observed frequency for each y_i, while for continuous data we can produce a rough estimate of the cumulative distribution on the basis of the observed $\boldsymbol{y}$ by calculating

$$\hat{F}(Y \leq y_{(k)}|\boldsymbol{y}) = \frac{\sum_{i=1}^{k} \frac{1}{2}\left(\hat{f}(y_{(i)}|\boldsymbol{y}) + \hat{f}(y_{(i-1)}|\boldsymbol{y})\right)(y_{(i)} - y_{(i-1)})}{\sum_{i=1}^{n} \frac{1}{2}\left(\hat{f}(y_{(i)}|\boldsymbol{y}) + \hat{f}(y_{(i-1)}|\boldsymbol{y})\right)(y_{(i)} - y_{(i-1)})} \tag{10.6}$$

for $k = 0, 1, \ldots, n, n + 1$, where $y_{(k)}$ is the kth ordered observation and $y_{(0)}, y_{(n+1)}$ are additional low and high values used to calculate the lower and upper tails of the distribution. Here we have set $y_{(0)} = \min(\boldsymbol{y}) - \text{SD}(\boldsymbol{y})$ and $y_{(n+1)} = \max(\boldsymbol{y}) + \text{SD}(\boldsymbol{y})$ with $\text{SD}(\boldsymbol{y})$ the sample standard deviation of the data $\boldsymbol{y}$. The $\hat{F}(Y \leq y_{(k)}|\boldsymbol{y})$ estimated above must be compared visually with the corresponding observed cumulative probability relative frequency of $Y < y_{(i)}$, which is equal to i/n.

Example 10.2: Manchester United's goals (continued).   In the WinBUGS model code of the Poisson model used for Example 10.2, we have additionally defined node ppo using the syntax

```
for (i in 1:nf){
    ppo[i] <- exp( -lambda + y[i]*log(lambda) - logfact(y[i]) )
}
```

which calculates the probability under the simple Poisson model for all observed values $y_1, y_2, \ldots, y_{nf}$ under the sampled parameter values in each MCMC cycle. The probability function is written in log-scale form and then exponentiated:

$$f(y|\lambda) = e^{-\lambda + y \log(\lambda) - \log(y!)} .$$

This is a usual practice in statistical computation, which is used to avoid arithmetic overflows, and it is also recommended when defining nodes within WinBUGS model codes.

Posterior means of ppo have been calculated after 2000 iterations (and an additional 1000 iterations as burnin) and are given in Table 10.15. Comparison of estimated PPOs and observed relative frequencies is provided in Figure 10.7. As expected, PPOs using this approach are similar to the posterior means of the frequencies in predicted data replicated in Section 10.3.1. For scored goals, differences between PPOs and observed relative frequencies are small, with the exception of values for $y = 3$, while for conceded goals differences seem to be more severe, indicating that an alternative distribution may be more appropriate.

Table 10.15   Estimated posterior predictive ordinates (PPO) and observed relative frequencies for scored and conceded goals in home games of Manchester United for premiership season 2006–2007 (Example 10.2)

	0	1	2	3	4	5
Scored goals						
Observed relative frequencies	0.11	0.16	0.21	0.32	0.16	0.05
Estimated PPO (posterior mean)	0.10	0.22	0.26	0.20	0.12	0.06
Conceded goals						
Observed relative frequencies	0.42	0.53	0.05	0.00	0.00	0.00
Estimated PPO (posterior mean)	0.54	0.32	0.10	0.02	0.00	0.00

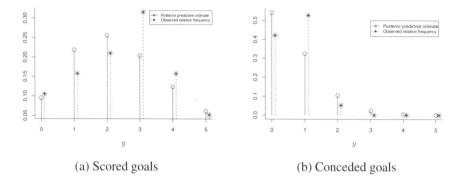

(a) Scored goals   (b) Conceded goals

Figure 10.7   Estimated posterior predictive ordinates (solid line) and observed relative frequencies (dashed line) for scored and conceded goals in home games of Manchester United for premiership season 2006–2007 (Example 10.2).

Example 10.3: Simulated normal data (continued).   Returning to Example 10.3, we have to define the node ppo using the syntax

```
for (i in 1:n){
    log.ppo[i] <- -0.5 * log( 2*3.14 ) + 0.5*log(tau)
                  -0.5*tau*(y[i]-mu)*(y[i]-mu)
    ppo[i] <- exp( log.ppo[i] )
}
```

in order to calculate the posterior predictive ordinates PPO_i for $i = 1, 2, \ldots, n$. Two additional PPOs are also calculated using the syntax

```
ysmall<-ranked(y[],1)-sd(y[])
log.ppo[n+1] <- - 0.5 * log( 2*3.14 ) + 0.5 * log(tau)
                - 0.5 * tau * (ysmall-mu) * (ysmall-mu)
ppo[n+1]<- exp( log.ppo[n+1] )
yhigh<-ranked(y[],n)+sd(y[])
log.ppo[n+2] <- - 0.5 * log( 2*3.14 ) + 0.5 * log(tau)
                - 0.5 * tau * (yhigh-mu) * (yhigh-mu)
ppo[n+2]<- exp( log.ppo[n+2] )
```

where ysmall and yhigh are the values $y_{(0)}$ and $y_{(n+1)}$ defined one standard deviation away from the minimum and maximum observed values in each dataset. These values are used to more accurately compute probabilities at the tails of the distribution required in calculation of the cumulative frequencies (10.6).

Estimated cumulative function of the predictive distribution compared with the observed one is provided in Figure 10.8, while the estimated density function of the predictive distribution is depicted in Figure 10.9 for all three datasets. From Figure 10.8 we observe that for dataset 1, the simple normal model fits the data satisfactorily. For dataset 2 we observe differences between the predictive and the observed distribution, which are even more severe for the third dataset. Therefore, improvement of the model may be needed for the second dataset, while for the third dataset the adopted model is clearly problematic and needs to be revised.

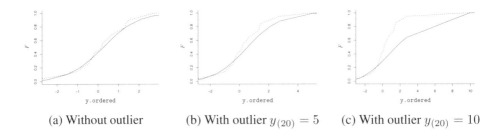

(a) Without outlier   (b) With outlier $y_{(20)} = 5$ (c) With outlier $y_{(20)} = 10$

Figure 10.8   Observed (dashed line) and estimated predictive (solid line) cumulative distribution for Example 10.3.

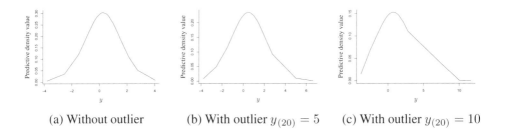

(a) Without outlier   (b) With outlier $y_{(20)} = 5$ (c) With outlier $y_{(20)} = 10$

Figure 10.9   Estimated predictive density function for Example 10.3.

10.3.5   Checking individual observations using residuals

In addition to the quantities calculated in the previous sections, we facilitate residuals for checking the fit of each observation and the identification of outliers.

Residual values can be based on the deviations of the data from the mean of the model; see Spiegelhalter et al. (1996a, pp. 43–47) for an example. Hence we define the simple (unstandardized) residual as

$$r_i = y_i - E(Y_i | \boldsymbol{\theta})$$

and its standardized version by dividing it by the standard deviation under the adopted model:

$$r_i^s = \frac{r_i}{\text{SD}(Y_i | \boldsymbol{\theta})} = \frac{y_i - E(Y_i | \boldsymbol{\theta})}{\sqrt{\text{Var}(Y_i | \boldsymbol{\theta})}} .$$

Finally, we can use the tail area probability

$$p_i^r = P(r_i^{\text{rep}} > r_i | \boldsymbol{y}) = P(y_i^{\text{rep}} > y_i | \boldsymbol{y}),$$

where r_i^{rep} is the residual value, based on the predictive/replicated values y_i^{rep}. The value

$$\min\left(p_i^r, 1 - p_i^r\right) = \min\left\{P(y_i^{\text{rep}} > y_i | \boldsymbol{y}), \ 1 - P(y_i^{\text{rep}} > y_i | \boldsymbol{y})\right\}$$

can be interpreted as the probability of "getting a more extreme observation." Details concerning the use of these quantities can be found in the classic BUGS manual (Spiegelhalter et al., 1996a, pp. 40–47).

Further residual values can be calculated using ordered statistics or frequencies. Here we additionally propose calculation of the ordered residual statistics

$$r_{(i)}^o = r_{(i)} - F^{-1}\left(\frac{i}{n}\right)$$

or their corresponding ith ranked p-value

$$p_i^{r_o} = P\left(r_{(i)}^{\text{rep}} > r_{(i)}|\boldsymbol{y}\right) = P\left(y_{(i)}^{\text{rep}} > y_{(i)}|\boldsymbol{y}\right).$$

These quantities measure differences between the expected and the observed ordered statistics and can also trace differences between the observed and assumed distributions of Y.

Example 10.3 (continued): Residual analysis in the simple normal example.

For the generated data of Example 10.3, we have calculated standardized residual values and their corresponding p-values using the following syntax:

```
for (i in 1:n){
    # ordered replicated/predicted data
    ordered.y.rep[i]<- ranked( y.rep[1:n], i )
    ordered.y[i]     <- ranked( y[1:n], i )
    # standardized residual
    rs[i] <- (y[i]-mu)*sqrt(tau)
    # tail area probability
    pr[i] <- step(y.rep[i]-y[i])
    # tail area probability for ordered data
    pro[i]<- step(ordered.y.rep[i]-ordered.y[i])
}
```

Standardized residuals with absolute posterior mean values higher than one are presented in Table 10.16, accompanied by detailed posterior summaries and their corrsponding tail area probabilities. Error bars of the residual values (produced using the compare menu of WinBUGS) are provided in Figure 10.10. Reference lines at values of -2 and 2 are also added in order to trace outlier values. Finally, in Table 10.17 we also present the distribution of $\min(p_i^r, 1 - p_i^r)$ and $\min(p_i^{r_o}, 1 - p_i^{r_o})$. The first must be compared with the corresponding expected probabilities under the model under consideration.

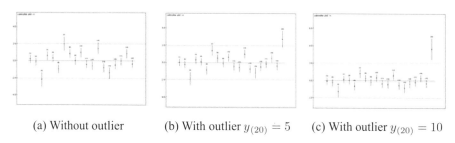

(a) Without outlier   (b) With outlier $y_{(20)} = 5$ (c) With outlier $y_{(20)} = 10$

Figure 10.10   Error bars for standardized residuals of Example 10.3.

For dataset 1, only two absolute values are close to the value of 2 with error bars including this value. Their corresponding probability values (0.026 and 0.039) are low but nonnegligible. Moreover, from Table 10.17, we observe no differences between expected and observed frequencies of the residual values. Probabilities based on order statistics are all high and thus do not indicate major differences.

Table 10.16   Posterior summaries for standardized residuals in Example 10.3[a]

Dataset 1

node	mean	sd	MC error	2.5%	median	97.5%	Pr	minPr	pro
rs[3]	-2.171	0.4461	0.01222	-3.076	-2.155	-1.333	0.974	0.026	0.675
rs[15]	-1.376	0.3335	0.00917	-2.054	-1.36	-0.734	0.895	0.105	0.506
rs[13]	1.445	0.3416	0.01211	0.8127	1.446	2.197	0.087	0.087	0.457
rs[7]	1.951	0.4124	0.01437	1.169	1.957	2.864	0.039	0.039	0.436

Dataset 2

rs[3]	-1.815	0.3916	0.01055	-2.625	-1.808	-1.057	0.957	0.043	0.481
rs[15]	-1.201	0.3118	0.00801	-1.838	-1.185	-0.578	0.860	0.140	0.346
rs[7]	1.368	0.3055	0.00824	0.766	1.377	1.970	0.086	0.086	0.495
rs[20]	2.716	0.4902	0.01413	1.763	2.711	3.717	0.011	0.011	0.123

Dataset 3

rs[3]	-1.278	0.3211	0.00831	-1.939	-1.264	-0.638	0.886	0.114	0.177
rs[20]	3.625	0.6304	0.01841	2.437	3.632	4.901	0.002	0.002	0.020

[a]Only cases with $|r_i^s| > 1$ are presented.

Table 10.17   Frequency tabulation of residual tail area probabilities for datasets of Example 10.3[a]

	$(0, 0.025]$	$(0.025, 0.05]$	$(0.05, 0.1]$	$(0.1, 0.2]$	$(0.2, 0.5]$
Frequency tabulation of $\min(p_i, 1 - p_i)$					
Dataset 1	0	2	1	3	13
Dataset 2	1	1	1	2	15
Dataset 3	1	0	0	1	18
Expected	1	1	2	4	12
Frequency tabulation of $\min(p_i^{r_o}, 1 - p_i^{r_o})$					
Dataset 1	0	0	0	0	19
Dataset 2	0	0	0	1	19
				(20)	
Dataset 3	1	0	3	8	8
	(20)		(8, 10, 16)		

[a] Values in parentheses indicate observations with corresponding probabilities.

For the second dataset, results are slightly worse. Again, only two absolute values are close to the value of 2, with the corresponding error bars including this value. For observation $y_{(20)} = 5$ p^r is estimated equal to 0.011, which is low but nonnegligible. Moreover, from Table 10.17, we observe no differences in the expected and observed frequencies of the residual values. Probabilities based on order statistics are all high except for the artificial value of $y_{(20)} = 5$, for which $p^{ro} = 0.123$, indicating a possible outlier.

Finally, for the third dataset, observation $y_{(20)} = 10$ is clearly spotted as an outlier since its error bar is far away from the value of 2; see Figure 10.10c. Moreover, both tail area probabilities are very low ($p^r = 0.002$ and $p^{ro} = 0.02$), indicating an outlier. The first states that the probability of observing a value higher than 10 under the fitted model is equal to 0.002, while the latter states that the probability of observing the maximum value in a sample of 20 observations higher than 10 under the fitted model is equal to 0.02. Moreover, from Table 10.17 we can see that the distribution of p^r is not as close as the expected one (the majority of observations have very high values in comparison to the expected ones), while three additional observations (indexed as 8, 10, 16) were also spotted with $p^{ro} < 0.10$. This may indicate an overall failure of the model resulted by the influential point $y_{(20)} = 10$.

10.3.6   Checking structural assumptions of the model

Posterior p-values can be used for checking the structural assumptions of the fitted model. For example, we can check whether the skewness and the kurtosis of the predictive and actual data are in agreement (which is particularly useful for normal models), or whether the assumption of equal mean and variance in Poisson models is valid.

To be more specific, we can calculate

$$g_1(\boldsymbol{y}) \;=\; \widehat{\text{skewness}}(\boldsymbol{y}) = \frac{1}{n}\sum_{i=1}^{n}\left(\frac{y_i - \bar{y}}{\text{SD}(\boldsymbol{y})}\right)^3$$

$$g_2(\boldsymbol{y}) \;=\; \widehat{\text{kurtosis}}(\boldsymbol{y}) = \frac{1}{n}\sum_{i=1}^{n}\left(\frac{y_i - \bar{y}}{\text{SD}(\boldsymbol{y})}\right)^4 - 3$$

and their posterior p-values $p_{g_k} = P\left(g_k(\boldsymbol{y}^{\text{rep}}) > g_k(\boldsymbol{y})\right)$. Alternatively, we may use their Bayesian versions on the basis of the standardized residual values defined in Section 10.3.5 (Spiegelhalter et al., 1996a, pp. 43–47)

$$g_1^B(\boldsymbol{y},\boldsymbol{\theta}) \;=\; \text{skewness}(\boldsymbol{y},\boldsymbol{\theta}) = \frac{1}{n}\sum_{i=1}^{n}\left(\frac{y_i - E(y_i|\boldsymbol{\theta})}{\sqrt{\text{Var}(y_i|\boldsymbol{\theta})}}\right)^3 = \frac{1}{n}\sum_{i=1}^{n}(r_i^s)^3$$

$$g_2^B(\boldsymbol{y},\boldsymbol{\theta}) \;=\; \text{kurtosis}(\boldsymbol{y},\boldsymbol{\theta}) = \frac{1}{n}\sum_{i=1}^{n}\left(\frac{y_i - E(y_i|\boldsymbol{\theta})}{\sqrt{\text{Var}(y_i|\boldsymbol{\theta})}}\right)^4 - 3 = \frac{1}{n}\sum_{i=1}^{n}(r_i^s)^4 - 3$$

and their corresponding p-values $p_{g_k^B} = P\left(g_k^B(\boldsymbol{y}^{\text{rep}}) > g_k^B(\boldsymbol{y})\right)$.

Similarly, for Poisson models we can calculate the sample dispersion index $\text{DI}(\boldsymbol{y}) = \bar{y}/\text{SD}(\boldsymbol{y})^2$ and its corresponding p-value

$$p_{\text{DI}} = P\left(\text{DI}(\boldsymbol{y}^{\text{rep}}) > \text{DI}(\boldsymbol{y})\right).$$

We can select or construct any statistic to check for the plausibility of any assumption of interest. The selection of such a statistic heavily depends on the problem and the model at hand.

Example 10.2 (continued): Testing for over or underdispersion in Manchester United data. We use the $\text{DI}(y)$ to test the assumption that the mean is equal to the variance. Hence we define the following in WinBUGS :

```
# calculation of replicated DI
di.rep <- mean( y.rep[] )/ pow( sd(y.rep[]), 2)
# calculation of observed di (tabulated data)
n        <- sum(fx[])                 # sample size
bar.y <- inprod( fx[], y[] )/n # sample mean
# sample variance components
for (i in 1:nf){ ss[i]<- fx[i] * pow( y[i] - bar.y, 2 ) }
var.y <- sum(ss[])/(n-1)              # sample variance
di.obs <- bar.y/var.y                 # observed DI
di.diff <- di.rep - di.obs            # DI difference
di.p     <- step(di.diff)             # DI p-value
```

Note that the mean and the variance for the actual data ware calculated using expressions

$$\bar{y} = \frac{1}{n}\sum_{i=1}^{n_f} f_{y_i} y_i, \;\; \text{SD}(y)^2 = \frac{1}{n-1}\sum_{i=1}^{n_f} f_{y_i}(y_i - \bar{y})^2, \;\; n = \sum_{i=1}^{n_f} f_{y_i}$$

(since the data are given in a tabulated format), where n_f is the number of (different) observed values of y and f_y is the observed frequency of y.

Results are given in Table 10.18 indicating that for the scored goals the assumption of equal variance and mean is plausible (p-value $= 0.29$, observed DI $= 1.26$, posterior expected DI for predicted data $= 1.12$). On the other hand, for the conceded goal the observed dispersion index (1.77) is far away from what is expected under the simple Poisson model (posterior mean 1.09, 95% posterior interval ranging from 0.6 to 1.78) with low posterior value (equal to 0.03), indicating that an overdispersed model (e.g., the negative binomial distribution) might better describe the distribution of these data.

Table 10.18 Dispersion index statistics for scored and conceded goals in home games of Manchester United for premiership season 2006–2007 (Example 10.2)[a]

| Goals | Dispersion index statistics | | | |
	Observed	Replicated	Difference	p-value
Scored	1.26	1.12 (0.57, 2.18)	-0.14 (-0.69, 0.92)	0.294
Conceded	1.77	1.09 (0.61, 1.78)	-0.68 (-1.16, 0.01)	0.030

[a]Replicated statistics are produced from y^{ep} and represent expected DI under the assumed model; posterior means (95% posterior intervals) are provided for columns 3 and 4 (replicated and difference, respectively).

Example 10.3 (continued): Checking for the skewness and the kurtosis of the model. Returning to the simulated data of Example 10.3, we calculate the p-values on the basis of the sample kurtosis and skewness coefficients as defined above. In order to specify g_1 and g_2 (and their p-values) in WinBUGS , we use the following syntax:

```
mean.y<-mean(y[])
s.y    <-sd(y[])
mean.yrep<-mean(y.rep[])
s.yrep    <-sd(y.rep[])
```

```
for (i in 1:n){
    m3[i]        <- pow( (y[i]-mean.y)/s.y, 3)
    m4[i]        <- pow( (y[i]-mean.y)/s.y, 4)
    m3.rep[i] <- pow( (y.rep[i]-mean.yrep)/s.yrep, 3)
    m4.rep[i] <- pow( (y.rep[i]-mean.yrep)/s.yrep, 4)
}
g[1] <- sum(m3[])/n
g[2] <- sum(m4[])/n-3
g.rep[1] <- sum(m3.rep[])/n
g.rep[2] <- sum(m4.rep[])/n-3
g.p[1]<-step( g.rep[1]-g[1] )
g.p[2]<-step( g.rep[2]-g[2] )
```

and their Bayesian versions, by the following syntax:

```
for (i in 1:n){
    mb3[i]        <- pow( (y[i]-mu)*sqrt(tau), 3)
    mb4[i]        <- pow( (y[i]-mu)*sqrt(tau), 4)
    mb3.rep[i] <- pow( (y.rep[i]-mu)*sqrt(tau), 3)
    mb4.rep[i] <- pow( (y.rep[i]-mu)*sqrt(tau), 4)
}
gb[1] <- sum(mb3[])/n
gb[2] <- sum(mb4[])/n-3
gb.rep[1] <- sum(mb3.rep[])/n
gb.rep[2] <- sum(mb4.rep[])/n-3
gb.p[1]<-step( gb.rep[1]-gb[1] )
gb.p[2]<-step( gb.rep[2]-gb[2] )
```

In this syntax we first calculate these elements used in the summations (defined in vectors m3, m4, mb3, and mb4) and then the desired measures using the sums of the above vectors.

Results for the three datasets are summarized in Table 10.19. As we can see, the Bayesian p-values are more conservative, but inference is similar whatever measure we used. For the first dataset all p-values are close to 0.5, indicating that the observed and expected data have

Table 10.19 Skewness and kurtosis statistics for simulated normal data of Example 10.3[a]

		Observed		Replicated	p-value
		Skewness statistics			
Dataset 1	g_1	-0.083		0.009 (-0.902,0.940)	0.568
	g_1^B	-0.104	(-1.592,1.484)	-0.012 (-1.839,1.837)	0.531
Dataset 2	g_1	0.782		0.018 (-0.804,0.824)	0.034
	g_1^B	0.784	(-0.502,2.579)	-0.019 (-1.836,1.898)	0.217
Dataset 3	g_1	2.372		0.018 (-0.804,0.824)	0.000
	g_1^B	2.412	(0.509,5.742)	-0.019 (-1.836,1.898)	0.040
		Kurtosis statistics			
Dataset 1	g_2	-0.370		-0.558 (-1.478,1.222)	0.291
	g_2^B	0.328	(-2.376,6.821)	0.048 (-2.432,6.016)	0.462
Dataset 2	g_2	0.952		-0.562 (-1.453,1.149)	0.034
	g_2^B	1.547	(-2.132,9.262)	0.020 (-2.507,6.023)	0.331
Dataset 3	g_2	6.555		-0.562 (-1.453,1.149)	0.000
	g_2^B	7.579	(-1.104,26.59)	0.020 (-2.507,6.023)	0.096

[a] Replicated statistics are produced from y^{rep} and represent expected skewness and kurtosis coefficients under the assumed model; values represent posterior means (and 95% posterior intervals).

the same skewness and kurtosis. For the second dataset, p-values for both g_1 and g_2 are equal to 0.034, indicating deviations between observed and expected data. For g_1^B and g_2^B, p-values are much higher (0.22 and 0.33, respectively), indicating minor deviations from the assumptions of the model. Finally, for the last dataset, all p-values are small (lower than 0.001 for g_1 and g_2, 0.04 and 0.096 for g_1^B and g_2^B, respectively) indicating that the data have shape statistics different from those for the corresponding measures of the assumed model.

The same approach can be used for any distribution. Moreover, these tests are useful for indirectly checking the assumption of the residual's normality in normal models; see, for example, Spiegelhalter et al. (1996a, pp. 45–46) and Section 10.5 of the current book.

10.3.7   Checking the goodness-of-fit of a model

For the evaluation of the goodness-of-fit of a model, χ^2 and the deviance measures defined in Section 10.1.2 and their corresponding p-values are the most frequent statistics used. Calculation of the χ^2 within WinBUGS can be attained by defining nodes chisq.obs and chisq.rep, which will calculate $\chi^2(y, \theta^{(t)})$ and $\chi^2(y^{\text{rep},(t)}, \theta^{(t)})$ in each iteration t of the MCMC algorithm. Their difference $\chi^2(y^{\text{rep},(t)}, \theta^{(t)}) - \chi^2(y, \theta^{(t)})$ must be monitored as well as their corresponding posterior p-value, given by the posterior mean of a node p.chisq defined as

```
chisq.p <- step( chisq.rep-chisq.obs)
```

which is a binary variable taking value 1 when $\chi^2(y^{\text{rep},(t)}, \theta^{(t)}) > \chi^2(y, \theta^{(t)})$ and zero otherwise. Similar to this is the syntax for specification of the deviance or other related function. Note that the deviance (for observed data) is calculated internally in WinBUGS (termed as *deviance*) and, therefore, we do not need to define it in the WinBUGS model code.

Poor fit is indicated for models with p-values close to zero since the observed statistic will be far away from what is expected under the assumed model.

Example 10.2 (continued): Goodness-of-fit measures for Manchester United data.   For the simple Poisson Example 10.2, we calculate χ^2 using the following syntax:

```
#
# chisq for predictive values
for (i in 1:n.rep){
        chisq.rep.vec[i] <- pow(y.rep[i]-lambda,2)/lambda}
chisq.rep <- sum( chisq.rep.vec[] )
#
# chisq for observed values
for (k in 1:nf){chisq.obs.vec[k] <- pow(y[k]-lambda,2)/lambda}
chisq.obs <- inprod( fx[], chisq.obs.vec[] )
chisq.diff <- chisq.rep - chisq.obs
chisq.p    <- step(chisq.diff)
```

In this syntax, nodes chisq.rep.vec and chisq.obs.vec are initially used to calculate the χ^2 elements for each observation involved in its summation. Then the inprod command is used to calculate the final χ^2 value for the observed values. Using similar logic, we can define the corresponding values for the deviance measure using the following syntax:

```
#
# deviance for replicated/predictive values
for (i in 1:n.rep){
        dev.rep.vec[i] <- -lambda + y.rep[i]*log(lambda)
```

```
                            - logfact(y.rep[i])
   }
   dev.rep <- -2*sum( dev.rep.vec[] )
   # deviance for observed values
   for (k in 1:nf){
         dev.obs.vec[k] <- -lambda+y[k]*log(lambda)-logfact(y[k])
   }
   dev.obs  <- -2*inprod( fx[], dev.obs.vec[] )
   dev.diff <- dev.rep - dev.obs
   dev.p    <- step(dev.diff)
```

Finally, we may consider as a goodness-of-fit measure the rescaled deviance, which is given by the deviance of the current model after removing the deviance of the saturated model, that is, the Poisson model mean equal to each observation y_i. This is a measure of the distance between the model estimates and the data. The rescaled version of the deviance for the simple Poisson model is given by

$$
\begin{aligned}
\text{Deviance}^r(\boldsymbol{y}) &= -2\sum_{i=1}^{n}\log\frac{f(y_i|\lambda)}{f(y_i|\lambda=y_i)} = -2\sum_{i=1}^{n}\left((y_i-\lambda)+y_i\log\frac{\lambda}{y_i}\right) \\
&= 2n(\lambda-\bar{y})-2\sum_{i=1}^{n}y_i\log\frac{\lambda}{y_i}.
\end{aligned}
$$

This measure cannot be calculated if at least one observation is equal to zero. To avoid this problem, in the calculation of the rescaled deviance we may substitute y_i by $y_i+\epsilon$, where ϵ is a small positive constant (e.g., in the following computations we have used $\epsilon = 10^{-13}$). Using this approach, we can define the rescaled deviance by the WinBUGS syntax

```
   e<-0.0000000000001
   for (i in 1:n.rep){
         devr.rep.vec[i]<-   (y.rep[i]+e-lambda)
                         + (y.rep[i]+e)*log(lambda/(y.rep[i]+e))
   }
   dev.rep <- -2*sum( dev.rep.vec[] )
   for (k in 1:nf){
         devr.obs.vec[k]<-y[k]+e-lambda+(y[k]+e)*log(lambda/(y[k]+e))
   }
   devr.rep  <- -2*sum( devr.rep.vec[] )
   devr.obs  <- -2*inprod( fx[], devr.obs.vec[] )
   devr.diff <- devr.rep - devr.obs
   devr.p    <- step(devr.diff)
```

For the Manchester United data, we have estimated posterior p-values 0.65 and 0.54 for χ^2 and deviance, respectively, for the goal scored, while the corresponding values for the conceded goals are higher (0.82 and 0.60). The probabilities for the rescaled deviance discussed above are similar (p-values 0.55 and 0.90, respectively). None of the results indicate any problem concerning the goodness-of-fit of the model. Figures of the χ^2 and deviance differences for the two sets of data are given in Figures 10.11 and 10.12.

Finally, we can use a χ^2 statistic to compare the predicted and observed frequencies using the statistic

$$
\chi_f^2(\boldsymbol{y},\boldsymbol{\theta}) = \sum_{i=1}^{n}\frac{\left[f_{y_i}-E(f_{y_i}|\boldsymbol{\theta})\right]^2}{\text{Var}(f_{y_i}|\boldsymbol{\theta})} = n\sum_{i=1}^{n}\frac{\left[f_{y_i}-f(y_i|\boldsymbol{\theta})\right]^2}{f(y_i|\boldsymbol{\theta})\left[1-f(y_i|\boldsymbol{\theta})\right]},
$$

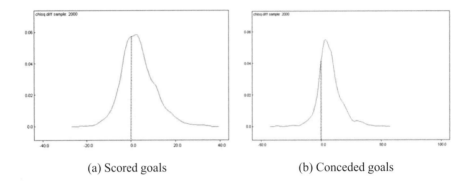

(a) Scored goals   (b) Conceded goals

Figure 10.11   Differences in χ^2 statistics for Manchester United data (Example 10.2).

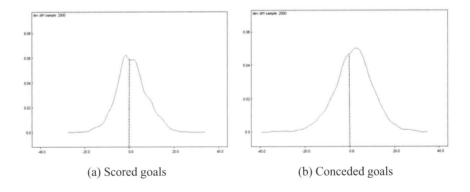

(a) Scored goals   (b) Conceded goals

Figure 10.12   Differences in deviance statistics for Manchester United data (Example 10.2).

where we have used the assumption of multinomial sampling since the size of vector n is fixed by design. The expected frequency of each y under the model is given by its probability function $f(y|\theta)$.

In order to calculate this measure and its corresponding p-values in WinBUGS , we first calculate the relative frequencies (see Section 10.3.1 for details) and then add the following commands:

```
for(i in 1:nf){
    f[i] <- exp( -lambda + y[i]*log(lambda) - logfact(y[i]) )
    chisq.freq.obs.vec[i]<-pow(fx[i]/n - f[i],2)/(f[i]*(1-f[i]))
    chisq.freq.rep.vec[i]<-pow(rel.freq[i]-f[i],2)/(f[i]*(1-f[i]))
}
chisq.freq.obs  <- n*sum(chisq.freq.obs.vec[])
chisq.freq.rep  <- n*sum(chisq.freq.rep.vec[])
chisq.freq.p    <- n*step( chisq.freq.rep-chisq.freq.obs )
```

The p-values based on χ^2 for frequencies are equal to 0.76 and 0.26 for scored and conceded goals, respectively. The latter indicates that, for conceded goals, differences between observed and expected frequencies exist but are not so severe as to reject the simple Poisson model.

Example 10.3 (continued): Goodness-of-fit measures for simulated normal data.

For testing the simple normal model, we propose using χ^2 measures for comparison of the cumulative frequencies rather than the actual frequencies as in the Poisson example above.

For this reason we may define a χ^2-based statistic measuring the difference between observed and expected frequencies for given cutpoints y'_k $(k = 1, \ldots n'$ with $y'_k > y'_{k-1})$ of the assumed distribution. The χ^2 statistic is given by

$$
\begin{aligned}
\chi_f^2(\boldsymbol{y}, \boldsymbol{\theta}, \boldsymbol{y}') &= \sum_{k=1}^{n'+1} \frac{\left[F_{y'_{k-1}, y'_k} - E\left(F_{y'_{k-1}, y'_k} | \boldsymbol{\theta} \right) \right]^2}{\mathrm{Var}\left(F_{y'_{k-1}, y'_k} | \boldsymbol{\theta} \right)} \\
&= n \sum_{k=1}^{n'+1} \frac{\left[F_{y'_k} - F_{y'_{k-1}} - F(y'_k|\boldsymbol{\theta}) + F(y'_{k-1}|\boldsymbol{\theta}) \right]^2}{\left(F(y'_k|\boldsymbol{\theta}) - F(y'_{k-1}|\boldsymbol{\theta}) \right)\left(1 - F(y'_k|\boldsymbol{\theta}) + F(y'_{k-1}|\boldsymbol{\theta}) \right)}
\end{aligned}
$$

where $F_{y'_{k-1}, y'_k} = F_{y'_k} - F_{y'_{k-1}}$ is the observed relative frequency of observations with values between y'_k and y'_{k-1} and

$$
F(y'_0|\boldsymbol{\theta}) = 0, \quad F(y'_{n'+1}|\boldsymbol{\theta}) = 1 \quad \text{and} \quad F_{y'_{n'+1}} = 1 .
$$

This expression was based in the assumption that $F_{y'_{k-1}, y'_k}$ will follow multinomial distribution success probabilities given by the probability of $y \in (y'_{k-1}, y'_k]$ under the assumed model. For the normal model, we consider the $k/(n'+1)$ quantiles of the standardized normal distribution, denoted by z_k, and define y'_k as $y'_k = \mu + z_k/\sqrt{\tau}$. The expression is simplified to

$$
\chi_f^2(\boldsymbol{y}, \boldsymbol{\theta}, \boldsymbol{z}) = n \sum_{k=1}^{n'+1} \frac{\left[F_{\mu+\sigma z_k} - F_{\mu+\sigma z_{k-1}} - \Phi(z_k) + \Phi(y'_{k-1}) \right]^2}{\left(\Phi(z_k) - \Phi(z_{k-1}) \right)\left(1 - \Phi(z_k) + \Phi(z_{k-1}) \right)},
$$

where $\Phi(x)$ is the cumulative distribution function of the standardized normal distribution.

An important drawback of this statistic is its sensitivity to the selection of cutpoints y'_k. Moreover, in the case where n' is large, then a considerable number of observed frequencies $F_{y'_{k-1},y'_k}$ will be equal to zero. For this reason, this statistic requires large samples to enable the use of numerous cutpoints and for accurate estimation.

To avoid the observed sensitivity, we may use a χ^2 statistic based on the cumulative frequencies. We expect the p-values resulting from the corresponding statistic to be more robust since F_{y_k} will not degenerate to zero as the number of cutpoints n' increases. Hence this version of χ^2 can be calculated by

$$\chi_F^2(\boldsymbol{y},\boldsymbol{\theta},\boldsymbol{y}') \;=\; \sum_{i=1}^{n'+1} \frac{\left[F_{y'_k} - E\!\left(F_{y'_k}|\boldsymbol{\theta}\right)\right]^2}{\mathrm{Var}\!\left(F_{y'_k}|\boldsymbol{\theta}\right)} \;=\; n\sum_{k=1}^{n'+1} \frac{\left[F_{y'_k} - F\!\left(y'_k|\boldsymbol{\theta}\right)\right]^2}{F\!\left(y'_k|\boldsymbol{\theta}\right)\left(1 - F\!\left(y'_k|\boldsymbol{\theta}\right)\right)},$$

which will be simplified to

$$\chi_F^2(\boldsymbol{y},\boldsymbol{\theta},\boldsymbol{z}) \;=\; n\sum_{k=1}^{n'+1} \frac{\left[F_{\mu+\sigma\,z_k} - \Phi\!\left(z_k\right)\right]^2}{\Phi\!\left(z_k\right)\left(1 - \Phi\!\left(z_k\right)\right)}.$$

Finally, in order to avoid the use of arbitrary cutpoints, we may construct a χ^2 statistic based directly on the cumulative frequencies

$$\chi_F^2(\boldsymbol{y},\boldsymbol{\theta}) \;=\; n\sum_{i=1}^{n} \frac{\left[F_{y_i} - E\!\left(F_{y_i}|\boldsymbol{\theta}\right)\right]^2}{\mathrm{Var}\!\left(F_{y_i}|\boldsymbol{\theta}\right)} \;=\; n\sum_{i=1}^{n} \frac{\left[F_{y_i} - F\!\left(y_i|\boldsymbol{\theta}\right)\right]^2}{F\!\left(y_i|\boldsymbol{\theta}\right)\left(1 - F\!\left(y_i|\boldsymbol{\theta}\right)\right)}.$$

For the normal case that we consider here, the χ^2 statistic will be given by

$$\chi_F^2(\boldsymbol{y},\boldsymbol{\theta}) \;=\; n\sum_{i=1}^{n} \frac{\left[F_{y_i} - \Phi\!\left((y_i-\mu)/s\right)\right]^2}{\Phi\!\left((y_i-\mu)/s\right)\left[1 - \Phi\!\left((y_i-\mu)/s\right)\right]}.$$

Finally, the Kolmogorov–Smirnov statistic (KS)

$$\mathrm{KS}(\boldsymbol{y},\boldsymbol{\theta}) \;=\; \max_{i\in\{1,\dots,n\}} \left|F_{y_i} - F(y_i|\boldsymbol{\theta})\right|$$

can be adopted in order to evaluate the difference between the predictive and the observed cumulative frequencies.

The original χ^2 as defined in the previous section cannot be used here since in the normal model we fit both the mean and the variance, and hence the abovementioned measure will not be able to trace deficiencies of the model; see Section 10.5 for more details.

P-values are summarized in Table 10.20 for all three datasets. Model code for $\chi_F^2(\boldsymbol{y},\boldsymbol{\theta})$, $\mathrm{KS}(\boldsymbol{y},\boldsymbol{\theta})$, $\chi_f^2(\boldsymbol{y},\boldsymbol{\theta},\boldsymbol{z})$, and $\chi_F^2(\boldsymbol{y},\boldsymbol{\theta},\boldsymbol{z})$ statistics is presented in Tables 10.21 and 10.22. No p-values indicate any important deviation between predicted and observed data. For the third dataset, specific p-values [based on $\chi_F^2(\boldsymbol{y},\boldsymbol{\theta},\boldsymbol{z})$ and KS] are relatively low (compared to the rest), indicating a possible problem in the overall fit of the model. Finally, in the current example, $\chi_f^2(\boldsymbol{y},\boldsymbol{\theta},\boldsymbol{z})$ is quite sensitive to the selection of cutpoints failing to trace possible differences, while $\chi_F^2(\boldsymbol{y},\boldsymbol{\theta},\boldsymbol{z})$ is quite robust over a variety of different number of cutpoints.

Table 10.20   P-values for comparison of differences between predictive and observed frequencies

	$\chi_f^2(\boldsymbol{y}, \boldsymbol{\theta}, \boldsymbol{z})$			$\chi_F^2(\boldsymbol{y}, \boldsymbol{\theta}, \boldsymbol{z})$			$\chi_F^2(\boldsymbol{y}, \boldsymbol{\theta})$	$KS(\boldsymbol{y}, \boldsymbol{\theta})$
	$n' = 9^a$	$n' = 19^b$	$n' = 39^c$	$n' = 9^a$	$n' = 19^b$	$n' = 39^c$		
Dataset 1	0.834	0.805	0.904	0.635	0.642	0.650	0.603	0.661
Dataset 2	0.737	0.778	0.839	0.486	0.527	0.526	0.590	0.509
Dataset 3	0.233	0.464	0.564	0.118	0.126	0.128	0.240	0.144

[a] 10% percentiles.
[b] 5% percentiles.
[c] 2.5% percentiles.

Table 10.21   WinBUGS code for computation of $\chi_F^2(\boldsymbol{y}, \boldsymbol{\theta})$ and Kolmogorov–Smirnov statistic

```
#
# 2nd version of chi^2_F
for (i in 1:n){
    F.obs2[i] <- rank( ordered.y[], i )/n
    F.rep2[i] <- rank( ordered.y.rep[], i )/n
    F.exp.obs2[i] <- phi( (ordered.y[i]-mu)*sqrt(tau) )
    F.exp.rep2[i] <- phi( (ordered.y.rep[i]-mu)*sqrt(tau) )
    chisq.F2.obs.vec[i] <- pow(F.obs2[i]-F.exp.obs2[i],2)/(F.exp.obs2[
        i]*(1-F.exp.obs2[i]))
    chisq.F2.rep.vec[i] <- pow(F.rep2[i]-F.exp.rep2[i],2)/(F.exp.rep2[
        i]*(1-F.exp.rep2[i]))
}
# chisq values
chisq.F2.obs <-  sum( chisq.F2.obs.vec[] )
chisq.F2.rep <-  sum( chisq.F2.rep.vec[] )
# chisq p-value
chisq.F2.p <- step( chisq.F2.rep - chisq.F2.obs )
# KS statistic
#
for (i in 1:n){
    F.diff.obs[i] <- abs( F.obs2[i] - F.exp.obs2[i] )
    F.diff.rep[i] <- abs( F.rep2[i] - F.exp.rep2[i] )
}
ks.obs <-  ranked( F.diff.obs[], n )
ks.rep <-  ranked( F.diff.rep[], n )
# chisq p-value
ks.p <- step( ks.rep - ks.obs )
```

Table 10.22 WinBUGS code for computation of $F_{y'_{i-1}, y_i}$ (denoted by f), $F_{y'_i}$ (denoted by F), $\chi_f^2(y, \theta, z)$ and $\chi_F^2(y, \theta, z)^a$

```
e<-0.0000001  # precision parameter
# zcut = cutpoints for N(0,1)
# calculation of rescaled cutpoints
# ycut = y' (cutpoints for y)
for (i in 1:ncut){ ycut[i] <- mu + zcut[i]*s }
#
# calculation of observed cumulative frequencies
# (for actual and replicated data)
for (i in 1:ncut){
for (j in 1:n){
        bin.freq.obs[i,j] <- step( ycut[i]-ordered.y[j] )
        bin.freq.rep[i,j] <- step( ycut[i]-ordered.y.rep[j] )
    }
  F.obs[i] <- sum( bin.freq.obs[i,1:n] )/n
  F.rep[i] <- sum( bin.freq.rep[i,1:n] )/n
  F.exp[i] <- phi( zcut[i] )
}
#
# calculation of frequencies values within interval y_k', y_{k-1}'
f.exp[1] <- F.exp[1]
f.obs[1] <- F.obs[1]
f.rep[1] <- F.rep[1]
for (i in 2:ncut ){
    f.obs[i] <- F.obs[i]-F.obs[i-1]
    f.rep[i] <- F.rep[i]-F.rep[i-1]
    f.exp[i] <- F.exp[i]-F.exp[i-1]
}
f.obs[ncut+1] <- 1 - F.obs[ncut]
f.rep[ncut+1] <- 1 - F.rep[ncut]
f.exp[ncut+1] <- 1 - F.exp[ncut]
#
for (i in 1:(ncut+1)){
    # setting zero expected frequencies equal to e
    f.exp2[i] <- f.exp[i] + e*equals(f.exp[i],0)
    chisq.f.obs.vec[i]<-pow(f.obs[i]-f.exp2[i],2)/(f.exp2[i]*(1-f.exp2
        [i]))
    chisq.f.rep.vec[i]<-pow(f.rep[i]-f.exp2[i],2)/(f.exp2[i]*(1-f.exp2
        [i]))
}
# chisq values
chisq.f.obs <-  sum( chisq.f.obs.vec[] )
chisq.f.rep <-  sum( chisq.f.rep.vec[] )
# chisq p-value
chisq.f.p <- step( chisq.f.rep - chisq.f.obs )
# 1st version of chi^2_F
for (i in 1:(ncut)){
    F.exp2[i] <- F.exp[i] + e*equals(F.exp[i],0)
    chisq.F.obs.vec[i]<-pow(F.obs[i]-F.exp2[i],2)/(F.exp2[i]*(1-F.exp2
        [i]))
    chisq.F.rep.vec[i]<-pow(F.rep[i]-F.exp2[i],2)/(F.exp2[i]*(1-F.exp2
        [i]))
}
# chisq values
chisq.F.obs <-  sum( chisq.F.obs.vec[] )
chisq.F.rep <-  sum( chisq.F.rep.vec[] )
# chisq p-value
chisq.F.p <- step( chisq.F.rep - chisq.F.obs )
```

^{a}ncut and zcut correspond to n', z and are loaded via a data list format.

10.4 USING CROSS-VALIDATION PREDICTIVE DENSITIES FOR MODEL CHECKING, EVALUATION, AND COMPARISON

All predictive diagnostics presented above have the disadvantage of double usage of the data. In this way, the predictive performance is overestimated. On the other hand, cross-validation offers a well defined tool to evaluate the predictive ability of a model. In this section we focus on the leave-one-out (CV-1) cross-validation. We illustrate estimation of the cross-validatory conditional predictive ordinate and discuss generation of random values from the appropriate CV-1 predictive distribution. After the simulation of such a sample, CV-1 predictive diagnostics can be calculated using a procedure similar to that followed for the full posterior predictive distribution described in the previous sections.

10.4.1 Estimating the conditional predictive ordinate

Estimation of the cross-validatory predictive ordinate can be performed in a straightforward manner using the MCMC output generated from the full posterior density. This can be achieved by the property

$$
\begin{aligned}
\left[f(y_i|\boldsymbol{y}_{\backslash i})\right]^{-1} &= \frac{f(\boldsymbol{y}_{\backslash i})}{f(\boldsymbol{y})} = \int \frac{f(\boldsymbol{y}_{\backslash i}|\boldsymbol{\theta})f(\boldsymbol{\theta})}{f(\boldsymbol{y})}d\boldsymbol{\theta} = \int \frac{1}{f(y_i|\boldsymbol{\theta})}\frac{f(\boldsymbol{y}|\boldsymbol{\theta})f(\boldsymbol{\theta})}{f(\boldsymbol{y})}d\boldsymbol{\theta} \\
&= \int \frac{1}{f(y_i|\boldsymbol{\theta})}f(\boldsymbol{\theta}|\boldsymbol{y})d\boldsymbol{\theta} = E_{\boldsymbol{\theta}|\boldsymbol{y}}\left[\frac{1}{f(y_i|\boldsymbol{\theta})}\right].
\end{aligned}
$$

Thus, we can estimate the inverse predictive ordinate CPO_i from the sample mean of the inverse density/probability function evaluated at y_i for each $\boldsymbol{\theta}^{(t)}$ generated from the full posterior distribution. A Monte Carlo estimate for CPO_i is then given by

$$
\widehat{CPO}_i = \left(\frac{1}{T'}\sum_{t=1}^{T'}\frac{1}{f\left(y_i|\boldsymbol{\theta}^{(t)}\right)}\right)^{-1},
$$

which is the harmonic mean of the density (or probability) distribution function evaluated at y_i for each $\boldsymbol{\theta}^{(t)}$ for $t = 1, 2, \ldots, T'$. Further details can be found in Gelfand and Dey (1994, p. 511), Spiegelhalter et al. (1996a, p. 41), and Gelfand (1996, pp. 154–155).

The preceding estimate can be easily obtained in WinBUGS by defining a deterministic–logical node that will enable calculation of $\left[f\left(y_i|\boldsymbol{\theta}^{(t)}\right)\right]^{-1}$. Its posterior mean provides an estimate for CPO_i^{-1}, and therefore an estimate of CPO_i is directly obtained by the inverse of these quantities. Thus, having calculated node ppo as illustrated in Section 10.3.4, we only need to add the command

```
icpo[i] <- 1/ppo[i]
```

in order to calculate its posterior mean and obtain an estimate of the inverse of the conditional predictive ordinate.

Since, for discrete data, CPO_i can be used to estimate the probability of observing y_i in the future when you have already observed $\boldsymbol{y}_{\backslash i}$, it has a direct interpretation. Comparison of CPO_i with the corresponding relative frequencies estimated from $\boldsymbol{y}_{\backslash i}$ provides a better picture of the model's predictive ability.

For continuous data, computation of $F(y_{(i)}|\boldsymbol{y}_{\backslash i})$ in a manner similar to that in Section 10.3.4 is computationally expensive and is not recommended.

Outliers can be easily traced for large values of ICPOs. In order to have a feeling of the scale of such quantities, we note that the inverse density for the 0.05, 0.02, 0.01, and 0.005 quantiles of the standardized normal are equal to 9.7, 20.6, 37.5, and 69.1, respectively. Assuming approximate normality, ICPO values larger than 40 can be considered as possible outliers and higher than 70 as extreme values.

Example 10.2: Manchester United's goals (continued). MCMC-based estimates of CPOs for Example 10.2 are provided in Table 10.23 and Figure 10.13. Higher differences are observed for the conceded goals as in the comparison of the PPOs in Section 10.3.4.

Table 10.23 Estimated leave-one-out conditional predictive ordinates (CPO) and observed relative frequencies for scored and conceded goals in home games of Manchester United for premiership season 2006–2007 (Example 10.2)[a]

	0	1	2	3	4	5
Scored goals						
Observed relative frequencies						
Actual	0.11	0.16	0.21	0.32	0.16	0.05
Leave-one-out	0.06	0.11	0.17	0.28	0.11	0.00
Estimated CPO (inverse posterior mean)	0.08	0.21	0.26	0.20	0.12	0.05
Estimated ICPO (posterior mean)	11.79	4.76	3.92	4.95	8.53	18.78
Conceded goals						
Observed relative frequencies						
Actual	0.42	0.53	0.05	0.00	0.00	0.00
Leave-one-out	0.39	0.50	0.00	—	—	—
Estimated CPO (inverse posterior mean)	0.52	0.32	0.09	—	—	—
Estimated ICPO (posterior mean)	1.91	3.12	11.19	—	—	—

[a] Leave-one-out observed frequencies are calculated after removing one observation with the corresponding value.

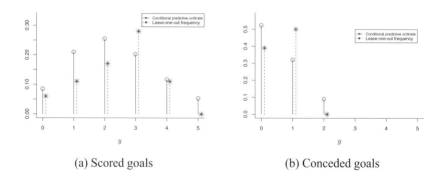

(a) Scored goals   (b) Conceded goals

Figure 10.13 Estimated conditional predictive ordinates (solid lines) and leave-one-out observed relative frequencies (dashed lines) for scored and conceded goals in home games of Manchester United for premiership season 2006–2007 (Example 10.2).

Example 10.3: Simulated normal data (continued). Using the simple rule of thumb proposed above, in Table 10.24 we present values with ICPO> 10. Note that one value (y_3) was indicated as a possible outlier in dataset 1, while in datasets 2 and 3 only y_{20} was flagged as an extremely outlying value. Especially for the last dataset, ICPO is extremely large, indicating that this observation cannot be actually observed under the assumed model. We also present the corresponding inverse PPOs, which have similar but milder behavior. Using inverse PPOs we can trace y_{20} only in the last dataset, where the extreme value of 10 was added.

Table 10.24 Estimated conditional and posterior predictive densities of extreme values for each dataset of Example 10.3; only for values with ICPO> 10

Observation	Dataset 1		Dataset 2		Dataset 3
	y_3	y_7	y_3	y_{20}	y_{20}
Rank	1	19	1	20	20
Value	-2.53	2.76	-2.53	5.00	10.00
ICPO	54.54	25.30	13.21	275.50	60700.0
IPPO	13.11	9.41	6.02	27.13	99.7
CPO	0.018	0.040	0.076	0.004	1.6×10^{-5}
PPO	0.076	0.106	0.166	0.039	0.010

10.4.2 Generating values from the leave-one-out cross-validatory predictive distributions

Generation of values from the leave-one-out cross-validation densities $f(y_i|\mathbf{y}_{\backslash i})$ can be achieved in a straightforward manner by generating values from the corresponding full predictive densities with data $\mathbf{y}_{\backslash i}$. Thus we need to rerun n times the MCMC algorithm (and the corresponding WinBUGS code) for all possible leave-one-out datasets $\mathbf{y}_{\backslash i}$ (for $i = 1, 2, \ldots, n$).

Gelfand (1996) described an alternative approach in order to generate a sample $(\boldsymbol{\theta}'^{(t)}; t = 1, 2, \ldots, T')$ from the leave-one-out posterior density $f(\boldsymbol{\theta}|\mathbf{y}_{\backslash i})$. He proposed resampling with replacement from the MCMC sample $(\boldsymbol{\theta}^{(t)}; t = 1, 2, \ldots, T')$ generated from the posterior distribution. For each $\boldsymbol{\theta}^{(t)}$ the probability of selection w_t is given by

$$w_t = \frac{\left(f(y_i|\mathbf{y}_{\backslash i}, \boldsymbol{\theta}^{(t)}) \right)^{-1}}{\sum_{k=1}^{T'} \left(f(y_i|\mathbf{y}_{\backslash i}, \boldsymbol{\theta}^{(t)}) \right)^{-1}}.$$

Usually random variables Y_i are assumed to be independent given the parameter vector $\boldsymbol{\theta}$. Thus, the resampling weights are simplified to

$$w_t = \frac{\left(f(y_i|\boldsymbol{\theta}^{(t)}) \right)^{-1}}{\sum_{k=1}^{T'} \left(f(y_i|\boldsymbol{\theta}^{(t)}) \right)^{-1}}.$$

Unfortunately, this approach cannot be performed within WinBUGS in a straightforward manner. Hence, in order to implement this approach, we need to generate a sample $\theta^{(t)}$ from the full posterior distribution using WinBUGS, export the data in another statistical program, and finally perform the resampling as well as further calculations in this program. Nevertheless, the abovementioned unnormalized weights can be provided by the `icpo` node defined in the previous section. Once we have generated samples from the corresponding leave-one-out density, we can calculate corresponding residuals or statistics for the implementation of other predictive checks.

Although the leave-one-out cross validation approach avoids double usage of the data, we frequently use the full posterior predictive distribution which is a good approximation of the leave-one-out distribution (Carlin and Louis, 2000, pp. 205–206), provided that the size of data is not small. In this way we avoid extensive computation, making predictive inference straightforward.

10.5   ILLUSTRATION OF A COMPLETE PREDICTIVE ANALYSIS: NORMAL REGRESSION MODELS

In this section we focus on implementation of the predictive model checks in normal regression models. When fitting normal models, we are interested in checking for

1. The structural assumptions of the model (independence, normality, and homoscedasticity of errors)

2. Possible outliers or observations that are rarely observed from the assumed model

3. The goodness-of-fit of the model

All the proposed checks are illustrated using Example 5.1.

10.5.1   Checking structural assumptions of the model

The structural assumptions of the model are the independence of errors, normality, and homoscedasticity (i.e., constant error variance across all observations). In order to identify problems concerning these assumptions, we use Bayesian p-values based on measures used in classical significance tests.

The independence of errors can be checked using the Durbin–Watson statistic

$$\mathrm{DW}(\boldsymbol{y}) = \frac{\sum_{i=2}^{n}(r_i - r_{i-1})^2}{\sum_{i=1}^{n} r_i^2},$$

which is essentially an estimate of the first order autocorrelation (i.e., for lag equal to one) of the errors. In order to be meaningful, observations must be arranged in a chronological order of collection. In this way we can identify dependencies due to the design or the sampling scheme of the study; for example, sequential sampling of people visiting a specific place may result in collecting data from different members of the same family, which induces dependencies in the sample that are due to similar beliefs.

Normality can be checked by

1. Using the approaches proposed in Section 10.3.6, that is, using various versions of χ^2 statistics and the KS statistic.

2. Testing for skewness and symmetry of the errors as described in Section 10.3.7.

3. Calculating the number of observations with standardized residual values outside intervals $(-2, 2)$ and $(-3, 3)$ and compared with the 5% and 1% expected under the normality assumption. In this way, we may identify possible problems at the tails of the distribution.

4. Visual evaluation of

   a. The 95% error bars of the cumulative frequencies $\hat{F}^{\mathrm{rep}}$ of the replicated/predictive values versus the observed cumulative frequency $\hat{F}_i$ for each residual value (as in Section 10.3.2), where

   $$\hat{F}_i^{\mathrm{rep}} = \frac{1}{n} \sum_{i=k}^{n} I\left(r_k^{\mathrm{rep}} \leq r_i\right) \ \text{ and } \ \hat{F}_i = \frac{1}{n} \sum_{i=k}^{n} I\left(r_k \leq r_i\right).$$

   b. The 95% error bars of the ordered predictive standardized residual values $r_{(i)}^{\mathrm{rep}}$ versus the posterior means of ordered standardized residuals $r_{(i)}$ (as in Section 10.3.3).

These measures focus on different aspects of the fitted distribution and thus may identify different problems concerning the normality of errors.

Finally, to test homoscedasticity of errors, we propose dividing the sample into equal parts on the basis of covariates or the estimated mean in each iteration. Then variances must be calculated within each subsample and compared with the overall variance using the Levene test for the equality of variances. Hence we use the test statistic

$$W = \frac{(n - K) \sum_{i=1}^{n} (\overline{Z}_{g_i} - \overline{Z})^2}{(K - 1) \sum_{i=1}^{n} (Z_i - \overline{Z}_{g_i})^2} \ \text{ with } \ Z_i = |r_i - \overline{r}_{g_i}|,$$

where $r_i = y_i - \mu_i$ is the residual for the i observation, $\overline{r}_g$ is the mean residual value for the g group, K is the number of subsamples into which we have divided our data, $\overline{Z}_g$ is the mean of the gth subsample, and g_i is the subsample indicator for the ith subject. We propose using four samples (split in the quantiles of the generated mean) or more if the sample size allows for further spliting.

10.5.2   Detailed checks based on residual analysis

Residual analysis can be performed in much the same way as described in Section 10.3.5. Additionally, PPO and CPO can be used to identify problematic cases. We can use the following techniques for the residual values:

1. Error bars of the cumulative frequencies $\hat{F}^{\mathrm{rep}}$ of the replicated/predictive values vs. the observed cumulative frequency $\hat{F}_i$ for each residual value (as in list item 4a in Section 10.5.1).

2. Error bars of the ordered predictive standardized residual values $r_{(i)}^{\mathrm{rep}}$ versus the posterior means of ordered standardized residuals $r_{(i)}$ (as in list item 4b in Section 10.5.1).

3. Error bars of standardized residuals versus the observation index.

4. Error bars of standardized residuals versus the posterior means of parameters μ_i or values of the explanatory variables.

5. Calculation of the probability of more extreme observations.

6. Calculation of PPO and CPO. Study values with ICPO> 10 and consider as possible outliers observations with ICPO> 40.

All these techniques can be used to trace residuals and evaluate the overall fit of the model. To be more specific, calculations 5 and 6 focus on the identification of outliers while plots of items 1 and 2 evaluate normality, a plot of item 3 identifies autocorrelated errors, while a graph of item 4 traces problems of heteroscedasticity.

10.5.3   Overall goodness-of-fit of the model

In order to check the overall goodness-of-fit, we can calculate the R^2 measures (as defined in Section 5.2.3) to identify the decrease of the error variance due to the explanatory variables included in the model. A rule of thumb indicates that models with acceptable predictive ability must have R^2 higher than 0.7.

A simple check of the importance of each model coefficient can be performed using the posterior tail area probability of value; hence we calculate the quantity

$$\min \left\{ f(\beta_j > 0 \,|\, \boldsymbol{y}), f(\beta_j < 0 \,|\, \boldsymbol{y}) \right\}.$$

Low values of this tail area probability indicate that the corresponding posterior distribution is far away from zero. Therefore the corresponding coefficient will be important for the model. This approach is only indicative and by no means should be considered as a formal model comparison or variable selection method. Such methods are briefly presented in the next chapter.

10.5.4   Implementation using WinBUGS

Calculation of the residual values: Standardized residual values based on actual and predictive values are calculated using the following syntax:

```
for (i in 1:n){
    r[i]        <- (y[i] - mu[i])*sqrt(tau)
    y.rep[i]  ~ dnorm( mu[i], tau )
    r.rep[i] <- (y.rep[i] - mu[i])*sqrt(tau)
}
```

Most of the model checks are based on standardized residual values.

Calculation of the Durbin–Watson statistic: The DW statistic and the corresponding p-value can be calculated in WinBUGS using the following syntax

```
dw.vec1[1] <- 0.0
dw.vec2[1] <- pow( r[1], 2)
dw.rep.vec1[1] <- 0.0
dw.rep.vec2[1] <- pow( r.rep[1], 2)
for ( i in 2:n){
    dw.vec1[i] <- pow( r[i]-r[i-1], 2)
    dw.vec2[i] <- pow( r[i], 2)
    dw.rep.vec1[i] <- pow( r.rep[i]-r.rep[i-1], 2)
    dw.rep.vec2[i] <- pow( r.rep[i], 2)
```

```
        }
    dw.obs <- sum ( dw.vec1 [] )/sum ( dw.vec2 [] )
    dw.rep <- sum ( dw.rep.vec1 [] )/sum ( dw.rep.vec2 [] )
    dw.p <- step ( dw.rep  - dw.obs )
```

where `r[i]` and `r.rep[i]` are the standardized residuals calculated using the actual and predicted values.

Calculation of the Levene statistic: Calculation of the Levene statistic is more complicated than the previous statistics. In the data format we define the number of groups using the deterministic node K and the ranks of the cutpoints in a vector node `ranksK`. In the following syntax we first calculate the ranks of μ_i (`ranksmu`), then the group membership of each observation (`group[i]`), and finally the means in each group (`group.index`) using a binary matrix. Each group mean is then calculated using the equation

$$\overline{Z}_k = \frac{\sum_{i=1}^n G_{ik} y_i}{\sum_{i=1}^n G_{ik}} = \frac{\sum_{i=1}^n I(g_i = k) y_i}{\sum_{i=1}^n I(g_i = k)} = \frac{1}{n} \sum_{i=1}^n I(g_i = k) y_i,$$

where G_{ik} and g_i correspond to the WinBUGS nodes `group.index[i,k]` and `group[i]`, respectively. The WinBUGS code for calculation of this statistic is given in Table 10.25.

Calculation of the proportion of extreme residual values: Calculation of the percentage of standardized residual values that lie outside an interval of type $(-z, z)$ are easily calculated using the expression

$$
\begin{aligned}
P\big(r \notin (-z, z)\big) &= \sum_{i=1}^n I\big(r_i^s \leq -z\big) + \sum_{i=1}^n I\big(r_i^s \geq z\big) \\
&= \sum_{i=1}^n I\big(-z - r_i^s 0 \geq 0\big) + \sum_{i=1}^n I\big(r_i^s - z \geq 0\big) \\
&= \sum_{i=1}^n \Big\{ I\big(-z - r_i^s 0 \geq 0\big) + I\big(r_i^s - z \geq 0\big) \Big\}. \quad (10.7)
\end{aligned}
$$

The binary indicator function $I(x \geq 0)$ corresponds to the WinBUGS command `step(x)`. Hence, in WinBUGS we first calculate each element of the summation (10.7) using the syntax

```
    vec[i] <- step( -z-r[i] ) + step( r[i]-z )
```

and then the proportion of residuals that lie outside the desired interval by

```
    prop <- mean(vec [])
```

Under the normality assumption, we expect 5% and 1% to lie outside intervals $(-2, 2)$ and $(-3, 3)$. To make the comparison more general, we can also calculate the corresponding proportion of the replicated residuals r_i^{rep} under the assumed model and calculate the corresponding p-value as usual.

The full code for the intervals $(-2, 2)$ and $(-3,3)$ is as follows:

```
    # percentage of stand. residuals outside (-2,2) and (-3,3)
    for (i in 1:n){
        p95.vec[i] <-step(r[i]-2)+step(-2-r[i])
        p99.vec[i] <-step(r[i]-3)+step(-3-r[i])
        p95.rep.vec[i] <-step(r.rep[i]-2)+step(-2-r.rep[i])
```

Table 10.25   WinBUGS code for implementation of Levene's statistic.

```
# levene test
# ------------
for (i in 1:n){
    # calculation of the ranks of mu's
    ranksmu[i] <- rank( mu[], i )
    # binary indicators for y_i < cut.y[i]+1
    for (k in 1:K){
        group.temp[i,k] <- step(ranksmu[i]-ranksK[k]-1) }
    # group indicators for cut.y[i-1] < y_i <= cut.y[i]
    group[i] <- sum( group.temp[i,1:K] )+1
    # binary indicators for each group
    for (k in 1:K){ group.index[i,k] <- equals( group[i], k ) }
}
# calculation of group means for y and y.rep
for (k in 1:K){
    barr.obs[k] <- inprod( r[],   group.index[1:n,k] )
                   / sum(group.index[1:n,k])
    barr.rep[k] <- inprod( r.rep[], group.index[1:n,k] )
                   / sum(group.index[1:n,k])
}
# calculation of z[i] for y and y.rep
for (i in 1:n){
    z.obs[i] <- abs( r[i]     - barr.obs[ group[i] ] )
    z.rep[i] <- abs( r.rep[i] - barr.rep[ group[i] ] )
}
# calculation of group means for z.obs and z.rep
for (k in 1:K){
    barz.obs[k] <- inprod( z.obs[],  group.index[1:n,k] )
                   / sum(group.index[1:n,k])
    barz.rep[k] <- inprod( z.rep[],  group.index[1:n,k] )
                   / sum(group.index[1:n,k])
}
# overall means for z's
grandmean.obs <- mean(z.obs[])
grandmean.rep <- mean(z.rep[])


for (i in 1:n){
    lev.obs.vec1[i]<-pow(barz.obs[group[i]] - grandmean.obs ,2)
    lev.rep.vec1[i]<-pow(barz.rep[group[i]] - grandmean.rep ,2)
    lev.obs.vec2[i]<-pow(z.obs[i] - barz.obs[group[i]], 2)
    lev.rep.vec2[i]<-pow(z.rep[i] - barz.rep[group[i]], 2)
}
levenes.obs<-(n-K)*sum(lev.obs.vec1[])/((K-1)*sum(lev.obs.vec2[]))
levenes.rep<-(n-K)*sum(lev.rep.vec1[])/((K-1)*sum(lev.rep.vec2[]))
levenes.p  <-step(levenes.rep-levenes.obs)
```

```
      p99.rep.vec[i] <-step(r.rep[i]-3)+step(-3-r.rep[i])
}
p95.obs <- mean(p95.vec[])
p99.obs <- mean(p99.vec[])
p95.rep <- mean(p95.rep.vec[])
p99.rep <- mean(p99.rep.vec[])
p95.p <- step(p95.rep-p95.obs)
p99.p <- step(p99.rep-p99.obs)
```

10.5.5 An Illustrative example

Example 10.4. Model checking for the soft drink delivery times data (Example 5.1 continued). Returning to the normal regression example (Example 5.1), tail area probabilities of the model checks suggested above are presented in Table 10.26. The proportion of residual values lying outside $(-2, 2)$ and $(-3, 3)$ intervals is given in Table 10.27. Figures 10.14 and 10.15 provide information concerning individual residual values as well as the overall fit of the model. Figure 10.16 offers a visual representation of individual residual values. Possible outliers with ICPO> 10 are given in Table 10.28, while R^2 values are listed in Table 10.29.

Table 10.26 Posterior p-values and tail area probabilities for Example 10.4

Assumption	Statistic	Node	p-value
Independence of errors	Durbin–Watson	dw.p	0.970
Homoscedasticity	Levene (5 groups)	levenes.p	0.077
Normality	χ^2-based tests	chisq.F.p	0.325
		chisq.F2.p	0.374
		chisq.f.p	0.236
	Kolomogov–Smirnov	ks.p	0.303
Symmety	Skewness coefficient	m3.p	0.497
Kurtosis	Kurtosis coefficient	m4.p	0.419
Tail area probabilities	β_0	p.beta0	0.979
	β_1	p.beta1	1.000
	β_2	p.beta2	0.999

Concerning the normality assumption, no important problems exist since all corresponding p-values are reasonably high. The p-values for testing for symmetry and the mesokurticity of the distribution are very close to 0.5 indicating no major deviations from these two assumptions. This is similar to the scenarios in Figures 10.14 and 10.15, where small deviations are observed between expected and observed values.

The proportion of observations outside $(-2, 2)$ is found equal to 5.3%, which is very close to the expected one (p-value 0.622), indicating no problem concerning the 95% tail area probability. Regarding interval $(-3, 3)$, 0.63% observations is found outside this interval, which is much higher than the corresponding expected percentage (p-value 0.853), possibly indicating an outlier.

The Durbin–Watson posterior p-value is high (0.97), indicating strong error autocorrelations (DW posterior mean 1.17, expected posterior mean under the assumed model, 1.92).

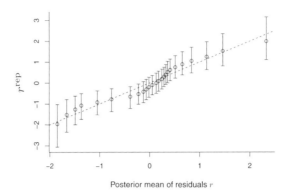

Figure 10.14   95% posterior error bars of ordered predictive standardized residuals $r_{(i)}^{\text{rep}}$ versus posterior means of ordered standardized residuals $r_{(i)}$ for Example 10.4.

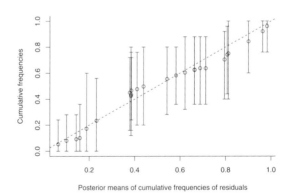

Figure 10.15   95% posterior error bars of cumulative predictive frequencies $\hat{F}_i^{\text{rep}}$ versus cumulative frequencies $\hat{F}_i$ for Example 10.4.

Table 10.27   Proportion of residuals outside $(-2, 2)$ and (-3,3) and their corresponding p-values for Example 10.4

node	mean	sd	MC error	2.5%	median	97.5%	start	sample
p95.obs	0.05324	0.04643	0.00111	0.0	0.04	0.2	1001	1000
p95.rep	0.04824	0.04259	0.001294	0.0	0.04	0.16	1001	1000
p95.p	0.622	0.4849	0.01389	0.0	1.0	1.0	1001	1000
p99.obs	0.00632	0.01459	4.838E-4	0.0	0.0	0.04	1001	1000
p99.rep	0.00296	0.01062	3.483E-4	0.0	0.0	0.04	1001	1000
p99.p	0.853	0.3541	0.01137	0.0	1.0	1.0	1001	1000

The corresponding picture of possible sequential correlations is depicted in Figure 10.16a where we can spot batches of residuals with similar values. Nevertheless, such checks may be nonsense if the ordering of the collected observations has no sequential or time related interpretation.

Similar problems are observed in the homogeneity of variance diagnostics. Levene statistic posterior p-value, after splitting the sample in five groups (of five observations each) according to the order of parameters μ_i, is equal to 0.077, suggesting possible differences in the residuals' dispersion. The picture is similar in Figure 10.16b, where the cluster of the last five observations has clearly higher dispersion than do the remaing values.

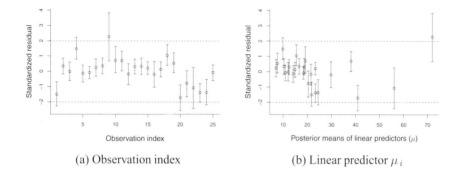

(a) Observation index   (b) Linear predictor μ_i

Figure 10.16   95% posterior error bars of standardized residuals r_i^s versus (a) observation index i and (b) posterior mean of μ_i for Example 10.4.

Finally, individual residual checks, have indicated 13 observations (about 52% of the total observations) with inverse conditional predictive ordinate higher than the threshold value of 10 (ICPO> 10). The large number of observations is alarming, indicating possible problems in the normality of error assumption. Nevertheless, only observations 20 and 9 are spotted with extremely large values (> 40), equal to 58.3 and 2703, respectively. The latter can certainly be considered as an outlier according to the values of all diagnostic checks presented in Table 10.28 (posterior mean of standardized residual 2.24, probability of more extreme 0.033, IPPO $= 52.2$).

Finally, R^2 values are high, close to 1.0, indicating a reasonably good fit of the model. Moreover, tail area probabilities $P(\beta_j > 0)$ ($j = 0, 1, 2$), given in Table 10.29, are high, indicating that the posterior distributions of all parameters are far away from zero.

To summarize, small deviations from normality are present, and autocorrelated and heteroscedastic errors might be more appropriate for this model. Two values are traced as possible outliers, and one of them has extremely high diagnostic values. Finally, the overall fit of the model is good, with both explanatory variables having large impact on the response variable Y (delivery time). Possible use of the logarithms of the delivery time might correct problems in heteroscedasticity of errors and improve the fit of the model. Checks for independence may not have any practical meaning since the order of the values may not represent any actual time sequence.

Table 10.28   Individual residual statistics for Example 10.4[a]

Observation index (i)	Standardized residual (r_i^s)	PEO	PPO	CPO	IPPO	ICPO
19	0.54	0.294	0.100	0.096	9.97	10.42
11	0.67	0.268	0.093	0.088	10.76	11.37
10	0.72	0.247	0.088	0.077	11.38	13.03
21	-0.80	0.207	0.084	0.074	11.96	13.56
18	1.04	0.149	0.069	0.062	14.48	16.24
23	-1.40	0.102	0.045	0.040	22.23	24.90
24	-1.39	0.112	0.047	0.035	21.27	28.45
4	1.49	0.078	0.041	0.032	24.34	31.07
22	-1.13	0.183	0.063	0.032	15.81	31.07
1	-1.53	0.077	0.039	0.028	25.61	35.16
20	-1.76	0.049	0.028	0.017	35.44	58.34
9	2.24	0.033	0.019	0.000	52.22	2703.00

[a] Values with ICPO> 10 only are presented.

Key: PEO = probability of more extreme observation given by $\min\{p_i^r, 1 - p_i^r\}$ with $p_i^r = P(Y_i^{\text{rep}} > Y_i | \boldsymbol{y}))$; IPPO = inverse of PPO (posterior predictive ordinate); ICPO = inverse of CPO (conditional predictive ordinate).

Table 10.29   Posterior summaries for R^2 measures for Example 10.4

node	mean	sd	MC error	2.5%	median	97.5%	start	sample
RB2	0.9556	0.01512	6.105E-4	0.9166	0.9585	0.9766	1001	1000
RB2adj	0.9516	0.01649	6.66E-4	0.9091	0.9547	0.9745	1001	1000

10.5.6 Summary of the model checking procedure

In this section, we have illustrated the full procedure for predictive model checking in normal models as presented in Section 10.5. It is evident that a wide variety of approaches is available for model checking. For this reason it is essential to summarize them as a brief guideline for future use:

1. Select appropriate measures and check the structural assumptions of the model (see Section 10.3.6).

2. Perform outlier analysis using residual checks and by examining extreme PPOs and CPOs (see Sections 10.3.4, 10.3.5, and 10.4.1).

3. Check the overall goodness-of-fit of the model using an appropriate measure (see for examples in Section 10.3.7).

4. Revise the model if checks 1–3 indicate that the model is not valid (and rerun the previous analysis).

Since the procedure is time-consuming, we suggest that users identify which measures they prefer and try to always implement the same measures for checking the model. Graphical representations should also be performed especially in the residual analysis or when comparing the goodness-of-fit of individual values, since they offer valuable information. Finally, cross-validation measures (such as CPO described in this chapter) must also be reported when the sample size is relatively small.

10.6 DISCUSSION

In this chapter, we have described the importance and the practical implementation of the predictive distribution in Bayesian inference using WinBUGS . First, predictions for future response data and estimation of missing response values are illustrated in detail using simple but descriptive examples. The use of the predictive distribution for checking the model fit and for the identification of outliers or surprising observations is further illustrated. Posterior p-values and residual analysis have been also described in detail, while the leave-one-out cross-validation predictive distribution has also been discussed. The chapter closes with a detailed implementation of the described approaches in normal regression models.

A further interesting area for implementation of predictive inference can be found in problems related to the evaluation and construction of performance ranking tables. In such cases we are interested in constructing league and ranking tables, which are particularly useful for the evaluation of institutions (hospitals, universities), individuals (students, athletes), or athletic teams. The predictive distribution can be easily used to rank institutions according to the collected data or even predict the future ranking in a sports contest; see Karlis and Ntzoufras (2008) for an implementation in association football (soccer) data.

The predictive checks described or proposed in this chapter are by no means exhaustive. Many additional approaches exist in Bayesian literature since it is a highly active area of research. Each Bayesian researcher can use different statistics to check whether an assumption of interest is plausible under the light of the observed data. Moreover, the calibration of posterior p-values and related predictive checks is a very important issue that needs further investigation since no clear threshold points exist indicating low and high values (Sellke et al., 2001). Another interesting approach is the idea of predictive checks using the MCMC output of different models running in parallel. In this way, comparison of two models m_1 and m_2 is feasible since we can check the distribution of a statistic when either of the two models under consideration are true; more details can be found in Congdon (2005b, 2006).

Furthermore, predictive distributions have also been used for model comparison or selection. Extensive work on the topic has been published by Proffessor Ibrahim and his associates; see Ibrahim and Laud (1994), Laud and Ibrahim (1995), Hoeting and Ibrahim (1998), and Meyer and Laud (2002) for some indicative examples. Decision theoretic approaches to model selection and averaging have been also illustrated by Guti érrez-Peña and Walker (2001) and Walker et al. (2001), respectively.

In the next and final chapter, we focus on the comparison of different models rather than checking for the adequacy of a single model. The purely Bayesian approach is briefly described by introducing the notions of prior predictive distribution, Bayes factor, and posterior model probabilities. Bayesian variable selection using MCMC and model averaging are also described and illustrated using a simple example. Finally, we focus on three information criteria: the Akaike (AIC), the Bayesian (BIC), and the deviance (DIC) criteria. We illustrate them using simple examples in WinBUGS .

Problems

10.1 Let us consider the number of outstanding claims counts (i.e., contracts with pending claims) for Example 10.1 of Section 10.2.2.

Year	1	2	B 3	4	5	6	7	Partial total
1989	527,003	183,260	90,539	50,471	37,875	10,110	21,165	920,423
1990	594,059	237,289	99,640	52,390	50,723	39,028		1,073,129
1991	810,593	257,122	87,318	72,538	81,336			1,308,907
A 1992	1,012,181	339,664	156,531	181,253				1,689,629
1993	1,459,202	448,651	188,698					2,096,551
1994	1,689,751	563,683						2,253,434
1995	1,698,534							1,698,534

Source: Ntzoufras and Dellaportas (2002). Copyright 2002 by the Society of Actuaries, Schaumburg, Illinois. Reprinted with permission.

  a) Consider a Poisson log-linear model with linear predictor similar to the model used in Section 10.2.2 to estimate the number of contracts that will be paid in the missing part of the triangle.

  b) The last column (total) contains the total number of outstanding claims. Use this information to model the claims counts assuming a multinomial distribution.

10.2 Check whether the fitted distributions are appropriate for the data of Problem 8.1.

10.3 Check whether the fitted distributions are appropriate for the data of Problem 8.6.

10.4 Check whether the Poisson distribution is appropriate for the data of Problem 8.7.

10.5 Perform residual analysis for the models for
  a) Problems 5.1 – 5.8 of Chapter 5.
  b) Problems 6.3 – 6.9 of Chapter 6.
  c) Problems 7.1 – 7.10 of Chapter 7.
  d) Trace outliers and extreme values.
  e) Check whether residuals have a normal (or approximately normal) distribution.

10.6 Check whether the assumptions of the normal model hold for
  a) Problems 5.1 – 5.8 of Chapter 5.
  b) Problems 6.3 – 6.9 of Chapter 6.

10.7 Implement goodness-of-fit tests for the models implemented in Problems 7.1 – 7.10 of Chapter 7.

10.8 Estimate the cross-validatory predictive measures for the data of Problems 8.1 and 8.6.

CHAPTER 11

BAYESIAN MODEL AND VARIABLE EVALUATION

11.1 PRIOR PREDICTIVE DISTRIBUTIONS AS MEASURES OF MODEL COMPARISON: POSTERIOR MODEL ODDS AND BAYES FACTORS

Bayesian comparison of two competing models m_1 and m_2 (or the corresponding hypotheses H_1 and H_2) is performed via the posterior model probabilities $f(m_k|\boldsymbol{y})$ and their corresponding ratio

$$\text{PO}_{12} = \frac{f(m_1|\boldsymbol{y})}{f(m_2|\boldsymbol{y})} = \frac{f(\boldsymbol{y}|m_1)}{f(\boldsymbol{y}|m_2)} \times \frac{f(m_1)}{f(m_2)} = B_{12} \times \frac{f(m_1)}{f(m_2)}, \qquad (11.1)$$

which is termed the *posterior model odds* of model m_1 versus model m_2. B_{12} is the Bayes factor of model m_1 versus model m_2 defined as the ratio of the "marginal" likelihoods $f(\boldsymbol{y}|m_1)$ and $f(\boldsymbol{y}|m_2)$. Hence we can summarize the preceding expression by

$$\text{Posterior model odds} = \text{Bayes factors} \times \text{prior model odds},$$

where *prior model odds* is the ratio of the prior model probabilities $f(m_1)$ and $f(m_2)$. The marginal likelihood $f(\boldsymbol{y}|m)$ for $m \in \{m_1, m_2\}$ is given by

$$f(\boldsymbol{y}|m) = \int f(\boldsymbol{y}|\boldsymbol{\theta}_m, m) f(\boldsymbol{\theta}_m|m) d\boldsymbol{\theta}_m,$$

where $f(\boldsymbol{y}|\boldsymbol{\theta}_m, m)$ is the likelihood of model m with parameters $\boldsymbol{\theta}_m$ and $f(\boldsymbol{\theta}_m|m)$ is the prior of $\boldsymbol{\theta}_m$ under model m. This definition of the *marginal likelihood* coincides with the

Bayesian Modeling Using WinBUGS, by Ioannis Ntzoufras
Copyright ©2009 John Wiley & Sons, Inc.

prior predictive density as defined in the previous chapter; see Eq. (10.1). Hence the Bayes factor is the ratio of the prior predictive densities under the compared models. Kass and Raftery (1995) call $f(\boldsymbol{y}|m)$ as the *predictive probability of the data* under model m, that is, the probability of obtaining the actually observed data before any data are available under the assumption that model m is the true stochastic mechanism generating the observed data.

The Bayes factor is of prominent importance within Bayesian theory since equal prior model probabilities are considered as a default choice when no information is available concerning the model structure. Hence, model comparison and model evaluation are frequently based solely on Bayes factors. If we consider the model comparison as a hypothesis testing problem where interest lies in evaluating the null hypothesis H_0 (corresponding to a model m_0) against the alternative H_1 (corresponding to a model m_1), then both the posterior model odds PO_{10} and the corresponding Bayes factor B_{10} evaluate the evidence *against* the null hypothesis, which is familiar to classical significance tests. On the other hand, PO_{01} and B_{01} evaluate the evidence *in favor* of the null hypothesis, which is not feasible in classical significance tests. To summarize, using posterior model odds and Bayes factors, we can

- Evaluate the evidence in favor of the null hypothesis

- Compare two or more non-nested models

- Draw inferences without ignoring model uncertainty

- Determine which set of explanatory variables gives better predictive results

Suggested interpretation of the Bayes factor is provided by Kass and Raftery (1995); see also Table 11.1.

Table 11.1 Bayes factor interpretation according to Kass and Raftery (1995)

$\log(B_{10})$	B_{10}	Evidence against H_0
$0-1$	$1-3$	Negligible
$1-3$	$3-20$	Positive
$3-5$	$20-150$	Strong
>5	>150	Very strong

Note that the integral involved in computation of the prior predictive density and Bayes factors is analytically tractable only in certain restricted examples, and therefore asymptotic approximations or Monte Carlo methods are frequently used instead. Moreover, these densities are sensitive to the dispersion of the prior distributions $f(\boldsymbol{\theta}_{m_k}|m_k)$. As a consequence, when large prior dispersion is specified for parameters that differ across compared models, then more parsimonious models are supported irrespective of which data we observe. For the same reason, flat improper prior distribution cannot be used since the integrals involved in the computation of Bayes factors are not tractable (i.e., tend to infinity). These problems are widely known as the *Lindley–Bartlett* or *Jeffreys paradox* (Lindley, 1957; Bartlett, 1957). Posterior model odds and Bayes factors cannot be generally calculated within Win-BUGS unless sophisticated approaches are used.

The difficulties described above have initiated a wide discussion concerning the use of Bayes factor in Bayesian inference [see, e.g., Gelfand (1996)], resulting in numerous versions of Bayes factor as well as different alternative approaches for model comparison and checking. Some of the most popular Bayes factor versions developed are the posterior, fractional, intrinsic and pseudo Bayes factors; see Aitkin (1991), O'Hagan (1995), Berger and Pericchi (1996a; 1996b), and Geisser and Eddy (1979), respectively.

11.2 SENSITIVITY OF THE POSTERIOR MODEL PROBABILITIES: THE LINDLEY–BARTLETT PARADOX

Lindley (1957) reported a surprising behavior of the Bayes factor and, consequently, of the posterior odds. Let us consider the model $Y_i \sim N(\theta, \sigma^2)$ with σ^2 known in which we wish to test the following simple hypotheses: $H_0 : \theta = \theta_0$ versus $H_1 : \theta \neq \theta_0$. Let us consider the usual normal prior distribution for θ under the null hypothesis centered around the value of the alternative H_0. Hence we consider $f(\theta|H_1) \sim N(\theta_0, \sigma_\theta^2)$. Then the posterior model odds are given by

$$
\text{PO}_{01} = \frac{f(H_0)}{f(H_1)} \sqrt{1 + n \frac{\sigma_\theta^2}{\sigma^2}} \exp \left\{ -\frac{1}{2\sigma^2} \left[\sum_{i=1}^n (y_i - \theta_0)^2 - \sum_{i=1}^n (y_i - \overline{y})^2 - \frac{n(\overline{y} - \theta_0)^2}{1 + n\sigma_\theta^2/\sigma^2} \right] \right\}.
$$

The resulting posterior odds depends on the sample size n and $\overline{y}$. Lindley considered samples being at the limit of rejection area of the usual significance test of $100q\%$ significance level. In these samples, $\overline{y} = \theta_0 \pm z_{q/2}\sigma/\sqrt{n}$, resulting in posterior odds equal to

$$
\text{PO}_{01} = \frac{f(H_0)}{f(H_1)} \sqrt{1 + n \frac{\sigma_\theta^2}{\sigma^2}} \exp \left\{ -\frac{1}{2} \frac{n\sigma_\theta^2}{n\sigma_\theta^2 + \sigma^2} Z_{q/2}^2 \right\},
$$

where z_q is the q quantile of the standardized normal distribution. From here, we can observe that when n increases, then the posterior odds also increase and tend to infinity for a given significance level q supporting the simpler hypothesis in contrast to classical significance tests, which reject the null hypothesis for sufficiently large n. This leads to a paradox since for sufficiently large samples, Bayesian methods and significance tests support different hypotheses.

Bartlett (1957) extended this paradox by observing that the prior variance used in the Bayes factor also affects the supporting hypothesis. He observed that the posterior Bayes factor in favor of H_0 is sensitive to the magnitude of the prior variance σ_θ^2. This behavior also appears for the Bayes factors of any two nested models/hypotheses, in which case results are sensitive to the variances of the additional parameters. This phenomenon is of much concern since the usual improper priors cannot be determined because of unknown constants involved in the computation of posterior odds while large variance priors fully support the simplest model. For these reasons, improper priors cannot be used and a large prior variance must be carefully selected in order to avoid activating the paradox described above. Although Bartlett (1957) noted this behavior, the term "Lindley's paradox" is used for any case where Bayesian and significance tests result in contradictive evidence (Shafer, 1982). The term "Bartlett paradox" was used by a few researchers such as Kass and Raftery (1995), while others refer to this phenomenon as "Jeffreys paradox" (Lindley, 1980; Berger and Delampady, 1987) or "Jeffreys-Lindley's paradox" (Robert, 1993). Lindley's paradox is discussed in detail by Shafer (1982).

Although Lindley (1993) noted that the sensitivity of the Bayes factor is natural, this drawback resulted in a series of publications that attempted to resolve the problem by finding good reference priors for model selection. In this category we may include Bayes factor variants (posterior, fractional and intrinsic); see Aitkin (1991), O'Hagan (1995), and Berger and Pericchi (1996a; 1996b), respectively.

11.3 COMPUTATION OF THE MARGINAL LIKELIHOOD

Various of alternative methods have been proposed in the literature in order to accurately estimate or approximate the marginal likelihood. Here we briefly present the most important ones and focus attention on the simpler ones that can be obtained via WinBUGS . Detailed review of the related methods can be found in Kass and Raftery (1995) and Gamerman and Lopes (2006, chap. 7).

11.3.1 Approximations based on the normal distribution

The most popular approximation of the marginal likelihood is the so called Laplace approximation based on the normal distribution. This approximation results in

$$f(\boldsymbol{y}|m) \approx (2\pi)^{d_m/2} \big|\widetilde{\boldsymbol{\Sigma}}_m\big|^{1/2} f(\boldsymbol{y}|\widetilde{\boldsymbol{\theta}}_m, m) f(\widetilde{\boldsymbol{\theta}}_m|m), \tag{11.2}$$

where $\widetilde{\boldsymbol{\theta}}_m$ is the posterior mode of the parameters of model m and $\widetilde{\boldsymbol{\Sigma}}_m = \left(\mathbf{H}_m(\widetilde{\boldsymbol{\theta}}_m) \right)^{-1}$, with $\mathbf{H}_m(\widetilde{\boldsymbol{\theta}}_m)$ being equal to the minus of the second derivative matrix of the log-posterior density $\log f(\boldsymbol{\theta}|\boldsymbol{y}, m)$ evaluated at the posterior mode $\widetilde{\boldsymbol{\theta}}_m$.

To avoid analytic calculation of $\widetilde{\boldsymbol{\Sigma}}_m$ and $\widetilde{\boldsymbol{\theta}}_m$, we may use the Laplace–Metropolis estimator proposed by Raftery (1996b) and Lewis and Raftery (1997). Using this approach, we estimate $\widetilde{\boldsymbol{\theta}}_m$ and $\widetilde{\boldsymbol{\Sigma}}_m$ from the output of a MCMC algorithm by the posterior mean and variance–covariance matrix of the simulated values, respectively. Hence the Laplace–Metropolis estimator is given by

$$f(\boldsymbol{y}|m) \approx \widehat{f}_1(\boldsymbol{y}|m) = (2\pi)^{d_m/2} \big|\boldsymbol{S}_m\big|^{1/2} f(\boldsymbol{y}|\overline{\boldsymbol{\theta}}_m, m) f(\overline{\boldsymbol{\theta}}_m|m) \tag{11.3}$$

with

$$\overline{\boldsymbol{\theta}}_m = \frac{1}{T} \sum_{t=1}^{T} \boldsymbol{\theta}_m^{(t)} \text{ and } \boldsymbol{S}_m = \frac{1}{T-1} \sum_{t=1}^{T} \left(\boldsymbol{\theta}_m^{(t)} - \overline{\boldsymbol{\theta}}_m \right) \left(\boldsymbol{\theta}_m^{(t)} - \overline{\boldsymbol{\theta}}_m \right)^T .$$

Note that this approximation works efficiently when the posterior distributions are symmetric. For this reason, substituting the posterior mode by the corresponding mean will not affect results.

11.3.2 Sampling from the prior: A naive Monte Carlo estimator

Since the marginal likelihood is given by

$$f(\boldsymbol{y}|m) = \int f(\boldsymbol{y}|\boldsymbol{\theta}_m, m) f(\boldsymbol{\theta}_m|m) d\boldsymbol{\theta}_m,$$

a straightforward estimate can be provided by

$$\widehat{f}_2(\boldsymbol{y}|m) = \frac{1}{T} \sum_{t=1}^{T} f(\boldsymbol{y}|\boldsymbol{\theta}_m^{*(t)}, m) \text{ with } \boldsymbol{\theta}_m^{*(t)} \sim f(\boldsymbol{\theta}_m|m) \text{ for } t = 1, \ldots, T .$$

Note that in this last equation, we use a sample from the prior distribution denoted by $\boldsymbol{\theta}_m^{*(t)}$ (for $t = 1, \ldots, T$) instead of the usual posterior sample denoted by $\boldsymbol{\theta}_m^{(t)}$ (for $t = 1, \ldots, T$). Unfortunately, this estimator is quite inefficient, especially when the prior distribution $f(\boldsymbol{\theta}_m|m)$ considerably differs from the posterior $f(\boldsymbol{\theta}_m|\boldsymbol{y}, m)$, for example, when the prior is relatively flat in comparison to the posterior. In such a case, the likelihood values for the majority of the generated (from the prior) values $\boldsymbol{\theta}_m^{(t)}$ will be almost zero and hence minimal contribution to the summation of the estimate above. This results in large standard errors of the estimate and very slow convergence to the true value (Kass and Raftery, 1995).

11.3.3   Sampling from the posterior: The harmonic mean estimator

We can construct a simple Monte Carlo estimate with values simulated from the posterior distribution by considering the following equation:

$$\int \frac{1}{f(\boldsymbol{y}|\boldsymbol{\theta}_m, m)} f(\boldsymbol{\theta}_m|\boldsymbol{y}, m) d\boldsymbol{\theta}_m = \int \frac{1}{f(\boldsymbol{y}|\boldsymbol{\theta}_m, m)} \frac{f(\boldsymbol{y}|\boldsymbol{\theta}_m, m) f(\boldsymbol{\theta}_m|m)}{f(\boldsymbol{y}|m)} d\boldsymbol{\theta}_m = \frac{1}{f(\boldsymbol{y}|m)}.$$

Hence an estimate can be obtained on the basis of the harmonic mean of the likelihoods calculated in each step of an MCMC algorithm given by

$$\widehat{f}_3(\boldsymbol{y}|m) = \left(\frac{1}{T} \sum_{t=1}^{T} \left\{ f(\boldsymbol{y}|\boldsymbol{\theta}_m^{(t)}, m) \right\}^{-1} \right)^{-1}.$$

This estimator, introduced by Newton and Raftery (1994), is called the *harmonic mean estimator*. Although this estimator is simple, it is quite unstable and sensitive to small likelihood values and hence is not recommended; see Raftery (1996b, p. 169) and Raftery et al. (2007) for discussions on the efficiency of this estimator. Raftery et al. (2007) have proposed methodology for improving the efficiency of this estimator.

The estimator derived above can be extended to Gelfand and Dey's (1994) *generalized harmonic mean estimator*, which is based on the following identity:

$$\begin{aligned}
\frac{1}{f(\boldsymbol{y}|m)} &= \int \frac{1}{f(\boldsymbol{y}|m)} g(\boldsymbol{\theta}_m) d\boldsymbol{\theta}_m = \int \frac{1}{f(\boldsymbol{y}|\boldsymbol{\theta}_m, m) f(\boldsymbol{\theta}_m|m)} f(\boldsymbol{\theta}_m|\boldsymbol{y}, m) g(\boldsymbol{\theta}_m) d\boldsymbol{\theta}_m \\
&= \int \frac{g(\boldsymbol{\theta}_m)}{f(\boldsymbol{y}|\boldsymbol{\theta}_m, m) f(\boldsymbol{\theta}_m|m)} f(\boldsymbol{\theta}_m|\boldsymbol{y}, m) d\boldsymbol{\theta}_m \\
&= E_{f(\boldsymbol{\theta}_m|y, m)} \left\{ \frac{g(\boldsymbol{\theta}_m)}{f(\boldsymbol{y}|\boldsymbol{\theta}_m, m) f(\boldsymbol{\theta}_m|m)} \right\}.
\end{aligned}$$

Hence an estimator is given by the harmonic mean of $W = g(\boldsymbol{\theta}_m) / \left[f(\boldsymbol{y}|\boldsymbol{\theta}_m, m) f(\boldsymbol{\theta}_m|m) \right]$ evaluated over values $\boldsymbol{\theta}_m^{(1)}, \ldots, \boldsymbol{\theta}_m^{(T)}$ generated from the posterior distribution using an MCMC algorithm, and hence the generalized harmonic mean estimator is given by

$$\widehat{f}_4(\boldsymbol{y}|m) = \left(\frac{1}{T} \sum_{t=1}^{T} w^{(t)} \right)^{-1} = \left(\frac{1}{T} \sum_{t=1}^{T} \frac{g(\boldsymbol{\theta}_m^{(t)})}{f(\boldsymbol{y}|\boldsymbol{\theta}_m^{(t)}, m) f(\boldsymbol{\theta}_m^{(t)}|m)} \right)^{-1}. \qquad (11.4)$$

A special case of this estimator is the harmonic mean estimator for $g(\boldsymbol{\theta}_m) = f(\boldsymbol{\theta}_m|m)$. The "importance" density g must be selected carefully and must be close to the posterior density. In such a case, inflation of the components of the summation involved in the estimator derived above is avoided since small likelihood values will be moderated by low small values of the importance density. A multivariate normal or t distribution with mean and variance set equal to the posterior mean and variance (estimated from an MCMC algorithm) usually provides an accurate estimate.

11.3.4 Importance sampling estimators

Sampling from an arbitrary importance sampling density $g(\boldsymbol{\theta})$ results in the estimate

$$\widehat{f}_5(\boldsymbol{y}|m) = \frac{1}{T}\sum_{t=1}^{T} \frac{f\left(\boldsymbol{y}|\boldsymbol{\theta}_m^{*(t)}, m\right)f\left(\boldsymbol{\theta}_m^{*(t)}|m\right)}{g\left(\boldsymbol{\theta}_m^{*(t)}\right)} \quad \text{with } \boldsymbol{\theta}_m^{*(t)} \sim g\left(\boldsymbol{\theta}_m^{*(t)}\right) \text{ for } t = 1, 2, \ldots, T$$

since

$$
\begin{aligned}
f(\boldsymbol{y}|m) &= \int f(\boldsymbol{y}|\boldsymbol{\theta}_m, m)f(\boldsymbol{\theta}_m|m)d\boldsymbol{\theta}_m = \int \frac{f(\boldsymbol{y}|\boldsymbol{\theta}_m, m)f(\boldsymbol{\theta}_m|m)}{g(\boldsymbol{\theta}_m)}g(\boldsymbol{\theta}_m)d\boldsymbol{\theta}_m \\
&= E_g\left[\frac{f(\boldsymbol{y}|\boldsymbol{\theta}_m, m)f(\boldsymbol{\theta}_m|m)}{g(\boldsymbol{\theta}_m)}\right].
\end{aligned}
$$

When the sampling density is known up to a constant [i.e., $g(\boldsymbol{\theta}_m) = C\,g^*(\boldsymbol{\theta}_m) \propto g^*(\boldsymbol{\theta}_m)$], then the estimator is slightly changed to

$$\widehat{f}_6(\boldsymbol{y}|m) = \frac{\sum_{t=1}^{T}\left\{f\left(\boldsymbol{y}|\boldsymbol{\theta}_m^{*(t)}, m\right)f\left(\boldsymbol{\theta}_m^{*(t)}|m\right)\left[g^*\left(\boldsymbol{\theta}_m^{*(t)}\right)\right]^{-1}\right\}}{\sum_{t=1}^{T}f\left(\boldsymbol{\theta}_m^{*(t)}|m\right)\left[g^*\left(\boldsymbol{\theta}_m^{*(t)}\right)\right]^{-1}}$$

since

$$
\begin{aligned}
C &= \int C f(\boldsymbol{\theta}_m|m)d\boldsymbol{\theta}_m = \int \frac{g(\boldsymbol{\theta}_m)}{g^*(\boldsymbol{\theta}_m)}f(\boldsymbol{\theta}_m|m)d\boldsymbol{\theta}_m = \int \frac{f(\boldsymbol{\theta}_m|m)}{g^*(\boldsymbol{\theta}_m)}g(\boldsymbol{\theta}_m)d\boldsymbol{\theta}_m \\
&= E_g\left[\frac{f(\boldsymbol{\theta}_m|m)}{g^*(\boldsymbol{\theta}_m)}\right].
\end{aligned}
$$

Different selection of the importance sampling density results in different estimates: the naive estimator for $g(\boldsymbol{\theta}_m) = f(\boldsymbol{\theta}_m|m)$ (importance density = prior), the harmonic mean estimator for $g(\boldsymbol{\theta}_m) = f(\boldsymbol{\theta}_m|\boldsymbol{y}, m)$ (importance density = posterior), and Newton and Raftery's (1994) estimator for $g(\boldsymbol{\theta}_m) = wf(\boldsymbol{\theta}_m|m) + (1-w)f(\boldsymbol{\theta}_m|\boldsymbol{y}, m)$ for $0 < w < 1$ (importance density = mixture of the prior and the posterior density).

11.3.5 Bridge sampling estimators

Another effective Monte Carlo estimator of the marginal likelihood was introduced by Meng and Wong (1996) on the basis of a simulation sampling scheme called *bridge sampling*. Using this approach, we can write

$$\frac{\int h(\boldsymbol{\theta}_m)f(\boldsymbol{\theta}_m|m)f(\boldsymbol{y}|\boldsymbol{\theta}_m, m)g(\boldsymbol{\theta}_m)d\boldsymbol{\theta}_m}{\int h(\boldsymbol{\theta}_m)g(\boldsymbol{\theta}_m)f(\boldsymbol{\theta}_m|\boldsymbol{y}, m)d\boldsymbol{\theta}_m}$$

$$= \frac{\int h(\boldsymbol{\theta}_m)f(\boldsymbol{\theta}_m|m)f(\boldsymbol{y}|\boldsymbol{\theta}_m, m)g(\boldsymbol{\theta}_m)d\boldsymbol{\theta}_m}{\int h(\boldsymbol{\theta}_m)g(\boldsymbol{\theta}_m)\left\{f(\boldsymbol{y}|\boldsymbol{\theta}_m, m)f(\boldsymbol{\theta}_m)/f(\boldsymbol{y}|m)\right\}d\boldsymbol{\theta}_m} = f(\boldsymbol{y}|m) \Leftrightarrow$$

$$f(\boldsymbol{y}|m) = \frac{E_{g(\theta)}\left\{h(\boldsymbol{\theta}_m)f(\boldsymbol{\theta}_m|m)f(\boldsymbol{y}|\boldsymbol{\theta}_m,m)\right\}}{E_{f(\theta_m|y,m)}\left\{h(\boldsymbol{\theta}_m)g(\boldsymbol{\theta}_m)\right\}},$$

where $g(\boldsymbol{\theta}_m)$ is a proposal distribution and $h(\boldsymbol{\theta}_m)$ is an arbitrary bridge function. A Monte Carlo estimator is given by

$$\widehat{f}_7(\boldsymbol{y}|m) = \frac{\frac{1}{T_1}\sum_{t=1}^{T_1} h\left(\boldsymbol{\theta}_m^{*(t)}\right)f\left(\boldsymbol{\theta}_m^{*(t)}|m\right)f\left(\boldsymbol{y}|\boldsymbol{\theta}_m^{*(t)},m\right)}{\frac{1}{T_2}\sum_{t=1}^{T_2} h\left(\boldsymbol{\theta}_m^{(t)}\right)g\left(\boldsymbol{\theta}_m^{(t)}\right)},$$

where $\boldsymbol{\theta}_m^{*(1)},\ldots,\boldsymbol{\theta}_m^{*(t)},\ldots\boldsymbol{\theta}_m^{*(T_1)}$ is a sample from the proposal distribution $g(\boldsymbol{\theta}_m)$ and $\boldsymbol{\theta}_m^{(1)},\ldots,\boldsymbol{\theta}_m^{(t)},\ldots\boldsymbol{\theta}_m^{(T_2)}$ is a sample from the posterior distribution $f(\boldsymbol{\theta}_m|\boldsymbol{y},m)$ (usually taken from an MCMC algorithm).

Obviously the efficiency of this estimator depends on selection of the pɪ ·posal distribution g and the bridge function h. The proposal distribution must be close to the target posterior distribution. The closer are the two distributions, the more efficient is the estimator. Moreover, the function h plays the role of the bridge that links the two distributions. For this reason, it must support both distributions. According to Meng and Wong (1996), an optimal function (under the mean square error) is the bridge function $h(\boldsymbol{\theta}_m) = T_2/\{T_2 g(\boldsymbol{\theta}_m) + T_1 f(\boldsymbol{\theta}_m|\boldsymbol{y},m)\}$ which depends on the marginal likelihood. Hence an estimate of the marginal likelihood can be obtained using the iterative scheme proposed by Meng and Wong (1996). Finally, a further generalization of the method was proposed by Gelman and Meng (1998) using a series of functions called *path*, while the corresponding estimator was called the *path sampling estimator* (Gamerman and Lopes, 2006). Although this method is quite efficient, the required specification of both the proposal distribution and the bridge function make this approach quite unattractive especially for inexperienced users.

11.3.6 Chib's marginal likelihood estimator

Among the most popular estimators for calculation of the marginal likelihood are the estimators proposed by Chib (1995) based on the Gibbs output and their extension by Chib and Jeliazkov (2001) using the output of Metropolis–Hastings algorithm. Both these estimators are also called *candidate's estimators* and are based on the identity

$$f(\boldsymbol{y}|m) = \frac{f(\boldsymbol{y}|\boldsymbol{\theta}_m,m)f(\boldsymbol{\theta}_m|m)}{f(\boldsymbol{\theta}_m|\boldsymbol{y},m)} \tag{11.5}$$

for all data $\boldsymbol{y}$ and for any value of the parameters $\boldsymbol{\theta}_m$ (Besag, 1989). Any value $\boldsymbol{\theta}_m^*$ will provide the correct value for the marginal likelihood. The problem is that the posterior distribution is not available and hence must be estimated. Hence an estimate of the posterior predictive ordinate $\widehat{f}(\boldsymbol{\theta}_m^*|\boldsymbol{y},m)$ at a point of interest $\boldsymbol{\theta}_m^*$ will be required to estimate the marginal likelihood

$$\widehat{f}(\boldsymbol{y}|m) = \frac{f(\boldsymbol{y}|\boldsymbol{\theta}_m^*,m)f(\boldsymbol{\theta}_m^*|m)}{\widehat{f}(\boldsymbol{\theta}_m^*|\boldsymbol{y},m)}.$$

When the full conditionals are known, then Gibbs sampling is facilitated. In this case the posterior distribution can be rewritten as

$$f(\boldsymbol{\theta}_m|\boldsymbol{y},m) = f(\theta_1|\boldsymbol{y},m)f(\theta_2,\ldots,\theta_d|\theta_1,\boldsymbol{y},m)$$

$$= \ f(\theta_1|\boldsymbol{y},m)f(\theta_2|\theta_1,\boldsymbol{y},m)f(\theta_3,\ldots,\theta_d|\theta_1,\theta_2,\boldsymbol{y},m)$$

$$= \ f(\theta_1|\boldsymbol{y},m)\prod_{j=2}^{d} f(\theta_j|\theta_1,\ldots,\theta_{j-1},\boldsymbol{y},m),$$

where d is the dimension of the parameter vector $\boldsymbol{\theta}_m = (\theta_1,\ldots,\theta_d)^T$. These conditional distributions can also be written as

$$f(\theta_1|\boldsymbol{y},m) \ = \ \int\cdots\int f(\theta_1,\theta_2,\ldots,\theta_d|\boldsymbol{y},m)d\theta_2\cdots d\theta_d$$

$$= \ \int\cdots\int f(\theta_1|\theta_2,\ldots,\theta_d,\boldsymbol{y},m)f(\theta_2,\ldots,\theta_d|\boldsymbol{y},m)d\theta_2\cdots d\theta_d$$

and

$$f(\theta_j|\theta_1,\ldots,\theta_{j-1},\boldsymbol{y},m)$$

$$= \ \int\cdots\int f(\theta_j,\theta_{j+1},\ldots,\theta_d|\theta_1,\ldots,\theta_{j-1},\boldsymbol{y},m)d\theta_{j+1}\cdots d\theta_d$$

$$= \ \int\cdots\int f(\theta_j|\boldsymbol{\theta}_{m,\backslash j},\boldsymbol{y},m)f(\theta_{j+1},\ldots,\theta_d,|\theta_1,\ldots,\theta_{j-1},\boldsymbol{y},m)d\boldsymbol{\theta}_{m,\backslash j},$$

where $\boldsymbol{\theta}_{m,\backslash j} = (\theta_1,\ldots,\theta_{j-1},\theta_{j+1},\ldots,\theta_d)$. Estimates can be obtained from the Gibbs output using the following simple estimators

$$\widehat{f}(\theta_1|\boldsymbol{y},m) \ = \ \frac{1}{T}\sum_{t=1}^{T} f(\theta_1^{(t)}|\theta_2^{(t)},\ldots,\theta_d^{(t)},\boldsymbol{y},m)$$

$$\widehat{f}(\theta_j|\theta_1,\ldots,\theta_{j-1},\boldsymbol{y},m) \ = \ \frac{1}{T}\sum_{t=1}^{T} f(\theta_j^{(t)}|\theta_1^{(t)},\ldots,\theta_{j-1}^{(t)},\theta_{j+1}^{(t)},\ldots,\theta_d^{(t)},\boldsymbol{y},m)$$

since the full conditional posterior densities are available in this sampling scheme.

Chib (1995) proposed to estimate the marginal likelihood using

$$\widehat{f}(\boldsymbol{\theta}_m^*|\boldsymbol{y},m) \ = \ \widehat{f}(\theta_1^*|\boldsymbol{y},m)\widehat{f}(\theta_2^*,\ldots,\theta_d^*|\theta_1^*,\boldsymbol{y},m)$$

$$= \ \widehat{f}(\theta_1^*|\boldsymbol{y},m)\widehat{f}(\theta_2^*|\theta_1^*,\boldsymbol{y},m)\widehat{f}(\theta_3^*,\ldots,\theta_d^*|\theta_1^*,\theta_2^*,\boldsymbol{y},m)$$

$$= \ \widehat{f}(\theta_1^*|\boldsymbol{y},m)\prod_{j=1}^{d}\widehat{f}(\theta_j^*|\theta_1^*,\ldots,\theta_{j-1}^*,\boldsymbol{y},m)$$

with

$$\widehat{f}(\theta_1^*|\boldsymbol{y},m) \ = \ \frac{1}{T}\sum_{t=1}^{T} f(\theta_1^{*(t)}|\theta_2^{(t)},\ldots,\theta_d^{(t)},\boldsymbol{y},m)$$

$$\widehat{f}(\theta_j|\theta_1^*,\ldots,\theta_{j-1}^*,\boldsymbol{y},m) \ = \ \frac{1}{T}\sum_{t=1}^{T} f(\theta_j^{*(t)}|\theta_1^{*(t)},\ldots,\theta_{j-1}^{*(t)},\theta_{j+1}^{(t)},\ldots,\theta_d^{(t)},\boldsymbol{y},m).$$

These formulations indicate that $\widehat{f}(\theta_1^*|\boldsymbol{y},m)$ will be estimated directly from the Gibbs sampler used to produce a sample from the posterior distribution while each of the remaining conditional posterior predictive ordinates $\widehat{f}(\theta_j | \theta_1^*, \ldots, \theta_{j-1}^*, \boldsymbol{y}, m)$ will be estimated from the same sampling scheme but restricted to the selected values $\theta_1^*, \ldots, \theta_{j-1}^*$. Hence, in order to estimate the posterior predictive ordinate, we need to run our Gibbs sampling algorithm d separate times. Selection of $\boldsymbol{\theta}_m^*$ is also important. Chib (1995) proposed using values close the center of the distribution; hence the posterior mean, median, or mode are usually good choices.

Chib and Jeliazkov (2001) implemented the same idea to the Metropolis–Hastings algorithm. Chen (2005) also proposed an extension of the method used when latent variables are available and that avoids some of the drawbacks of Chib's estimators. Details concerning the implementation of the method using Matlab can be found in Congdon (2006b, sec. 2.3). Calculation of the marginal likelihood using WinBUGS is possible, but it is not recommended (at least from the author of this book) since the user does not know the sampling scheme implemented by the program and, furthermore, calculation of the full conditional posterior distributions and multiple runs of the algorithm are required.

11.3.7 Additional details and further reading

Review, comparison of several marginal likelihood estimators, and illustrative examples can be found in Gamerman and Lopes (2006). Congdon (2006b) provides details and examples for the implementation of several method using Matlab. Several other methods have been proposed in the literature for estimation of the marginal likelihood. As we have already mentioned, Raftery et al. (2007) have proposed a method to stabilize the harmonic mean estimator, Gelman and Meng (1998) extended the bridge sampling estimator by proposing the path sampling estimator, Chib and Jeliazkov (2005) extended their method for posterior samples generated by the accept—reject Metropolis—Hastings algorithm, and Chen (2005) proposed a method that can be considered as an extension of Chib's estimator.

Neal (2001) used ideas from importance sampling and simulated annealing to construct an estimator of the marginal likelihood. Furthermore, Friel and Pettit (2008) proposed an estimator based on power posteriors by combining ideas from simulated annealing and path sampling.

These approaches entail running each model under comparison and estimating the marginal likelihood of each model separately. In real life model selection problems and especially in variable selection models, the set of models under consideration is usually large, and hence such methods cannot be implemented since a high computational burden is required. In such cases, we need to use an algorithm that can also work as a model search algorithm and will be able to trace the "good" models having high posterior model probabilities. Such methods are based on extensions of usual MCMC methods such as the reversible jump MCMC (Green, 1995); for more details, review, and comparison of such methods, see Han and Carlin (2001), Dellaportas et al. (2002), and Sisson (2005). In Section 11.5, we concentrate on the variable selection problem, presenting Gibbs-based methods that can be easily implemented in WinBUGS. A brief description of the reversible jump MCMC algorithm is also provided in Section 11.9

11.4 COMPUTATION OF THE MARGINAL LIKELIHOOD USING WinBUGS

In this section we illustrate the computation of the marginal likelihood models under consideration using the output generated using WinBUGS. More specifically, in this section we illustrate the Laplace–Metropolis estimator, the harmonic mean estimator, and the generalized harmonic mean estimator. Two simple examples with conjugate prior setups were used in which the marginal likelihood was also available analytically. This allows us to present and compare the efficiency of the illustrated methods since the actual marginal likelihood value is known. The harmonic mean estimator fails even in simple cases presented for illustration in this chapter. The other two estimators give reasonable results, at least within the family of the generalized linear models.

To estimate the Laplace–Metropolis estimator, we proceed with the following steps

1. Produce an MCMC sample in WinBUGS.

2. Estimate from the MCMC output produced by WinBUGS:

   a. The posterior means of the parameters of interest denoted by $\overline{\boldsymbol{\theta}}_m$

   b. The posterior standard deviations of the parameters of interest denoted by $s_{\theta_m} = (s_1, \ldots, s_d)$

   c. The posterior correlation between the parameters of interest (from Inference>Correlations) menu, denoted by R_{θ_m}

3. Calculate the expression

$$\log \widehat{f}(\boldsymbol{y}|m) \;=\; \frac{1}{2} d_m \log(2\pi) + \frac{1}{2} \log \left| R_{\theta_m} \right| + \sum_{j=1}^{d_m} \log s_j$$
$$+ \sum_{i=1}^{n} \log f(y_i|\overline{\boldsymbol{\theta}}_m, m) + \log f(\overline{\boldsymbol{\theta}}_m|m),$$

where s_j are the posterior standard deviations of θ_j parameter estimated from the MCMC output.

The harmonic mean estimator can be obtained by calculating the likelihood in each iteration. In order to do so, we first calculate the log-likelihood term ℓ_i for each observation i using the node log.like[i] (for $i = 1, 2, \ldots, n$) and then the inverse likelihood inv.like by the expression $\exp\left(-\sum_{i=1}^{n} \ell_i\right)$ using the WinBUGS syntax

```
inv.loglike <- exp( -sum( log.like[1:n] ) )
```

Finally, the estimate of the marginal likelihood is given by the inverse of the posterior mean of node inv.like:

$$\widehat{f}(\boldsymbol{y}|m) = 1/\overline{\texttt{inv.like}}.$$

Similarly, we calculate the generalized harmonic mean estimate of the marginal likelihood. We calculate node w.like by the expression

$$w^{(t)} = \frac{g\left(\boldsymbol{\theta}_m^{(t)}\right)}{f\left(\boldsymbol{y}|\boldsymbol{\theta}_m^{(t)}, m\right) f\left(\boldsymbol{\theta}_m^{(t)}|m\right)}$$

and then estimate the marginal likelihood by (11.4), that is, the inverse of the posterior mean of w.like:

$$\widehat{f}(\boldsymbol{y}|m) = 1/\overline{\texttt{w.like}}.$$

11.4.1 A beta–binomial example

Example 11.1. Kobe Bryant's field goals in NBA (Example 1.4 revisited). Here we reconsider Kobe Bryant's field goals in NBA (Example 1.4). Two models are fitted: the one with equal success probabilities across all seasons (model m_0) and the one with different success probabilities (model m_1). Hence this is equivalent to testing $H_0 : \pi_k = \pi$ for all $k \in \{1999, \dots, 2006\}$ versus the alternative that $H_0: \pi_k \neq \pi_l$ for all $k \neq l; k, l \in \{1999, \dots, 2006\}$

A simple beta–binomial model is adopted in which the marginal likelihood can be calculated analytically.

Analytic computation of the marginal likelihoods. Under model m_1, the success probabilities for each year/season are the parameters of interest $\boldsymbol{\theta}_{m_1} = \boldsymbol{\pi} = (\pi_{1999}, \dots, \pi_{2006})$, while the log-likelihood is given by

$$f(\boldsymbol{y}|\boldsymbol{\theta}_{m_1} = \boldsymbol{\pi}, m_1) = \prod_{i=1999}^{2006} \frac{N_i!}{y_i!(N_i - y_i)} \pi_i^{y_i}(1 - \pi_i)^{N_i - y_i} \ .$$

We consider a set of independent (conjugate) beta prior distributions with parameters a_i and b_i:

$$f(\boldsymbol{\theta}_{m_1} = \boldsymbol{\pi}|m_1) = \prod_{i=1999}^{2006} f_B(\pi_i; a_i, b_i).$$

The marginal posterior likelihood for model m_1 is given by

$$
\begin{aligned}
f(\boldsymbol{y}|m_1) &= \int f(\boldsymbol{y}|\boldsymbol{\pi}, m_1) f(\boldsymbol{\pi}|m_1) d\boldsymbol{\pi} \\
&= \prod_{i=1999}^{2006} \frac{N_i!}{y_i!(N_i - y_i)} \frac{\Gamma(a_i + b_i)}{\Gamma(a_i)\Gamma(b_i)} \frac{\Gamma(y_i + a)\Gamma(N_i - y_i + b_i)}{\Gamma(N_i + a_i + b_i)} \ .
\end{aligned}
$$

Similarly, for model m_0 we have a common parameter $\boldsymbol{\theta}_{m_0} = \pi$

$$f(\boldsymbol{y}|\boldsymbol{\theta}_{m_0} = \pi, m_0) = \prod_{i=1999}^{2006} \frac{N_i!}{y_i!(N_i - y_i)} \pi^{y_i}(1 - \pi)^{N_i - y_i}$$

with prior

$$f(\boldsymbol{\theta}_{m_0} = \pi|m_0) = f_B(\pi; a, b),$$

resulting in marginal likelihood

$$
\begin{aligned}
f(\boldsymbol{y}|m_0) &= \int f(\boldsymbol{y}|\pi, m_0) f(\pi|m_0) d\pi \\
&= \left\{ \prod_{i=1999}^{2006} \frac{N_i!}{y_i!(N_i - y_i)} \right\} \frac{\Gamma(a + b)}{\Gamma(a)\Gamma(b)} \frac{\Gamma(\sum_{i=1999}^{2006} y_i + a)\Gamma(N - \sum_{i=1999}^{2006} y_i + b)}{\Gamma(N + a + b)} \ .
\end{aligned}
$$

Finally, the Bayes factor of model m_0 versus model m_1 is given by the ratio of the two marginal likelihoods above. Here we consider the uniform distribution as a prior (i.e.,

$a_i = b_i = 1$ and $a = b = 1$). In this case the marginal likelihoods are equal to

$$\log f(\boldsymbol{y}|m_1) = -\sum_{i=1999}^{2006} (N_i + 1) = -58.0228$$

$$\log f(\boldsymbol{y}|m_0) = \log\left[\left\{\prod_{i=1999}^{2006} \frac{N_i!}{y_i!(N_i - y_i)}\right\} \frac{\Gamma(\sum_{i=1999}^{2006} y_i + 1)\Gamma(N - \sum_{i=1999}^{2006} y_i + 1)}{\Gamma(N+2)}\right]$$

$$= -39.2308,$$

resulting in a log-Bayes factor equal to $\log B_{01} = -39.2308 + 58.0228 = 18.792$, which provides support in favor of model m_0.

Implementation in WinBUGS.   Illustration is described for model m_1. Implementation for model m_0 is similar, but one common parameter π is used to model the success probabilities instead of a vector.

To calculate the harmonic mean estimator, we need to define the node `log.like`

```
for (t in 1:8){
    log.like[t]<- loggam(N[t]+1) -loggam(y[t]+1) -loggam(N[t]-y[t]+1)
                        +y[t]*log(p[t]) +(N[t]-y[t])*log(1-p[t])
}
```

and then set

```
inv.like <- exp( - sum(log.like[1:8]) )
```

to calculate the inverse of the likelihood in each iteration. The harmonic mean estimator is given by the inverse of the posterior mean of `inv.like`.

The procedure for the generalized inversed distribution is similar to the one described above. We need to define a reasonably well-behaved function, $g(\boldsymbol{\pi})$. Here we consider that

$$\pi_i \sim N\left(\frac{y_i}{N_i}, \frac{(1 - y_i/N_i)y_i/N_i}{N_i}\right).$$

To calculate the generalized harmonic mean estimator under this importance function we use the following syntax

```
for(t in 1:8){
    ps[t]  <- y[t]/N[t]
    qs[t]  <- 1-ps[t]
    var[t] <-ps[t]*qs[t]/N[t]
    log.g[t]  <- -0.5*log( 2*3.14 ) - 0.5*log( var[t] )
                -0.5*(p[t]-ps[t])*(p[t]-ps[t])/var[t]
    log.prior[t]<-0.0 # uniform prior
    wl[t] <- log.g[t] - log.like[t] - log.prior[t]
}
wlike <- exp( sum( wl[1:8] )
```

in addition to the log-likelihood defined as described above. An estimate of the marginal likelihood is given by the inverse of the posterior mean of node `wlike`.

Finally, in order to calculate the Laplace–Metropolis estimator, we estimate the posterior means and standard deviations from an MCMC run. Then we calculate the posterior correlation matrix by the menu `Inference>Correlations`; results for model m_1 are provided in Table 11.2.

Then we calculate the marginal likelihood in an external software such as R. After importing the MCMC-based estimated posterior summaries (denoted by p, s, and R, respectively),

Table 11.2 Posterior summaries and correlation of success probabilities for model m_1 in Example 11.1

π	Posterior		Posterior correlations							
	Mean	SD	π_{1999}	π_{2000}	π_{2000}	π_{2000}	π_{2000}	π_{2000}	π_{2000}	π_{2000}
π_{1999}	0.468	0.0145	1.000							
π_{2000}	0.464	0.0131	0.141	1.000						
π_{2001}	0.469	0.0125	0.138	0.175	1.000					
π_{2002}	0.451	0.0113	0.176	0.152	0.185	1.000				
π_{2003}	0.438	0.0146	0.130	0.149	0.132	0.149	1.000			
π_{2004}	0.433	0.0138	0.123	0.138	0.138	0.143	0.119	1.000		
π_{2005}	0.450	0.0107	0.175	0.187	0.191	0.199	0.142	0.161	1.000	
π_{2006}	0.472	0.0170	0.107	0.124	0.113	0.143	0.111	0.108	0.125	1.000

we calculate the logarithm of the Laplace estimate for the marginal likelihood using the R commands

```
dm <-8 # number of parameters
(dm/2)*log(2*3.14)+sum(log(s))+0.5*log(det(R))+sum(dbinom(y,n,p,log=
    TRUE))
```

The full code for model m_1 is given in Table 11.3. Code for model m_0 is similar, but p[t] is substituted by a single p.

Table 11.3 WinBUGS code for estimation of marginal likelihood for model m_1 in Example 11.1

```
model {
    for (t in 1:YEARS){
        y[t]    ~ dbin( p[t], N[t] )
        p[t]~dbeta( 1, 1)
        log.like[t] <- loggam( N[t]+1 )- loggam( y[t]+1 )
                        - loggam( N[t] - y[t]+1 )
                        + y[t]*log( p[t] ) + (N[t]-y[t])*log( 1-p[t] )
    }
    inv.like <- exp( - sum(log.like[1:YEARS]) )
    for(t in 1:YEARS){
        ps[t] <- y[t]/N[t]
        qs[t] <- 1-ps[t]
        var[t] <-ps[t]*qs[t]/N[t]
        log.g[t] <- -0.5*log( 2*3.14 ) - 0.5*log( var[t] )
                    - 0.5*(p[t]-ps[t])*(p[t]-ps[t])/var[t]
        log.prior[t]<-0.0
        wl[t] <- log.g[t] - log.like[t] - log.prior[t]
    }
    wlike <- exp( sum( wl[1:YEARS] )
}
```

Results. Results based on 5000 iterations (and an additional 1000 iterations as burnin period) are presented in Table 11.4. We observe that the generalized harmonic mean and the Laplace–Metropolis estimators are very close to the true values while the harmonic mean fails to provide a reasonably accurate estimate of the marginal likelihoods, especially for model m_1. From Figure 11.1 we can also see that the latter is quite unstable since the ergodic estimators do not seem to stabilize.

Table 11.4 Estimated log–marginal likelihood and Bayes factor for Example 11.1

		m_0	m_1	$\log B_{01}$
Inverse likelihood	inv.like	3.917×10^{15}	8.168×10^{16}	—
Weighted inverse likelihood	wlike	1.091×10^{17}	1.585×10^{25}	—
Log–marginal likelihood	Harmonic mean	-35.904	-38.942	3.038
	Generalized harmonic mean	-39.231	-58.025	18.794
	Laplace–Metropolis	-39.218	-58.203	18.985
	True value	-39.231	-58.023	18.792

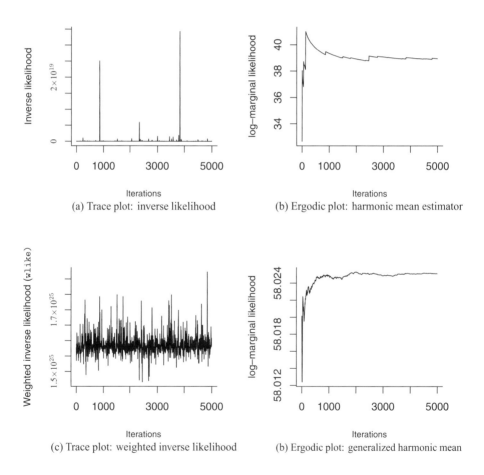

(a) Trace plot: inverse likelihood

(b) Ergodic plot: harmonic mean estimator

(c) Trace plot: weighted inverse likelihood

(b) Ergodic plot: generalized harmonic mean

Figure 11.1 Trace and ergodic plots for simple and generalized harmonic mean estimators of the marginal likelihood for model m_1.

11.4.2   A normal regression example with conjugate normal–inverse gamma prior

Example 11.2. Soft drink delivery times data (Example 5.1 revisited). Here we reconsider Example 5.1, in which simple normal models with two covariates are considered, resulting in a total of four possible models under consideration.

Analytic computation of the marginal likelihoods. Let us consider two models, m_0 and m_1, given by

$$y \sim N\left(X_m \beta_m, \sigma^2 I_n\right),$$

where $m \in \{m_0, m_1\}$, n is the sample size, β_m is the $P_m \times 1$ vector of model parameters, X_m is the $n \times P_m$ design (or data) matrix of model m, P_m is the number of parameters involved in the linear predictor of model m, $N(\mu, \Sigma)$ is the multivariate normal distribution with mean μ and variance covariance matrix Σ, and I_n is an $n \times n$ identity matrix.

We adopt the conjugate prior distribution given by

$$f(\beta_m | \sigma^2, m) \sim N\left(\mu_{\beta_m}, c^2 V_m \sigma^2\right) \tag{11.6}$$

and a gamma prior with parameters a and b for the residual variance $f(\sigma^2) \sim \mathrm{IG}(a, b)$.

The resulting marginal likelihood for model m is given by

$$f(y|m) = \frac{\Gamma(a + n/2)}{\Gamma(a)} (2\pi)^{-n/2} b^a c^{-P_m} \left(\frac{|\widetilde{\Sigma}_m|}{|V_m|}\right)^{1/2} \left(\frac{1}{2}\mathrm{SS}_m + b\right)^{-n/2-a},$$

while the corresponding posterior model odds for comparison between models m_0 and m_1 are provided by the expression

$$\mathrm{PO}_{01} = c^{P_{m_1} - P_{m_0}} \left(\frac{|V_{m_0}|}{|V_{m_1}|}\right)^{-1/2} \left(\frac{|\widetilde{\Sigma}_{m_0}|}{|\widetilde{\Sigma}_{m_1}|}\right)^{1/2} \left(\frac{\mathrm{SS}_{m_0} + 2b}{\mathrm{SS}_{m_1} + 2b}\right)^{-n/2-a} \frac{f(m_0)}{f(m_1)}, \tag{11.7}$$

where SS_m are the posterior sum of squares

$$\mathrm{SS}_m = y^T y - \widetilde{\beta}_m^T \widetilde{\Sigma}_m^{-1} \widetilde{\beta}_m + c^{-2} \mu_{\beta_m}^T V_m^{-1} \mu_{\beta_m}, \tag{11.8}$$

$$\widetilde{\beta}_m = \widetilde{\Sigma}_m \left(X_m^T X_m \widehat{\beta}_m + c^{-2} V_m^{-1} \mu_{\beta_m}\right), \tag{11.9}$$

$$\widetilde{\Sigma}_m^{-1} = X_m^T X_m + c^{-2} V_m^{-1}, \tag{11.10}$$

μ_{β_m} and V_m are the prior mean vector and a prior matrix associated with the prior variance–covariance matrix of the parameter vector β_m respectively, and $\widetilde{\beta}_m$ and $\widetilde{\Sigma}_m$ are the corresponding posterior measures.

An alternative expression of the posterior sum of squares is given by Atkinson (1978) and Pericchi (1984), where

$$\mathrm{SS}_m = \mathrm{RSS}_m + \left(\widehat{\beta}_m - \mu_{\beta_m}\right)^T \left[\left(X_m^T X_m\right)^{-1} + c^2 V_m\right]^{-1} \left(\widehat{\beta}_m - \mu_{\beta_m}\right), \tag{11.11}$$

and $\mathrm{RSS}_m = y^T y - \widehat{\beta}_m^T X_m^T X_m \widehat{\beta}_m$ is the usual residual sum of squares. This quantity is the sum of two measures: a goodness of fit and a measure of distance between maximum likelihood estimates and the prior mean.

Implementation and results.   In this example we consider an independent normal prior distribution with $V_m = I_{P_m}$ and select various values for $c^2 \in \{10, 100, 1000\}$. For the inverse gamma distribution we set $a = b = 0.001$. Results based on the analytic computation of the marginal likelihoods are provided in Table 11.5. The second model (m_2), with covariate the number of cases that the employee needs to load on the machine, is the one supported by prior setups within the range of $c^2 \in [10, 1000]$. Lindley–Bartlett paradox is depicted in this table since the marginal likelihoods and the posterior model probabilities of the simpler models increase in contrast to the more complicated model m_4 when c^2 increases. The effect of the covariate measuring the number cases is strong, resulting in posterior model probabilities that are quite robust for a wide range of values for the prior parameter c^2. The Lindley–Bartlett paradox eventually appears for large prior variances eliminating the actual data evidence and finally fully supporting the simplest (constant) model (see results for $c^2 = e^{700}$).

Table 11.5   Marginal log–likelihood and posterior model probabilities for models under consideration in soft drink data (Example 5.1)

	c^2	m_1 (constant)	m_2 (cases)	m_3 (distance)	m_4 (cases + distance)
Marginal Log–likelihood	10	-113.65	-84.93	-102.28	-86.10
	100	-114.70	-87.18	-104.55	-89.51
	1000	-115.84	-89.48	-106.85	-92.96
	e^{700}	-462.38	-782.57	-799.94	-1132.59
Posterior probabilities	10	2.6×10^{-13}	0.7627	2.2×10^{-8}	0.2373
	100	1.0×10^{-12}	0.9108	2.6×10^{-8}	0.0892
	1000	3.5×10^{-12}	0.9700	2.8×10^{-8}	0.0300
	e^{700}	1.000	0.0000	0.000	0.0000

The Laplace–Metropolis, the harmonic mean, and the generalized harmonic mean estimates of the marginal likelihoods for $c^2 = 1000$ are presented in Table 11.6. Code for the four models under consideration are available in the book's Webpage. For the generalized harmonic mean estimates, we used the posterior means and standard deviations from each model in order to specify independent normal importance distributions for parameters β_j and a simple gamma for the precision parameter. These were similar to the results obtained when a multivariate normal importance function was used for all parameters. In all cases studied, the harmonic mean estimator proved inaccurate and unstable in contrast to the other two estimates, which provided accurate results. Note that the accuracy of the generalized harmonic mean estimator depends heavily on the choice of the importance function. In this example, the simple strategy of considering posterior summaries to construct independent importance functions has proved quite effective.

Note that, because of the large values the exponents involved in the harmonic and the generalized harmonic mean estimator, it was not possible to calculate the inverse and the weighted inverse likelihood (`inv.like` and `wlike`) directly in WinBUGS. For this reason, in the example presented here, we subtracted a large value from each component before taking the exponent of the log-likelihood (here we have used $C = 90$). Hence, instead of

Table 11.6 Estimated marginal log–likelihoods and posterior model probabilities for $c^2 = 1000$ for soft drink data (Example 5.1)

	Estimator	m_1 (constant)	m_2 (cases)	m_3 (distance)	m_4 (cases + distance)
Marginal log–likelihood	Harmonic mean	-105.36	-73.28	-86.80	-68.78
	Generalized harmonicmean	-115.84	-89.43	-104.72	-92.57
	Laplace–Metropolis	-115.85	-89.35	-106.73	-92.77
	True value	-115.84	-89.48	-106.85	-92.96
Posterior probabilities	Harmonic mean	1.3×10^{-16}	0.0110	1.5×10^{-8}	0.9890
	Generalized harmonic mean	3.2×10^{-12}	0.9585	2.2×10^{-7}	0.0415
	Laplace–Metropolis	3.0×10^{-12}	0.9683	2.7×10^{-8}	0.0317
	True value	3.5×10^{-12}	0.9701	2.8×10^{-8}	0.0299

calculating

$$\texttt{inv.like} = \exp\left(-\sum_{i=1}^{n}\ell_i\right) \text{ and } \texttt{wlike} = \exp\left(g(\boldsymbol{\theta}_m) - \sum_{i=1}^{n}\ell_i - f(\boldsymbol{\theta}_m|m)\right)$$

in each iteration, we have calculated

$$\texttt{inv.like*} = \exp\left(-\sum_{i=1}^{n}\ell_i - C\right),$$

$$\texttt{wlike*} = \exp\left(g(\boldsymbol{\theta}_m) - \sum_{i=1}^{n}\ell_i - f(\boldsymbol{\theta}_m|m) - C\right),$$

and then obtained the estimates of the log–marginal likelihood by

$$\texttt{inv.like} = -\log(\texttt{inv.like*}) - C \text{ and } \texttt{wlike} = -\log(\texttt{wlike*}) - C,$$

where ℓ_i is $f(y_i|\boldsymbol{\theta}_m, m)$ for $i = 1, \ldots, n$.

11.5 BAYESIAN VARIABLE SELECTION USING GIBBS-BASED METHODS

In GLM-type models, which we have focused on in this introductory book, the main model selection problem is the selection of the covariates involved in the linear predictor. Even when the number of covariates is moderate, the number under consideration is large (e.g., for $p = 20$, it is equal to $2^{20} = 1,048,576$), making it difficult to estimate the marginal likelihoods of all models using the methods described in the previous section. In fact, we need to implement algorithms that efficiently search the model space, focus on the most probable a posteriori models, and provide estimates of their posterior model probabilities. This can be achieved by the methods described in this section. Here we focus on the Gibbs variable selection introduced by Dellaportas et al. (2002). Other Gibbs-based methods such as the stochastic search variable (George and McCulloch, 1993) selection and the Kuo and Mallick (1998) sampler are discussed and compared. All these methods can be implemented in WinBUGS ; for additional details and examples, see Dellaportas et al. (2000) and Ntzoufras (2002).

In variable selection, the set of models $\mathcal{M}$ under consideration can be represented by a vector of binary indicators $\boldsymbol{\gamma} \in \{0, 1\}^p$. This vector of binary indicators reveals which of the p possible sets of covariates are present in the model. For example, for a generalized linear model, the linear predictor can now be written as

$$\eta = \sum_{j=0}^{p} \gamma_j \boldsymbol{X}_j \boldsymbol{\beta}_j, \tag{11.12}$$

where $\boldsymbol{X}$ and $\boldsymbol{\beta}$ are the design matrix and the parameter vector of the full model (including all p available covariates in the linear predictor), respectively; $\boldsymbol{X}_j$ is the vector or matrix with the elements $\boldsymbol{X}$ that correspond to the coefficients $\boldsymbol{\beta}_j$ of the jth covariate under consideration; while $\boldsymbol{X}_0 = \mathbf{1}_n$ is the column that corresponds to the constant term β_0. There is a one-to-one transformation connecting the usual model indicator m with the vector of variable indicators $\boldsymbol{\gamma}$. One way to transform $\boldsymbol{\gamma}$ to a unique model indicator m, is to consider $\boldsymbol{\gamma}$ as a number in the binary numerical system and transform it to the corresponding number in the decimal numerical system; see Section 11.6 for more details.

Finally, we consider the partition of $\boldsymbol{\beta}$ into $(\boldsymbol{\beta}_{\boldsymbol{\gamma}}, \boldsymbol{\beta}_{\backslash \boldsymbol{\gamma}})$ corresponding to those components of $\boldsymbol{\beta}$ that are included ($\gamma_j = 1$) or not included ($\gamma_j = 0$) in the model. Hence the vector $\boldsymbol{\beta}_{\boldsymbol{\gamma}}$ corresponds to the active parameters of the model (i.e., $\boldsymbol{\beta}_m$), while $\boldsymbol{\beta}_{\backslash \boldsymbol{\gamma}}$ corresponds to the remaining parameters, which are not included in the model defined by $\boldsymbol{\gamma}$.

Before proceeding to a description of the Gibbs-based variable selection methods, we briefly provide details concerning a set of priors that can be used for Bayesian variable selection.

11.5.1 Prior distributions for variable selection in GLM

The specification of the prior distribution when no prior information is available is very important in variable selection in order to allow the data to determine which variables are important for the model without activating the Lindley–Bartlett paradox. Here we present a family of prior distributions based on the Zellner's g-prior (Zellner, 1986). We describe an interpretation based on imaginary data and the power prior of Ibrahim and Chen (2000) and Chen et al. (2000). The prior is presented for normal, Poisson, and binomial data, but it can be specified for the rest of the models using similar arguments.

Zellner's (1986) g-prior is simply defined if we specify $\boldsymbol{V}_m = \left(\boldsymbol{X}_m^T \boldsymbol{X}_m \right)^{-1}$ in the conjugate prior (11.6):

$$\boldsymbol{\beta}_m | \sigma^2, m \sim N \left(\boldsymbol{\mu}_{\beta_m}, c^2 \left(\boldsymbol{X}_m^T \boldsymbol{X}_m \right)^{-1} \sigma^2 \right) . \tag{11.13}$$

This prior can be interpreted using the power prior of Ibrahim and Chen (2000) and Chen et al. (2000). Let us assume imaginary data $\boldsymbol{y}^*$ obtained under the same design matrix $\boldsymbol{X}_m$. Then we set our prior proportional to a power of the likelihood obtained by the imaginary data:

$$f(\boldsymbol{\beta}_m | \boldsymbol{y}^*, m) \propto \left(f(\boldsymbol{y}^* | \boldsymbol{\beta}_m, m) \right)^{1/c^2} .$$

This prior accounts for information equal to n/c^2 data points. In the normal regression model, the above prior results in

$$\boldsymbol{\beta}_m | \boldsymbol{y}^*, m \sim N \left(\left(\boldsymbol{X}_m^T \boldsymbol{X}_m \right)^{-1} \boldsymbol{X}_m^T \boldsymbol{y}^*, \ c^2 \left(\boldsymbol{X}_m^T \boldsymbol{X}_m \right)^{-1} \sigma^2 \right)$$

which is the Zellner g-prior for imaginary data equal to $y^* = X_m \mu_{\beta_m}$.

If no prior information is available, it is logical to center our prior beliefs for β_m around zero and hence set $\mu_{\beta_m} = 0$. Furthermore, a default choice for c^2 (denoted by g in Zellner's original paper) is to set $c^2 = n$, which corresponds to adding prior information equal to one data point, resulting to the prior

$$\beta_m | m \sim N\left(0, \, n\left(X_m^T X_m\right)^{-1} \sigma^2\right). \tag{11.14}$$

In this way we a priori support the simplest model (by setting the means equal to zero) but in a minimal way, since our prior accounts only for one data point in the final analysis. This approach is also sensible in terms of the parsimony principle. Posterior model odds (and Bayes factors) penalize the model likelihood for deviations of the actual data from the prior distribution (Raftery, 1996a, eq. 12). Since the prior presented above can be generated using a set of minimally weighted imaginary data that fully support the constant model, it provides a sensible a priori support of more parsimonious models.

Similar arguments were used by Fouskakis et al. (2008) to adopt the prior of Ntzoufras et al. (2003)

$$f(\beta_m | m) = N\left(\mu_{\beta_m}, n\left[\mathcal{I}(\beta_m)\right]^{-1}\right) \tag{11.15}$$

for the logistic regression case, where $\mathcal{I}(\beta_m)$ is the information matrix

$$\mathcal{I}(\beta_m) = X_m^T W_m X_m.$$

Here W_m is a diagonal matrix that in the binomial case (McCullagh and Nelder, 1989) takes the form

$$W_m = \text{diag}\left(N_i \pi_i (1 - \pi_i)\right).$$

This is the *unit information prior* introduced by Kass and Wasserman (1995), which corresponds to adding one data point to the data. Here we use this prior as a base, but we specify π_i in the information matrix according to our prior information. In this manner we avoid (even minimal) reuse of the data in the prior.

Similarly to the normal regression case, when little prior information is available, a reasonable choice for the prior mean of β_m is $\mu_{\beta_m} = 0$. This corresponds to the assumption that a reasonable prior estimate (when no information is available) for all fitted probabilities of the binomial regression model is $\pi_i = \frac{1}{2}$. With this choice and binary data (i.e., $N_i = 1$ for all i), Eq. (11.15) simplifies to

$$f(\beta_m | m) = N\left(0, 4n\left(X_m^T X_m\right)^{-1}\right), \tag{11.16}$$

where n is the total number of Bernoulli experiments. This prior distribution can also be motivated using the power prior approach of Chen et al. (2000). After observing the design matrix X_m for any model m, we consider a set of imaginary data $y_i^* = N_i$ and $N_i^* = 2N_i$ for $i = 1, \ldots, n$ that assigns probabilities $\frac{1}{2}$ for all binomial observations i and therefore supports the simplest (constant) model. We consider a prior that is generated using the likelihood of these imaginary data,

$$f(\beta_m | y^*, m) \quad \propto \quad \left(\prod_{i=1}^n \pi_i^{N_i} (1 - \pi_i)^{N_i}\right)^{1/(2N)},$$

where $N = \sum_{i=1}^{n} N_i$. Note when binary data are considered, then $N_i = 1$ for all $i = 1, \ldots, n$ resulting to the setup of Fouskakis et al. (2008). Using this prior, the posterior becomes

$$f(\boldsymbol{\beta}_m | \boldsymbol{y}, m) \propto \prod_{i=1}^{n} \pi_i^{y_i + 1/(2N)} (1 - \pi_i)^{N_i (1 + 1/N) - \left[y_i + N_i/(2N) \right]};$$

therefore this is equivalent to obtaining information from $\sum_{i=1}^{n} (N_i + N_i/n) = (N + 1)$ Bernoulli experiments, instead of N when using a flat prior. Thus the proposed prior (11.17) introduces to the posterior distribution additional information equivalent to one data point.

Using a Laplace approximation to (11.17) [see, e.g., Bernardo and Smith (1994, p. 286)], we obtain

$$\boldsymbol{\beta}_m | \boldsymbol{y}^*, m \overset{\text{approx}}{\sim} N\left(\widehat{\boldsymbol{\beta}}_m, 2N \, \mathcal{I}(\widehat{\boldsymbol{\beta}}_m)^{-1} \right),$$

where $\widehat{\boldsymbol{\beta}}_m$ is the maximum likelihood estimate if the imaginary data $\boldsymbol{y}_i^*$ were observed and $\mathcal{I}(\widehat{\boldsymbol{\beta}}_m)$ is the observed information matrix given by

$$\mathcal{I}(\widehat{\boldsymbol{\beta}}_m) = \boldsymbol{X}_m^T \, \text{diag}\left(2N_i \, \widehat{\pi}_i^* (1 - \widehat{\pi}_i^*) \right) \boldsymbol{X}_m,$$

in which $\widehat{\pi}_i^* = \left(1 + \exp(-\boldsymbol{X}_{(i)} \widehat{\boldsymbol{\beta}}_m) \right)^{-1}$ is the fitted success probability for all i under model m when observing data $\boldsymbol{y}^*$. According to these imaginary data, $\widehat{\boldsymbol{\beta}}_m = \boldsymbol{0}$ and $\widehat{\pi}_i = \frac{1}{2}$ for all i, yielding $\mathcal{I}(\widehat{\boldsymbol{\beta}}_m) = \frac{1}{2} \left(\boldsymbol{X}_m^T \boldsymbol{X}_m \right)$ and therefore leading to the prior given by (11.16).

Similar arguments can be used for the Poisson case. If all imaginary data y_i^* are equal to one, then the corresponding prior that accounts for one additional data points is given by

$$\boldsymbol{\beta}_m | m \sim N\left(\boldsymbol{0}, n(\boldsymbol{X}_m^T \boldsymbol{X}_m)^{-1} \right).$$

This prior can be informative; for the constant parameter, we might use a vague prior for β_0 (if it is included in all models) and use the prior formulated above for the parameters of the covariates, namely, $\boldsymbol{\beta}_{\backslash 0, m} | m \sim N\left(\boldsymbol{0}, n(\boldsymbol{x}_m^T \boldsymbol{x}_m)^{-1} \right)$, where $\boldsymbol{\beta}_{\backslash 0, m}$ is the parameter vector $\boldsymbol{\beta}_m$ without the constant parameter β_0 and $\boldsymbol{x}_m$ is the data matrix $\boldsymbol{X}_m$ after removing the first column that corresponds to the constant parameter.

Using similar arguments, we may use an empirical Bayes approach by estimating the posterior means $\widetilde{\mu}_\theta^2$ and the variance–covariance matrix $\widetilde{\boldsymbol{\Sigma}}_\theta$ multiplied by the sample size n. In this way we assume that we have imaginary data such as the ones already observed, which account for one additional data point. Hence the effect of the double usage of data is low and simplifies the implementation of Bayesian variable selection.

In model selection the uniform prior on model space it is typically used by setting

$$f(m) = \frac{1}{|\mathcal{M}|} \quad \text{for all } m \in \mathcal{M}.$$

When using the variable selection indicators $\boldsymbol{\gamma}$, this prior is equivalent to specifying independent Bernoulli prior distributions with inclusion probability equal to $\frac{1}{2}$:

$$\gamma_j \sim \text{Bernoulli}\left(\frac{1}{2} \right) \quad \text{for all } j = 1, \ldots, p.$$

Hence, in some cases, it is sensible to set $f(\gamma_j|\gamma_{\setminus j}) = f(\gamma_j)$ (where the subscript $\setminus j$ denotes all elements of a vector except the jth) , whereas in other cases (e.g., hierarchical or graphical log-linear models), it is required that $f(\gamma_j|\gamma_{\setminus j})$ depends on $\gamma_{\setminus j}$ (Chipman, 1996).

Although, the prior for γ with each term present or absent independently with probability $\frac{1}{2}$ may be considered noninformative in the sense that it gives the same weight to all possible models, George and Foster (2000) argue that this prior can be considered as informative since it puts more weight on models of size close to $p/2$ supporting a priori overparametrized and complicated models.

11.5.2   Gibbs variable selection

Gibbs variable selection (GVS) was introduced by Dellaportas et al. (2002). In order to setup GVS, we need to specify the prior distribution for (γ, β) with the following hierarchical structure: $f(\gamma, \beta) = f(\gamma)f(\beta|\gamma)$. If we consider the partition of β into $(\beta_\gamma, \beta_{\setminus\gamma})$, then the prior $f(\beta|\gamma)$ may be partitioned into model parameter prior $f(\beta_\gamma|\gamma)$ and pseudoprior $f(\beta_{\setminus\gamma}|\beta_\gamma, \gamma)$.

The full conditional posterior distributions for the model parameters are given by

$$f(\beta_\gamma|\beta_{\setminus\gamma}, \gamma, y) \quad \propto \quad f(y|\beta, \gamma)f(\beta_\gamma|\gamma)f(\beta_{\setminus\gamma}|\beta_\gamma, \gamma) \qquad (11.17)$$
$$f(\beta_{\setminus\gamma}|\beta_\gamma, \gamma, y) \quad \propto \quad f(\beta_{\setminus\gamma}|\beta_\gamma, \gamma), \qquad (11.18)$$

and for the variable indicator γ_j by

$$\gamma_j|\beta, \gamma_{\setminus j}, y \sim \text{Bernoulli}\left(\frac{O_j}{1 + O_j}\right) \qquad (11.19)$$

with

$$O_j = \frac{f(\gamma_j = 1|\gamma_{\setminus j}, \beta, y)}{f(\gamma_j = 0|\gamma_{\setminus j}, \beta, y)} = \frac{f(y|\beta, \gamma_j = 1, \gamma_{\setminus j})}{f(y|\beta, \gamma_j = 0, \gamma_{\setminus j})} \frac{f(\beta|\gamma_j = 1, \gamma_{\setminus j})}{f(\beta|\gamma_j = 0, \gamma_{\setminus j})} \frac{f(\gamma_j = 1, \gamma_{\setminus j})}{f(\gamma_j = 0, \gamma_{\setminus j})}, \qquad (11.20)$$

where $\gamma_{\setminus j}$ denotes all terms of γ except γ_j.

Note that the full conditional posterior distribution of the active model parameters β_γ in (11.17) also depends on the pseudoprior $f(\beta_{\setminus\gamma}|\beta_\gamma, \gamma)$, which seems awkward, at least on first glance at of this expression. Although this dependence may be useful, especially when correlated variables are considered as possible covariates, we may avoid it by assuming that the actual model parameters β_γ and the inactive parameters $\beta_{\setminus\gamma}$ are a priori independent. This plausible assumption implies that $f(\beta_{\setminus\gamma}|\beta_\gamma, \gamma) = f(\beta_{\setminus\gamma}|\gamma)$, simplifying the conditionals posterior distributions (11.17) and (11.18) to

$$f(\beta_\gamma|\beta_{\setminus\gamma}, \gamma, y) \quad \propto \quad f(y|\beta, \gamma)f(\beta_\gamma|\gamma)$$
$$f(\beta_{\setminus\gamma}|\beta_\gamma, \gamma, y) \quad \propto \quad f(\beta_{\setminus\gamma}|\gamma)$$

The algorithm is further simplified when assuming prior conditional independence of all β_j terms for each model γ. This is a rather restrictive assumption that may influence posterior model probabilities. Nevertheless, it is realistic in the case where the data matrix X has orthogonal columns X_j (appearing in 11.12) and the imposed priors are intended to be noninformative. Then, each prior for $\beta_j|\gamma$ consists of a mixture of two densities. The first, $f(\beta_j|\gamma_j = 1, \gamma_{\setminus j})$, is the true prior for the parameter, whereas the second, $f(\beta_j|\gamma_j =$

$0, \gamma_{\setminus j}$), is a pseudoprior. For example, considering normal prior and pseudoprior for each β_j results in a prior setup that is a mixture of two normal distributions given by

$$f(\boldsymbol{\beta}_j|\boldsymbol{\gamma}_j) = \gamma_j N(0, \boldsymbol{\Sigma}_j) + (1 - \gamma_j)N(\overline{\mu}_j, \overline{\boldsymbol{S}}_j). \tag{11.21}$$

The prior with $f(\boldsymbol{\beta}_j|\boldsymbol{\gamma}) = f(\boldsymbol{\beta}_j|\gamma_j)$ potentially makes the method less efficient, but it is appropriate in datasets where $\boldsymbol{X}$ is orthogonal. If prediction, rather than inference, is of primary interest, then $\boldsymbol{X}$ may always be chosen or appropriately transformed to be orthogonal [see, e.g., Clyde et al. (1996)]. Under this simplified prior setup, the full conditional posterior distribution is now given by

$$f(\boldsymbol{\beta}_j|\boldsymbol{\gamma}, \boldsymbol{\beta}_{\setminus j}, \boldsymbol{y}) \propto \begin{cases} f(\boldsymbol{y}|\boldsymbol{\gamma}, \boldsymbol{\beta})N(0, \boldsymbol{\Sigma}_j) & \gamma_j = 1 \\ N(\overline{\mu}_j, \overline{\boldsymbol{S}}_j) & \gamma_j = 0 \end{cases},$$

and a clear difference between this and SSVS is that the pseudoprior $f(\boldsymbol{\beta}_j|\gamma_j = 0)$ does not affect the posterior distribution and may be chosen as a "linking density" to increase the efficiency of the sampler. Possible choices of $\overline{\mu}_j$ and $\overline{\boldsymbol{S}}_j$ may be obtained from a pilot run of the full model. An alternative is to set $\overline{\mu}_j$ equal to zero and set the proposal variance low but proportional to the actual prior variance $\overline{\boldsymbol{S}}_j = k_j^{-2}\boldsymbol{\Sigma}_j$ in a fashion similar to the prior setup used in the SSVS algorithm, which is described in Section 11.5.3. Empirical results have shown that this setup is efficient in specific cases and can be a good alternative when parameters vary across models, and therefore an overall proposal cannot be specified from a single pilot run from the model with all covariates (Ntzoufras, 1999a).

11.5.3   Other Gibbs-based methods for variable selection

Stochastic search variable selection.   Stochastic search variable selection (SSVS) was introduced by George and McCulloch (1993) for linear regression models and has been adapted for generalized linear models (George et al., 1996; George and McCulloch, 1997), Poisson log-linear models (Ntzoufras et al., 2000), and multivariate regression models (Brown et al., 1998). It has also been implemented in genetics-related problems [see, e.g., Oh et al. (2003) and Yi et al. (2003)].

The difference between SSVS and GVS described above is that the parameter vector $\boldsymbol{\beta}$ retains its full dimension p under all models since the linear predictor is

$$\boldsymbol{\eta} = \sum_{j=0}^{p} \boldsymbol{X}_j \boldsymbol{\beta}_j. \tag{11.22}$$

Therefore $\boldsymbol{\eta} = \boldsymbol{X}\boldsymbol{\beta}$ for all models, where $\boldsymbol{X}$ contains all the potential explanatory variables. The indicator variables γ_j are involved in the modeling process through a mixture prior of the type

$$\boldsymbol{\beta}_j|\gamma_j \sim \gamma_j N(0, \boldsymbol{\Sigma}_j) + (1 - \gamma_j)N(0, k_j^{-2}\boldsymbol{\Sigma}_j) \tag{11.23}$$

for prespecified prior parameters k_j and prior variance–covariance matrix $\boldsymbol{\Sigma}_j$. The prior parameters k_j and $\boldsymbol{\Sigma}_j$ in (11.23) are chosen so that when $\gamma_j = 0$ (i.e., when the covariate is "absent" from the linear predictor), the prior distribution for $\boldsymbol{\beta}_j$ ensures that $\boldsymbol{\beta}_j$ is constrained to be close to zero.

The full conditional posterior distributions for $\boldsymbol{\beta}_j$ are given by

$$f(\boldsymbol{\beta}_j|\boldsymbol{y}, \boldsymbol{\gamma}, \boldsymbol{\beta}_{\setminus j}) \propto f(\boldsymbol{y}|\boldsymbol{\gamma}, \boldsymbol{\beta})f(\boldsymbol{\beta}_j|\gamma_j),$$

while for the variable indicators γ_j, the full conditional posterior distribution is again a Bernoulli distribution with success probability $O_j/(1 + O_j)$ as in the corresponding Gibbs variable selection step (11.18) with

$$O_j = \frac{f(\gamma_j = 1|\boldsymbol{y}, \boldsymbol{\gamma}_{\backslash j}, \boldsymbol{\beta})}{f(\gamma_j = 0|\boldsymbol{y}, \boldsymbol{\gamma}_{\backslash j}, \boldsymbol{\beta})} = \frac{f(\boldsymbol{\beta}|\gamma_j = 1, \boldsymbol{\gamma}_{\backslash j})}{f(\boldsymbol{\beta}|\gamma_j = 0, \boldsymbol{\gamma}_{\backslash j})} \frac{f(\gamma_j = 1, \boldsymbol{\gamma}_{\backslash j})}{f(\gamma_j = 0, \boldsymbol{\gamma}_{\backslash j})}. \tag{11.24}$$

If we use the prior distributions for $\boldsymbol{\beta}$ defined by (11.23) and assume that $f(\gamma_j = 0, \boldsymbol{\gamma}_{\backslash j}) = f(\gamma_j = 1, \boldsymbol{\gamma}_{\backslash j})$ for all j, then

$$\frac{f(\gamma_j = 1|\boldsymbol{y}, \boldsymbol{\gamma}_{\backslash j}, \boldsymbol{\beta})}{f(\gamma_j = 0|\boldsymbol{y}, \boldsymbol{\gamma}_{\backslash j}, \boldsymbol{\beta})} = k_j^{-d_j} \exp\left(-\frac{1}{2}(1 - k_j^2)\boldsymbol{\beta}_j^T \boldsymbol{\Sigma}_j^{-1} \boldsymbol{\beta}_j\right), \tag{11.25}$$

where d_j is the dimension of $\boldsymbol{\beta}_j$.

Posterior model probabilities calculated by this approach are slightly different from the corresponding probabilities calculated using the traditional posterior model probabilities, assuming that specific parameters β_j are constraint to zero and therefore are fully eliminated from the model. The resulting posterior model probabilities are heavily dependent on the choice of the prior parameters k_j^2 and $\boldsymbol{\Sigma}_j$. One way of specifying these parameters is by setting $\boldsymbol{\Sigma}_j$ large (for $\gamma_j = 1$) as in Section 11.5.1. Then, k_j^2 can be chosen by considering the value of $|\boldsymbol{\beta}_j|$ at which the densities of the two components of the prior distribution are equal. This can be considered as the smallest value of $|\boldsymbol{\beta}_j|$ at which the term is considered to be significant in practice as proposed by George and McCulloch (1993). See also Ntzoufras et al. (2000) for specification of the prior in the Poisson log-linear interaction models with multidimensional $\boldsymbol{\beta}_j$.

Using unconditional prior distribution: The Kuo–Mallick sampler. Kuo and Mallick (1998) approach is similar to GVS since the linear predictor has the form (11.12). They considered a prior distribution for the model parameters $f(\boldsymbol{\beta})$ independent of the model structure $\boldsymbol{\gamma}$, resulting in the prior

$$f(\boldsymbol{\beta}, \boldsymbol{\gamma}) = f(\boldsymbol{\beta})f(\boldsymbol{\gamma}) = f(\boldsymbol{\beta}_{\boldsymbol{\gamma}}|\boldsymbol{\beta}_{\backslash \boldsymbol{\gamma}})f(\boldsymbol{\beta}_{\backslash \boldsymbol{\gamma}})f(\boldsymbol{\gamma}).$$

Therefore, the full conditional posterior distributions are given by

$$f(\boldsymbol{\beta}_j|\boldsymbol{y}, \boldsymbol{\gamma}, \boldsymbol{\beta}_{\backslash j}) \propto \begin{cases} f(\boldsymbol{y}|\boldsymbol{\gamma}, \boldsymbol{\beta})f(\boldsymbol{\beta}_j|\boldsymbol{\beta}_{\backslash j}) & \gamma_j = 1 \\ f(\boldsymbol{\beta}_j|\boldsymbol{\beta}_{\backslash j}) & \gamma_j = 0 \end{cases} \tag{11.26}$$

while, similarly to the corresponding step of GVS and SSVS, variable indicators γ_j conditional on the remaining parameters are a posteriori Bernoulli distributed with success probability $O_j/(1 + O_j)$ [see Eq. (11.18)] with

$$O_j = \frac{f(\gamma_j = 1|\boldsymbol{y}, \boldsymbol{\gamma}_{\backslash j}, \boldsymbol{\beta})}{f(\gamma_j = 0|\boldsymbol{y}, \boldsymbol{\gamma}_{\backslash j}, \boldsymbol{\beta})} = \frac{f(\boldsymbol{y}|\gamma_j = 1, \boldsymbol{\gamma}_{\backslash j}, \boldsymbol{\beta})}{f(\boldsymbol{y}|\gamma_j = 0, \boldsymbol{\gamma}_{\backslash j}, \boldsymbol{\beta})} \frac{f(\gamma_j = 1, \boldsymbol{\gamma}_{\backslash j})}{f(\gamma_j = 0, \boldsymbol{\gamma}_{\backslash j})}. \tag{11.27}$$

The main strength of this approach is that it can be implemented in a straightforward manner. It is necessary only to specify the usual prior on $\boldsymbol{\beta}$ (for the full model), and the conditional prior distributions $f(\boldsymbol{\beta}_j|\boldsymbol{\beta}_{\backslash j})$ replace the pseudopriors required by GVS. However, this is also a drawback in specific cases, since it does not allow for any flexibility of the method and, thereby for improvement of the efficiency of the sampler.

Discussion and comparison of the variable selection algorithms.   The three Gibbs sampling variable selection methods presented above can be easily summarized by inspecting the conditional probabilities (11.24), (11.27), and, in particular, (11.20).

In SSVS, $f(y|\beta,\gamma)$ is not a function of γ, and so the first ratio on the right side of (11.20) is absent in (11.24). For the "unconditional priors approach" of Kuo and Mallick (1998), the second term on the right side of (11.20) is absent in (11.27) as β and γ are a priori independent. For Gibbs variable selection, both likelihood and prior appear in the variable selection step.

The key differences between the methods are their requirements in terms of prior and/or linking densities. GVS requires linking densities whose sole function is to enhance the efficiency of the sampler. The prior parameters in SSVS all have an impact on the posterior, and therefore the densities cannot really be regarded as linking densities. The simplest method, described by Kuo and Mallick (1998), does not require one to specify anything other than the usual priors for the model parameters of the full model. The association between the variable selection Gibbs samplers is summarized in Table 11.7.

Table 11.7   Components of full conditional posterior odds for inclusion of term j (O_j) in each variable selection algorithm

Method	η	O_j PSR$_j$	LR$_j$	PR$_j$
SSVS	$X\beta$			$\checkmark$
KM	$\sum \gamma_j X_j \beta_j$		$\checkmark$	
GVS	$\sum \gamma_j X_j \beta_j$	$\checkmark$	$\checkmark$	$\checkmark$

Key: PSR = Pseudoprior Ratio; LR = Likelihood Ratio; PR = Prior Density Ratio; SSVS = stochastic search variable selection; KM = Kuo–Mallick method; GVS = Gibbs variable selection.

11.6   POSTERIOR INFERENCE USING THE OUTPUT OF BAYESIAN VARIABLE SELECTION SAMPLERS

The main focus in Bayesian variable selection is to estimate the maximum a posteriori (MAP) model. This can be estimated by a simple frequency tabulation of the model indicator m or of the joint distribution γ. Usually for every indicator γ we assign the following model indicator

$$m(\gamma) = \sum_{j=k}^{p} \gamma_j 2^{j-k},$$

which transforms the γ to a unique decimal number, where $k = 1$ when the constant term is included in all models under consideration and $k = 0$ otherwise.

Hence we estimate the posterior model probabilities $f(m|y)$ simply using

$$\widehat{f}(m|y) = \frac{1}{T-B} \sum_{t=B+1}^{T} I(m^{(t)} = m),$$

where T and B are the total and burnin iterations of the algorithm and $m^{(t)}$ is the model indicator value at iteration t. Note that only for the best models we can accurately estimate the posterior model probabilities. Posterior model odds can be estimated using the ratios of the posterior model probabilities calculated by the MCMC output. These estimates are accurate only when both compared models have been visited in a sufficient number of iterations by the algorithm. An alternative is to use the approach of Ntzoufras et al. (2005) and calibrate prior model probabilities in such a way that all models under consideration are visited sufficiently often. Then, the Bayes factor and the true posterior model odds are calculated by some inverse calculations based on definition of the posterior model odds (11.1); see also in Katsis and Ntzoufras (2005) for implementation using WinBUGS .

From the MCMC output, we can estimate the posterior inclusion probabilities of each variable

$$f(\gamma_j = 1|\boldsymbol{y}) = \sum_{\boldsymbol{\gamma}_{\backslash j} \in \{0,1\}^{p-1}} f(\gamma_j = 1, \boldsymbol{\gamma}_{\backslash j}|\boldsymbol{y})$$

using the simple estimator

$$\widehat{f}(\gamma_j = 1|\boldsymbol{y}) = \frac{1}{T - B} \sum_{t=B+1}^{T} I(\gamma_j^{(t)} = 1) \,.$$

Using these inclusion probabilities, we can trace the *median probability model* (MP), which includes all variables with $f(\gamma_j = 1|\boldsymbol{y}) > 0.5$. The MP model has better predictive performance than the MAP model under certain conditions; for model details, see Barbieri and Berger (2004).

An important task in variable selection is to identify a reasonable solution when the number of covariates p is large, especially in comparison to the sample size n. For large p, the model space becomes large, rendering the efficient implementation of the above algorithm difficult. For example, with 20, 50, and 100 variables under consideration, the model space includes more than 10^6, 10^{15}, and 10^{30} models. In such cases, it is rather unrealistic to expect any algorithm to accurately calculate the posterior probabilities of even the best models. Nevertheless, the algorithms described above can work as model search algorithms, which can identify the most probable models. Moreover, the posterior inclusion probabilities can be estimated efficiently. Thus we can follow the strategy of Fouskakis et al. (2008) and reduce the model space by removing variables with very low marginal inclusion probabilities — say, for example, with values lower than 0.2.

Concerning the model parameters $\boldsymbol{\beta}_{\boldsymbol{\gamma}}$, the posterior distributions can $f(\boldsymbol{\beta}_{\boldsymbol{\gamma}}|\boldsymbol{\gamma}, \boldsymbol{y})$ can be estimated but only for models that the algorithm visited in a sufficient number of iterations.

In specific cases we do not wish to select a specific model, but make inference or predictions by considering model uncertainty also in our analysis. Hence we may obtain the model averaged posterior density of a quantity of interest ξ by

$$f(\xi|\boldsymbol{y}) = \sum_{m \in \mathcal{M}} f(m|\boldsymbol{y}) f(\xi|m, \boldsymbol{y}) \,.$$

For example, ξ might be a future observation y^{rep}.

Note that the marginal posterior distribution $f(\boldsymbol{\beta}|\boldsymbol{y})$ (i.e., under all models) does not have any meaning unless the parameters have the same interpretation across all models. In GLMs, this is true only for the case where $\boldsymbol{X}$ is close to orthogonality.

11.7   IMPLEMENTATION OF GIBBS VARIABLE SELECTION IN WinBUGS USING AN ILLUSTRATIVE EXAMPLE

Here we illustrate the implementation of GVS in WinBUGS using a simple normal example. Additional examples (including binomial and Poisson data) and illustration of SSVS and the Kuo and Mallick (1998) sampler can be found in Ntzoufras et al. (2000).

> **Example 11.3. Dellaportas et al. (2002) simulated data.** Here we consider for illustration purposes the first simulated dataset due to Dellaportas et al. (2002) with $p = 15$ covariates and sample size equal to $n = 50$; data are available in the book's Website with kind permission of Springer science and business media. Covariates X_j (for $j = 1, \ldots, 15$) were generated from a standardized normal distribution, while the response variable was generated from
>
> $$Y_i \sim N(X_{i4} + X_{i5}, (2.5)^2) \text{ for } i = 1, 2, \ldots, 50.$$

For this examples we use the following set of prior distributions:

1. The prior originally used by Dellaportas et al. (2002): $\beta_j \sim N(0, 100)$ for $j = 0, 1, 2, \ldots, 15$ and $\sigma^2 \sim \text{IG}(10^{-4}, 10^{-4})$ for σ^{-2}.

2. Zellner's g-prior with zero mean and $c^2 = n$ given by (11.14).

3. Empirical Bayes independent prior distribution accounting for one data point. We set $\beta_j \sim N(\tilde{\beta}_j, n\tilde{S}^2_{\beta_j})$ for $j = 0, \ldots, 15$, where $\tilde{\beta}_j$ and $\tilde{S}^2_{\beta_j}$ are the posterior mean and variance of β_j estimated by the full model.

Implementation in WinBUGS.   The likelihood is specified as usual in WinBUGS. The difference is that we now need to incorporate the binary indicators γ_j in the linear predictor. Hence we can write

```
for (i in 1:n){
    y[i] ~ dnorm( mu[i], tau)
    mu[i] <- gamma0*beta0 + x[i,1]*gamma[1]*beta[1] + ...
                          + x[i,15]*gamma[15]*beta[15]
}
```

This expresion can be written in WinBUGS in a more general way by avoiding the long summation involved in the linear predictor by using the command inprod. Therefore, the syntax can be written as

```
for (j in 1:p){ gb[j] <- gamma[j]*beta[j] }
for (i in 1:n){
    y[i] ~ dnorm( mu[i], tau )
    mu[i] <- gamma0 * beta0 + inprod( x[i,1:p], gb[1:p] )
}
```

In this code, node gb corresponds to the active values of $\boldsymbol{\beta}_{\setminus 0}$ under the current model since its elements assume values equal to β_j when $\gamma_j = 1$ and values equal to zero otherwize (i.e., when the corresponding variable is excluded from the model). All parameters (including β_0) can be incorporated in a single node (e.g., B) using the following syntax for the likelihood definition:

```
for (i in 1:n){
    X[i,1] <- 1.0
    for (j in 1:p){
```

```
            X[i,j+1]<-x[i,j]
    }
}
B[1]<-beta0
g[1]<-gamma0
for (j in 1:p){
    B[j+1]<-beta[j]
    g[j+1]<-gamma[j]
}
for (j in 1:(p+1)){ gb[j] <- g[j]*B[j] }
for (i in 1:n){
        y[i] ~ dnorm( mu[i], tau )
        mu[i] <- inprod( X[i,], gb[] )
}
```

In this code, X and x are the design matrices including or excluding the constant term. This setup should be used when a multivariate normal prior is used for $\boldsymbol{\beta}$ as in the second prior setup here: Zellner's g-prior with zero mean and $c^2 = n$.

The prior for the inclusion indicators are easily defined here since $\gamma_j \sim \text{Bernoulli}(0.5)$ for all $j = 0, 1, \ldots, p$.

Specification of the prior and the proposal densities in WinBUGS when we use independent priors are used (as in prior setups 1 and 3) is relatively easy since we need to specify that

$$\beta_j \sim N(\mu_{\gamma_j,\beta_j}, S^2_{\gamma_j,\beta_j})$$

with

$$\mu_{\gamma_j,\beta_j} = \gamma_j \mu_{\beta_j} + (1 - \gamma_j)\overline{\mu}_{\beta_j} \text{ and } S^2_{\gamma_j,\beta_j} = \gamma_j S^2_{\beta_j} + (1 - \gamma_j)\overline{S}^2_{\beta_j},$$

where μ_{β_j} and $S^2_{\beta_j}$ are the prior mean and variance of β_j while $\overline{\mu}_{\beta_j}$ and $\overline{S}^2_{\beta_j}$ are the corresponding proposal (or pseudoprior) parameters, respectively. Here, the proposal parameters were set equal to the posterior mean $\widetilde{\mu}_{\beta_j}$ and variance $\widetilde{S}^2_{\beta_j}$ of each β_j estimated by a pilot MCMC run of the full model. For the first prior setup, we set $\mu_{\beta_j} = 0$ and $S^2_{\beta_j} = 100$, resulting in

$$\mu_{\gamma_j,\beta_j} = (1 - \gamma_j)\widetilde{\mu}_{\beta_j} \text{ and } S^2_{\gamma_j,\beta_j} = \gamma_j 100 + (1 - \gamma_j)\widetilde{S}^2_{\beta_j},$$

which can be modeled in WinBUGS using the syntax

```
gamma0~dbern(0.5)
for (j in 1:p){ gamma[j]~dbern(0.5)  }
beta0 ~ dnorm( mb0, taub0)
mb0 <- (1-gamma0) * prop.mean.beta0
taub0 <- gamma0*0.01 + (1-gamma0) /pow(prop.sd.beta0 ,2)
for (j in 1:p){
    beta[j] ~ dnorm( mb[j], taub[j])
    mb[j] <- (1-gamma0) * prop.mean.beta[j]
    taub[j]<- gamma[j]*0.01 + (1-gamma[j])/pow(prop.sd.beta[j],2)
}
tau~dgamma( 0.01, 0.01)
```

where prop.mean.beta[j] and prop.sd.beta[j] are the proposal means $\overline{\mu}_{\beta_j}$ and standard deviations $\overline{S}_{\beta_j}$ of β_j. For the empirical Bayes (third prior setup), the mean and variance of the mixture are given by

$$\mu_{\gamma_j,\beta_j} = \widetilde{\mu}_{\beta_j} \text{ and } S^2_{\gamma_j,\beta_j} = \{\gamma_j n + (1 - \gamma_j)\}\widetilde{S}^2_{\beta_j},$$

which can specified in WinBUGS using the syntax

```
gamma0~dbern(0.5)
for (j in 1:p){ gamma[j]~dbern(0.5)   }
beta0  ~ dnorm( mb0, taub0)
mb0 <- (1-gamma0) * prop.mean.beta0
taub0 <- gamma0*0.01 + (1-gamma0)/pow(prop.sd.beta0, 2)
for (j in 1:p){
     beta[j]  ~ dnorm( mb[j], taub[j])
     mb[j] <- prop.mean.beta[j]
     taub[j]<- ( gamma[j]/n + (1-gamma[j]) )/pow(prop.sd.beta[j],2)
}
tau~dgamma(0.01, 0.01)
```

In order to specify Zellner's g-prior, we need to work on the whole parameter vector $\boldsymbol{\beta}$, which in the following code is denoted by B. Hence we use a multivariate normal prior distribution with mean $\boldsymbol{\mu}_{\gamma,\beta}$ and precision $\boldsymbol{T}_{\gamma,\beta}$:

$$\boldsymbol{\beta} \sim N_p\left(\boldsymbol{\mu}_{\gamma,\beta}, \boldsymbol{T}_{\gamma,\beta}^{-1}\right),$$

where $\boldsymbol{\beta}$ is denoted by B in the WinBUGS model code. The elements of the mean $\boldsymbol{\mu}_\beta$ are defined as in the case of independent prior distributions by

$$\mu_{\gamma,\beta,j} = (1 - \gamma_j)\overline{\mu}_{\beta_j},$$

while for the prior precision matrix $\boldsymbol{T}$, each element T_{jk} is equal to the elements of matrix $c^{-2}\sigma^{-2}\boldsymbol{X}^T\boldsymbol{X}$ in the case where both variables X_j and X_k are included in the model. When at least one of them is not included in the model, then $T_{jk} = 0$ for $j \neq k$. Finally, diagonal elements T_{jj} denote the pseudoprior precision for $\gamma_j = 0$ that is, when X_j is excluded from the model. Hence we set

$$T_{jk} = \frac{\gamma_j\gamma_k}{n\sigma^2}[\boldsymbol{X}^T\boldsymbol{X}]_{jk} + (1 - \gamma_j\gamma_k)I(j = k)\overline{S}_{\beta_j}^2$$

for $i, j = 1, 2, \ldots, p$ and $c^2 = n$. This mixture prior can be specified in WinBUGS using the following syntax:

```
for (j in 1:(p+1)){ g[j]~dbern(0.5)   }
B[1:(p+1)]  ~ dmnorm( mean.beta[1:(p+1)], T[ 1:(p+1), 1:(p+1)] )
tau~dgamma( 0.01, 0.01)
for (j in 1:(p+1) ){ mean.beta[j] <-  (1-g[j])*prop.mean.beta[j] }
for (j in 1:(p+1) ){ for (k in 1:(p+1) ){
     T[j,k] < - g[j]*g[k]*tau*XTX[j,k]/n
               + ( 1- g[j]*g[k])*equals(j,k)*pow(prop.sd.beta[k],-2)
}}
gamma0 <-g[1]
beta0 <-B[1]
for (j in 1:p){
     beta[j]    <- B[j+1]
     gamma[j] <- g[j+1]
}
```

Before we conclude this section, we describe how we can obtain in WinBUGS a uniquely defined model indicator m for $\boldsymbol{\gamma}$ and how we can calculate posterior model probabilities for specific models of interest. To transform $\boldsymbol{\gamma}$ to a model indicator m, we use the simple formula that converts numbers defined in the binary numerical system to the corresponding numbers in the decimal numerical system. Hence we use the equation

$$m(\boldsymbol{\gamma}) = \sum_{j=0}^{p} \gamma_j 2^j,$$

which in WinBUGS is expressed using the syntax

```
for (j in 1:p){ mindex[j] <- pow(2,j)}
model <- gamma0 + inprod( gamma[], mindex[] )
```

In this code, WinBUGS model is used insted of m.

The model indicator m and the corresponding WinBUGS node, model, can then be exported to a statistical software using the command CODA in order to obtain its frequency tabulation and, therefore, estimates of the posterior model probabilities. Nevertheless, posterior probabilities of specific models can also be calculated within WinBUGS using the command equals. In this way, we can trace whether a specific model is visited at each iteration of the MCMC run. For the current example, we monitor models $X_4 + X_5$ and $X_4 + X_5 + X_{12}$, which are the true and the maximum a posteriori ones according to Dellaportas et al. (2002) analysis, using the syntax

```
pmodel[1] <- equals( model, pow(2,4)+pow(2,5) )
pmodel[2] <- equals( model, pow(2,4)+pow(2,5)+pow(2,12) )
```

This code uses a set binary indicators pmodel to identify which iteration m (model in WinBUGS) takes values equal to $2^4 + 2^5 = 48$ or $2^4 + 2^5 + 2^{12} = 4144$ that correspond to the target models $X_4 + X_5$ and $X_4 + X_5 + X_{12}$.

When the number of all models is small (e.g., for $p = 3$, the models under consideration are only 8), then all posterior model probabilities can be calculated using this approach. But in the case of a large model space, this approach is not recommended since it will slow down WinBUGS because of the large amount of values stored by using these binary model indicators.

Results. Posterior variable inclusion probabilities and posterior model probabilities are presented in Tables 11.8 and 11.9 from MCMC runs of 21,000 iterations after discarding the

Table 11.8   Posterior variable inclusion probabilities for simulated data of Dellaportas et al. (2002)

	Prior 1[a]		Prior 2[b]		Prior 3[c]	
	$f(\gamma_j = 1|\boldsymbol{y})$	MC error	$f(\gamma_j = 1|\boldsymbol{y})$	MC error	$f(\gamma_j = 1|\boldsymbol{y})$	MC error
γ_0	0.042	0.0045	0.134	0.0025	0.039	0.0016
γ_1	0.031	0.0012	0.128	0.0025	0.106	0.0022
γ_2	0.039	0.0014	0.136	0.0027	0.113	0.0023
γ_3	0.033	0.0013	0.127	0.0024	0.101	0.0018
γ_4	0.970	0.0024	0.992	0.0001	0.990	0.0001
γ_5	0.999	0.0001	1.000	0.0000	1.000	0.0001
γ_6	0.046	0.0016	0.155	0.0028	0.128	0.0025
γ_7	0.037	0.0015	0.138	0.0028	0.117	0.0023
γ_8	0.041	0.0015	0.133	0.0023	0.105	0.0025
γ_9	0.044	0.0014	0.168	0.0027	0.138	0.0027
γ_{10}	0.043	0.0015	0.141	0.0029	0.115	0.0021
γ_{11}	0.048	0.0015	0.184	0.0030	0.147	0.0027
γ_{12}	0.338	0.0033	0.615	0.0040	0.545	0.0034
γ_{13}	0.038	0.0014	0.137	0.0024	0.106	0.0024
γ_{14}	0.042	0.0014	0.137	0.0023	0.104	0.0021
γ_{15}	0.076	0.0019	0.277	0.0037	0.243	0.0032

[a] Independent $\beta_j \sim N(0, 100)$ used by Dellaportas et al. (2002).
[b] Zellner's g-prior with zero mean and $c^2 = n$ given by (11.14).
[c] Empirical Bayes prior of type $\beta_j \sim N(\widetilde{\beta}_j, n\widetilde{S}^2_{\beta_j})$, where $\widetilde{\beta}_j$ and $\widetilde{S}^2_{\beta_j}$ are the posterior mean and variance of β_j estimated by the full model.

Table 11.9   Posterior model probabilities and odds for simulated data of Dellaportas et al. (2002)

Rank	m	m_k	Model	$f(m|y)$	$\text{PO}_{m_1 m_k}$
			Prior 1[a]		
1	48	m_1	$X_4 + X_5$	0.3664	1.00
2	4,144	m_2	$X_4 + X_5 + X_{12}$	0.1854	1.98
3	32,816	m_3	$X_4 + X_5 + X_{15}$	0.0292	12.55
4	560	m_4	$X_4 + X_5 + X_9$	0.0196	18.69
5	112	m_5	$X_4 + X_5 + X_6$	0.0178	20.58
6	16,432	m_6	$X_4 + X_5 + X_{14}$	0.0176	20.82
7	2,096	m_7	$X_4 + X_5 + X_{11}$	0.0172	21.30
8	1,072	m_8	$X_4 + X_5 + X_{10}$	0.0157	23.34
9	8,240	m_9	$X_4 + X_5 + X_{13}$	0.0150	24.43
10	49	m_{10}	$X_0 + X_4 + X_5$	0.0149	24.59
			Prior 2[b]		
1	4,144	m_2	$X_4 + X_5 + X_{12}$	0.0679	0.67
2	48	m_1	$X_4 + X_5$	0.0453	1.00
3	36,912	m_{11}	$X_4 + X_5 + X_{12} + X_{15}$	0.0252	1.80
4	6,192	m_{12}	$X_4 + X_5 + X_{11} + X_{12}$	0.0176	2.57
5	32,816	m_3	$X_4 + X_5 + X_{15}$	0.0158	2.87
6	4,208	m_{13}	$X_4 + X_5 + X_6 + X_{12}$	0.0118	3.84
7	12,336	m_{14}	$X_4 + X_5 + X_{12} + X_{13}$	0.0116	3.91
8	4,656	m_{15}	$X_4 + X_5 + X_9 + X_{12}$	0.0115	3.94
9	4,272	m_{16}	$X_4 + X_5 + X_7 + X_{12}$	0.0114	3.97
10	5,168	m_{17}	$X_4 + X_5 + X_{10} + X_{12}$	0.0112	4.04
			Prior 3[c]		
1	4,144	m_2	$X_4 + X_5 + X_{12}$	0.1014	0.88
2	48	m_1	$X_4 + X_5$	0.0896	1.00
3	36,912	m_{11}	$X_4 + X_5 + X_{12} + X_{15}$	0.0312	2.87
4	32,816	m_3	$X_4 + X_5 + X_{15}$	0.0277	3.23
5	6,192	m_{12}	$X_4 + X_5 + X_{11} + X_{12}$	0.0207	4.33
6	4,656	m_{15}	$X_4 + X_5 + X_9 + X_{12}$	0.0151	5.93
7	560	m_4	$X_4 + X_5 + X_9$	0.0142	6.31
8	5,168	m_{17}	$X_4 + X_5 + X_{10} + X_{12}$	0.0138	6.49
9	4,208	m_{13}	$X_4 + X_5 + X_6 + X_{12}$	0.0136	6.59
10	112	m_5	$X_4 + X_5 + X_6$	0.0133	6.74

[a] Independent $\beta_j \sim N(0, 100)$ used by Dellaportas et al. (2002).
[b] Zellner's g-prior with zero mean and $c^2 = n$ given by (11.14).
[c] Empirical Bayes prior of type $\beta_j \sim N(\widetilde{\beta}_j, n\widetilde{S}^2_{\beta_j})$, where $\widetilde{\beta}_j$ and $\widetilde{S}^2_{\beta_j}$ are the posterior mean and variance of β_j estimated by the full model.

initial 1000 iterations as burnin. Results under the three different prior setups differ because of the Lindley–Bartlett paradox described in a previous section. We managed to reproduce the results of Dellaportas et al. (2002) under prior setup 1; see also in Ntzoufras (1999a, pp. 110–111). Under the first prior setup the true model $X_4 + X_5$ has posterior model probability 0.366, which is about twice as high as the corresponding probability of model $X_4 + X_5 + X_{12}$. Under this prior setup, more parsimonious models are supported than the ones obtained by using Zellner's g-prior with zero mean and $c^2 = n$ (prior setup 2) and the empirical prior setup (setup 3). Note that under all prior setups used in this illustration, variables X_4 and X_5 must be included in the model since their posterior inclusion probabilities are close to one.

In a second stage, following the approach of Fouskakis et al. (2008), we can eliminate variables with posterior inclusion probabilities lower than 0.20 and rerun the MCMC algorithm. With this approach, we keep variables X_4, X_5, and X_{12} for first prior setup and the same variables including variable X_{15} under prior setups 2 and 3. Posterior model probabilities in this reduced model space are presented in Table 11.10.

Table 11.10 Posterior model probabilities and odds for simulated data of Dellaportas et al. (2002) in reduced model spaces

Model	m	Posterior model probability			Posterior model odds[d]		
		Prior 1[a]	Prior 2[b]	Prior 3[c]	Prior 1[a]	Prior 2[b]	Prior 3[c]
$X_4 + X_5$	4	0.6505	0.2987	0.3503	1.00	1.00	1.00
$X_4 + X_5 + X_{12}$	8	0.3265	0.4338	0.4118	1.99	0.69	0.85
X_5	3	0.0127	0.0017	0.0013	51.22	175.71	269.46
$X_5 + X_{12}$	7	0.0102	0.0025	0.0035	63.77	119.48	100.09
$X_4 + X_5 + X_{15}$	12	–	0.1032	0.1055	–	2.89	3.32
$X_4 + X_5 + X_{12} + X_{15}$	16	–	0.1568	0.1239	–	1.90	2.83

[a] Independent $\beta_j \sim N(0, 100)$ used by Dellaportas et al. (2002).
[b] Zellner's g-prior with zero mean and $c^2 = n$ given by (11.14).
[c] Empirical Bayes prior of type $\beta_j \sim N(\widetilde{\beta}_j, n\widetilde{S}^2_{\beta_j})$, where $\widetilde{\beta}_j$ and $\widetilde{S}^2_{\beta_j}$ are the posterior mean and variance of β_j estimated by the full model.
[d] Each model is compared with the true model $X_4 + X_5$.

11.8 THE CARLIN–CHIB METHOD

Carlin and Chib (1995) proposed a Gibbs sampling strategy used to generate from the posterior distribution $f(m, \boldsymbol{\beta}_m | \boldsymbol{y})$. In order to do this, we need to consider the joint distribution the model indicator m and parameter vectors $\boldsymbol{\beta}_{m'}$ for all models m' under consideration, that is, for $\{\boldsymbol{\beta}_{m'}$ for all $m' \in \mathcal{M}\}$. Therefore, we need to define prior distributions $f(m)$ and $f(\boldsymbol{\beta}_{m'}|m)$ for all $m, m' \in \mathcal{M}$. Distributions $f(\boldsymbol{\beta}_m | m)$ are the usual prior distributions while $f(\boldsymbol{\beta}_{m'}|m)$ with $m' \neq m$ are called *pseudopriors* or *linking densities*, which are used to facilitate the Gibbs sampler and do not influence the posterior distribution.

The conditional posterior distributions required for the Gibbs sampler are

$$f(\boldsymbol{\beta}_{m'}|m, \{\boldsymbol{\beta}_s : s \in \mathcal{M} \setminus \{m'\}\}, \boldsymbol{y},) \propto \begin{cases} f(\boldsymbol{y}|m, \boldsymbol{\beta}_m)f(\boldsymbol{\beta}_m|m) & m' = m \\ f(\boldsymbol{\beta}_{m'}|m) & m' \neq m \end{cases} \quad (11.28)$$

$$f(m|\{\boldsymbol{\beta}_s : s \in \mathcal{M}\}, \boldsymbol{y}) = \frac{A_m}{\sum\limits_{s \in \mathcal{M}} A_s}, \qquad (11.29)$$

where

$$A_m = f(\boldsymbol{y}|m, \boldsymbol{\beta}_m)\left\{\prod_{s \in \mathcal{M}} f(\boldsymbol{\beta}_s|m)\right\} f(m), \quad \forall \ m \in \mathcal{M}.$$

Therefore, when $m' = m$, we generate from the usual conditional posterior for model m, and when $m' \neq m$ we generate from the corresponding pseudoprior, $f(\boldsymbol{\beta}_{m'}|m)$. The model indicator m is generated as a discrete random variable using (11.29).

The pseudopriors have no influence on $f(\boldsymbol{\beta}_m|m)$, the marginal posterior distribution of interest. They act only as linking densities. Their careful specification is essential for the mobility and the efficiency of the sampler. Ideally, $f(\boldsymbol{\beta}_{m'}|m \neq m')$ must resemble the marginal posterior distribution $f(\boldsymbol{\beta}_{m'}|m', \boldsymbol{y})$; see Carlin and Chib (1995) for proposed strategies to achieve this.

The flexibility of this method is due to the ability to specify and tune these pseudopriors which assist the sampler move efficiently across models of different dimension. The number of pseudopriors can also be perceived as a drawback, especially in problems where the model space $\mathcal{M}$ is large since the specification of efficient pseudopriors may be time-consuming and cumbersome. A further drawback of the method is the requirement for generating parameter vectors $\boldsymbol{\beta}_{m'}$ for all models under consideration $m' \in \mathcal{M}$ at each stage of the sampler. This may be avoided by using a Metropolis–Hastings step to generate m as proposed by Dellaportas et al. (2002). The description of this sampler is outside the scope of the current book; for more details, we refer the reader to the work of Dellaportas et al. (2002).

An example of the implementation of this method is provided in Spiegelhalter et al. (1996c, pp. 47–59, pines example) and in Spiegelhalter et al. (2003c, pp. 38–39; pines example). In pines example, both models are normal with different mean and variance. Hence the overall model is simply expressed as a mixture of two normal distributions with different parameters. When the models under comparison are of different distributional form, then the zeros trick must be used as described in Section 8.1. Also note that the binary indicators approach used in variable selection be can also adopted here to simplify the problem of model specification. For an example comparing different distributions using WinBUGS, see Katsis and Ntzoufras (2005).

11.9 REVERSIBLE JUMP MCMC (RJMCMC)

Reversible jump Markov chain Monte Carlo (Green, 1995) is a flexible MCMC sampling strategy for generating observations from the joint posterior distribution $f(m, \boldsymbol{\beta}_m|\boldsymbol{y})$. The method is a natural extension of the simple Metropolis–Hastings algorithm and is based on creating a Markov chain that can "jump" between models with parameter spaces of different dimensions.

Supposing that the current state of the Markov chain is $(m, \boldsymbol{\beta}_m)$, where $\boldsymbol{\beta}_m$ has dimension d_m, one version of the procedure would be as follows:

- Propose a new model m' with probability $j(m, m')$.

- Generate $\boldsymbol{u}$ from a specified proposal density $q(\boldsymbol{u}|\boldsymbol{\beta}_m, m, m')$.

- Set $(\boldsymbol{\beta}'_{m'}, \boldsymbol{u}') = h_{m,m'}(\boldsymbol{\beta}_m, \boldsymbol{u})$, where $h_{m,m'}$ is a specified invertible function for which $h_{m',m} = h_{m,m'}^{-1}$. This equation implies that $d_m + d(\boldsymbol{u}) = d_{m'} + d(\boldsymbol{u}')$, where d_m and $d(\boldsymbol{u})$ are the dimensions for the parameters of model m and vector $\boldsymbol{u}$, respectively.

- Accept the proposed move to model m' with probability

$$\alpha = \min\left(1, \frac{f(\boldsymbol{y}|\boldsymbol{\beta}'_{m'}, m')f(\boldsymbol{\beta}'_{m'}|m')f(m')j(m', m)q(\boldsymbol{u}'|\boldsymbol{\beta}'_{m'}, m', m)}{f(\boldsymbol{y}|\boldsymbol{\beta}_m, m)f(\boldsymbol{\beta}_m|m)f(m)j(m, m')q(\boldsymbol{u}|\boldsymbol{\beta}_m, m, m')} \left| \frac{\partial h(\boldsymbol{\beta}_m, \boldsymbol{u})}{\partial(\boldsymbol{\beta}_m, \boldsymbol{u})} \right| \right).$$
(11.30)

There are many variations or simpler versions of reversible jump that can be applied in specific model selection problems. In particular, if all parameters of the proposed model are generated from a proposal distribution, then $(\boldsymbol{\beta}'_{m'}, \boldsymbol{u}') = (\boldsymbol{u}, \boldsymbol{\beta}_m)$ with $d_m = d(\boldsymbol{u}')$ and $d_{m'} = d(\boldsymbol{u})$, and the Jacobian term in (11.30) is one. This version of the reversible jump can be used for jumping between models for which no appropriate parameter transformation exists. When model m is nested to model m', then a natural transformation function $h_{m,m'}$ may be the identity function such that $d(\boldsymbol{u}') = 0$ and $\boldsymbol{\beta}_{m'} = (\boldsymbol{\beta}_m, \boldsymbol{u})$; see, for example, Dellaportas and Forster (1999). Finally, if $m' = m$, then the corresponding step is the same as in a standard Metropolis–Hastings step. More details concerning RJMCMC can be found in Han and Carlin (2001), Dellaportas et al. (2002) and Sisson (2005).

The WinBUGS jump interface, recently developed by Dave Lunn, can be used for implementing variable selection and automatic spline models. The interface is available at the WinBUGS development site:

http://www.winbugs-development.org.uk/main.html.

For more details, see Lunn et al. (2005), Lunn et al. (2006) and the interface's manual and examples.

11.10 USING POSTERIOR PREDICTIVE DENSITIES FOR MODEL EVALUATION

The posterior Bayes factor (PBF) is a natural variation of the Bayes factor based on the ratio of the posterior predictive densities of the observed data. Hence the PBF of model m_1 versus m_2 is defined as

$$\mathrm{PBF}_{12} = \frac{f(\boldsymbol{y}|\boldsymbol{y}, m_1)}{f(\boldsymbol{y}|\boldsymbol{y}, m_2)}.$$

Although PBF allows the use of improper priors, it has been strongly criticized for its double usage of the data, leading to violation of the likelihood principle. Moreover, it does not correspond to any purely Bayesian measure as the posterior model probabilities $f(m|\boldsymbol{y})$ and their corresponding model odds. Finally, PBF supports more complicated models than do Bayes factors because of the double use of data.

For this reason, it is more sensible to use the leave-one-out cross-validatory predictive density to construct an overall measure of model evaluation and comparison. Hence we may calculate the *pseudo-Bayes factor* (Geisser and Eddy, 1979; Gelfand and Dey, 1994), given by

$$\mathrm{PSB}_{12} = \prod_{i=1}^{n} \frac{f(y_i|\boldsymbol{y}_{\backslash i}, m_1)}{f(y_i|\boldsymbol{y}_{\backslash i}, m_2)} = \prod_{i=1}^{n} \frac{\mathrm{CPO}_i(m_1)}{\mathrm{CPO}_i(m_2)}$$

where $\mathrm{CPO}_i(m)$ is the leave-one-out cross-validation predictive density for y_i under model m. Computation of the pseudo Bayes factor can be achieved by calculating the cross-validation predictive likelihood, defined as

$$\ell_{\mathrm{CV}}^{\mathrm{pred}}(m) = \prod_{i=1}^{n} f(y_i|\boldsymbol{y}_{\backslash i}, m) = \prod_{i=1}^{n} \mathrm{CPO}_i(m).$$

Computation of each $\mathrm{CPO}_i(m)$ is described in Section 10.4.1. Alternatively, we may use the negative cross-validatory log-likelihood (Spiegelhalter et al., 1996a, p. 42), given by

$$\mathrm{LS}_{\mathrm{CV}}(m) = -\log \ell_{\mathrm{CV}}^{\mathrm{pred}}(m) = -\sum_{i=1}^{n} \log \mathrm{CPO}_i(m) \qquad (11.31)$$

also referred as the *cross-validatory predictive log-score* (Draper and Krnjajić, 2006).

   Although generation of the replicated/predicted data $\boldsymbol{y}^{\mathrm{rep}}$ from the full predictive density and computation of CPOs in the leave-one-out cross-validation is straightforward, sampling from all n leave-one-out cross-validatory predictive densities is not as straightforward as sampling from the full predictive density (Gelfand, 1996). A simple but computationally expensive method is to run the MCMC sampler n times by omitting one observation each time from the sample. To simplify computations, we can approximate the leave-one-out cross-validatory conditional predictive ordinate $f(y_i|\boldsymbol{y}_{\backslash i}, m)$ by the full posterior predictive ordinate $f(y_i|\boldsymbol{y}, m)$ when the sample size n is large [see, e.g., Carlin and Louis (2000, pp. 205–206)]. Hence, for the negative cross validation log-likelihood we can write

$$\mathrm{LS}_{\mathrm{CV}}(m) \approx -\sum_{i=1}^{n} \log f(y_i|\boldsymbol{y}, m) = -\sum_{i=1}^{n} \log \mathrm{PPO}_i(m) = \mathrm{LS}(m).$$

The logarithmic score $\mathrm{LS}(m)$ is simpler to calculate using a single MCMC run; for details, see Section 10.3.4. Note that $\mathrm{LS}(m)$ is based on the full predictive density and does not correspond to the same supported models as in posterior Bayes factor. In the logarithmic score we first calculate all PPO_i using expression (10.3) and then consider their product. Therefore, from each MCMC run we calculate the posterior means of $f(y_i|\boldsymbol{\theta}_m^{(t)}, m)$ over all iterations $t = 1, \ldots, T$ and then consider the product of these posterior means. On the other hand, for the posterior Bayes factor, we calculate the full likelihood (i.e., we consider the product) and then integrate out over the predictive density; for details, see the appendix in Draper and Krnjajić (2006).

   Other related methods are the approaches of the fractional and intrinsic Bayes factors introduced by O'Hagan (1995) and Berger and Pericchi (1996a,b). Both of these approaches try to avoid the double usage of the data and appropriately define the prior distribution in order to avoid the Lindley–Bartlett paradox. In the first approach, a fraction of the likelihood is used to construct the prior distribution and the remaining fraction to calculate Bayes factors and evaluate models, while in the latter a minimal training sample is used to construct the prior distribution and the remaining data to calculate the Bayes factor. The selection of the training sample is an important task in the second approach. This led to the proposal of median and mean intrinsic Bayes factors (and other variations), which have different properties and provide different values for comparison of the same models or hypotheses.

11.10.1 Estimation from an MCMC output

For calculation of the posterior Bayes factor, we need to calculate the posterior predictive density evaluated at the observed data y:

$$f(y|y, m) = \int f(y|\theta_m, m) f(\theta_m|y, m) d\theta_m.$$

This can be estimated in a straightforward manner by the posterior mean of the likelihood over all sampled parameter values from the output of an MCMC sampler. Hence, an estimate is given by

$$\widehat{f}(y|y, m) = \frac{1}{T} \sum_{t=1}^{T} f(y|\theta_m^{(t)}, m) = \overline{f(y|\theta_m, m)},$$

where $\theta_m^{(t)}$, $t = 1, \ldots, T$ is a sample from the posterior distribution. In WinBUGS this can be estimated in a straightforward manner by defining a node L for the likelihood given by

$$L = \exp\left(\sum_{i=1}^{n} \ell_i\right),$$

where $\ell_i = \log f(y_i|\theta_m, m)$ is the log-likelihood component for each observation i. Users must be careful with overflows and underflows of this node. For this reason, L may be divided by a constant term C to avoid such problems.

Computation of the pseudo Bayes factors and the leave-one-out cross-validation likelihood can be achieved by estimating $\text{CPO}_i(m)$ for $i = 1, \ldots, n$ from an MCMC output (i.e., within WinBUGS) as described in Section 10.4.1 and then calculate their product outside WinBUGS. Hence in WinBUGS we define the node `ICPO[i]` as the inverse of the likelihood and estimate CPOs by the inverse of their means:

$$\widehat{\text{CPO}}_i(m) = \left[\overline{\text{ICPO}}_i(m)\right]^{-1}.$$

The leave-one-out cross-validation likelihood is then estimated by

$$\widehat{\ell}_{\text{CV}}^{\text{pred}}(m) = \prod_{i=1}^{n} \widehat{\text{CPO}}_i(m) = \left(\prod_{i=1}^{n} \overline{\text{ICPO}}_i(m)\right)^{-1}.$$

The above quantity is frequently close to zero and, for this reason, its logarithm is usually reported instead expressed by

$$\widehat{\text{LS}}_{\text{CV}}(m) = -\log \widehat{\ell}_{\text{CV}}^{\text{pred}}(m) = -\log\left(\prod_{i=1}^{n} \widehat{\text{CPO}}_i(m)\right) = \sum_{i=1}^{n} \log \overline{\text{ICPO}}_i(m)$$

which is called the *negative cross-validatory log-likelihood* for model m.

Finally, the logarithmic score can be calculated in a software outside WinBUGS by the expression

$$\widehat{\text{LS}}(m) = -\sum_{i=1}^{n} \log \widehat{\text{PPO}}_i(m) = -\sum_{i=1}^{n} \log \overline{\text{like[i]}}$$

where $\overline{\text{like[i]}}$ is the posterior mean of the likelihood component for observation i corresponding to the MCMC estimate of $\text{PPO}_i(m)$; for details, see Section 10.3.4.

11.10.2   A simple example in WinBUGS

Example 11.4. Soft drink delivery times data (Example 5.1 revisited). Let us reconsider Example 5.1. Two covariates are available, and hence four models can be fitted. Here we calculate the posterior and the pseudo-Bayes factors between the models under consideration as well as the negative cross-validatory log-likelihood and the logarithmic score for each model.

The full WinBUGS code for calculation of the predictive likelihood L, the inverse CPO (ICPO), and PPO is available at the book's Website. Results are presented in Table 11.11, where we present the logarithm of the predictive likelihood, the cross-validatory predictive log-score $LS_{CV}(m)$, the log-score $LS(m)$, and the logarithms of the posterior and pseudo Bayes factor of each model compared to the one with both covariates in the model (model m_4). Finally, the last two columns correspond to the model weights based on the posterior and pseudo Bayes factors. All results have been generated with the same prior as in Section 11.4.2 with $c^2 = 10$, using 1000 iterations as burnin and an additional 5000 iterations as a final sample.

Table 11.11   Results based on the posterior predictive measures for soft drink data (Example 11.4)

				Logarithm of		Model weights	
Model (m)	$\log f(\boldsymbol{y}\|\boldsymbol{y}, m)$	$LS_{CV}(m)$	$LS(m)$	$\overline{PBF}_{m,m_4}$	PSB_{m,m_4}	w_{PBF}	w_{PsBF}
m_1 (constant)	-104.3	-107.9	-102.7	-39.5	-34.8	0.0000	0.0000
m_2 (cases)	-71.2	-78.4	-69.4	-6.5	-5.4	0.0015	0.0046
m_3 (distance)	-84.7	-88.9	-83.7	-20.0	-15.9	0.0000	0.0000
m_4 (cases + distance)	-64.8	-73.0	-63.1	0.0	0.0	0.9985	0.9954

Key: PBF_{m,m_4} and PSB_{m,m_4} = posterior and pseudo-Bayes factors of model m versus model m_4; w_{PBF} and w_{PsBF} = weights based on posterior and pseudo-Bayes factors calculated by $w_{PBF}(m) = PBF_{m,m_4} / \sum_{k=1}^{4} PBF_{m_k, m_4}$ and $w_{PSB}(m) = PSB_{m,m_4} / \sum_{k=1}^{4} PSB_{m_k, m_4}$.

The model with both covariates is fully supported. In particular, the weights of both posterior and pseudo-Bayes factors are concentrated (> 0.99) in this model, which is less parsimonious than the corresponding one supported by the analysis using the usual posterior model probabilities and the corresponding Bayes factors. Note that these measures are not sensitive to the prior variance controlled by c^2. Results for both $c^2 = 10$ and $c^2 = 1000$ are identical, and therefore we present results only for the first prior choice.

11.11   INFORMATION CRITERIA

The two most popular information criteria used in Bayesian statistics are: the Bayes information criterion (BIC), (Schwarz, 1978) and the Akaike information criterion (AIC), (Akaike, 1973). More recently, the deviance information criterion was introduced by Spiegelhalter et al. (2002), as an extension of AIC; for details, see Section 6.4.3. In simple one-stage models, AIC and DIC are identical. Differences occur in hierarchical and latent variable models where DIC uses the number of "effective" parameters instead of the actual number of parameters used by AIC. Here we also focus on the Bayesian versions of AIC and BIC as described by Brooks (2002).

11.11.1   The Bayes information criterion (BIC)

The Bayes information criterion (BIC) is based on the criterion originally introduced by Schwarz (1978), given by

$$S_{01} = \log(f(\boldsymbol{y}|\widehat{\boldsymbol{\theta}}_{m_1}, m_1) - \log(f(\boldsymbol{y}|\widehat{\boldsymbol{\theta}}_{m_0}, m_0) - \frac{1}{2}(d_{m_1} - d_{m_0})\log(n), \qquad (11.32)$$

where n is the sample size, $\widehat{\boldsymbol{\theta}}_m$ are the maximum likelihood estimates of parameters $\boldsymbol{\theta}_m$ of model m and d_m is the dimension of $\boldsymbol{\theta}_m$. The main property of the Schwarz criterion S it that

$$\frac{S_{01} - \log(B_{10})}{\log(B_{10})} \to 0 \quad \text{when} \quad n \to \infty, \qquad (11.33)$$

and therefore it can be used to obtain a rough approximation of the log–Bayes factor under a wide family of prior distributions; see Kass and Wasserman (1995) and Kass and Raftery (1995) for more details and Kuha (2004) for a nice review and description of BIC. The BIC value for a model m is defined as

$$\text{BIC}(m) = D(\widehat{\boldsymbol{\theta}}_m, m) + d_m \log n$$

where $D(\widehat{\boldsymbol{\theta}}_m, m)$ is the deviance measure of model m as defined in Section 4.2.5. From this,it is evident that BIC is a penalized deviance (or log-likelihood) measure with penalty equal to $\log n$ for each parameter estimated by the model. From this last equation, the connection between BIC and the Schwarz criterion (11.32) is clear and can be expressed as

$$S_{01} = -\frac{1}{2}\Big\{\text{BIC}(m_0) - \text{BIC}(m_1)\Big\},$$

whereas from (11.33) we obtain

$$-2\log B_{01} \approx \text{BIC}(m_0) - \text{BIC}(m_1) = \Delta\text{BIC}_{01} \qquad (11.34)$$

for large n. Since BIC is related to the Bayes factor, we can obtain approximate posterior model probabilities by the expression

$$f(m|\boldsymbol{y}) \approx \frac{\exp\left(-\frac{1}{2}\text{BIC}(m)\right)}{\sum\limits_{m'\in\mathcal{M}} \exp\left(-\frac{1}{2}\text{BIC}(m')\right)} \qquad (11.35)$$

assuming the uniform prior distribution for all models under consideration. Moreover, we may use Table 11.1 (Kass and Raftery, 1995) to interpret differences between models. Hence a rule of thumb as follows: If the BIC difference between two models is lower than 2, then we cannot discriminate between the two compared models. BIC differences from $2 - 6, 6 - 10$ and higher than 10 express positive, strong, and very strong evidence in favor of the model with the lower BIC value.

   In the expression of BIC, although as sample size n we frequently use the dimension of the $\boldsymbol{y}$ vector, Raftery (1996a) defines it as the dimension of $\boldsymbol{y}$ in normal models, as the sum of all Bernoulli trials in binomial models and as the sum of all counts in Poisson models. For further details on the Schwarz approximation, see Kass and Wasserman (1995), Raftery (1996a), and Pauler (1998).

11.11.2 The Akaike information criterion (AIC)

The AIC statistic was introduced by Akaike (1973) as an approximation of the expected Kullback–Leibler distance between a true model and an estimated model. In his subsequent research, Akaike (1974) extends his work to time series models and proposes using the minimum AIC to select a model.

Generally the AIC and BIC have different motivations. While BIC is a rough approximation of the log–Bayes factor, AIC is an approximately unbiased estimator of the expected Kullback–Leibler (KL) distance between true and estimated models and supports models that have predictive performance equivalent to the true performance. Since AIC is one of the approximately unbiased estimators the KL distance, a wide variety of other estimators have been proposed in the literature; see Kuha (2003) for AIC and related methods.

The Akaike information criterion is defined as

$$\mathrm{AIC}(m) = D(\widehat{\boldsymbol{\theta}}_m, m) + 2d_m,$$

and therefore it is also a penalized deviance measure with penalty equal to 2 for each estimated parameter. The penalty induced by AIC is lower than the corresponding one imposed by BIC for a reasonable sample size ($n > 7$), and hence AIC supports less parsimonious models than does BIC. AIC is closely related to the C_p criterion of Mallows (1973), R^2_{adj} used in normal regression models and results obtained using the leave-one-out cross-validation method (Stone, 1977); see also Kuha (2004) for a related discussion.

Model weights based on AIC can be obtained by the expression

$$f(m|\boldsymbol{y}) \approx \frac{\exp\left(-\frac{1}{2}\mathrm{AIC}(m)\right)}{\sum_{m' \in \mathcal{M}} \exp\left(-\frac{1}{2}\mathrm{AIC}(m')\right)}, \tag{11.36}$$

with the Bayesian interpretation that it will correspond to posterior model odds when the prior model weights

$$f(m) \propto \exp\left(\frac{d_m}{2}\{\log(n) - 1\}\right)$$

are adopted and BIC can be considered as a good approximation of the Bayes factor (Burnham and Anderson, 2004).

Burnham and Anderson (2004) suggest that all models with AIC difference < 2 from the best one must be reported as equally "good" models having substantial support (evidence) against the remaining ones. Moreover, models with AIC difference ranging between 4 and 7 are weakly supported by the data, while models with difference > 10 are not supported at all. This rule of thumb is comparable to the one proposed by Kass and Raftery (1995) for the Bayes factor and can be motivated by the Bayesian interpretation of AIC described in the previous paragraph.

Finally, the deviance information criterion (DIC), more recently introduced by Spiegelhalter et al. (2002), can be considered as the Bayesian analog of AIC. Its justification is similar to that for AIC, but the expectations used in its derivation are now with respect to parameters instead of sampling distributions used in the original derivation of the latter. The importance of DIC is that it can be directly calculated from an MCMC output and can be applied in a much wider variety of models, including hierarchical and latent variable models, where the number of estimated parameters is unclear.

11.11.3   Other criteria

A wide variety of penalized likelihood or deviance criteria is available in the statistical literature. The alphabet of such information criteria (IC) is described by Kuha (2004) and includes AIC, BIC, and DIC (already discussed); for more details, see Ntzoufras (1999a, sec. 2.3), Kuha (2004), Konishi and Kitagawa (2008), and references cited therein.

Generally, most information criteria minimize the quantity

$$\text{IC}(m) = D(\widehat{\boldsymbol{\theta}}_m, m) + d_m F, \tag{11.37}$$

where F is the penalty imposed on the deviance measure for each additional parameter used in the model. Different penalty functions result in different criteria such as for $F = 2$ and $F = \log(n)$, for which we obtain AIC and BIC, respectively.

If we wish to compare two models, m_0 and m_1, then we select the one that has a lower value of IC, and therefore we may use the corresponding difference ΔIC_{01} between the IC values for the two compared models

$$\Delta\text{IC}_{01} = D(\widehat{\boldsymbol{\theta}}_{m_0}, m_0) - D(\widehat{\boldsymbol{\theta}}_{m_1}, m_1) - (d_{m_1} - d_{m_0})F. \tag{11.38}$$

Without loss of generality, we assume that $d(m_0) < d(m_1)$. Note that if $\Delta\text{IC}_{01} < 0$ we select model m_0 and if $\Delta\text{IC}_{01} > 0$, we select model m_1. We can generalize the abovementioned criterion difference by substituting the expression $[d(m_1) - d(m_0)]F$ by a more complicated penalty function ψ.

Shao (1993) divides the information criteria into two major categories: (1) criteria that are asymptotically valid under the assumption that a true model exists and (2) criteria that are asymptotically valid under the assumption that a true model does not exist. Generally, information criteria with the penalty F fixed as $n \to \infty$ (such as AIC) and criteria with $F \to \infty$ as $n \to \infty$ (such as BIC) are two differently behaving categories of criteria, usually referred as AIC-like and BIC-like criteria. Criteria that attempt to compromise between the two approaches also exist. The main argument in favor of BIC-like criteria is that the existence of a true model is doubtful, and even if it does exists, we may prefer to select a simpler model that approximates sufficiently the true one.

All information and model selection criteria consider both the goodness of fit of the model and the parsimony principle. The first is measured by the log-likelihood ratio and the latter, by the number of model parameters. In other words, they aim to facilitate selection of the model, which describes the data as accurately as possible but with the simplest possible model structure. Each information criterion differs in the weight that it gives to each of these two principles.

Comparison of information criteria, posterior odds, and likelihood ratios and their connections are provided by Atkinson (1981) and Chow (1981). Note that Bayes factor variants and Bayesian predictive or utility-based criteria are (in most cases) equivalent to information criteria. For example, the predictive criterion L_m of Ibrahim and Laud (1994) is equivalent to an information criterion (11.38) with penalty function

$$F = \frac{n}{d_{m_1} - d_{m_0}} \log\left(\frac{n - d_{m_0} - 2}{n - d_{m_1} - 2}\right),$$

while the posterior Bayes factor is approximately equal to a criterion (11.38) with penalty equal to $F = \log(2)$ (Aitkin, 1991; O'Hagan, 1995). Other interesting criteria have been introduced by George and Foster (2000) using empirical Bayes methods, Bernardo (1999) using Bayesian decision theory, and Gelfand and Ghosh (1998) using a predictive loss approach.

11.11.4   Calculation of penalized deviance measures from the MCMC output

In terms of the Bayesian approach, we can calculate the posterior distribution of any penalized deviance measure of the form

$$\text{IC}(m) = D(\boldsymbol{\theta}_m, m) + d_m F$$

from the MCMC output and then obtain posterior summaries as usual.

All information criteria of type (11.37), including AIC and BIC, can be approximately estimated from an MCMC output using the minimum deviance value over the generated parameter values $\boldsymbol{\theta}^{(t)}$ for $t = 1, \ldots, T$. Hence

$$\text{IC}(m) \approx \min_{t=1,\ldots,T} D(\boldsymbol{\theta}_m^{(t)}, m) + d_m F \ .$$

Note that this value may be close to the correct minimum value but it will never be exactly equal to the correct minimum value since MCMC is a sampling, and not an optimization, algorithm. Nevertheless, the minimum deviance value obtained from an MCMC output with a large number of generated iterations usually provides sufficiently accurate results. Even more accurate results can be obtained if we use the posterior mean or median obtained from a posterior distribution with a flat prior distribution instead of the maximum-likelihood estimates used in the formal definition of IC (Raftery, 1996b). Hence we may estimate $\text{IC}(m)$ with

$$\widetilde{\text{IC}}(m) = D(\overline{\boldsymbol{\theta}}_m, m) + d_m F,$$

where $\overline{\boldsymbol{\theta}}_m$ is the posterior mean under model m. The posterior mean generally offers an accurate approximation, provided that the prior information is vague and the resulting posterior distribution is relatively symmetric to the mean.

Another estimate of of the minimum deviance (and hence of AIC and BIC) can be obtained via calculation of the posterior Bayes factor. As we have described in Section 11.10, the PBF is simply given as a ratio of the posterior predictive densities of two competing models which can be estimated by the posterior mean of the likelihood $\overline{f(\boldsymbol{y}|\boldsymbol{\theta}_m, m)}$ from each model's MCMC output. Hence we may obtain an estimate of the minimum deviance by

$$\widetilde{\min} D(\boldsymbol{\theta}_m, m) = -2 \log \overline{f(\boldsymbol{y}|\boldsymbol{\theta}_m^*, m)} - d_m \log(2)$$

using the result of O'Hagan (1995) which expresses the PBF as an information criterion with penalty equal to $\log(2)$. Then an information criterion of type (11.37) can be estimated by

$$\widetilde{\text{IC}}(m) = \widetilde{\min} D(\boldsymbol{\theta}_m, m) + d_m F \ .$$

Alternatively, we can obtain the whole posterior distribution of AIC or BIC as proposed by Brooks (2002) and compare, for example, their 95% posterior credible intervals to decide whether two models can be considered of equal value. Model weights can also be calculated for any posterior summary of AIC, BIC, or DIC, in order to obtain comparable measures inside the zero–one interval. Nevertheless, these weight do not have any theoretical justification.

11.11.5   Implementation in WinBUGS

Implementation in WinBUGS is straightforward. The sample from the posterior distribution of AIC and BIC can be obtained using the simple syntax

```
Deviance <- -2*sum( log.like[1:n] )
AIC <- Deviance + dm*2
BIC <- Deviance + dm*log(n)
```

Although the deviance node is automatic, specified internally in WinBUGS, it cannot be used inside the model code to define other nodes or variables. For this reason we redefine a Deviance node as minus twice the sum of all log-likelihood components. The minimum value of nodes AIC and BIC can be obtained after calculating the minimum deviance from the MCMC output in a statistical program outside WinBUGS. Similarly, after calculating the posterior means (or medians) using WinBUGS, the mean/median-based AIC and BIC estimates can be calculated using software other than WinBUGS.

11.11.6 A simple example in WinBUGS

Example 11.5. Soft drink delivery times data (Example 5.1 revisited).
Here we consider again Example 5.1 and compare the four models under consideration using the AIC and BIC values.

The full WinBUGS code for calculation of AIC and BIC is available at the book's Website. MCMC-based estimates of the minimum deviance for each model are presented in Table 11.12, while the corresponding posterior summaries are given in Table 11.13. Similarly, estimates of the minimum values of AIC and BIC for each model are given in Table 11.14, while their posterior distribution is depicted in Figures 11.2 and 11.3. All results have been generated with the same prior as in Section 11.4.2 with $c^2 = 1000$, using 1000 iterations as burnin and an additional 5000 iterations as a final sample. The minimum deviance value was estimated using 20,000 and 100,000 iterations in order to monitor the improvement in estimated minimum deviance, based on the minimum value of the sample.

From Table 11.12 we can see that the minimum value from an MCMC sample approaches sufficiently the actual minimum deviance; see the first column of Table 11.12. Although longer chains increase the accuracy of the estimate, even in samples of 5000 iterations, the correct values were calculated with accuracy of one decimal digit. Estimates based on the posterior mean and median are also very close to the actual minimum deviance value. Finally, although the PBF-based approach provides estimates close to the actual AIC and BIC values, they are less accurate than the estimates using the previous approaches.

Table 11.12 Minimum deviance values estimated from MCMC output for models under comparison for soft drink data (Example 11.5)

Model (m)	d_m	$D(\widehat{\theta}_m, m)$	Min(5K)	Min(20K)	Min(100K)	Mean	Median	PBF
m_1 (constant)	2	207.049	207.049	207.049	207.049	207.049	207.058	207.263
m_2 (cases)	3	140.395	140.401	140.401	140.395	140.399	140.408	140.428
m_3 (distance)	3	167.421	167.431	167.424	167.423	167.422	167.430	167.407
m_4 (cases + distance)	4	126.829	126.872	126.848	126.847	126.843	126.852	126.792

Key: Min(5K), Min(20K), Min(100K) correspond to minimum deviance values from MCMC output of length equal to 5000, 20,000, and 100,000 iterations, respectively (with an additional 1000 iterations discarded as burnin). Estimates based on the mean, median, and PBF were obtained using an MCMC chain of 5000 iterations (and 1000 iterations burnin).

Table 11.13   Deviance posterior summaries for models under comparison for soft drink data (Example 11.5); MCMC results based on 5000 iterations plus 1000 iterations as burnin

Model (m)	Minimum	2.5% percentile	Mean	97.5% percentile
m_1 (constant)	207.05	207.11	209.08	214.81
m_2 (cases)	140.40	140.61	143.46	150.13
m_3 (distance)	167.43	167.63	170.40	176.85
m_4 (cases + distance)	126.87	127.34	130.79	137.95

Table 11.14 presents estimates of minimum AIC and BIC. All estimates indicate that the model with both variables is the best one. From Figures 11.2 and 11.3 we can clearly see that the posterior distributions of AIC and BIC for this model are clearly lower than those of remaining models, indicating that this model is better than the rest. We obtain a similar picture of posterior weights using Eqs. (11.35) and (11.36). For AIC, model 4 is supported with weight equal to 0.997 using any of the estimates of Table 11.14, while model 2 has weight equal to 0.003. Similar model weights are obtained using the 25%, 50%, and 75% posterior percentiles of AIC (0.9964, 0.9952, and 0.9939, respectively). Finally, model 4 is also supported with high model weight (equal to 0.994) from BIC, which is slightly lower than the corresponding one of AIC because of the increased penalty used by BIC, supporting in this way more parsimonious models in comparison with AIC.

Table 11.14   AIC and BIC estimates for soft drink data (Example 11.5); MCMC results based on 5000 iterations plus 1000 iterations as burnin

Model (m)	MLE	Min	Post. mean	Post. median	PBF
Akaike information criterion					
m_1 (constant)	211.049	211.049	211.049	211.058	211.263
m_2 (cases)	146.395	146.401	146.399	146.408	146.428
m_3 (distance)	173.421	173.431	173.422	173.430	173.407
m_4 (cases + distance)	134.829	134.872	134.843	134.852	134.792
Bayes information criterion					
m_1 (constant)	213.487	213.487	213.487	213.496	213.700
m_2 (cases)	150.052	150.058	150.056	150.065	150.085
m_3 (distance)	177.078	177.088	177.079	177.087	177.064
m_4 (cases + distance)	139.704	139.747	139.718	139.727	139.667

Key: Min = estimates based on the minimum deviance of an MCMC sample; MLE, Post. mean, Post.median = estimates based on the maximum-likelihood estimates, posterior means, and posterior medians of the model parameters; PBF = estimates obtained indirectly via calculation of posterior Bayes factor.

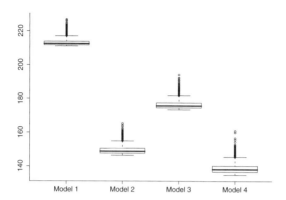

Figure 11.2 Posterior error bars of AIC measures for models under comparison for soft drink data (Example 5.1).

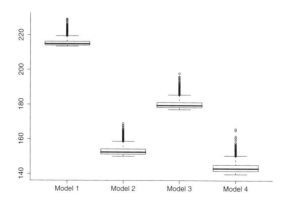

Figure 11.3 Posterior error bars of BIC measures for models under comparison for soft drink data (Example 11.5).

11.12   DISCUSSION AND FURTHER READING

In this chapter we have briefly described Bayesian model selection methods with more focus on variable selection. We have described how to estimate the marginal likelihood of a model and the corresponding posterior model odds of two competing models. Variable selection methods are also described and illustrated using simple examples. Transdimensional methods for Bayesian model selection are briefly described. Finally, we conclude this chapter with illustration of model comparison methods based on posterior predictive distributions and on AIC and BIC criteria. This chapter is by no means exhaustive on the topic. It provides only a short introduction to the topic with emphasis on the implementation using WinBUGS. Advanced WinBUGS users can concentrate more on variable selection following details described by Dellaportas et al. (2000) and Ntzoufras (2002). Implementation of more complicated model comparisons using Carlin and Chib's (1995) sampler can be found in Spiegelhalter et al. (2003c) and Katsis and Ntzoufras (2005). Implementation of reversible jump MCMC in WinBUGS can be achieved via JUMP interface; for more details, see Lunn et al. (2005, 2006) and the manual for this interface.

Generally, implementation of variable and model selection methods is more complicated than simple estimation of model parameters and is recommended only for experienced users and scientists who have thoroughly comprehended at least all basic notions of Bayesian modeling presented in the previous chapters of this book.

Problems

11.1   Estimate the marginal log-likelihood for the regression models including and excluding the constant for the model implemented in Problem 5.1.

11.2   Estimate the posterior model odds of the model used in Problem 5.2 versus the constant model using the methods for estimating the marginal likelihoods.

11.3   Estimate the marginal likelihoods and the corresponding posterior weights of the models used in Problem 6.6.

11.4   For the data of the WinBUGS example seeds (Spiegelhalter et al., 2003a) used in Problem 7.6, estimate the marginal likelihoods of models with and without the dose and the log-dose. Also calculate the corresponding model weights. Use the prior suggested in the current chapter.

11.5   For the data of the WinBUGS example beetles (Spiegelhalter et al., 2003b) used in Problem 7.9
   a) Estimate the marginal likelihoods of models with and without the drug's concentration.
   b) Estimate model for the logit, probit, and clog–log links.
   c) Calculate the corresponding model weights.

11.6   Use zero–one dummy variables to fit the model of Problem 5.5 and the variable selection methods to identify which groups can be considered equal. Use Zellner's g-prior with $c^2 = n$ and an independence prior with prior variances equal to 100.

11.7   Apply the variable selection methods for the models implemented in Problems 6.5 and 6.9.

11.8   Apply the Gibbs-based methods for variable selection for the models fitted in Problems 7.1 and 7.2.

11.9   Apply the Gibbs-based methods for variable selection for the models used in Problems 7.7 and 7.8.

11.10   For Problems 8.1 and 8.2, calculate the marginal likelihoods for the models with different distributional assumptions. Which distribution is more appropriate?

11.11   Use the zero–ones trick and the approach of Katsis and Ntzoufras (2005) to compare the different distributions in Problems 8.1 and 8.2.

11.12   Implement variable selection methods on the survival data of Problems 8.3 and 8.4.

11.13   Calculate the posterior Bayes factors and the negative log-likelihood for the models used in Problems 11.1–11.4.

11.14   Calculate AIC and BIC measures for the models fitted in Problems 11.1–11.4.

APPENDIX A

MODEL SPECIFICATION VIA DIRECTED ACYCLIC GRAPHS: THE DOODLE MENU

A.1 INTRODUCTION: STARTING WITH DOODLE

DOODLE is a graphical interface of WinBUGS in which we can represent our model via a directed acyclic graph (DAG) and then automatically generate the corresponding Win-BUGS code from this graph. The benefit of DOODLE is that a user, who is unfamiliar with programming, can build a model without having to write the corresponding model code.

In order to construct a DAG from a model, we need to specify

- The nodes representing the variables of the model

- The edges representing dependencies between the variables induced by the model

Nodes and variables are depicted by rectangular or oval boxes depending on their type, while edges are depicted using uni-directional arrows. The head of the arrow points on the dependent variable, namely, the variable defined by the node or variable from which the arrow starts.

To complete the model specification in WinBUGS, we further need to denote the repeating structures, that is, actions or procedures that need to be repeated in a similar manner. An example of repeating structure in a statistical model is the likelihood, where every observation is assumed to follow the same distribution with similar parameter pattern, for example, $Y_i \sim N(\mu_i, \sigma^2)$ for $i = 1, \ldots, n$. These repeating structures are called *panels* in DOODLE.

Bayesian Modeling Using WinBUGS, by Ioannis Ntzoufras
Copyright ©2009 John Wiley & Sons, Inc.

In order to start drawing a DAG in WinBUGS, we select `Doodle > New...` from the menu bar. Then, a dialog box asks the user to specify the dimensions of the window in which the DAG is drawn. After specifying the desired dimensions, click OK, and a window for drawing the graphical representation of the model will appear.

Note that when the DAG is completed, we can generate the WinBUGS model code by selecting `Write Code` from the DOODLE menu.

A.2 NODES

A stochastic node is drawn when clicking (the left mouse button) in any position of the DOODLE window. This node is used to represent a variable of the model. When the node is highlighted, the properties of the specific node appear on the upper part of the window.

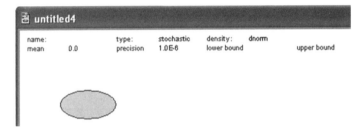

The most important properties of a node are the following:

- **Name**: Insert the name of the node or variable here.

- **Type**: Set the type of the node. You can select from three types: stochastic, logical, and constant. *Stochastic* refers to random variables, *logical* refers to nodes that have been specified using deterministic expressions while *constant* refers to fixed quantities. Note that logical nodes can be either random or fixed, depending on the variables defining them.

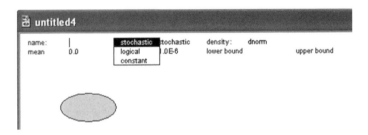

To select the type of node, click on the word `type` at its properties. Then a small menu with the available node types will appear.

The following node properties are type-specific:

1. For the *stochastic* node we need to specify

    a. **Density**: Set the density name from a prespecified list. Click on the word `density` for the menu to appear.

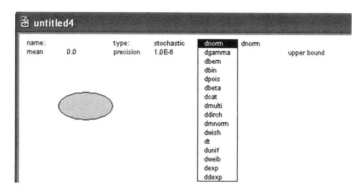

b. **Parameters**: Type the values of the density parameters. For example, for the normal density you have to define the mean and the precision.

c. **Upper and lower bounds**: Set up the upper and lower bounds of the distribution in case of censoring.

2. For the *logical* node we need to specify

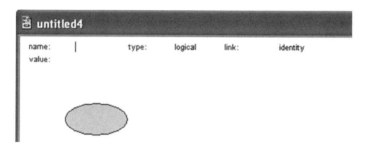

a. **Link**: Set the link function $g(x)$ for the logical node x. There are five available choices:

    (1) Identity: $g(x) = x$
    (2) Logit: $g(x) = \log(x/(1-x))$
    (3) Probit: $g(x) = \Phi^{-1}(x)$; where $\Phi^{-1}(x)$ is the inverse of the cumulative probability function of the standardized normal distribution
    (4) Cloglog: $g(x) = \log(-\log(1-x))$
    (5) log: $g(x) = \log(x)$

Click on the word link for the menu to appear.

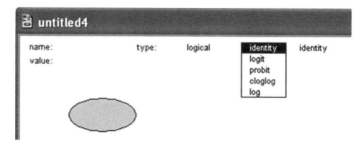

   b. **Value**: Value or expression from which node x is defined. For example, if η is the value or expression of this option, we define x by the expression $g(x) = \eta$.

3. Finally, for a *constant* node we do not need to specify any additional value. The constant node is depicted using a rectangular box and its value is defined in the data section.

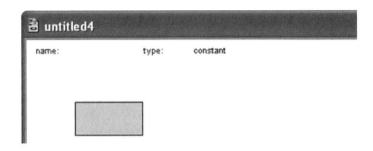

The following operations pertain to nodes:

- **Highlighting or changing the properties of a node**: Just left-click on the desired node. Its properties will appear on the upper part of the working window.

- **Moving a node**: Click on a node and then move it by holding the left mouse button.

- **Removing a node**: Highlight the node and then press simultaneously the Control (Ctrl) and Delete (Del) keys.

A.3   EDGES

Edges of a DAG are depicted using arrows starting at one node and ending at another one. They define conditional associations of the type $Y|X$ if Y is stochastic, and of the type $Y = g(X)$ if Y is a logical or deterministic node.

   Two nodes are connected with an edge by the following procedure

- **Connect two nodes using an edge or arrow**: Highlight the terminal node (where the arrow will point) and then, while holding the Control key, click on the starting node (where the arrow starts).

An edge is deleted if we repeat the same procedure as above on two nodes that are already connected with an edge or arrow.

A.4   PANELS

Panels are rectangular frames representing a set of identical repeating operations (equivalent to the for syntax). To create a panel, hold the Ctrl key and click on the point where you want it to appear.

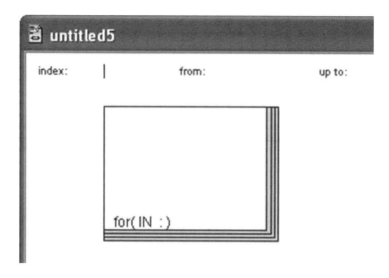

The panel properties appear on the upper part of the window when the panel is selected or highlighted. In the properties of the panel, the user needs to specify the index name and its range of values. The *index* refers to a counter that controls how many times an operation within the panel is repeated. Any node included in panels must depend on the index of the surrounding panel.

The full set of operations related to a panel are as follows:

- **Create a panel**: Ctrl key and left-click mouse button.

- **Highlight/select a panel**: Left-click on the thick edges.

- **Delete a panel**: After highlighting the panel, press Ctrl and Del keys.

- **Move a panel**: After highlighting the panel, press and hold the left mouse button on the thick edge of the panel; move the panel using the mouse (while holding the left mouse button). Minor adjustments can be made using cursor keys.

- **Resize a panel**: After highlighting the panel, press and hold the left mouse button on the corner of the thick edge of the panel; change the size and the shape of the panel using the mouse (while holding the left mouse button).

A.5   A SIMPLE EXAMPLE

Let us consider the following simple model

$$
\begin{aligned}
Y_i &\sim N(\mu_i, \tau^{-1}) \\
\mu_i &= a + bX_i \text{ for } i = 1, \ldots, n \\
a &= N(\mu_a, 100) \\
b &\sim N(0, 100) \\
\tau &\sim \text{gamma}(0.1, 0.1) \\
\sigma &= 1/\sqrt{\tau} \\
\mu_a &= 0.0 \\
n &= 10
\end{aligned}
$$

In this model we have the following nodes:

1. Y_i for $i = 1, \ldots, n$: stochastic nodes
2. μ_i for $i = 1, \ldots, n$: logical nodes
3. τ: stochastic node
4. X_i for $i = 1, \ldots, n$: constant nodes
5. n: constant node
6. a, b: stochastic nodes
7. X_i for $i = 1, \ldots, n$: constant nodes
8. σ: logical node
9. μ_a: constant node

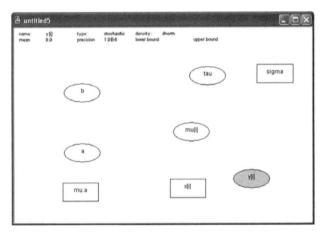

The relationships (edges in the graph) are defined as follows

1. $Y_i | \mu_i, \tau$ results in $(\mu_i, \tau) \rightarrow Y_i$.
2. μ_i is a function of a, b and X_i hence $(X_i, a, b) \rightarrow \mu_i$.
3. a depends on μ_a; hence $a \rightarrow \mu_a$.
4. σ is a transformation of τ hence $\tau \rightarrow \sigma$.

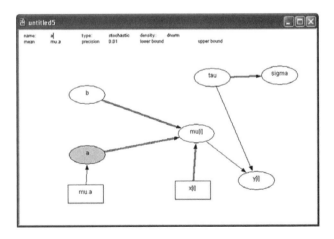

We further need one panel with index i moving from 1 to n that is used to specify the likelihood of the model. Nodes Y_i, X_i, and μ_i must be embedded in this panel.

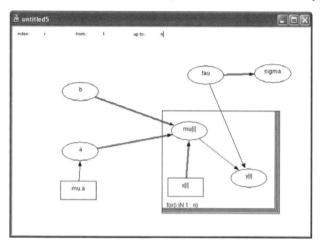

Table 1.1 provides all details used for each parameter of the above model.

Table 1.1 Detailed description of node options used in DAG of Appendix A

Node	Name	Type	Distribution[a]	Parameters[a]		Value[b]
Y_i	y[i]	Stochastic	dnorm	mean: mu[i]	precision: tau	—
μ_i	mi[i]	Logical	—	—	—	a+b*x[i]
τ	tau	Stochastic	dgamma	shape: 0.1	scale: 0.1	—
X_i	x[i]	Constant	—	—	—	—
a	a	Stochastic	dnorm	mean: mu.a	precision: 0.01	—
b	b	Stochastic	dnorm	mean: 0.0	precision: 0.01	—
σ	sigma	Logical	—	—	—	1/sqrt(tau)
μ_a	mu.a	Constant	—	—	—	—

[a] Options for stochastic nodes.
[b] Options for logical nodes.

Finally, we generate the WinBUGS code of the model by the menu `Doodle>Write Code`, which results in the following sequence:

```
model;
{
   for( i in 1 : n ) {
       y[i]  ~ dnorm(mu[i],tau)
   }
   for( i in 1 : n ) {
       mu[i] <- a + b * x[i]
   }
   tau ~ dgamma( 0.1, 0.1)
   sigma <- 1 / sqrt(tau)
   a  ~ dnorm(mu.a,0.01)
   b  ~ dnorm( 0.0,0.01)
}
```

Further examples of DAGs can be found in WinBUGS examples (Spiegelhalter et al. 2003a–c), where each example is accompanied by the corresponding graph.

APPENDIX B

THE BATCH MODE: RUNNING A MODEL IN THE BACKGROUND USING SCRIPTS

B.1 INTRODUCTION

After having set up the model, we frequently need to perform the same analysis multiple times. The use of dialog menus described in Chapter 3 requires interaction by the user each time we run a model. Alternatively, the user may be facilitated by the batch mode in which a set of commands (script) performs the same operations as those performed via menus. The batch mode is the corresponding background running mode in the classic BUGS. A list of the available commands is provided in the WinBUGS manual (Spiegelhalter et al., 2003*d*). In the following we illustrate the most common commands using a simple example.

Four different types of files must be prepared in order to run a procedure in the batch mode:

1. A file with the model code and definition

2. One (or more) files including the data

3. One (or more) files including the initial values of the parameters

4. A script file with the commands that correspond to the WinBUGS menus and dialog boxes. The user can run more than one script file with the assistance of the `script` command.

Bayesian Modeling Using WinBUGS, by Ioannis Ntzoufras
Copyright ©2009 John Wiley & Sons, Inc.

All files can be saved in either text or WinBUGS format, with `.txt` or `.odc` extensions, respectively.

Finally, in the last version of WinBUGS, the shortcut `BackBUGS` runs directly all commands contained in the file `script.odc`, which is located in the directory of WinBUGS.

B.2 BASIC COMMANDS: COMPILING AND RUNNING THE MODEL

Lets us consider the simple example of Section 4.1. We first need to set up the files needed for the script mode. We save the model code under the name `model.odc`, the data under the name `data.odc`, and the initial values with the name `initial.odc` in a directory with path `PATHNAME`. Here we have considered the path `C:/myexample/`. In order to run a WinBUGS full model in script session, you need to

1. Open a log file (or window) where the WinBUGS output will be saved using the command `display('log')`.

2. Check the model's syntax, load the data, compile the model, and finally load the initial values using the commands

```
# check the syntax of the model for
# model.odc in directory PATHNAME
check('PATHNAME/model.odc')
# load data from file data.odc
data('PATHNAME/data.odc')
# compile one chain
compile(1)
# provide initial values for the first chain
inits(1, 'PATHNAME/initial.odc')
# generate random initial values for unitialized chains
gen.inits()
```

3. Update the chain with burnin iterations (1000 iterations here) using the command `update(1000)`.

4. Set the monitored nodes using the command `set(nodename)`.

5. Use the command `set.dic()` to start calculating DIC value.

6. Update the chain with the iterations that will be kept for analysis.

7. Produce posterior analysis. For example, the command `stats(nodename)` calculates posterior summaries for `nodename` from the values sampled in step 6, while the command `stats(*)` provides posterior summaries for all monitored nodes.

8. The command `dic.stats()` calculates the DIC value.

9. The command `CODA` saves the sampled values for a node in a file or in a window. The syntax is `CODA(nodename,'filename')`. If instead of the name of a specific node we use the asterisk (*), then the values of all monitored nodes are saved. The command saves at least two files: `filename1.txt` and `filenameIndex.txt`. The first file includes the sampled values for the first chain. If two or more chains are generated, then the corresponding values are stored with different numbers. The format of a CODA file contains only two columns. The first refers to the iteration

number and the second, to the sampled value. If more than one node is saved, then the values of the nodes are stacked sequentially. The ordering of the saved nodes is denoted in the second file (`filenameIndex.txt`), which has three columns: the node name and the position (starting line and the ending line) of the sampled values of each node. Finally, if the `filename` is left blank, then all values appear in separate windows instead of being saved them to a file.

The full syntax for the example used here is given as follows:

```
# open log file
display('log')
# check the syntax of the model file (model.odc)
check('C:/myexample/model.odc')
# load data from file data.odc
data('C:/myexample/data.odc')
# compile one chain
compile(1)
# provide initial values for the first chain
inits(1, 'C:/myexample/initial.odc')
# generate random initial values for unitialized chains
# here it is not needed
gen.inits()
#
# generate 1000 iterations as burnin
update(1000)
#
# monitor the following parameters/nodes
set(pi)
set(R)
#
# monitor dic values
dic.set()
#
# Produce Dynamic Trace plot for monitored nodes
trace(*)
#
# refresh plot and chain every 10 iterations
refresh(10)
# Update the chain by 5000 iterations
update(5000)
# Calculate the (posterior) statistics for monitored parameters
stats(*)
# Calculate the DIC value
dic.stats()
# see Trace plot for sampled values of monitored nodes
history(*)
# Plot posterior density estimates
density(*)
# Plot autocorrelation plots for sampled values
autoC(*)
# Plot of quantiles
quantiles(*)
# Open windows with sampled values in CODA format
coda(*,'C:/myexample/CODAoutput')
# Save the log file with the name "ExampleLog.odc"
save('C:/myexample/exampleLog')
```

APPENDIX C
CHECKING CONVERGENCE USING
CODA/BOA

C.1 INTRODUCTION

The convergence of an MCMC algorithm is an important issue for the correct estimation of the posterior distribution of interest. A problem with MCMC methods is that convergence cannot always be diagnosed as clearly as in optimization methods. The user must specify both the length of the burnin period and the size of the MCMC output that will be used for the posterior analysis (i.e., the number of iterations/observations needed to keep for analysis). A secondary, but also important, problem is specification of the thinning interval, that is, the number of iterations we need to discard until two successive observations become independent. Graphical methods for monitoring convergence have been described in Section 2.2.2.4. Furthermore, low Monte Carlo errors are also essential for precise estimation of the posterior distribution. More formal diagnostic tests have been developed in the literature and can be implemented using CODA [convergence diagnosis and output analysis software for Gibbs sampling analysis (Best et al., 1996; Plummer et al., 2006)] or BOA [Bayesian output analysis Smith (2005; 2007)] packages; both are available in R.

In this appendix we briefly describe the diagnostics used by CODA and BOA and illustrate their use with a simple example.

C.2   A SHORT HISTORICAL REVIEW

According to Plummer et al. (2006), the original version of CODA was written for Splus as a companion to a review by Cowles and Carlin (1996). CODA was later adopted by the BUGS team to serve as a companion to the classic BUGS (Spiegelhalter et al., 1996a), which had limited output facilities. Under this team, versions 0.30 (Best et al., 1996) and 0.40 (Best et al., 1997) were released. With the development of R and its growing popularity, CODA was rewritten in 2000 by Brian Smith (Smith, 2005) to form the package known as Bayesian output analysis (BOA). In parallel, the CODA package was also developed by Martyn Plummer in collaboration with the original CODA team (Plummer et al., 2006).

Note that both packages are menu driven, with CODA also providing the possibility to separately call each diagnostic, making it extremely appealing to experienced users who wish to embody such commands in their own MCMC code.

In the following, we briefly present the four diagnostics used by CODA/BOA in R and illustrate them (in R) using a simple example.

C.3   DIAGNOSTICS IMPLEMENTED BY CODA/BOA

CODA provides four diagnostic tests suggested respectively by Geweke (1992), Gelman and Rubin (1992), Raftery and Lewis (1992), and Heidelberger and Welch (1992). The same tests are also implemented in BOA, but appear in a different order at the corresponding diagnostics menu. Details of these diagnostics can be found in Best et al. (1996) and Cowles and Carlin (1996).

C.3.1   The Geweke diagnostic

Geweke (1992) suggested a diagnostic test for checking the convergence of the mean of each parameter separately from the sampled values of a single chain. To construct this test, he proposed viewing the set of simulated values, obtained by the MCMC output, as a time series. This diagnostic applies a simple Z test to check whether the means estimated from two different subsamples of the total MCMC output are equal. These subsamples refer to observations coming from the beginning and the end of the generated chain. Hence, if the means at the beginning and the end of the total MCMC output are rejected, then convergence of the chain cannot be assumed. CODA/BOA compare by default the initial 10% and the last 50% of the total iterations. For the asymptotic variance of the sample mean, an estimate from spectral density theory is adopted. Hence, for a set of generated values

$$\theta_1^{(1)}, \ldots, \theta_1^{(T)}$$

of a parameter of interest θ, the sample mean $\bar{\theta}$ is calculated, followed by calculation of the spectral density $S_\theta(\omega)$ for these time series. Then, the standard error of the mean is given by $\sqrt{S_\theta(0)/T}$ and the corresponding Z diagnostic by

$$Z = \frac{\bar{\theta}^A - \bar{\theta}^B}{\sqrt{S_\theta^A(0)/T_A + S_\theta^B(0)/T_B}},$$

which will asymptotically follow the standardized normal distribution, where $\bar{\theta}^A$, $\bar{\theta}^B$ are the sample means for the two subsamples described above: T_A, T_B are the corresponding sizes; and $S_\theta^A(0)/T_A$, $S_\theta^B(0)/T_B$ the corresponding variances of the sample means. This expression can also be used for any function $g(\theta)$.

Since Z asymptotically follows the standardized normal distribution, values that lie in its tails provide an indication of nonconvergence (Best et al., 1996). To be more specific, parameters with $|Z| > 2$ indicate differences in the means between the first and the last set of iterations and hence nonconvergence. Nevertheless, because of the type I error of classical significance tests, in multiparameter models, we allow a 5% of the calculated Zs to lie outside this range.

C.3.2   The Gelman–Rubin diagnostic

The Gelman–Rubin diagnostic (Gelman and Rubin, 1992) involves checking the convergence of the chain using two or more samples generated in parallel. This test is an ANOVA-type diagnostic, calculating a shrinking factor R as described in Section 4.3.3. Values of R close to one indicate convergence.

C.3.3   The Raftery–Lewis diagnostic

The third diagnostic test was proposed by Raftery and Lewis (1992). It can be applied to output coming from a single chain, and it focuses on achieving a prespecified degree of accuracy of specific quantiles rather than the convergence of the mean. The default measure in CODA is the 2.5% percentile estimated with accuracy 0.005 and probability 0.95. CODA reports $N_{\min}$, M, B, and I; more specifically:

- $N_{\min}$ is the minimum number of iterations required to estimate the quantile of interest with the prespecified accuracy under the assumption of independence (i.e., with zero autocorrelation).

- N is the total number of iterations that the chain must run.

- M is the number of burnin iterations.

- I is the *dependence factor* given by $I = N/N_{\min}$, which indicates the relative increase of the total sample due to autocorrelations. If I is equal to one, then the generated values are independent. On the other hand, values greater than 5 often indicate a problematic behavior; for details, see Best et al. (1996). In general I can be considered as a rough estimate of the required thinning interval.

Note that if the current number of generated observations T is lower than $N_{\min}$, then an error is printed and the remaining values are not estimated. This diagnostic is very detailed, giving valuable information regarding the length of the sample.

C.3.4   The Heidelberger–Welch diagnostic

The last convergence diagnostic was proposed by Heidelberger and Welch (1992). It is used for the analysis of single chains from univariate observations. The test consists of two parts and is based on ideas from Brownian bridge theory. This diagnostic tests whether stationarity of the Markov chain is attained using the values from an MCMC output (null hypothesis). If this hypothesis is rejected, then the first 10% of the total iterations is discarded and the test is repeated on the remaining sample. The procedure is repeated by dropping additional 10% of the sample until the test of stationarity is not rejected or more than 50% of the sample is discarded. In the latter case, the null hypothesis is rejected and more iterations are required. If the test is not rejected (i.e., if it is passed following the terminology of CODA/BOA), then the number of iterations that have been used to pass the test, the discarded iterations, and the Cramer–von Mises statistic are reported. In the second part of the diagnostic, the halfwidth test is implemented. The portion of the chain that passed the stationarity test is treated as time series, from which we estimate the spectral density at zero $S(0)$. Then, the asymptotic standard error of the mean is given by $\sqrt{S(0)/N_p}$ as in the Geweke (1992) diagnostic; where N_p is the length of the retained chain. If the halfwidth of the 95% confidence interval of the mean, evaluated with the asymptotic standard error, is less than ε times the sample mean, the halfwidth test is passed, where ε is a small fraction with CODA default value equal to 0.1. In the opposite case, the halfwidth test reports failure and a longer chain must be considered to achieve the required precision of the parameter of interest.

C.3.5 Final remarks

All convergence diagnostics work like "alarms" that detect certain problems concerning the convergence of the chain. Since the focus of each diagnostic is different, to ensure convergence all tests must be passed (not rejected). Nevertheless, the required precision of the quantiles of the diagnostic of Raftery and Lewis (1992) must be adjusted according to the scaling of each variable.

C.4 A FIRST LOOK AT CODA/BOA

Here we describe the main menus of CODA/BOA and present the required actions to import the data and obtain each diagnostic test. Additional descriptive measures and plots are available in both packages; for details, see Best et al. (1996) and Smith (2005), respectively.

C.4.1 CODA

After downloading, installing, and loading the CODA package in R, we call the main menu by the command

```
codamenu()
```

The following text menu appears where two selections are available: (1) importing data from a WinBUGS output file (obtained using CODA option in the Inference>Samples tool) or (2) data with the conventional CODA format (usually saved from previous CODA session).

```
CODA startup menu

1: Read BUGS output files
2: Use an mcmc object
3: Quit
```

For the first selection, an index file (with the positions of each stored variable) and at least one output file must be provided by the user. In the second selection, the name of an R object must be provided. This object must have the same format as data saved after a CODA session. The data of a CODA session are saved by default with the name coda.dat.

This object is a list with the data of each chain stored in a matrix (each row is an iteration and each column a variable/node).

Once the data are imported in CODA, the following menu appears on the screen

```
1: Output Analysis
2: Diagnostics
3: List/Change Options
4: Quit
```

Selection of the second option provides a list of available diagnostics

```
CODA Diagnostics Menu

1: Geweke
2: Gelman and Rubin
3: Raftery and Lewis
4: Heidelberger and Welch
5: Autocorrelations
6: Cross-Correlations
7: List/Change Options
8: Return to Main Menu
```

Options 1–4 apply each of the diagnostics described in the previous section, while options 5 and 6 provide some graphical details concerning autocorrelations and correlations across different parameters. For more details regarding the additional available analysis and options of CODA, see the CODA manual (Best et al., 1996) available through the WinBUGS Webpage.

Exiting CODA will save the data of the current session under the name coda.dat. Store this file in another object in case you need to use it again.

C.4.2 BOA

There are several differences between the menus of BOA and CODA. In the following we present the main menus of BOA. Note that detailed documentation of BOA (including examples) is provided in Smith (2005; 2007).

After downloading, installing, and loading the BOA package in R, we call the main menu with the command

```
boa.menu()
```

Then the following initial menu of BOA appears

```
BOA MAIN MENU
*************

1: File     >>
2: Data     >>
3: Analysis >>
4: Plot     >>
5: Options  >>
6: Window   >>
```

To import WinBUGS files, select 1 (File) and from the next menu, select 3 (Import Data):

```
Selection: 1

FILE MENU
=========
```

```
1: Back
2: ----------------------+
3: Import Data          >> |
4: Load Session            |
5: Save Session            |
6: Exit BOA                |
7: ----------------------+

Selection: 3

IMPORT DATA MENU
----------------

1: Back
2: --------------------------+
3: CODA Output Files         |
4: Flat ASCII File           |
5: Data Matrix Object        |
6: View Format Specifications |
7: Options...                |
8: --------------------------+

Selection:
```

WinBUGS output files are loaded with option 3 (CODA Output Files), after we specify the working directory in 7 (Options...). In BOA, unlike CODA, output and index files must have the same names with different extensions for each file: .out for the output files and .ind for the index file. Return to the BOA MAIN MENU by typing 1, enter twice, and then select 3 (analysis) to obtain the analysis menu:

```
Selection: 3

ANALYSIS MENU
=============

1: Back
2: --------------------------+
3: Descriptive Statistics >> |
4: Convergence Diagnostics >> |
5: Options...                |
6: --------------------------+
```

Select 4 to obtain the convergence diagnostics menu

```
CONVERGENCE DIAGNOSTICS MENU
----------------------------

1: Back
2: ----------------------+
3: Brooks, Gelman & Rubin |
4: Geweke                 |
5: Heidelberger & Welch   |
6: Raftery & Lewis        |
7: ----------------------+
```

The diagnostics displayed above are the same as in CODA but in a different order (they are arranged alphabetically).

In BOA, data are not stored by default in an R object as in CODA. We can store the whole BOA session by the Save Session option of the File Menu (option 1 in the main menu). By this option a list object is stored with extensive details concerning the current

BOA session. The data are available in `boaobject$chain$master`, which is a list like the one saved by CODA; where `boaobject` is the name of the R object where the BOA session is stored.

C.5   A SIMPLE EXAMPLE

The output of Example 1.4 as implemented in Section 4.1 is used in the following to briefly illustrate the use of CODA for obtaining the diagnostics of Geweke (1992), Raftery and Lewis (1992) and Heidelberger and Welch (1992). Illustration of the Gelman–Rubin diagnostic (Gelman and Rubin, 1992) is omitted since it can be implemented in WinBUGS , as illustrated in Section 4.3.3. In the following we assume that we have generated one chain. The output file is saved under the name `CODAoutput1.txt` and the index file, under the name `CODAoutputIndex.txt`.

C.5.1   Illustration in CODA

Firstly, we load the data using the following procedure

1. Load the CODA package and then call the main menu with the command `codamenu()`.

2. In the menu

   ```
   > codamenu()
   CODA startup menu

   1: Read BUGS output files
   2: Use an mcmc object
   3: Quit
   ```

   select 1 (and press Enter):

   ```
   Selection: 1
   ```

3. In the following request

   ```
   Enter CODA index file name
   (or a blank line to exit)
   ```

   insert the name of the index file

   ```
   1: CODAoutputIndex.txt
   ```

   and then insert the name of the output file:

   ```
   Enter CODA output file names , separated by return key
   (leave a blank line when you have finished)
   1: CODAoutput1.txt
   2:
   ```

   The package expects a second output file, unless Enter is pressed without any other information. If the data are successfully loaded, then a list of the loaded nodes and the corresponding number of iterations will appear on the screen

   ```
   Abstracting R[2] ... 5000 valid values
   Abstracting R[3] ... 5000 valid values
   Abstracting R[4] ... 5000 valid values
   Abstracting R[5] ... 5000 valid values
   ```

```
Abstracting R[6] ... 5000 valid values
Abstracting R[7] ... 5000 valid values
Abstracting R[8] ... 5000 valid values
Abstracting pi[1] ... 5000 valid values
Abstracting pi[2] ... 5000 valid values
Abstracting pi[3] ... 5000 valid values
Abstracting pi[4] ... 5000 valid values
Abstracting pi[5] ... 5000 valid values
Abstracting pi[6] ... 5000 valid values
Abstracting pi[7] ... 5000 valid values
Abstracting pi[8] ... 5000 valid values
Checking effective sample size ...OK
```

followed by CODA Main Menu:

```
CODA Main Menu

1: Output Analysis
2: Diagnostics
3: List/Change Options
4: Quit
```

The data are now loaded and are ready to use. From the main menu, select 2 and then the desired model diagnostic from the following menu:

```
Selection: 2

WCODA Diagnostics Menu

1: Geweke
2: Gelman and Rubin
3: Raftery and Lewis
4: Heidelberger and Welch
5: Autocorrelations
6: Cross-Correlations
7: List/Change Options
8: Return to Main Menu
```

Results of the Geweke (1992) diagnostic are obtained by selecting the first option:

```
Selection: 1

GEWEKE CONVERGENCE DIAGNOSTIC (Z-score)
==========================================

Iterations used = 1001:6000
Thinning interval = 1
Sample size per chain = 5000

$CODAoutput1.txt

Fraction in 1st window = 0.1
Fraction in 2nd window = 0.5

    R[2]      R[3]      R[4]      R[5]      R[6]      R[7]      R[8]     pi[1]
-0.11145  -0.17586  -0.26609   0.60018   1.11425  -1.49937  -0.96585  -0.07320
   pi[2]     pi[3]     pi[4]     pi[5]     pi[6]     pi[7]     pi[8]
-0.20697  -0.62976  -1.18390   0.02331   2.03886   0.02998  -1.54755


Geweke plots menu
```

```
1: Change window size
2: Plot Z-scores
3: Change number of bins for plot
4: Return to Diagnostics Menu

Selection:
```

From this output we observe that all Z values are within -2 and 2, indicating no differences in the means for the first and last sets of iterations. Selecting the second option provides a graphical picture of Z scores.

Returning to the diagnostics menu (option 4) and selecting the Raftery–Lewis diagnostic (Raftery and Lewis, 1992) provides the following output:

```
Selection: 3

RAFTERY AND LEWIS CONVERGENCE DIAGNOSTIC
==========================================

Iterations used = 1001:6000
Thinning interval = 1
Sample size per chain = 5000

$CODAoutput1.txt

Quantile (q) = 0.025
Accuracy (r) = +/- 0.005
Probability (s) = 0.95
```

	Burn-in (M)	Total (N)	Lower bound (Nmin)	Dependence factor (I)
R[2]	36	40968	3746	10.90
R[3]	22	23208	3746	6.20
R[4]	16	17023	3746	4.54
R[5]	12	12894	3746	3.44
R[6]	10	10532	3746	2.81
R[7]	8	8602	3746	2.30
R[8]	5	5577	3746	1.49
pi[1]	13	13827	3746	3.69
pi[2]	8	9404	3746	2.51
pi[3]	12	14242	3746	3.80
pi[4]	8	9658	3746	2.58
pi[5]	12	14242	3746	3.80
pi[6]	8	10112	3746	2.70
pi[7]	4	5169	3746	1.38
pi[8]	4	4792	3746	1.28

```
1: Change parameters
2: Return to diagnostics menu
```

From this CODA output display, we conclude that a larger sample of size equal to 40,968 iterations (which is the maximum of all N) is required to obtain the prespecified accuracy for the 2.5% quantile. No additional burnin is needed since M is generally low (here 1000 iterations have been already removed). I for some parameters is high, indicating high autocorrelations for these parameters and that a thinning interval may be required to make observations independent. For example, here we may finally consider to $N \times I' = 3746 \times 11 = 41,206$ iterations with thinning interval equal to $I' = 11$, finally keeping

3746 independent observations; where I' is the maximum I rounded to the closest larger integer number: $I' = round(\max\{I\} + 0.5)$.

Returning to the diagnostics menu (option 2) and requesting for the Heidelberg–Welch diagnostic (Heidelberger and Welch, 1992), we find the following output printed on the screen:

```
Selection: 4

HEIDELBERGER AND WELCH STATIONARITY AND INTERVAL HALFWIDTH TESTS
================================================================

Iterations used = 1001:6000
Thinning interval = 1
Sample size per chain = 5000

Precision of halfwidth test = 0.1

$CODAoutput1.txt

        Stationarity start      p-value
        test            iteration
R[2]    passed          1        0.551
R[3]    passed          1        0.891
R[4]    passed          1        0.603
R[5]    passed          1        0.511
R[6]    passed          1        0.760
R[7]    passed          1        0.243
R[8]    passed          1        0.798
pi[1]   passed          1        0.534
pi[2]   passed          1        0.633
pi[3]   passed          1        0.973
pi[4]   passed          1        0.119
pi[5]   passed          1        0.971
pi[6]   passed          1        0.193
pi[7]   passed          1        0.607
pi[8]   passed          1        0.791

        Halfwidth Mean  Halfwidth
        test
R[2]    passed    0.990 0.006362
R[3]    passed    1.013 0.003643
R[4]    passed    0.963 0.002748
R[5]    passed    0.970 0.003706
R[6]    passed    0.990 0.004002
R[7]    passed    1.040 0.003074
R[8]    passed    1.051 0.001837
pi[1]   passed    0.469 0.002081
pi[2]   passed    0.463 0.001228
pi[3]   passed    0.469 0.000781
pi[4]   passed    0.451 0.000831
pi[5]   passed    0.437 0.001141
pi[6]   passed    0.433 0.000953
pi[7]   passed    0.450 0.000454
pi[8]   passed    0.472 0.000450


1: Change precision
2: Return to diagnostics menu
```

This output indicates that all tests have been passed and convergence has been reached.

From the analysis above, some problems were traced using the Raftery–Lewis diagnostic. The user may argue that two out of three of the applied diagnostics have diagnosed convergence. Alternatively, if we wish to ensure convergence, we should obtain additional iterations according to the details of the Raftery–Lewis diagnostic.

Finally, after returning to the main menu and exiting CODA, all values of the MCMC output are stored in the R object coda.dat. We recommend saving it in a different object for future use. For example, we can store the data in an object named ex1coda with the command

```
ex1coda <- coda.dat
```

You may restart CODA using this object, by selecting the second option in the introductory menu

```
[1] "mcmc.list"
> ex1coda <- coda.dat
> codamenu()
CODA startup menu

1: Read BUGS output files
2: Use an mcmc object
3: Quit

Selection: 2
```

and then insert the name of the object

```
Enter name of saved object (or type "exit" to quit)
1: ex1coda
```

If the following message and the main CODA menu appear, then the data have been loaded successfully:

```
Checking effective sample size ...OK
CODA Main Menu

1: Output Analysis
2: Diagnostics
3: List/Change Options
4: Quit
```

C.5.2   Illustration in BOA

After downloading, installing, and loading the BOA package in R, we call the main menu using the command

```
boa.menu()
```

The initial menu of BOA gives the following options:

```
BOA MAIN MENU
*************

1: File     >>
2: Data     >>
3: Analysis >>
4: Plot     >>
5: Options  >>
6: Window   >>
```

To import WinBUGS files, select 1 (File) and then 3 (Import Data):

```
FILE MENU
=========

1: Back
2: ----------------------+
3: Import Data       >> |
4: Load Session         |
5: Save Session         |
6: Exit BOA             |
7: ----------------------+

Selection: 3

IMPORT DATA MENU
----------------

1: Back
2: --------------------------+
3: CODA Output Files         |
4: Flat ASCII File           |
5: Data Matrix Object        |
6: View Format Specifications |
7: Options...                |
8: --------------------------+

Selection:
```

We then specify the working directory using 7 (Options...):

```
Selection: 7

Data Parameters
===============

Files
-----
1) Working Directory: ""
2) ASCII File Ext:    ".txt"

Select parameter to change or press <ENTER> to continue
1: 1
Read 1 item

DESCRIPTION: Specified directory must not end with a slash

Enter new character string
1: C:/myexample
Read 1 item

IMPORT DATA MENU
----------------

1: Back
2: --------------------------+
3: CODA Output Files         |
4: Flat ASCII File           |
5: Data Matrix Object        |
6: View Format Specifications |
7: Options...                |
8: --------------------------+
```

```
Selection:
```

Then load the WinBUGS data using selection 3 (CODA Output Files). The files must be saved under the same name with different extensions: .out and .ind for the output and index file, respectively. In the following the original files were renamed boa.out and boa.ind:

```
Selection: 3

Enter filename prefix without the .ind or .out extension [Working
    Directory: "C:/myexample"]
1: boa
```

If the data are successfully loaded, the following message appears:

```
Read 1 item
Read 15 records
Read 75000 records
+++ Data successfully imported +++
```

This is followed by the IMPORT DATA MENU:

```
IMPORT DATA MENU
----------------

1: Back
2: -------------------------+
3: CODA Output Files        |
4: Flat ASCII File          |
5: Data Matrix Object       |
6: View Format Specifications |
7: Options...               |
8: -------------------------+

Selection:
```

Return to the BOA MAIN MENU by typing 1, press the Enter key twice, and then select 3 (analysis) to obtain the corresponding menu:

```
Selection: 3

ANALYSIS MENU
=============

1: Back
2: -------------------------+
3: Descriptive Statistics  >> |
4: Convergence Diagnostics >> |
5: Options...               |
6: -------------------------+
```

Select 4 to obtain the convergence diagnostics menu:

```
CONVERGENCE DIAGNOSTICS MENU
----------------------------

1: Back
2: ---------------------+
3: Brooks, Gelman & Rubin |
4: Geweke               |
```

```
5: Heidelberger & Welch   |
6: Raftery & Lewis         |
7: ----------------------+
```

These diagnostics are the same as in BOA arranged in alphabetical order. Results are the same as in CODA. For the Geweke (1992) diagnostic, p-values are additionally provided.

APPENDIX D

NOTATION SUMMARY

D.1 MCMC

- B: number of burnin iterations
- T: total number of iterations obtained by a simulation method
- $T' = T - B$: total number of iterations retained after discarding the burnin period
- K: number of batches used for estimation of Monte Carlo errors; also number of groups in implementation of Levene's test in Chapter 10
- L: lag/thin interval
- $\nu = T'/K$ (or T/K): size of each batch used for the estimation of Monte Carlo error
- $l_i = \log f(y_i|\boldsymbol{\theta})$: log-likelihood
- κ: number of generated chains
- $\hat{V} = \text{BSS}/T'$; $\hat{R} = \hat{V}/\text{WSS}$: Gelman–Rubin diagnostic
- WSS/BSS: within/between-chain sum-of-squares (Gelman–Rubin diagnostic)

D.2 SUBSCRIPTS AND INDICES

- i: index for subject/experimental unit ($i = 1, 2, \ldots, n$)

- j, l: indices for parameters/variables in linear predictor

- ℓ: index for levels in categorical variables taking values $\ell = 1, 2, ..., L_A$

- k: subject index within each level of a categorical variable

- Subscript i: index for subjects/observations

- Subscript $\backslash i$: index for defining a vector without the ith element

- Subscript (i): index for i row of a matrix (used for design matrices to extract covariate values for i individual/observation); also

    $\boldsymbol{\theta}_{(i)}$: parameters corresponding to i subject

    $\boldsymbol{X}_{(i)}$: i row of a matrix (used for design matrices to extract covariate values for i individual/observation)

    $\boldsymbol{y}_{(i)}$: element of ith order of vector $\boldsymbol{y}$

- Subscript j in a matrix $(\boldsymbol{X}_j)$: j column of matrix $\boldsymbol{X}$

- Subscript y, θ: random variable/parameter for which a measure (e.g., expected value or variance) is calculated

- Superscript (t): iteration number

D.3 PARAMETERS

- $\boldsymbol{\theta}$: parameter vector

- $\boldsymbol{\theta}_m$: parameter vector of model m

- $\boldsymbol{\theta}_m^{(t)}$: t observation of $\boldsymbol{\theta}_m$ generated from the posterior distribution

- $\boldsymbol{\theta}_m^{*(t)}$: t observation of $\boldsymbol{\theta}_m$ generated from the prior or an arbitrary distribution $g(\boldsymbol{\theta}_m)$

- $\boldsymbol{\theta}_{(i)}$: parameters for i individual/observation

- $G(\boldsymbol{\theta})$: function of parameter vector $\boldsymbol{\theta}$

- $h(\boldsymbol{\theta})$: link function

- $\boldsymbol{\beta}, \boldsymbol{\beta}_m$: parameters involved in the linear predictor of the model

- $\boldsymbol{\beta} = (\beta_0, \beta_1, \ldots, \beta_p)^T$

- $\boldsymbol{b} = (\beta_1, \ldots, \beta_p)^T$

D.4   RANDOM VARIABLES AND DATA

- n: sample size
- p: number of explanatory variables
- P: number of parameters in linear predictor (usually $P = p + 1$)
- p_m and P_m: the corresponding p and P for model m
- d or d_m: dimension of model m; in models where only linear predictor parameters are involved $d = P = p + 1$; normal models $d = P + 1 = p + 2$.
- $X_1, \ldots, X_p$: explanatory variables
- $x_{i1}, \ldots, X_{ip}$: data of i individual for explanatory variables $X_1, \ldots, X_p$
- $\boldsymbol{x}$: $n \times p$ matrix with elements x_{ij}
- $\boldsymbol{X} = \begin{bmatrix} \mathbf{1}_n, \boldsymbol{x} \end{bmatrix}$: design/data matrix of dimension $n \times P$ in which the first column has each element equal to one (i.e. is a vector of ones: $\mathbf{1}_n$), if the intercept is included in the model, and the remaining columns contain the values of each covariate X_j
- $\boldsymbol{X}_{(i)}, \boldsymbol{x}_{(i)}$: the ith row of matrices $\boldsymbol{X}$ and $\boldsymbol{x}$, respectively
- Y: response variable
- Y_i: response variable for i individual
- y_i: observed value of Y_i for i individual
- $\boldsymbol{y} = (y_1, \ldots, y_n)^T$ vector of length n of the response data
- $y_{(i)}$: ith larger (order) observed value of vector $\boldsymbol{y}$
- $\left(y_{(1)}, \ldots, y_{(n)} \right)^T$ vector of length n of the response data arranged in ascending order
- $\boldsymbol{y}_{\backslash i} = (y_1, \ldots, y_{i-1}, y_{i+1}, \ldots, y_n)^T$: vector of observed values without the ith element
- $E(Y), V(Y)$ or $\mathrm{Var}(Y), \mathrm{SD}(Y)$: expectation, variance, and standard deviation of Y:

    $E(\boldsymbol{\theta}), \mathrm{Var}(\boldsymbol{\theta}), \mathrm{SD}(\boldsymbol{\theta})$: corresponding prior measures
    $E(\boldsymbol{\theta}|\boldsymbol{y}), \mathrm{Var}(\boldsymbol{\theta}|\boldsymbol{y}), \mathrm{SD}(\boldsymbol{\theta}|\boldsymbol{y})$: corresponding posterior measures
    $\mathrm{SD}(\boldsymbol{y})$: sample standard deviation of vector $\boldsymbol{y}$

D.5   SAMPLE ESTIMATES

- $\bar{y}, s_y^2, \hat{\sigma}_y^2$: sample mean, unbiased, and biased sample estimates of variance
- s_y: sample standard deviation of y
- $\widehat{\theta}$ (hat): maximum-likelihood estimate of parameter θ
- SS: sum of squares used in calculation of posterior quantities in normal models

D.6   SPECIAL FUNCTIONS, VECTORS, AND MATRICES

- $I(x)$: indicator function taking value equal to one if x is true and zero otherwise

- $\boldsymbol{I}_k$: unit matrix of dimension $k \times k$

- π: mathematical constant equal to 3.14159

- $\mathbf{1}_k$: vector of length k with all elements equal to one

- $\mathbf{1}_{rc}, \mathbf{1}_{[r \times c]}$: matrix of dimension $r \times c$ with all elements equal to one

- $\mathbf{0}_k$: vector of length k with all elements equal to zero

- $\mathbf{0}_{rc}$: matrix of dimension $r \times c$ with all elements equal to zero

D.7   DISTRIBUTIONS

- beta(a, b): beta distribution with parameters a and b

- Bernoulli(π): Bernoulli distribution with success probability π

- binomial(π, N), $B(\pi, N)$: binomial distribution with success probability π and N replications of a Bernoulli experiment

- Dirichlet$(\boldsymbol{\alpha})$: Dirichlet distribution with parameter vector $\boldsymbol{\alpha}$

- $f_D(y; \boldsymbol{\theta})$: density or probability function evaluated at y of distribution D and parameter vector $\boldsymbol{\theta}$

- $f_G(y; a, b)$: density function of gamma distribution with parameters a and b evaluated at y

- $f_N(y; \mu, \sigma^2)$: density function of normal distribution with mean μ and variance σ^2 evaluated at y

- gamma(a, b): gamma distribution with parameters a and b (mean a/b and variance a/b^2)

- IG(a, b): inverse gamma distribution with parameters a and b

- $N(\mu, \sigma^2)$: univariate normal distribution with mean μ and variance σ^2

- $N_d(\boldsymbol{\mu}, \boldsymbol{\Sigma})$: d-dimensional normal distribution with mean $\boldsymbol{\mu}$ and variance $\boldsymbol{\Sigma}$

- NB(π, N): negative binomial distribution with success probability π

- NG(μ, c, a, b): normal gamma distribution with parameters μ, c, a, b

- NIG(μ, c, a, b): normal inverse gamma distribution with parameters μ, c, a, b

- Poisson(λ): Poisson distribution with parameter λ

- $U(a, b)$: continuous uniform distribution defined in the interval (a, b)

D.8   DISTRIBUTION-RELATED NOTATION

- a, b: parameters of gamma distribution with mean a/b and variance a/b^2; also in gamma related distributions:

    a_0, b_0: prior parameters

    $\widetilde{a}, \widetilde{b}$: posterior parameters

    $\overline{a}, \overline{b}$: proposal parameters

- c^2, c_θ^2: variance multiplicators (used in prior or proposal)

- $\mathcal{D}(\boldsymbol{\alpha})$: distribution $\mathcal{D}$ with parameter vector $\boldsymbol{\alpha}$

- μ, σ^2, τ: mean, variance, and precision of normal distribution

- μ_0, σ_0^2, τ: mean, variance, and precision of normal prior distribution

- $\mu_\theta, \widetilde{\mu}_\theta^2$: prior and posterior mean of θ

- $\widetilde{\mu}, \widetilde{\sigma}^2, \widetilde{\tau}$: mean, variance, and precision of normal posterior distribution

- $\overline{\mu}, \overline{\sigma}^2, \overline{\tau}$: mean, variance, and precision of normal proposal distribution

- $\boldsymbol{\mu}, \boldsymbol{\Sigma}, \boldsymbol{T}, \boldsymbol{R}$: mean vector, variance, precision, and correlation matrices of multivariate normal distribution

- $\boldsymbol{\mu}_\beta$: prior mean of $\boldsymbol{\beta}$

- $\boldsymbol{\mu}_0, \boldsymbol{\Sigma}_0$: prior mean vector and variance of a multivariate normal prior distribution

- $\widetilde{\boldsymbol{\mu}}, \widetilde{\boldsymbol{\Sigma}}$: posterior mean vector and variance matrix of multivariate normal posterior distribution

- $\overline{\boldsymbol{\mu}}, \overline{\boldsymbol{S}}$: proposal mean vector and variance matrix of a multivariate normal proposal distribution

- N: size in binomial distribution

- N_i: size in i binomial case

- π: success probability in binomial distribution

- π_i: success probability in i binomial case

- Π: mathematical constant equal to 3.14159

- $\sigma_\theta^2, \widetilde{\sigma}_\theta^2$: prior and posterior variances of θ

- σ_β^2: prior variance of β

- z_q: quantile of $q100\%$ of the normal distribution

- $\widetilde{\theta}$: posterior parameter θ

- $\widehat{\theta}$: maximum-likelihood estimate of θ

- $\overline{\theta}$: θ parameter of a proposal distribution

D.9 NOTATION USED IN ANOVA AND ANCOVA

- $y_{\ell k}$: k observation of ℓ group in ANOVA

- A, B, C: random factors used in ANOVA and ANCOVA models

- A_i, B_i, C_i: random variables for i subject/experimental unit

- a_i, b_i, c_i: data codes for level/category of factors A, B, and C of subject/experimental unit

- α_ℓ: effect of ℓ level of factor A

- i: index for subject/experimental unit ($i = 1, 2, \ldots, n$)

- j, l: indices for parameters/variables in linear predictor

- ℓ: index for levels in categorical variables taking values $\ell = 1, 2, ..., L_A$

- k: subject index within each level of a categorical variable

- μ_i: mean for i individual

- μ'_ℓ: mean for ℓ group

- L_A, L_B, L_C: levels of factors A, B, and C, respectively

- $D^A_{i\ell}, D^B_{i\ell}, D^C_{i\ell}$: dummy variables using the corner constraint for factors A, B, and C, respectively

- $D^{A,STZ}_{i\ell}, D^B_{i\ell}, D^C_{i\ell}$: dummy variables using the sum-to-zero constraint for factors A, B, and C, respectively

- δ_ℓ: interaction parameters between quantitative and qualitative variables in ANCOVA

D.10 VARIABLE AND MODEL SPECIFICATION

- γ_j: binary variable indicators indicating inclusion of X_j in the linear predictor of the model

- $\boldsymbol{\gamma} = (\gamma_1, \gamma_2, \ldots, \gamma_P)$: binary variable indicators indicating the inclusion of X_j in the linear predictor of the model (the length of $\boldsymbol{\gamma}$ may be equal to p or P depending on the model specification)

D.11 DEVIANCE INFORMATION CRITERION (DIC)

- $\mathrm{DIC}(m)$: deviance measure of model m

- $D(\boldsymbol{\theta}_m, m) = -2 \log f(\boldsymbol{y}|\boldsymbol{\theta}_m, m)$: deviance measure of model m and parameters $\boldsymbol{\theta}_m$

- $\overline{D(\boldsymbol{\theta}_m, m)}$: posterior mean of deviance evaluated by an MCMC sample

- $p_m = \overline{D(\boldsymbol{\theta}_m, m)} - D(\bar{\boldsymbol{\theta}}_m, m)$: number of "effective" parameters for model m

D.12 PREDICTIVE MEASURES

- $\boldsymbol{y}^{\mathrm{rep}} = (y_1^{\mathrm{rep}}, \ldots, y_n^{\mathrm{rep}})^T$ vector of length n of the predictive/replicated response data

- CV-1: leave-one-out cross-validation

- PPO_i, CPO_i: posterior and conditional predictive ordinates for individual/observation i

- r_i: residual of ith observation

- r_i^s: standardized residual of ith observation

- $r_{(i)}^o$: residual of ith ordered observation $y_{(i)}$

REFERENCES

Abramowitz, M. and Stegun, I. (1974), *Handbook of Mathematical Functions*, Dover, New York.

Agarwal, D., Gelfand, A. and Citron-Pousty, S. (2002), "Zero-inflated models with application to spatial count data", *Environmental and Ecological Statistics* **9**, 341–355.

Agresti, A. (1990), *Categorical Data Analysis*, Wiley-Interscience, New York.

Agresti, A. (2002), *Categorical Data Analysis*, 2nd ed., Wiley-Interscience, Hoboken, NJ.

Agresti, A. and Hitchcock, D. (2005), "Bayesian inference for categorical data analysis", *Statistical Methods and Applications* **14**, 297–330.

Aguilar, O., Prado, R., Huerta, G. and West, M. (1999), "Bayesian inference on latent structure in time series (with discussion)", in J. Bernardo, J. Berger, A. Dawid, and A. Smith, eds., *Bayesian Statistics*, Vol. 6, Oxford University Press, pp. 3–26.

Aitkin, M. (1991), "Posterior Bayes factors", *Journal of the Royal Statistical Society B* **53**, 111–142.

Akaike, H. (1973), "Information theory and an extension of the maximum likelihood principle", in B. Petrov and F. Csaki, eds., *Proceedings of 2nd International Symposium on Information Theory*, Academiai Kiado, Budapest, pp. 267–281.

Akaike, H. (1974), "A new look at the statistical model identification", *IEEE Transactions on Automatic Control* **19**, 716–723.

Albert, J. and Chib, S. (1993), "Bayesian analysis of binary and polychotomous response data", *Journal of the American Statistical Association* **88**, 669–679.

Albert, J. and Chib, S. (1997), "Bayesian tests and model diagnostics in conditionally independent hierarchical models", *Journal of the American Statistical Association* **92**, 916–925.

Aldrich, J. and Nelson, F. (1984), *Linear Probability, Logit, and Probit Models*, Quantitative Applications in the Social Sciences, 07–045, Sage Publications, Inc., Los Angeles, CA.

Ameniya, T. (1981), "Quantitative response models: A survey", *Journal of Economics Literature* **19**, 1483–1536.

American Psychiatric Association (1987), *DSM-III- R: Diagnostic and Statistical Manual of Mental Disorders*, 3rd revised ed., APA, Washington, DC.

Andersen, P. and Gill, R. (1982), "Cox's regression model for counting processes: A large sample study", *Annals of Statistics* **10**, 1100–1120.

Anderson, H. (1986), "Metropolis, Monte Carlo and the MANIAC", *Los Alamos Science* pp. 96–107.

Aranda-Ordaz, F. (1981), "On two families of transformations to additivity for binary response data", *Biometrika* **68**, 357–363.

Atkinson, A. (1978), "Posterior probabilities for choosing a regression model", *Biometrika* **65**, 39–48.

Atkinson, A. (1981), "Likelihood ratios, posterior odds and information criteria", *Journal of Econometrics* **16**, 15–20.

Barbieri, M. and Berger, J. (2004), "Optimal predictive model selection", *Annals of Statistics* **32**, 870–897.

Bartholomew, D. and Knott, M. (1999), *Latent Variable Models and Factor Analysis*, Kendall's Library of Statistics, Vol. 7, 2nd ed., Hodder Arnold Publications, UK.

Bartlett, M. (1957), "Comment on D.V. Lindley's statistical paradox", *Biometrika* **44**, 533–534.

Basu, S. and Mukhopadhyay, S. (2000), "Binary response regression with normal scale mixture links", in D. Dey, S. Ghosh, and B. Mallick, eds., *Generalized Linear Models: A Bayesian Perspective*, Marcel Dekker, New York, pp. 231–241.

Baxter, M. and Stevenson, R. (1988), "Discriminating between the Poisson and negative binomial distributions: An application to goal scoring in association football", *Journal of Applied Statistics* **15**, 347–438.

Bayarri, M. and Berger, J. (2000), "P-values for composite null models (with discussion)", *Journal of the American Statistical Association* **95**, 1127–1142.

Bennet, J., Racine-Poon, A. and Wakefield, J. (1996), "MCMC for nonlinear hierarchical models", in W. Gilks, S. Richardson, and D. Spiegelhalter, eds., *Markov Chain Monte Carlo in Practice*, Chapman & Hall, Suffolk, UK, pp. 339–358.

Berger, J. and Delampady, M. (1987), "Testing precise hypotheses", *Statistical Science* **2**, 317–352.

Berger, J. and Pericchi, L. (1996*a*), "The intrinsic Bayes factor for linear models", in J. Bernardo, J. Berger, A. Dawid, and A. Smith, eds., *Bayesian Statistics*, Vol. 5, Oxford University Press, pp. 25–44.

Berger, J. and Pericchi, L. (1996*b*), "The intrinsic Bayes factor for model selection and prediction", *Journal of the American Statistical Association* **91**, 109–122.

Berkson, J. (1944), "Application of the logistic function to bio-assay", *Journal of the American Statistical Association* **39**, 357–365.

Berkson, J. (1951), "Why I prefer logits to probits", *Biometrics* **7**, 327–339.

Bernardo, J. (1999), "Nested hypothesis testing: The Bayesian reference criterion", in J. Bernardo, J. Berger, A. Dawid, and A. Smith, eds., *Bayesian Statistics*, Vol. 6, Oxford University Press, pp. 101–130.

Bernardo, J. and Smith, A. (1994), *Bayesian Theory*, Wiley, Chichester, UK.

Besag, J. (1989), "A candidate's formula: A curious result in Bayesian prediction", *Biometrika* **76**, 183.

Best, N., Cowles, M. and Vines, K. (1996), *CODA: Convergence Diagnostics and Output Analysis Software for Gibbs Sampling Output,* Version 0.30, MRC Biostatistics Unit, Institute of Public Health, Cambridge, UK.

Best, N., Cowles, M. and Vines, K. (1997), *CODA: Convergence Diagnostics and Output Analysis Software for Gibbs Sampling Output,* Version 0.40 (addendum to manual), MRC Biostatistics Unit, Institute of Public Health, Cambridge, UK.

Birch, M. W. (1963), "Maximum likelihood in three-way contingency tables", *Journal of the Royal Statistical Society B* **25**, 220–233.

Bliss, C. I. (1935), "The calculation of the dosage-mortality curve", *Annals of Applied Biology* **22**, 134–167.

Bohning, D. (1998), "Zero-inflated Poisson models and c.a.man: A tutorial collection of evidence", *Biometrical Journal* **40**, 833–843.

Bohning, D., Dietz, E., Schlattmann, P., Mendonca, L. and Kirchner, U. (1999), "The zero-inflated Poisson model and the decayed, missing and filled teeth index in dental epidemiology", *Journal of the Royal Statistical Society A* **162**, 195–209.

Box, G. and Tiao, G. (1973), *Bayesian Inference for Statistical Analysis*, Wiley Classics Library, Wiley, New York.

Brooks, S. (1998), "Markov chain Monte Carlo method and its application", *The Statistician* **47**, 69–100.

Brooks, S. (2002), "Discussion of the paper by Spiegelhalter, Best, Carlin and van der Linde", *Journal of the Royal Statistical Society B* **64**, 616–618.

Brooks, S. and Gelman, A. (1998), "Alternative methods for monitoring convergence of iterative simulations", *Journal of Computational and Graphical Statistics* **7**, 434–455.

Brooks, S. and Roberts, G. (1998), "Assessing convergence of Markov chain Monte Carlo algorithms", *Statistics and Computing* **8**, 319–335.

Brown, H. and Prescott, R. (2006), *Applied Mixed Models in Medicine*, Statistics in Practice, 2nd ed., Wiley, Chichester, UK.

Brown, P., Vannucci, M. and Fearn, T. (1998), "Multivariate Bayesian variable selection and prediction", *Journal of the Royal Statistical Society B* **60**, 627–641.

Browne, W., Subramanian, S., Jones, K. and Goldstein, H. (2005), "Variance partitioning in multilevel logistic models that exhibit overdispersion", *Journal of the Royal Statistical Society A* **168**, 599–613.

Burnham, K. and Anderson, D. (2004), "Multimodel inference: Understanding AIC and BIC in model selection", *Sociological Methods Research* **33**, 261–304.

Carlin, B. (1996), "Hierarchical longitudinal modelling", in W. Gilks, S. Richardson, and D. Spiegelhalter, eds., *Markov Chain Monte Carlo in Practice*, Chapman & Hall, Suffolk, UK, pp. 303–320.

Carlin, B. and Chib, S. (1995), "Bayesian model choice via Markov chain Monte Carlo methods", *Journal of the Royal Statistical Society B* **157**, 473–484.

Carlin, B. and Louis, T. (2000), *Bayes and Empirical Bayes Methods for Data Analysis*, Texts in Statistical Science, Chapman & Hall/CRC, New York.

Casella, G. and George, E. (1992), "Explaining the Gibbs sampler", *The American Statistician* **46**, 167–174.

Chen, D. (2007), "Bootstrapping estimation for estimating the relative potency in combinations of bioassays", *Computational Statistics and Data Analysis* **51**, 4597–4604.

Chen, M.-H. (2005), "Computing marginal likelihoods from a single MCMC output", *Statistica Neerlandica* **59**, 16–29.

Chen, M., Ibrahim, J. and Shao, Q. (2000), "Power prior distributions for generalized linear models", *Journal of Statistical Planning and Inference* **84**, 121–137.

Cheung, Y. (2002), "Zero-inflated models for regression analysis of count data: A study of growth and development", *Statistics in Medicine* **21**, 1461–1469.

Chib, S. (1995), "Marginal likelihood from the Gibbs output", *Journal of the American Statistical Association* **90**, 1313–1321.

Chib, S. and Greenberg, E. (1995), "Understanding the Metropolis–Hastings algorithm", *American Statistician* **49**, 327–335.

Chib, S. and Jeliazkov, I. (2001), "Marginal likelihood from the Metropolis–Hastings output", *Journal of the American Statistical Association* **96**, 270–281.

Chib, S. and Jeliazkov, I. (2005), "Accept–reject Metropolis-Hastings sampling and marginal likelihood estimation", *Statistica Neerlandica* **59**, 30–44.

Chipman, H. (1996), "Bayesian variable selection with related predictors", *Canadian Journal of Statistics* **24**, 17–36.

Chow, G. (1981), "A comparison of the information and posterior probability criteria for model selection", *Journal of Econometrics* **16**, 21–33.

Clayton, D. (1991), "A Monte Carlo method for Bayesian inference in frailty models", *Biometrics* **47**, 467–485.

Clayton, D. (1996), "Generalized linear mixed models", in W. Gilks, S. Richardson, and D. Spiegelhalter, eds., *Markov Chain Monte Carlo in Practice*, Chapman & Hall, Suffolk, UK, pp. 275–302.

Clyde, M., DeSimone, H. and Parmigiani, G. (1996), "Prediction via orthogonalized model mixing", *Journal of the American Statistical Association* **91**, 1197–1208.

Cohen, A. C. (1963), "Estimation in mixtures of discrete distributions", in *Proceedings of the International Symposium on Discrete Distributions*, Montreal, Canada, pp. 373–378.

Congdon, P. (2003), *Applied Bayesian Modelling*, Wiley Series in Probability and Statistics, Wiley, Chichester, UK.

Congdon, P. (2005a), *Bayesian Models for Categorical Data*, Wiley Series in Probability and Statistics, Wiley, Chichester, UK.

Congdon, P. (2005b), "Bayesian predictive model comparison via parallel sampling", *Computational Statistics and Data Analysis* **48**, 735–753.

Congdon, P. (2006a), "Bayesian model comparison via parallel model output", *Journal of Statistical Computation and Simulation* **76**, 149–165.

Congdon, P. (2006b), *Bayesian Statistical Modelling*, Wiley Series in Probability and Statistics, 2nd ed., Wiley, Chichester, UK.

Consul, P. (1989), *Generalized Poisson Distribution: Properties and Applications*, Marcel Decker, New York.

Consul, P. and Famoye, F. (1992), "Generalized Poisson regression model", *Communications in Statistics: Theory and Methods* **21**, 89–109.

Cowles, M. and Carlin, B. (1996), "Markov chain Monte Carlo convergence diagnostics: A comparative review", *Journal of the American Statistical Association* **91**, 883–904.

Cox, D. (1972), "Regression models and life-tables", *Journal of the Royal Statistical Society B* **34**, 187–220.

Cox, D. and Oakes, D. (1984), *Analysis of Survival Data*, Chapman & Hall, Cambridge, UK.

Cox, D. R. (1958), "Two further applications of a model for binary regression", *Biometrika* **45**, 562–565.

Czado, C. and Raftery, A. (2006), "Choosing the link function and accounting for link uncertainty in generalized linear models using Bayes factors", *Statistical Papers* **47**, 419–442.

Damien, P., Wakefield, J. and Walker, S. (1999), "Gibbs sampling for Bayesian non-conjugate and hierarchical models by using auxiliary variable", *Journal of the Royal Statistical Society B* **61**, 331–344.

Darby, S. (1980), "A Bayesian approach to parallel line bioassay", *Biometrika* **67**, 607–612.

de Leeuw, J. and Meijer, E. (2008), *Handbook of Multilevel Analysis*, Springer-Verlag, New York.

Dellaportas, P. and Forster, J. (1999), "Markov chain Monte Carlo model determination for hierarchical and graphical log-linear models", *Biometrika* **86**, 615–633.

Dellaportas, P., Forster, J. and Ntzoufras, I. (2000), "Bayesian variable selection using the Gibbs sampler", in D. Dey, S. Ghosh, and B. Mallick, eds., *Generalized Linear Models: A Bayesian Perspective*, Marcel Dekker, New York, pp. 271–286.

Dellaportas, P., Forster, J. and Ntzoufras, I. (2002), "On Bayesian model and variable selection using MCMC", *Statistics and Computing* **12**, 27–36.

Dellaportas, P. and Smith, A. (1993), "Bayesian inference for generalized linear and proportional hazards models via Gibbs sampling", *Journal of the Royal Statistical Society* C **42**, 443–460.

Dey, D., Ghosh, S. and Mallick, B. (2000), *Generalized Linear Models: A Bayesian Perspective*, Marcel Dekker, New York.

Dixon, M. and Coles, S. (1997), "Modelling association football scored and inefficiencies in football betting market", *Journal of the Royal Statistical Society* C **46**, 265–280.

Dobson, A. (2002), *An Introduction to Generalized Linear Models*, 2nd ed., Chapman & Hall, New York.

Draper, D. (1995), "Inference and hierarchical modeling in the social sciences", *Journal of Educational and Behavioral Statistics* **20**, 115–147.

Draper, D. and Krnjajić, M. (2006), *Bayesian Model Specification*, Technical Report, Department of Applied Mathematics and Statistics, Baskin School of Engineering, University of California, Santa Cruz.

Dunson, D. (2000), "Bayesian latent variable models for clustered mixed outcomes", *Journal of the Royal Statistical Society* B **62**, 355–366.

Dunson, D. and Herring, A. (2005), "Bayesian latent variable models for mixed discrete outcomes", *Biostatistics* **6**, 11–25.

Erkanli, A. (1994), "Laplace approximations for posterior expectation when the model occurs at the boundary of the parameter space", *Journal of the American Statistical Association* **89**, 205–258.

Evans, M. and Swartz, T. (1996), "Discussion of methods for approximating integrals in statistics with special emphasis on Bayesian integration problems", *Statistical Science* **11**, 54–64.

Fahrmeir, L. and Tutz, G. (1994), "Dynamic stochastic models for time-dependent ordered paired comparison system", *Journal of the American Statistical Association* **89**, 1438–1449.

Fahrmeir, L. and Tutz, G. (2001), *Multivariate Statistical Modelling Based on Generalized Linear Models*, Springer Series in Statistics, 2nd ed., Springer-Verlag, New York.

Famoye, F. and Singh, K. (2006), "Zero-inflated generalized Poisson regression model with an application to domestic violence data", *Journal of Data Science* **4**, 117–130.

Fernandez, C., Ley, E. and Steel, M. (2000), "Benchmark priors for Bayesian model averaging", *Journal of Econometrics* **100**, 381–427.

Feuewerger, A. (1979), "On some methods of analysis for weather experiments", *Biometrika* **66**, 665–668.

Fienberg, S. (1981), *The Analysis of Cross-Classified Categorical Data*, 2nd revised ed., The MIT Press, Cambridge, MA.

Fisher, R. A. (1922), "On the interpretation of chi-square from contingency tables, and the calculation of p", *Journal of the Royal Statistical Society* B **85**, 87–94.

Fouskakis, D., Ntzoufras, I. and Draper, D. (2008), "Bayesian variable selection using cost-adjusted BIC, with application to cost-effective measurement of quality of health care", *Annals of Applied Statistics* (to appear) .

Friel, N. and Pettitt, A. (2008), "Marginal likelihood estimation via power posteriors", *Journal of the Royal Statistical Society B* **70**, 589–607.

Futing Liao, T. (1994), *Interpreting Probability Models: Logit Probit, and Other Generalized Linear Models*, Quantitative Applications in the Social Sciences, 007–101, Sage Publications, Inc., Los Angeles, CA.

Gamerman, D. (1997), "Sampling from the posterior distribution in generalized linear mixed models", *Statistics and Computing* **7**, 57–68.

Gamerman, D. and Lopes, H. (2006), *Markov Chain Monte Carlo*, Texts in Statistical Science, 2nd ed., Chapman & Hall, New York.

Gan, N. (2000), *General Zero-Inflated Models and Their Applications*, PhD thesis, North Carolina State University, at Releigh.

Geisser, S. and Eddy, W. (1979), "A predictive approach to model selection", *Journal of the American Statistical Association* **74**, 153–160. Corrigenda in Vol. 75, p. 765.

Gelfand, A. (1996), "Model determination using sampling-based methods", in W. Gilks, S. Richardson, and D. Spiegelhalter, eds., *Markov Chain Monte Carlo in Practice*, Chapman & Hall, Suffolk, UK, pp. 145–161.

Gelfand, A. and Dey, D. (1994), "Bayesian model choice: Asymptotic and exact calculations", *Journal of the Royal Statistical Society B* **56**, 501–514.

Gelfand, A., Dey, D. and Chang, H. (1992), "Model determination using predictive distributions with implementation via sampling-based methods (with discussion)", in J. Bernardo, J. Berger, A. Dawid, and A. Smith, eds., *Bayesian Statistics*, Vol. 4, Oxford University Press, pp. 407–425.

Gelfand, A. and Ghosh, S. (1998), "Model choice: A minimum posterior predictive loss approach", *Biometrika* **85**, 1–13.

Gelfand, A., Hills, S., Racine-Poon, A. and Smith, A. (1990), "Illustration of Bayesian inference in normal data models using Gibbs sampling", *Journal of the American Statistical Association* **85**, 972–985.

Gelfand, A. and Smith, A. (1990), "Sampling-based approaches to calculating marginal densities", *Journal of the American Statistical Association* **85**. 398–409.

Gelfand, A., Smith, A. and Lee, T.-M. (1992), "Bayesian analysis of constrained parameter and truncated data problems using Gibbs sampling", *Journal of the American Statistical Association* **87**, 523–532.

Gelman, A. (2006), "Prior distributions for variance parameters in hierarchical models", *Bayesian Analysis* **1**, 515—-533.

Gelman, A., Carlin, J., Stern, H. and Rubin, D. (1995), *Bayesian Data Analysis*, Texts in Statistical Science, Chapman & Hall, London.

Gelman, A., Carlin, J., Stern, H. and Rubin, D. (2004), *Bayesian Data Analysis*, Texts in Statistical Science, 2nd ed., Chapman & Hall, London.

Gelman, A. and Hill, J. (2006), *Data Analysis Using Regression and Multilevel/Hierarchical Models*, Cambridge University Press, New York.

Gelman, A., Huang, Z., van Dyk, D. and Boscardin, W. J. (2008), "Using redundant parameters to fit hierarchical models", *Journal of Computational and Graphical Statistics* **17**, 95–122.

Gelman, A. and Meng, X.-L. (1996), "Model checking and model improvement", in W. Gilks, S. Richardson, and D. Spiegelhalter, eds., *Markov Chain Monte Carlo in Practice*, Chapman & Hall, Suffolk, UK, pp. 189–201.

Gelman, A. and Meng, X.-L. (1998), "Simulating normalizing constants: From importance sampling to bridge sampling to path sampling", *Statistical Science* **13**, 163–185.

Gelman, A., Meng, X.-L. and Stern, H. (1996), "Posterior predictive assessment of model fitness via realized discrepancies", *Statistica Sinica* **6**, 733–807.

Gelman, A. and Pardoe, I. (2006), "Bayesian measures of explained variance and pooling in multilevel (hierarchical) models", *Technometrics* **48**, 241–251.

Gelman, A. and Rubin, D. (1992), "Inference from iterative simulation using multiple sequences", *Statistical Science* **7**, 457–511.

Geman, S. and Geman, D. (1984), "Stochastic relaxation, Gibbs distributions and the Bayesian restoration of images", *IEEE Transactions on Pattern Analysis and Machine Intelligence* **6**, 721–741.

Genter, F. and Farewell, V. (1985), "Goodness-of-link testing in ordinal regression models", *Canadian Journal of Statistics* **13**, 37–44.

George, E. and Foster, D. (2000), "Calibration and empirical Bayes variable selection", *Biometrika* **87**, 731–748.

George, E. and McCulloch, R. (1993), "Variable selection via Gibbs sampling", *Journal of the American Statistical Association* **88**, 881–889.

George, E. and McCulloch, R. (1997), "Approaches for Bayesian variable selection", *Statistica Sinica* **7**, 339–373.

George, E., McCulloch, R. and Tsay, R. (1996), "Two approaches to Bayesian model selection with applications", in D. Berry, K. Chaloner, and J. Geweke, eds., *Bayesian Analysis in Statistics and Econometrics: Essays in Honor of Arnold Zellner*, Wiley, New York, pp. 339–348.

Geweke, J. (1992), "Evaluating the accuracy of sampling-based approaches to calculating posterior moments", in J. Bernardo, J. Berger, A. Dawid, and A. Smith, eds., *Bayesian Statistics*, Vol. 4, Claredon Press, Oxford, pp. 169–194.

Geyer, C. (1992), "Practical Markov chain Monte Carlo (with discussion)", *Statistical Science* **7**, 473–511.

Ghosh, S., Mukhopadhyay, P. and Lu, J. (2006), "Bayesian analysis of zero-inflated regression models", *Journal of Statistical Planning and Inference* **136**, 1360–1375.

Gilks, W. (1996), "Full conditional distributions", in W. Gilks, S. Richardson, and D. Spiegelhalter, eds., *Markov Chain Monte Carlo in Practice*, Chapman & Hall, Suffolk, UK, pp. 75–88.

Gilks, W., Richardson, S. and Spiegelhalter, D. (1996), *Markov Chain Monte Carlo in Practice*, Interdisciplinary Statistics, Chapman & Hall, Suffolk, UK.

Gilks, W. and Roberts, G. (1996), "Strategies for improving MCMC", in W. Gilks, S. Richardson, and D. Spiegelhalter, eds., *Markov Chain Monte Carlo in Practice*, Chapman & Hall, Suffolk, UK, pp. 89–110.

Gilks, W. and Wild, P. (1992), "Adaptive rejection sampling for Gibbs sampling", *Journal of the Royal Statistical Society* C **41**, 337–348.

Givens, G. and Hoeting, J. (2005), *Computational Statistics*, Wiley Series in Probability and Statistics, Wiley, Hoboken, NJ.

Glonek, G. (1996), "A class of regression models for multivariate categorical responses", *Biometrika* **83**, 15–28.

Goldstein, H., Browne, W. and Rasbash, J. (2002), "Partitioning variation in multilvevel models", *Understanding Statistics* **1**, 223–231.

Goodman, L. (1979), "Simple models for the analysis of association in cross-classifications having ordered categories", *Journal of the American Statistical Association* **74**, 537–552.

Goodman, L. (1981), "Association models and canonical correlation in the analysis of cross-classifications having ordered categories", *Journal of the American Statistical Association* **76**, 320–334.

Goodman, L. (1985), "The analysis of cross-classified data having ordered and/or unordered categories: Association models, correlation models and asymmetry models for contingency tables with or without missing entries", *Annals of Statistics* **13**, 10–69.

Green, P. (1995), "Reversible jump Markov chain Monte Carlo computation and Bayesian model determination", *Biometrika* **82**, 711–732.

Guerrero, V. and Johnson, R. (1982), "Use of the Box–Cox transformation with binary response models", *Biometrika* **69**, 309–314.

Gupta, P., Gupta, R. and Tripathi, R. (1996), "Analysis of zero-adjusted count data", *Computational Statistics and Data Analysis* **23**, 207–218.

Gutiérrez-Peña, E. and Walker, S. (2001), "A Bayesian predictive approach to model selection", *Journal of Statistical Planning and Inference* **93**, 259–276.

Hall, D. (2000), "Zero-inflated Poisson and binomial regression with random effects: A case study", *Biometrics* **56**, 1030–1039.

Han, C. and Carlin, B. (2001), "Markov chain Monte Carlo methods for computing Bayes factors: a comparative review", *Journal of the American Statistical Association* **96**, 1122–1132.

Hans, C., Dobra, A. and West, M. (2007), "Shotgun stochastic search for "large p" regression", *Journal of the American Statistical Association* **102**, 507–516.

Haro-López, R., Mallick, B. and Smith, A. (2000), "Binary regression using data adaptive robust link functions", in D. Dey, S. Ghosh, and B. Mallick, eds., *Generalized Linear Models: A Bayesian Perspective*, Marcel Dekker, New York, pp. 243–253.

Hastings, W. (1970), "Monte Carlo sampling methods using Markov chains and their applications", *Biometrika* **57**, 97–109.

Hedeker, D. and Gibbons, R. (2006), *Longitudinal Data Analysis*, Wiley-Interscience, Hoboken, NJ.

Heibron, D. (1994), "Zero-altered and other regression models for count data with added zeros", *Biometrical Journal* **36**, 531–547.

Heidelberger, P. and Welch, P. (1992), "Simulation run length control in the presence of an initial transient", *Operations Research* **31**, 1109–1144.

Higdon, D. (1998), "Auxiliary variable methods for Markov chain Monte Carlo with applications", *Journal of the American Statistical Association* **93**, 585–595.

Hoeting, J. and Ibrahim, J. (1998), "Bayesian predictive simultaneously variable and transformation selection in the linear model", *Journal of Computational Statistics and Data Analysis* **28**, 87–103.

Hoffmann-Jørgensen, J. (1994), *Probability with a View Towards Statistics,* Vol.1, Probability Series, Chapman & Hall, New York.

Hosmer, D. and Lemeshow, S. (2000), *Applied Logistic Regression*, 2nd ed., Wiley, New York.

Hosmer, D., Lemeshow, S. and May, S. (2008), *Applied Survival Analysis: Regression Modeling of Time to Event Data*, 2nd ed., Wiley, Hoboken, NJ.

Huard, D., Évin, G. and Favre, A.-C. (2006), "Bayesian copula selection", *Computational Statistics and Data Analysis* **51**, 809–822.

Hubble, E. (1929), "A relationship between distance and radial velocity among extra-galactic nebulae", in *Proceedings of the National Academy of Science*, Vol. 15, pp. 168–173, available at http://www.pnas.org/cgi/reprint/15/3/168.

Ibrahim, J. and Chen, M. (2000), "Power prior distributions for regression models", *Statistical Science* **15**, 46–60.

Ibrahim, J. and Laud, P. (1994), "A predictive approach to the analysis of designed experiments", *Journal of the American Statistical Association* **89**, 309–319.

Iliopoulos, G., Kateri, M. and Ntzoufras, I. (2007*a*), "Bayesian estimation of unrestricted and order-restricted association models for a two-way contingency table", *Computational Statistics and Data Analysis* **51**, 4643–4655.

Iliopoulos, G., Kateri, M. and Ntzoufras, I. (2007*b*), *Bayesian Model Comparison for the Order Restricted RC Association Model*, Technical Report, Department of Statistics, Athens University of Economics and Business, Athens, Greece, available at http://stat-athens.aueb.gr/~jbn/papers/paper18.htm.

Iliopoulou, K. (2004), *Schizotypy and Consumer Behavior* (in Greek), Master's thesis, Department of Business Administration, University of the Aegean, Chios, Greece, available at http://stat-athens.aueb.gr/~jbn/courses/diplomatikes/business/Iliopoulou(2004).pdf.

Jasra, A., Stephens, D. and Holmes, C. (2007), "Population-based reversible jump Markov chain Monte Carlo", *Biometrika* **94**, 787–807.

Johnson, N., Kotz, S. and Balakrishnan, N. (1997), *Discrete Multivariate Distributions*, Wiley, New York.

Kahn, M. (2005), "An exhalent problem for teaching statistics", *Journal of Statistics Education* **13**, available at www.amstat.org/publications/jse/v13n2/datasets.kahn.html.

Karlis, D. and Melikotzidou, L. (2005), "Multivariate Poisson regression with covariance structure", *Statistics and Computing* **15**, 255–265.

Karlis, D. and Ntzoufras, I. (2000), "On modelling soccer data", *Student* **3**, 229–244.

Karlis, D. and Ntzoufras, I. (2003*a*), "Analysis of sports data using bivariate Poisson models", *Journal of the Royal Statistical Society* D **52**, 381–393.

Karlis, D. and Ntzoufras, I. (2003*b*), "Bayesian and non-Bayesian analysis of soccer data using bivariate Poisson regression models", in *Proceedings of the Greek Statistical Institute*, Vol 16, pp. 605–612, available at http://www.stat-athens.aueb.gr/~jbn/tr/TR59_Greek_Soccer.ps.

Karlis, D. and Ntzoufras, I. (2005), "Bivariate Poisson and diagonal inflated bivariate Poisson regression models in R", *Journal of Statistical Software* **10**, 1–37.

Karlis, D. and Ntzoufras, I. (2006), "Bayesian analysis of the differences of count data", *Statistics in Medicine* **25**, 1885–1905.

Karlis, D. and Ntzoufras, I. (2008), "Bayesian modelling of football outcomes: Using the Skellam's distribution for the goal difference", *IMA Journal of Management Mathematics* (to appear) .

Karlis, D. and Tsiamyrtzis, P. (2008), "Exact Bayesian modeling for bivariate Poisson data and extensions", *Statistics and Computing* **18**, 27–40.

Kass, R., Carlin, B., Gelman, A. and Neal, R. (1998), "Markov chain Monte Carlo in practice: A roundtable discussion", *The American Statistician* **52**, 93–100.

Kass, R. and Raftery, A. (1995), "Bayes factors", *Journal of the American Statistical Association* **90**, 773–795.

Kass, R. and Wasserman, L. (1995), "A reference Bayesian test for nested hypotheses and its relationship to the Schwarz criterion", *Journal of the American Statistical Association* **90**, 928–934.

Kateri, M., Nicolaou, A. and Ntzoufras, I. (2005), "Bayesian inference for the RC(m) association model", *Journal of Computational and Graphical Statistics* **14**, 116–138.

Katsis, A. and Ntzoufras, I. (2005), "Testing hypotheses for the distribution of insurance claim counts using the Gibbs sampler", *Journal of Computational Methods in Sciences and Engineering* **5**, 201–214.

Knuiman, M. and Speed, T. (1988), "Incorporating prior information into the analysis of contingency tables", *Biometrics* **44**, 1061–1071.

Kocherlakota, S. and Kocherlakota, K. (1992), *Bivariate Discrete Distributions*, Marcel Dekker, New York.

Kolev, N., Anjos, U. and Mendez, B. (2006), "Copulas: A review and recent developments", *Stochastic Models* **22**, 617–660.

Konishi, S. and Kitagawa, G. (2008), *Information Criteria and Statistical Modeling*, Springer Series in Statistics, Springer-Verlag, New York.

Kuha, J. (2003), *Model Assessment and Model Choice: An Annotated Bibliography*, Technical Report, Department of Statistics and the Methodology Institute, London School of Economics, available at http://stats.lse.ac.uk/kuha/msbib/biblio/.

Kuha, J. (2004), "AIC and BIC: Comparisons of assumptions and performance", *Sociological Methods Research* **33**, 188–229.

Kuo, L. and Mallick, B. (1998), "Variable selection for regression models", *Sankhyā B* **60**, 65–81.

Kuonen, D. (1996), *Modelling the Success of Football Teams in the European Championships*, Technical Report, No. 96.1, Department of Mathematics, Swiss Federal Institute of Technology, Lausanne, Switzerland.

Kuonen, D. (1997), *Statistical Models for Knock-out Soccer Tournaments*, Technical Report, No. 97.3, Department of Mathematics, Swiss Federal Institute of Technology, Lausanne, Switzerland.

Kuonen, D. and Roehrl, A. (2000), "Was France's world cup win pure chance?", *Student* **3**, 153–166.

Lambert, D. (1992), "Zero-inflated Poisson regression, with applications to defects in manufacturing", *Technometrics* **34**, 1–14.

Lang, J. (1999), "Bayesian ordinal and binary regression models with a parametric family of mixture links", *Computational Statistics and Data Analysis* **31**, 59–87.

Laskey, K. and Myers, J. (2003), "Population Markov chain Monte Carlo", *Machine Learning* **50**, 175–196.

Laud, P. and Ibrahim, J. (1995), "Predictive model selection", *Journal of the Royal Statistical Society B* **57**, 247–262.

Lawson, A., Browne, W. and Vidal Rodeiro, C. (2003), *Disease Mapping with WinBUGS and MLwiN*, Statistics in Practice, Wiley, Hoboken, NJ.

Lee, A. (1997), "Modeling scores in the premier league: Is Manchester United really the best?", *Chance* **10**, 15–19.

Lewis, S. and Raftery, A. (1997), "Estimating Bayes factor via posterior simulation with the Laplace-Metropolis estimator", *Journal of the American Statistical Association* **92**, 648–655.

Li, C., Lu, J., Park, J., Kim, K. and Peterson, J. (1999), "Multivariate zero-inflated Poisson models and their applications", *Technometrics* **41**, 29–38.

Liang, F., Paulo, R., Molina, G., Clyde, M. and Berger, J. (2008), 'Mixtures of *g* priors for Bayesian variable selection', *Journal of the American Statistical Association* **103**, 410–423.

Lindley, D. (1957), "A statistical paradox", *Biometrika* **44**, 187–192.

Lindley, D. (1980), "L.J.Savage — his work in probability and statistics", *Annals of Statistics* **8**, 1–24.

Lindley, D. (1993), "On presentation of evidence", *Mathematical Scientist* **18**, 60–63.

Lindley, D. and Smith, A. (1972), "Bayes estimates for the linear model", *Journal of the Royal Statistical Society B* **34**, 1–41.

Lindsey, J. (1997), *Applying Generalized Linear Models*, Springer Texts in Statistics, Springer-Verlag, New York.

Liu, I. and Agresti, A. (2005), "The analysis of ordered categorical data: An overview and a survey of recent developments (with discussion)", *Test* **14**, 1–73.

Lunn, D. J., Best, N. and Whittaker, J. (2005), *Generic Reversible Jump MCMC Using Graphical Models*, Technical Report, No EPH-2005-01, Department of Epidemiology and Public Health, Imperial College, London, UK, available at `https://www1.imperial.ac.uk/resources/8b3cf549-039e-4f96-8bec-cab969a0695ceph-2005-01.pdf`.

Lunn, D. J., Whittaker, J. C. and Best, N. (2006), "A Bayesian toolkit for genetic association studies", *Genetic Epidemiology* **30**, 231–247.

Lynch, S. (2007), *Introduction to Applied Bayesian Statistics and Estimation for Social Scientists*, Statistics for Social and Behavioral Sciences, Springer-Verlag, New York.

Mackowiak, P. A., Wasserman, S. S. and Levine, M. M. (1992), "A critical appraisal of 98.6 degrees F, the upper limit of the normal body temperature, and other legacies of Carl Reinhold August Wunderlich", *Journal of the American Medical Association* **268**, 1578–1580.

Maher, M. (1982), "Modelling association football scores", *Statistica Neerlandica* **36**, 109–118.

Mallick, B. and Gelfand, A. (1994), "Generalized linear models with unknown number of components", *Biometrika* **81**, 237–245.

Mallows, C. (1973), "Some comments on C_p", *Technometrics* **15**, 661–675.

McCullagh, P. and Nelder, J. (1989), *Generalized Linear Models*, Monographs on Statistics and Applied Probability, Vol. 37, 2nd ed., Chapman & Hall, Cambridge, UK.

Mendoza, M. (1990), "A Bayesian analysis of the slope ratio bioassay", *Biometrics* **46**, 1059–1069.

Meng, X.-L. (1994), "Posterior predictive p-values", *Annals of Statistics* **22**, 1142–1160.

Meng, X.-L. and Wong, W. (1996), "Simulating ratios of normalizing constants via a simple identity: A theoretical exploration", *Statistica Sinica* **6**, 831–860.

Metropolis, N., Rosenbluth, A., Rosenbluth, M., Teller, A. and Teller, E. (1953), "Equations of state calculations by fast computing machine", *Journal of Chemical Physics* **21**, 1087–1092.

Metropolis, N. and Ulam, S. (1949), "The Monte Carlo method", *Journal of the American Statistical Association* **44**, 335–341.

Meyer, M. and Laud, P. (2002), "Predictive variable selection in generalized linear models", *Journal of the American Statistical Association* **97**, 859–871.

Møller, J. (1999), "Perfect simulation of conditionally specified models", *Journal of the Royal Statistical Society B* **61**, 251–264.

Montgomery, D. and Peck, E. (1992), *Introduction to Regression Analysis*, Wiley, New York.

Montgomery, D., Peck, E. and Vining, G. (2006), *Introduction to Linear Regression Analysis*, 4th ed., Wiley, Hoboken, NJ.

Movellan, J. (2006), *Tutorial on Multivariate Logistic Regression*, Machine Perception Laboratory, University of California, San Diego, available at the tutorial section of `http://mplab.ucsd.edu/wordpress/`.

Mwalili, S., Lesaffre, E. and Declerck, D. (2008), "The zero-inflated negative binomial regression model with correction for misclassification: An example in caries research", *Statistical Methods in Medical Research* **17**, 123–139.

Neal, P. and Roberts, G. (2008), "Optimal scaling for random walk Metropolis on spherically constrained target densities", *Methodology and Computing in Applied Probability* **10**, 277–297.

Neal, R. (1998), "Suppressing random walks in Markov chain Monte Carlo using ordered overrelaxation", in M. Jordan, ed., *Learning in Graphical Models*, Kluwer Academic Publishers, Dordrecht, pp. 205–230; also available at `http://www.cs.utoronto.ca/~radford/publications.html`.

Neal, R. (2001), "Annealed importance sampling", *Statistics and Computing* **11**, 125–139.

Neal, R. (2003), "Slice sampling", *The Annals of Statistics* **31**, 705–767.

Newton, M. and Raftery, A. (1994), "Approximate Bayesian inference with the weighted likelihood bootstrap", *Journal of the Royal Statistical Society B* **56**, 3–48.

Ntzoufras, I. (1999*a*), *Aspects of Bayesian Model and Variable Selection Using MCMC*, PhD thesis, Department of Statistics, Athens University of Economics and Business, Athens, Greece, available at http://stat-athens.aueb.gr/~jbn/publications.htm.

Ntzoufras, I. (1999*b*), "Discussion on Bayesian model averaging and model search strategies", in J. Bernardo, J. Berger, A. Dawid, and A. Smith, eds., *Bayesian Statistics*, Vol. 6, Oxford University Press, pp. 178–179.

Ntzoufras, I. (2002), "Gibbs variable selection using BUGS", *Journal of Statistical Software* **7**, 1–19.

Ntzoufras, I. and Dellaportas, P. (2002), "Bayesian modelling of outstanding liabilities incorporating claim count uncertainty (with discussion)", *North American Actuarial Journal* **6**, 113–128.

Ntzoufras, I., Dellaportas, P. and Forster, J. (2000), "Stochastic search variable selection for log-linear models", *Journal of Statistical Computation and Simulation* **68**, 23–38.

Ntzoufras, I., Dellaportas, P. and Forster, J. (2003), "Bayesian variable and link determination for generalized linear models", *Journal of Planning and Inference* **111**, 165–180.

Ntzoufras, I., Katsis, A. and Karlis, D. (2005), "Bayesian assessment of the distribution of insurance claim counts using reversible jump MCMC", *North American Actuarial Journal* **9**, 90–10.

O'Brien, S. and Dunson, D. (2004), "Bayesian multivariate logistic regression", *Biometrics* **60**, 739–746.

Oh, C., Ye, K., He, Q. and Mendell, N. (2003), "Locating disease genes using Bayesian variable selection with the Haseman-Elston method", *BMC Genetics* **4**, Supl.1 – S9, available at http://www.biomedcentral.com/1471-2156/4/s1/S69.

O'Hagan, A. (1995), "Fractional Bayes factors for model comparison", *Journal of the Royal Statistical Society B* **57**, 57.

Pauler, D. (1998), "The Schwarz criterion and related methods for normal linear models", *Biometrika* **85**, 13–27.

Pericchi, L. (1984), "An alternative to the standard Bayesian procedure for discrimination between normal linear models", *Biometrika* **71**, 575–586.

Pitt, M., Chan, D. and Kohn, R. (2006), "Efficient Bayesian inference for Gaussian copula regression models", *Biometrika* **93**, 537—554.

Plummer, M., Best, N., Cowles, K. and Vines, K. (2006), "CODA: Convergence diagnosis and output analysis for MCMC", *R News* **6**(1), 7–11, available at http://CRAN.R-project.org/doc/Rnews/Rnews_2006-1.pdf.

Powers, D. and Xie, Y. (1999), *Statistical Methods for Categorical Data Analysis*, Academic Press, San Diego, CA.

Pregibon, D. (1980), "Goodness of link tests for generalized linear models", *Journal of the Royal Statistical Society C* **29**, 15–24.

Prentice, R. (1976), "Generalization of the probit and logit methods for dose response curves", *Biometrics* **32**, 761–768.

Press, S. (1989), *Bayesian Statistics: Principles, Models and Applications*, Wiley, New York.

Propp, J. and Wilson, D. (1996), "Exact sampling with coupled Markov chains and applications to statistical mechanics", *Random Structures and Algorithms* **9**, 223–252.

Qiu, Z., Song, P. X.-K. and Tan, M. (2002), "Bayesian hierarchical models for multi-level repeated ordinal data using WinBUGS", *Journal of Biopharmaceutical Statistics* **12**, 121–135.

Raftery, A. (1996*a*), "Approximate Bayes factors and accounting for model uncertainty in generalized linear models", *Biometrika* **83**, 251–266.

Raftery, A. (1996*b*), "Hypothesis testing and model selection", in W. Gilks, S. Richardson, and D. Spiegelhalter, eds., *Markov Chain Monte Carlo in Practice*, Chapman & Hall, Suffolk, UK, pp. 163–188.

Raftery, A. and Lewis, S. (1992), "How many iterations in the Gibbs sampler?", in J. Bernardo, J. Berger, A. Dawid, and A. Smith, eds., *Bayesian Statistics*, Vol. 4, Claredon Press, Oxford, pp. 763–774.

Raftery, A., Madigan, D. and Hoeting, J. (1997), "Bayesian model averaging for linear regression models", *Journal of the American Statistical Association* **92**, 179–191.

Raftery, A., Newton, M., Satagopan, J. and Krivitsky, P. (2007), "Estimating the integrated likelihood via posterior simulation using the harmonic mean identity (with discussion)", in J. Bernardo, J. Bayarri, and J. Berger, eds., *Bayesian Statistics*, Vol. 8, Oxford University Press, pp. 1–45.

Raine, A. (1991), "The SPQ: A scale for the assessment of schizotypal personality based on dsm-iii-r criteria", *Schizophrenia Bulletin* **17**, 555–564.

Rasch, G. (1961), "On general laws and the meaning of measurement in psychology", in *Proceedings of 4th Berkeley Symposium on Mathematics, Statistics, and Probability*, University of California Press, Berkeley, pp. 321–333.

Reep, C. and Benjamin, B. (1968), "Skill and chance in association football", *Journal of the Royal Statistical Society* A **131**, 581–585.

Reep, C., Pollard, R. and Benjamin, B. (1971), "Skill and chance in ball games", *Journal of the Royal Statistical Society* A **134**, 623–629.

Ridout, M., Demetrio, C. and Hinde, J. (1998), "Models for count data with many zeros", in *Proceedings of 19th International Biometric Conference*, Cape Town, South Africa, pp. 179–190.

Rigby, R. and Stasinopoulos, D. (2005), "Generalized additive models for location, scale and shape (with discussion)", *Journal of the Royal Statistical Society* C **54**, 507–554.

Robert, C. (1993), "A note on Jeffreys-Lindley paradox", *Statistica Sinica* **3**, 601–608.

Robert, C. (2007), *The Bayesian Choice*, 2nd ed., Springer-Verlag, New York.

Robert, C. and Casella, G. (2004), *Monte Carlo Statistical Methods*, Springer Texts in Statistics, 2nd ed., Springer-Verlag, New York.

Roberts, G. (1996), "Markov chain concepts related to sampling algorithms", in W. Gilks, S. Richardson, and D. Spiegelhalter, eds., *Markov Chain Monte Carlo in Practice*, Chapman & Hall, Suffolk, UK, pp. 45–58.

Roberts, G., Gelman, A. and Gilks, W. (1997), "Weak convergence and optimal scaling of random walk Metropolis algorithms", *The Annals of Statistics* **7**, 110–120.

Roberts, G. and Rosenthal, J. (2001), "Optimal scaling for various Metropolis–Hastings algorithms", *Statistical Science* **16**, 351–367.

Rosner, B. (2005), *Fundamentals of Biostatistics*, 6th revised ed., Brooks Cole.

Rue, H. and Salvesen, O. (2000), "Prediction and retrospective analysis of soccer matches in a league", *Journal of the Royal Statistical Society* D **49**, 399–418.

Ryan, T. (1997), *Modern Regression Methods*, Wiley Series in Probability and Statistics, Wiley, New York.

Sahu, S., Dey, D. and Branco, M. (2003), "A new class of multivariate skew distributions with applications to Bayesian regression models", *The Canadian Journal of Statistics* **31**, 129–150.

Sarantinidis, M. (2003), *A Survey for the Development of Natural Products Shops: The Case of "Masticha Shop"* (in Greek), Master's thesis, Department of Business Administration, University of the Aegean, Chios, Greece, available at `http://stat-athens.aueb.gr/~jbn/courses/diplomatikes/business/Sarantinidis(2003).pdf`.

Schlesselman, J. (1982), *'Case-Control Studies: Design, Conduct, Analysis*, Monographs in Epidemiology and Biostatistics, Oxford University Press, New York.

Schwarz, G. (1978), "Estimating the dimension of a model", *Annals of Statistics* **6**, 461–464.

Scollnik, D. (1998), "On the analysis of the truncated generalized Poisson distribution using a Bayesian method", *ASTIN Bulletin* **28**, 135–152.

Scollnik, D. (2002), "Implementation of four models for outstanding liabilities in WinBUGS: A discussion of a paper by Ntzoufras and Dellaportas", *North American Actuarial Journal* **6**, 128–136.

Scott, D. W. (1992), *Multivariate Density Estimation*, Wiley, New York.

Sellke, T., Bayarri, M. and Berger, J. (2001), "Calibration of p-values for testing precise null hypotheses", *The American Statistician* **55**, 62–71.

Seltzer, M., Wong, W. and Bryk, A. (1996), "Bayesian analysis in applications of hierarchical models: Issues and methods", *Journal of Educational and Behavioral Statistics* **21**, 131–167.

Seshadri, V. (1993), *The Inverse Gaussian Distribution: A Case Study in Exponential Families*, Oxford Science Publications, UK.

Shafer, J. (1982), "Lindley's paradox (with discussion)", *Journal of the American Statistical Association* **77**, 325–334.

Shao, J. (1993), "Linear model selection by cross-validation", *Journal of the American Statistical Association* **88**, 486–494.

Shoemaker, A. L. (1996), "What's normal? – temperature, gender, and heart rate", *Journal of Statistics Education* **4**(2), available at `http://www.amstat.org/publications/jse/v4n2/datasets.shoemaker.html`.

Sisson, S. (2005), "Trans-dimensional Markov chains: A decade of progress and future perspectives", *Journal of the American Statistical Association* **100**, 1077–1089.

Skellam, J. (1946), "The frequency distribution of the difference between two Poisson variates belonging to different populations", *Journal of the Royal Statistical Society* A **109**, 296.

Smith, A. and Roberts, G. (1993), "Bayesian computation via the Gibbs sampler and related Markov chain Monte Carlo methods", *Journal of the Royal Statistical Society* B **55**, 3–23.

Smith, B. (2005), *(B)ayesian (O)utput (A)nalysis Program (BOA) Version 1.1.5 User's Manual*, Technical Report, Department of Public Health, The University of Iowa, available at `http://www.public-health.uiowa.edu/boa`.

Smith, B. (2007), "BOA: An R package for MCMC output convergence assessment and posterior inference", *Journal of Statistical Software* **21**, available at `http://www.jstatsoft.org/`.

Spiegelhalter, D., Abrams, K. and Myles, J. (2004), *Bayesian Approaches to Clinical Trials and Health-Care Evaluation*, Statistics in Practice, Wiley, Chichester, UK.

Spiegelhalter, D., Best, N., Carlin, B. and van der Linde, A. (2002), "Bayesian measures of model complexity and fit (with discussion)", *Journal of the Royal Statistical Society* B **64**, 583–639.

Spiegelhalter, D. and Smith, A. (1988), "Bayes factors for linear and log-linear models with vague prior information", *Journal of the Royal Statistical Society* B **44**, 377–387.

Spiegelhalter, D., Thomas, A., Best, N. and Gilks, W. (1996*a*), *BUGS 0.5: Bayesian Inference Using Gibbs Sampling Manual*, MRC Biostatistics Unit, Institute of Public Health, Cambridge, UK.

Spiegelhalter, D., Thomas, A., Best, N. and Gilks, W. (1996*b*), *BUGS 0.5: Examples Volume 1*, MRC Biostatistics Unit, Institute of Public Health, Cambridge, UK.

Spiegelhalter, D., Thomas, A., Best, N. and Gilks, W. (1996*c*), *BUGS 0.5: Examples Volume 2*, MRC Biostatistics Unit, Institute of Public Health, Cambridge, UK.

Spiegelhalter, D., Thomas, A., Best, N. and Lunn, D. (2003*a*), *WinBUGS Examples,* Vol. 1, MRC Biostatistics Unit, Institute of Public Health and Department of Epidemiology and Public Health, Imperial College School of Medicine, UK, available at `http://www.mrc-bsu.cam.ac.uk/bugs`.

Spiegelhalter, D., Thomas, A., Best, N. and Lunn, D. (2003*b*), *WinBUGS Examples,* Vol. 2, MRC Biostatistics Unit, Institute of Public Health and Department of Epidemiology and Public Health, Imperial College School of Medicine, UK, available at `http://www.mrc-bsu.cam.ac.uk/bugs`.

Spiegelhalter, D., Thomas, A., Best, N. and Lunn, D. (2003*c*), *WinBUGS Examples,* Vol. 3, MRC Biostatistics Unit, Institute of Public Health and Department of Epidemiology and Public Health, Imperial College School of Medicine, UK, available at `http://www.mrc-bsu.cam.ac.uk/bugs`.

Spiegelhalter, D., Thomas, A., Best, N. and Lunn, D. (2003*d*), *WinBUGS User Manual,* Version 1.4, MRC Biostatistics Unit, Institute of Public Health and Department of Epidemiology and Public Health, Imperial College School of Medicine, UK, available at `http://www.mrc-bsu.cam.ac.uk/bugs`.

Stanton, J. (2001), "Galton, Pearson, and the peas: A brief history of linear regression for statistics instructors", *Journal of Statistics Education* **9**(3).

Stone, M. (1977), "An asymptotic equivalence of choice of model by cross-validation and Akaike's criterion", *Journal of the Royal Statistical Society B* **39**, 44–47.

Stukel, T. (1988), "Generalized logistic models", *Journal of the American Statistical Association* **83**, 426–431.

Tanner, M. and Wong, W. (1987), "The calculation of the posterior distributions by data augmentation", *Journal of the American Statistical Association* **82**, 528–549.

Ter Berg, P. (1996), "A loglinear Lagrangian Poisson model", *ASTIN Bulletin* **26**, 123–129.

Tiao, G. and Zellner, A. (1964), "On the Bayesian estimation of multivariate regression", *Journal of the Royal Statistical Society B* **26**, 277–285.

Tierney, L. and Kadane, J. (1986), "Accurate approximations for posterior moments and marginal densities", *Journal of the American Statistical Association* **81**, 82–86.

Tierney, L., Kass, R. and Kadane, J. (1989), "Fully exponential Laplace approximations to expectations and variances of nonpositive functions", *Journal of the American Statistical Association* **84**, 710–716.

Tsionas, E. (2001), "Bayesian multivariate Poisson regression", *Communications in Statistics—Theory and Methods* **30**, 243–255.

Vehtari, A. and Lampinen, J. (2003), "Expected utility estimation via cross-validation", in J. Bernardo, M. Bayarri, J. Berger, A. Dawid, D. Heckerman, A. Smith, and M. West, eds., *Bayesian Statistics,* Vol. 7, Oxford University Press, pp. 701–710.

Wahlin, J. (2001), "Bivariate zip models", *Biometrical Journal* **43**, 147–160.

Wakefield, J. and Stephens, D. (2000), "Bayesian errors-in-variables modeling", in D. Dey, S. Ghosh, and B. Mallick, eds., *Generalized Linear Models: A Bayesian Perspective*, Marcel Dekker, New York, pp. 331 – 348.

Walker, S., Gutiérrez-Peña, E. and Muliere, P. (2001), "A decision theoretic approach to model selection", *Journal of the Royal Statistical Society D (The Statistician)* **50**, 31–39.

Weisberg, S. (2005), *Applied Linear Regression*, 3rd ed., Wiley-Interscience, New York.

West, M. and Harrison, P. (1997), *Bayesian Forecasting and Dynamic Models*, Kendall's Library of Statistics, Vol. 7, 2nd ed., Springer-Verlag, New York.

Woodworth, G. (2004), *Biostatistics: A Bayesian Introduction*, Wiley Series in Probability and Statistics, Wiley, Hoboken, NJ.

Yang, R. and Berger, J. (1996), *A Catalog of Noninformative Priors*, Technical Report, Institute of Statistics and Decision Sciences, Duke University, Durham, NC.

Yi, N. (2004), "A unified Markov chain Monte Carlo framework for mapping multiple quantitative trait loci", *Genetics* **167**, 967–975.

Yi, N., George, V. and Allison, D. (2003), "Stochastic search variable selection for identifying multiple quantitative trait loci", *Genetics* **164**, 1129–1138.

Zeger, S. and Karim, M. (1991), "Generalized linear models with random effects; a Gibbs sampling approach", *Journal of the American Statistical Association* **86**, 79–86.

Zellner, A. (1976), "Bayesian and non-Bayesian analysis of the regression model with multivariate Student-t error terms", *Journal of the American Statistical Association* **71**, 400–405.

Zellner, A. (1986), "On assessing prior distributions and Bayesian regression analysis using g-prior distributions", in P. Goel and A. Zellner, eds., *Bayesian Inference and Decision Techniques: Essays in Honor of Bruno de Finetti*, North-Holland, Amsterdam, pp. 233–243.

INDEX

WILEY SERIES IN COMPUTATIONAL STATISTICS